Lecture Notes in Computer Science 7219

Commenced Publication in 1973
Founding and Former Series Editors:
Gerhard Goos, Juris Hartmanis, and Jan van Leeuwen

Editorial Board

David Hutchison
Lancaster University, UK

Takeo Kanade
Carnegie Mellon University, Pittsburgh, PA, USA

Josef Kittler
University of Surrey, Guildford, UK

Jon M. Kleinberg
Cornell University, Ithaca, NY, USA

Alfred Kobsa
University of California, Irvine, CA, USA

Friedemann Mattern
ETH Zurich, Switzerland

John C. Mitchell
Stanford University, CA, USA

Moni Naor
Weizmann Institute of Science, Rehovot, Israel

Oscar Nierstrasz
University of Bern, Switzerland

C. Pandu Rangan
Indian Institute of Technology, Madras, India

Bernhard Steffen
TU Dortmund University, Germany

Madhu Sudan
Microsoft Research, Cambridge, MA, USA

Demetri Terzopoulos
University of California, Los Angeles, CA, USA

Doug Tygar
University of California, Berkeley, CA, USA

Gerhard Weikum
Max Planck Institute for Informatics, Saarbruecken, Germany

Youssef Hamadi   Marc Schoenauer (Eds.)

Learning and Intelligent Optimization

6th International Conference, LION 6
Paris, France, January 16-20, 2012
Revised Selected Papers

 Springer

Volume Editors

Youssef Hamadi
Microsoft Research
Cambridge CB3 0FB, UK
E-mail: youssefh@microsoft.com

Marc Schoenauer
INRIA Saclay, Université Paris Sud
91405 Orsay Cedex, France
E-mail: marc.schoenauer@inria.fr

ISSN 0302-9743   e-ISSN 1611-3349
ISBN 978-3-642-34412-1   e-ISBN 978-3-642-34413-8
DOI 10.1007/978-3-642-34413-8

Springer Heidelberg Dordrecht London New York

Library of Congress Control Number: 2012949747

CR Subject Classification (1998): F.2.2, I.2.8, G.1.6, F.1.1, G.2.2, J.3

LNCS Sublibrary: SL 1 – Theoretical Computer Science and General Issues

© Springer-Verlag Berlin Heidelberg 2012

This work is subject to copyright. All rights are reserved, whether the whole or part of the material is concerned, specifically the rights of translation, reprinting, re-use of illustrations, recitation, broadcasting, reproduction on microfilms or in any other way, and storage in data banks. Duplication of this publication or parts thereof is permitted only under the provisions of the German Copyright Law of September 9, 1965, in its current version, and permission for use must always be obtained from Springer. Violations are liable to prosecution under the German Copyright Law.
The use of general descriptive names, registered names, trademarks, etc. in this publication does not imply, even in the absence of a specific statement, that such names are exempt from the relevant protective laws and regulations and therefore free for general use.

Typesetting: Camera-ready by author, data conversion by Scientific Publishing Services, Chennai, India

Printed on acid-free paper

Springer is part of Springer Science+Business Media (www.springer.com)

Foreword

This LION conference (Learning and Intelligent OptimizatioN) was the sixth in a series of conferences that target the interface between optimization and machine learning, and the ever increasing success of these events bears witness to the growing interest of the scientific community in this research area today, as confirmed by the 109 submissions from 39 different countries that we received for this year's event. We would like to thank all of the authors for submitting some of their best work to LION 6.

Of the 109 submissions, there were 78 long papers and 21 short papers presenting original work, and 10 papers presenting work that had already been published. Due to this very high pressure, and the single-track format of the conference, we chose to give room to original works rather than works already published, regardless of the quality of the papers.

Out of these 99 original submissions, 24 papers were accepted as long papers (hence an acceptance rate of 31%), and 30 papers were accepted as short papers (19 that had been submitted as long papers, and 5 that had been submitted as short papers). All long papers were assigned to 3 independent reviewers, and all papers received at least 2 reviews. Note that the papers submitted to the special sessions were assigned by the special session chairs, except the ones that were authored by some of the session chairs. These were handled by the conference chairs, to ensure the anonymity of the reviewers (similarly, papers authored by one of the conference co-chairs were handled by the other co-chair and one member of the steering committee, unknown by the authors). We wish to heartily thank here all the reviewers (not anonymous any more, see next pages) for their hard and timely work, emphasizing the importance of such peer review, the best (if not only) way we know today to make a review process as fair as possible.

Because LION is a unique occasion for people from different research communities, the conference was single track (no parallel sessions) and the program left room for interaction among attendees with long coffee breaks. For the same reason, though the presentations of long papers (resp. short papers) were scheduled with 25minute (resp. 15minute) slots, the presentations themselves were not allowed more than 20minutes (resp. 12minutes), allowing time for questions and discussions. We want to thank here the session chairs, who were very strict on respecting these constraints, and thus made sure that the conference ran smoothly.

The final program of the conference also included 3 invited speakers, who presented forefront research results and frontiers, and 3 tutorial talks, which were crucial in bringing together the different components of the LION community. We wish to thank all these speakers who focused on different aspects of LION themes, and thus contributed to a better view and understanding of intelligent optimization at large.

Beside the authors, the reviewers, and the invited speakers, there are other people who made this event possible whom we wish to thank: Pierre-Louis Xech (Microsoft France), for arranging the venue at Microsoft France Technology Center, and smoothing out many small details that would otherwise have become incredibly time-consuming; Pietro Zanoni (Reactive Search Inc.), for setting up and diligently maintaining the conference Web site; Mireille Moulin (INRIA Saclay Île-de-France, Finance Department), for taking care of all the financial details with efficiency and flexibility; Esther Slamitz (INRIA Saclay Île-de-France, Events Department), for looking up and planning all local arrangements; Emmanuelle Perrot (INRIA Saclay Île-de-France, Communication Department), for providing many goodies, ... including the printing of the conference booklet; Chantal Girodon (INRIA Rocquencourt, Conferences & Seminars Office), for managing the registration system; and last but not least, Marie-Carol Lopes, for her tremendous help in gathering and formatting the material for these proceedings.

Finally, we would like to thank our sponsors, Microsoft Research, Microsoft France, and INRIA Saclay Île-de-France, for their financial support, which helped to keep the registration fees reasonable.

January 2012

Youssef Hamadi
Marc Schoenauer

Organization

LION Steering Committee

Roberto Battiti (Chair)	Università degli Studi di Trento, Italy
Christian Blum	Universitat Politècnica de Catalunya, Spain
Mauro Brunato	Università degli Studi di Trento, Italy
Martin Charles Golumbic	University of Haifa, Israel
Holger Hoos	University of British Columbia, Canada
Thomas Stützle	Université Libre de Bruxelles, Belgium
Benjamin W. Wah	The Chinese University of Hong Kong, China
Xin Yao	University of Birmingham, UK

Scientific Liaison with Springer

Thomas Stützle	Université Libre de Bruxelles, Belgium

Web Chair

Pietro Zanoni	Reactive Search SrL, Italy

Program Committee

Patrick Albert	IBM France
Ethem Alpaydin	Bogazici University, Turkey
Carlos Ansótegui	Universitat de Lleida, Spain
Josep Argelich	Universitat de Lleida, Spain
Anne Auger	INRIA Saclay Île-de-France, France
Ender Özcan	University of Nottingham, UK
Roberto Battiti	University of Trento, Italy
Mauro Birattari	Université Libre de Bruxelles, Belgium
Christian Blum	Universitat Politècnica de Catalunya, Spain
Lucas Bordeaux	Microsoft Research Cambridge, UK
Peter Bosman	CWI, Amsterdam, The Netherlands
Jürgen Branke	University of Warwick, UK
Dimo Brockoff	INRIA Lille Nord-Europe, France
Mauro Brunato	Università di Trento, Italy
Alba Cabiscol	Universitat de Lleida, Spain
Philippe Codognet	Japanese-French Laboratory for Informatics, and Université Paris 6, France
Carlos Coello Coello	CINVESTAV, Mexico

Pierre Collet	Université de Strasbourg, France
Carlos Cotta	Universidad de Malaga, Spain
Peter Demeester	KaHo Sint-Lieven, Belgium
Clarisse Dhaenens	University of Lille 1 and INRIA Lille Nord-Europe, France
Luca Di Gaspero	DIEGM - Università degli Studi di Udine, Italy
Karl F. Doerner	Johannes Kepler University Linz, Austria
Andries Engelbrecht	University of Pretoria, South Africa
Antonio J. Fernandez Leiva	Universidad de Málaga, Spain
Álvaro Fialho	Nokia Institute of Technology, Brazil
Valerio Freschi	University of Urbino, Italy
Cyril Furtlehner	INRIA Saclay Île-de-France, France
Ruben Ruiz Garcia	Universidad Politécnica de Valencia, Spain
Walter J. Gutjahr	University of Vienna, Austria
Nikolaus Hansen	INRIA Saclay Île-de-France, France
Jin-Kao Hao	University of Angers, France
Geir Hasle	SINTEF Applied Mathematics, Oslo, Norway
Federico Heras	University College Dublin, Ireland
Francisco Herrera	University of Granada, Spain
Tomio Hirata	Nagoya University, Japan
Holger Hoos	University of British Columbia, Canada
Frank Hutter	University of British Columbia, Canada
Matthew Hyde	University of Nottingham, UK
Mark Jelasity	University of Szeged, Hungary
Yaochu Jin	University of Surrey, UK
Laetitia Jourdan	LIFL and INRIA Lille Nord-Europe, France
Narendra Jussien	Ecole des Mines de Nantes, France
Zeynep Kiziltan	University of Bologna, Italy
Jiri Kubalik	Czech Technical University in Prague, Czech Republic
Arnaud Lallouet	Université de Caen, France
Frédéric Lardeux	Université d'Angers, France
Manuel López-Ibáñez	IRIDIA, CoDE, Université Libre de Bruxelles, Belgium
Jordi Levy	IIIA-CSIC, Spain
Chu-Min Li	Université de Picardie Jules Verne, France
Arnaud Liefooghe	Université Lille 1, France
Vittorio Maniezzo	University of Bologna, Italy
Felip Manya	IIIA-CSIC, Spain
Ruben Martins	INESC-ID Lisboa, Portugal
Bernd Meyer	Monash University, Australia
Zbigniew Michalewicz	University of Adelaide, Australia
Nicolas Monmarché	Université de Tours, France
Nysret Musliu	Vienna University of Technology, Austria
Amir Nakib	University of Paris-Est Créteil, France

Gabriela Ochoa	University of Nottingham, UK
Djamila Ouelhadj	University of Portsmouth, UK
Gisele Pappa	Universidade Federal de Minas Gerais, Brazil
Panos M. Pardalos	University of Florida, USA
Marcello Pelillo	University of Venice, Italy
Vincenzo Piuri	Università degli Studi di Milano, Italy
Jordi Planes	Universitat de Lleida, Spain
Günther R. Raidl	Vienna University of Technology, Austria
Celso C. Ribeiro	Universidade Federal Fluminense, Brazil
Florian Richoux	CNRS / University of Tokyo, Japan
Andrea Roli	Alma Mater Studiorum Università di Bologna, Italy
Wheeler Ruml	University of New Hampshire, USA
Thomas Runarsson	University of Iceland, Iceland
Ilya Safro	Argonne National Laboratory, USA
Lakhdar Sais	Université de Lens, France
Horst Samulowitz	IBM Research, USA
Frédéric Saubion	Université d'Angers, France
Pierre Savéant	Thalès Research & Technology, France
Andrea Schaerf	University of Udine, Italy
Michèle Sebag	CNRS - LRI, Université Paris-Sud, France
Yaroslav D. Sergeyev	Università della Calabria, Italy
Patrick Siarry	Université Paris-Est Creteil, France
Thomas Stützle	Université Libre de Bruxelles, Belgium
Ke Tang	University of Science and Technology of China, China
Olivier Teytaud	INRIA Saclay Île-de-France, France
Dirk Thierens	Utrecht University, The Netherlands
Jose Torres Jimenez	CINVESTAV, Mexico
Tamara Ulrich	ETH Zurich, Switzerland
Greet Vanden-Berghe	CODeS - KAHO Sint-Lieven, Belgium
Sébastien Vérel	Université de Nice, France
Stefan Voss	University of Hamburg, Germany
Toby Walsh	NICTA and UNSW, Australia
David L. Woodruff	University of California, Davis, USA
Shin Yoo	University College London, UK
Zhu Zhu	Université de Picardie, France

Invited Talks

Optimization problems and algorithms for the high-level control of dynamic systems

Gérard Verfaillie
ONERA, France

Abstract: The high-level control of dynamic systems, such as aircraft, airports, air traffic, or spacecraft, consists in deciding at each control step on which action(s) to be performed as a function of current observations and objectives. Successive decisions must entail that the dynamics of the controlled system satisfies user objectives as best as possible. To do so, a usual approach, inspired from the Model Predictive Approach in Automatic Control consists at each control step in (i) collecting current observations and objectives (ii) solving a deterministic planning problem over a given horizon ahead, (iii) extracting the first action from the best plan produced, (iv) applying it, and (v) considering the next step. From the optimization point of view, this implies to be able to solve quickly many successive similar planning problems over a sliding horizon, maybe not in an optimal way. I will try to present and illustrate this approach and to explain the potential impact of learning techniques.

Short bio:

Graduated from école Polytechnique (Paris) in 1971 and from SUPAéRO (French national engineering school in aeronautics and space, Computer science specialization, Toulouse) in 1985, Gérard Verfaillie is now Research supervisor at ONERA (The French Aerospace Lab). His research activity is related to models, methods, and tools for combinatorial optimization and constrained optimization, especially for planning and decision-making.

Autonomous Search

Frédéric Saubion
Université d'Angers, France

Abstract: Decades of innovations in combinatorial problem solving have produced better and more complex algorithms. These new methods are better since they can solve larger problems and address new application domains. They are also more complex, which means that they are hard to reproduce and often harder to fine tune to the peculiarities of a given problem. This last point has created a paradox where efficient tools became out of reach for practitioners. Autonomous search represents a new research field defined to precisely address the above challenge. Its major strength and originality consist in the fact that problem solvers can now perform self-improvement operations based on analysis of the performances of the solving process – including short-term reactive reconfiguration and long-term improvement through self-analysis of the performance, offline tuning and online control, and adaptive control and supervised control. Autonomous search "crosses the chasm" and provides engineers and practitioners with systems that are able to autonomously self-tune their performance while effectively solving problems. In this talk, we review existing works and we attempt to classify the different paradigms that have been proposed during past years to build more autonomous solvers. We also draw some perspectives and futures directions.

Short bio: Frédéric Saubion coheads the Metaheuristics, Optimization and Applications team at the Université d'Angers (France); his research topics include hybrid and adaptive evolutionary algorithms and applications of metaheuristics to various domains such as information retrieval, nonmonotonic reasoning and biology. www.info.univ-angers.fr/pub/saubion

Surrogate-Assisted Evolutionary Optimisation: Past, Present and Future

Yaochu Jin
Nature-Inspired Computing and Engineering Group, Department of Computing,
University of Surrey, UK

Abstract: Surrogate-assisted (or meta-model based) evolutionary computation uses efficient computational models, often known as surrogates or meta-models, for approximating the fitness function in evolutionary algorithms. Research on surrogate-assisted evolutionary computation began over a decade ago and has received considerably increasing interest in recent years. Very interestingly, surrogate-assisted evolutionary computation has found successful applications not only in solving computationally expensive single- or multi-objective optimization problems, but also in addressing dynamic optimization problems, constrained optimization problems and multi-modal optimization problems. This talk provides an up-to-date overview of the history and recent developments in surrogate-assisted evolutionary computation and suggests a few future trends in this research area.

Short bio: Yaochu Jin received the B.Sc., M.Sc., and Ph.D. degrees from Zhejiang University, China, in 1988, 1991, and 1996, respectively, and the Dr.-Ing. Degree from Ruhr University Bochum, Germany, in 2001. He is a Professor of Computational Intelligence and Head of the Nature Inspired Computing and Engineering (NICE) Group, Department of Computing, University of Surrey, UK. He was a Principal Scientist with the Honda Research Institute Europe in Germany. His research interests include understanding evolution, learning and development in biology and bio-inspired approaches to solving engineering problems. He (co)authored over 130 peer-reviewed journal and conference papers. He is an Associate Editor of BioSystems, IEEE Transactions on Neural Networks, IEEE Transactions on Systems, Man, and Cybernetics, Part C: Applications and Reviews, IEEE Transactions on Nanobioscience, and IEEE Computational Intelligence Magazine. He has delivered over ten Plenary/Keynote speeches at international conferences on multi-objective machine learning, computational modeling of neural development, morphogenetic robotics and evolutionary design optimization. He is the General Chair of the 2012 IEEE Symposium on Computational Intelligence in Bioinformatics and Computational Biology. He presently chairs the Intelligent System Applications Technical Committee of the IEEE Computational Intelligence Society. Professor Jin is a Fellow of BCS and Senior Member of IEEE.

Tutorial Talks

Addressing Numerical Black-Box Optimization: CMA-ES

Anne Auger and Nikolaus Hansen
INRIA Saclay Île-de-France

Abstract: Evolution Strategies (ESs) and many continuous domain Estimation of Distribution Algorithms (EDAs) are stochastic optimization procedures that sample a multivariate normal (Gaussian) distribution in the continuous search space, $\mathbb{R}^n$. Many of them can be formulated in a unified and comparatively simple framework. This introductory tutorial focuses on the most relevant algorithmic question: how should the parameters of the sample distribution be chosen and, in particular, updated in the generation sequence? First, two common approaches for step-size control are reviewed, one-fifth success rule and path length control. Then, Covariance Matrix Adaptation (CMA) is discussed in depth: rank-one update, the evolution path, rank-mu update. Invariance properties and the interpretation as natural gradient descent are touched upon. In the beginning, general difficulties in solving non-linear, non-convex optimization problems in continuous domain are revealed, for example non-separability, ill-conditioning and ruggedness. Algorithmic design aspects are related to these difficulties. In the end, the performance of the CMA-ES is related to other well-known evolutionary and non-evolutionary optimization algorithms, namely BFGS, DE, PSO,...

Short bios: Anne Auger is a permanent researcher at the French National Institute for Research in Computer Science and Control (INRIA). She received her diploma (2001) and PhD (2004) in mathematics from the Paris VI University. Before to join INRIA, she worked for two years (2004–2006) at ETH in Zurich. Her main research interest is stochastic continuous optimization including theoretical aspects and algorithm designs. She is a member of ACM-SIGECO executive committee and of the editorial board of Evolutionary Computation. She has been organizing the biannual Dagstuhl seminar "Theory of Evolutionary Algorithms" in 2008 and 2010. Nikolaus Hansen is researcher at The French National Institute for Research in Computer Science and Control (INRIA). He received a Ph.D. in civil engineering in 1998 from the Technical University Berlin under Ingo Rechenberg. Before joining INRIA, he has been working in applied artificial intelligence and in genomics, and he has been researching in evolutionary computation and computational science at the Technical University Berlin and the ETH Zurich. His main research interests are learning and adaptation in evolutionary computation and the development of algorithms applicable in practice. He has been a main driving force behind the development of CMA-ES over many years.

Intelligent Optimization with Submodular Functions

Andreas Krause
ETH Zurich, Switzerland

Abstract: In recent years, submodularity, a discrete analogue of convexity, has emerged as very useful in a variety of machine learning problems. Similar to convexity, submodularity allows to efficiently find provably (near-) optimal solutions. In this tutorial, I will introduce the notion of submodularity, discuss examples and properties of submodular functions, and review algorithms for submodular optimization. I will also cover recent extensions to the online (no-regret) and adaptive (closed-loop) setting. A particular focus will be on relevant applications such as active learning and optimized information gathering, ranking and algorithm portfolio optimization.

Short bio: Andreas Krause received his Diplom in Computer Science and Mathematics from the Technical University of Munich, Germany (2004) and his Ph.D. in Computer Science from Carnegie Mellon University (2008). He joined the California Institute of Technology as an assistant professor of computer science in 2009, and is currently assistant professor in the Department of Computer Science at the Swiss Federal Institute of Technology Zurich. His research is in adaptive systems that actively acquire information, reason and make decisions in large, distributed and uncertain domains, such as sensor networks and the Web. Dr. Krause is a 2010 Kavli Frontiers Fellow, and received an NSF CAREER award, the Okawa Foundation Research Grant recognizing top young researchers in telecommunications, as well as awards at several premier conferences (AAAI, KDD, IPSN, ICML, UAI) and the ASCE Journal of Water Resources Planning and Management.

Symmetry in Mathematical Programming

Leo Liberti
Ecole Polytechnique, Palaiseau, France

Abstract: This tutorial will introduce some basic concepts about group theory and how it applies to mathematical programming. We shall give an overview of the main existing research streams on this subjects, and then discuss the latest developments. We shall show how to put together existing computational tools (GAP, AMPL, CPLEX, Couenne, Rose, kept together using shell scripts) in order to automatically detect and exploit symmetry in a given mathematical programming instance.

Short bio: Leo Liberti received his PhD in 2004 from Imperial College, London. He then obtained a postdoctoral fellowship at Politecnico di Milano, and has been at LIX, Ecole Polytechnique ever since 2006, where he is an associate professor. He co-founded (and currently heads) the System Modelling and Optimization (SYSMO) team, he is co-director of the Optimization and Sustainable Development (OSD) Microsoft-CNRS sponsored chair, and is vice-president of the Computer Science department. He is Editor-in-Chief of 4OR, and holds associate editorships with several international journals (DAM, JOGO, ITOR, EURJCO, CMS). He has published more than 100 papers on mathematical programming and optimization techniques and applications.

Table of Contents

Long Papers

Short Papers

Iterative-Deepening Search
with On-Line Tree Size Prediction

Ethan Burns and Wheeler Ruml

University of New Hampshire
Department of Computer Science
{eaburns,ruml}@cs.unh.edu

Abstract. The memory requirements of best-first graph search algorithms such as A* often prevent them from solving large problems. The best-known approach for coping with this issue is iterative deepening, which performs a series of bounded depth-first searches. Unfortunately, iterative deepening only performs well when successive cost bounds visit a geometrically increasing number of nodes. While it happens to work acceptably for the classic sliding tile puzzle, IDA* fails for many other domains. In this paper, we present an algorithm that adaptively chooses appropriate cost bounds on-line during search. During each iteration, it learns a model of the search tree that helps it to predict the bound to use next. Our search tree model has three main benefits over previous approaches: 1) it will work in domains with real-valued heuristic estimates, 2) it can be trained on-line, and 3) it is able to make predictions with only a small number of training examples. We demonstrate the power of our improved model by using it to control an iterative-deepening A* search on-line. While our technique has more overhead than previous methods for controlling iterative-deepening A*, it can give more robust performance by using its experience to accurately double the amount of search effort between iterations.

1   Introduction

Best-first search is a fundamental tool for automated planning and problem solving. One major drawback of best-first search algorithms, such as A* [1], is that they store every node that is generated. This means that for difficult problems in which many nodes must be generated, A* runs out of memory. If optimal solutions are still required, however, iterative deepening A* (IDA*) [3] can often be used instead. IDA* performs a series of depth-first searches where each search expands all nodes whose estimated solution cost falls within a given bound. As with A*, the solution cost of a node n is estimated using the value $f(n) = g(n) + h(n)$ where $g(n)$ is the cost accrued along the current path from the root to n and $h(n)$ is a lower-bound on the cost that will be required to reach a goal node, which we call the *heuristic* value of n. After every iteration that fails to expand a goal, the bound is increased to the minimum f value of any node that was generated but not previously expanded. Because the heuristic

Y. Hamadi and M. Schoenauer (Eds.): LION 6, LNCS 7219, pp. 1–15, 2012.
© Springer-Verlag Berlin Heidelberg 2012

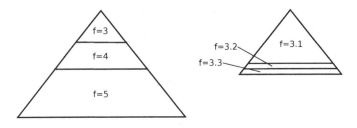

Fig. 1. Geometric versus non-geometric growth

estimator is defined to be a lower-bound on the cost-to-go and because the bound is increased by the minimum amount, any solution found by IDA* is guaranteed to be optimal. Also, since IDA* uses depth-first search at its core, it only uses an amount of memory that is linear in the maximum search depth. Unfortunately, it performs poorly on domains with few nodes per f layer as it will re-expand many interior nodes in order to expand only a very small number of new frontier nodes on each iteration.

One reason why IDA* performs well classic academic benchmarks like the sliding tiles puzzle and Rubik's cube is that both of these domains have a geometrically increasing number of nodes that fall within the successive iterations as the bound used for the search is increased by the minimum possible amount. This means that each iteration of IDA* will re-expand not only all of the nodes of the previous iterations but it will also expand a significant number of new nodes that were previously out-of-bounds. Sarkar et al. [10] show that, in a domain with this geometric growth, IDA* will expand $\mathcal{O}(n)$ nodes where n is the number of nodes expanded by A*. They also show, however, that in a domain that does not exhibit geometric growth, IDA* may expand as many as $\mathcal{O}(n^2)$ nodes. Figure 1 shows this graphically. The diagram on the left shows a tree with three f-layers each of an integer value and each layer encompasses a sufficient portion of the tree such that successive iterations of IDA* will each expand many new nodes that were not expanded previously. The right diagram in Figure 1, on the other hand, shows a tree with real-valued f layers where each layer contains only a very small number of nodes and therefore IDA* will spend a majority of its time re-expanding nodes that it has expanded previously. Because domains with real-valued edge costs tend to have many distinct f values, they fall within this later category in which IDA* performs poorly.

The main contribution of this work is a new type of model that can be used to estimate the number of nodes expanded in an iteration of IDA*. While the state-of-the-art approach to estimating search effort is able to predict the number of expansion with surprising accuracy in several domains it has two drawbacks: 1) it requires a large amount of off-line training to learn the distribution of heuristic values and 2) it does not extend easily to domains with real-valued heuristic estimates. Our new model, which we call an *incremental model*, is able to predict as accurately as the current state-of-the-art model for the 15-puzzle when trained off-line. Unlike the previous approaches, however, our incremental model can also handle domains with real-valued heuristic estimates. Furthermore, while the previous approaches require large amounts of off-line training, our model may be trained on-line during a search. We show that our model can be used to control an IDA* search by using information learned on completed iterations to determine a bound to use in the subsequent iteration. Our results show that our new model accurately predicts IDA* search effort. While IDA* guidance using our model tends to be expensive in terms of CPU time, the gain in accuracy allows the search to remain robust. Unlike the other IDA* variants which occasionally give very poor performance, IDA* using an incremental model is the only IDA* variant that can perform well over all of the domains used in our experiments.

2 Previous Work

Korf et al. [4] give a formula (henceforth abbreviated *KRE*) for predicting the number of nodes IDA* will expand with a given heuristic when searching to a given cost threshold. The KRE method uses an estimate of the heuristic value distribution in the search space to determine the percentage of nodes at a given depth that are *fertile*. A fertile node is a node within the cost bound of the current search iteration and hence will be expanded by IDA*.

The KRE formula requires two components: 1) the heuristic distribution in the search space and 2) a function for predicting the number of nodes at a given depth in the brute-force search tree. They showed that off-line random sampling can be used to learn the heuristic distribution. For their experiments, a sample size of ten billion states was used to estimate the distribution of the 15-puzzle. Additionally, they demonstrate that a set of recurrence relations, based on a special feature that they called the *type* of a node, can be used to find the number of nodes at a given depth in the brute-force search tree for a tiles puzzle or Rubik's cube. The node type used by the KRE method for the 15-puzzle is the location of the blank tile: on a side, in a corner, or in the middle. Throughout this paper, a node type can be any feature of a state that is useful for predicting information about its offspring. The results of the KRE formula using these two techniques gave remarkably accurate predictions when averaged over a large number of initial states for each domain.

Zahavi et al. [14] provide a further generalization of the KRE formula called Conditional Distribution Prediction (CDP). The CDP formula uses a conditional heuristic distribution to predict the number of nodes within a cost threshold. The formula takes into account more information than KRE such as the heuristic value and node type of the parent and grandparent of each node as conditions on the heuristic distribution. This extra information enables CDP to make predictions for individual initial states and to extend to domains with inconsistent heuristics. Using CDP, Zahavi et al. show that substantially more accurate predictions can be made on the sliding tiles puzzle and Rubik's cube given different initial states with the same heuristic value.

While the KRE and CDP formulas are able to give accurate predictions, their main drawback is that they require copious amounts of off-line training to estimate the heuristic distribution in a state space. Not only does this type of training take an excessive amount of time but it also does not allow the model to learn any instance-specific information. In addition, the implementation of these formulas as specified by Zahavi et al. [14] assumes that the heuristic estimates have integer values so that they can be used to index into a large multi-dimensional array. Many real-world domains have real-valued edge costs and therefore these techniques are not applicable in those domains.

2.1   Controlling Iterative Search

The problem with IDA* in domains with many distinct f values is well known and has been explored in past work. Vempaty et al. [12] present an algorithm called DFS*. DFS* is a combination of IDA* and depth-first search with branch-and-bound that sets the bounds between iterations more liberally than standard IDA*. While the authors describe a sampling approach to estimate the bound increase between iterations, in their experiments, the bound is simply increased by doubling.

Wah et al. [13] present a set of three linear regression models to control an IDA* search. Unfortunately, intimate knowledge of the growth properties of f layers in the desired domain is required before the method can be used. In many settings, such as domain-independent planning for example, this knowledge is not available in advance.

IDA* with Controlled Re-expansion (IDA*$_{CR}$) [10] uses a method similar to that of DFS*. IDA*$_{CR}$ uses a simple model of the search space that tracks the f values of the nodes that were pruned during the previous iteration and uses them to find a bound for the next iteration. IDA*$_{CR}$ uses a histogram to count the number of nodes with each out-of-bound f value during each iteration of search. When the iteration is complete, the histogram is used to estimate the f value that will double the number of nodes in the next iteration. The remainder of the search proceeds as in DFS*, by increasing the bound and performing branch-and-bound on the final iteration to guarantee optimality.

While IDA*$_{CR}$ is simple, the model that it uses to estimate search effort relies upon two assumptions about the search space to achieve good performance. The first is that the number of nodes that are generated outside of the bound must be at least the same as the number of nodes that were expanded. If there are an insufficient number of pruned nodes, IDA*$_{CR}$ sets the bound to the greatest pruned f value that it has seen. This value may be too small to significantly advance the search. The second assumption is that none of the children of the pruned frontier nodes of one iteration should fall within the bound on the next iteration. If this happens, then the next iteration may be much larger than twice the size of the previous. As we will see, this can cause the search to overshoot the optimal solution cost on its final iteration, giving rise to excessive search effort.

3   Incremental Models of Search Trees

To estimate the number of nodes that IDA* will expand when using a given cost threshold, we would like to know the distribution of f values in the search space. Assuming a consistent heuristic[1], all nodes with f values within the threshold will be expanded. If this distribution is given as a histogram that contains the number of nodes with each f value, then we can simply find the bound for which the number of nodes with f values less than the bound matches our desired value. Our new incremental model performs this task and has the ability to be trained both off-line with sampling and on-line during a search.

We will estimate the distribution of f values in two steps. In the first step, we learn a model of how the f values are changing from nodes to their offspring. In the second step, we extrapolate from the model of change in f values to estimate the overall distribution of f values. This means that our incremental model manipulates two main distributions: we call the first one the Δf distribution and the second one the f distribution. In the next section, we will describe the Δf distribution and give two techniques for learning it. We will then describe how the Δf distribution can be used to estimate the f distribution.

3.1   The Δf Distribution

The goal of learning the Δf distribution is to predict how the f values in the search space change between nodes and their offspring. The advantage of storing Δf values instead of storing the f values themselves is that it enables our model to extrapolate to portions of the search space for which it has no training data, a necessity when using the model on-line or with few training samples. We will use the information from the Δf distribution to build an estimate of the distribution of f values over the search nodes.

The CDP technique of Zahavi et al. [14] learns a conditional distribution of the heuristic value and node type of a child node c, conditioned on the node type and heuristic estimate of the parent node p, notated $P(h(c), t(c)|h(p), t(p))$. As described by [14], this requires indexing into a multi-dimensional array according to $h(p)$ and so the heuristic estimate must be an integer value. Our incremental model also learns a conditional distribution, however in order to handle real-valued heuristic estimates, our incremental model uses the integer valued search-space-steps-to-go estimate d of a node instead of its cost-to-go lower bound, h. In unit-cost domains, d, also known as the distance estimate, will typically be the same as h, however in domains with real-valued edge costs they will differ. d is typically easy to compute while computing h [11]. The distribution that is learned by the incremental model is $P(\Delta f(c), t(c), \Delta d(c)|d(p), t(p))$, that is, the distribution over the change in f value between a parent and child, the child

---

[1] A heuristic is consistent when the change in the h value between a node and its successor is no greater than the cost of the edge between the nodes. If the heuristic is not consistent then a procedure called pathmax [6] can be used to make it consistent locally along each path traversed by the search.

node type and the change in d estimate between a parent and child, given the distance estimate of the parent and the type of the parent node.

The only non-integer term used by the incremental model is $\Delta f(c)$. Our implementation uses a large multi-dimensional array of fixed-sized histograms over $\Delta f(c)$ values. Each of the integer-valued features is used to index into the array, resulting in a histogram of the $\Delta f(c)$ values. By storing counts, the model can estimate the branching factor of the search space by dividing the total number of nodes with a given d and t by the total number of their offspring. This branching factor will be used below to estimate the number of successors of a node when building the f distribution.

Zahavi et al. [14] found that it is often important to take into account information about the grandparent of a node for the distributions used in CDP. We accomplish this with the incremental model by rolling together the node types of the parent and grandparent into a single type. For example, on the 15-puzzle, if the parent state has the blank in the center and it was generated by a state with the blank on the side, then the parent type would be a *side–center* node. This allows us to use an array with the same dimensionality across domains that take different amounts of ancestry into account.

Learning Off-Line. We can learn an incremental Δf model off-line using the same method as with KRE and CDP. A large number of random states from a domain are sampled, and the children (or grandchildren) of each sampled state are generated. The change in distance estimate $\Delta d(c) = d(c) - d(p)$, node type $t(c)$ of the child node, node type $t(p)$ of the parent node, and the distance estimate $d(p)$ of the parent node are computed and a count of 1 is then added to the appropriate histogram for the (possibly real-valued) change in f, $\Delta f(c) = f(c) - f(p)$ between parent and child.

Learning On-Line. An incremental Δf model can also be learned on-line during search. Each time a node is generated, the $\Delta d(c)$, $t(c)$, $t(p)$ and $d(p)$ values are computed for the parent node p and child node c and a count of 1 is added to the corresponding histogram for $\Delta f(c)$, as in the off-line case. In addition, when learning a Δf model on-line, the *depth* of the parent node in the search tree is also known. We have found that this feature greatly improves accuracy in some domains (such as the vacuum domain described below) and so we always add it as a conditioning feature when learning an incremental model on-line.

Each iteration of IDA* search will expand a superset of the nodes expanded during the previous iteration. To avoid duplicating effort, our implementation tracks the bound used in the previous iteration and the model is only updated when expanding a node that would have been pruned on the previous iteration. Additionally, the search spaces for many domains form graphs instead of trees. In these domains, our implementation of depth-first search does cycle checking by using a hash table of all of the nodes along the current path. In order for our model to take this extra pruning into account, we only train the model on the successors of a node that pass the cycle detection.

Learning a Backed-Off Model. Due to data sparsity, and because the Δf model will be used to extrapolate information about the search space for which it may not have any training data, a backed-off version of the model may be needed that is conditioned on fewer features of each node. When querying the model, if there is no training data for a given set of features, the more general backed-off model is consulted instead. When learning a model on-line, because the model is learned on instance-specific data, we found that it was only necessary to learn a model that backs off the depth feature. When training off-line, however, we learn a series of two back-off models, first eliminating the parent node distance estimate and then eliminating both the parent distance and type.

3.2   The f Distribution

Our incremental model predicts a bound that will result in expanding the desired number of nodes for a given start state by estimating the distribution of f values of the nodes in the search space. The f value distribution of one search depth and the model of Δf are used to generate the f value distribution for the next depth. By beginning with the root node, which has a known f value, our procedure simulates the expansions of each depth layer to incrementally compute estimates of the f value distribution at the next layer. The accumulation of these depth-based f value distributions can then be used to make our prediction.

To increase accuracy, the distribution of f values at each depth is conditioned on node type t and distance estimate d. We begin our simulation with a model of depth 0 which is simply a count of 1 for $f = f(root)$, $t = t(root)$ and $d = d(root)$. Next, the Δf model is used to find a distribution over Δf, t and Δd values for the offspring of the nodes at each combination of t and d values at the current depth. By storing Δ values, we can compute $d(c) = d(p) + \Delta d(c)$ and $f(c) = f(p) + \Delta f(c)$ for each parent p with a child c. This gives us the number of nodes with each f, t and d value at the next depth of the search.

Because the Δf values may be real numbers, they are stored as histograms by our Δf model. In order to add $f(p) + \Delta f(c)$, we use a procedure called additive convolution [9,8]. Each node, i.e. every count in the histogram for the current layer, will have offspring whose f values differ according the Δf distribution. The additive convolution procedure sums the distribution of child f values for every count in the current layer's f histogram, resulting in a histogram of f values over all successors. More formally, the convolution of two histograms ω_a and ω_b, where ω_a and ω_b are functions from values to weights, is a histogram ω_c, where $\omega_c(k) = \sum_{i \in Domain(\omega_a)} \omega_a(i) \cdot \omega_b(k - i)$. By convolving the f distribution of a set of nodes with the distribution of the change in f values between these nodes and their offspring, we get the f distribution of the offspring.

Since the maximum depth of a shortest-path search tree is typically unknown, our simulation must use a special criterion to determine when to stop. With a consistent heuristic the f values of nodes will be non-decreasing along a path [7] and therefore the change in f stored in our model will always be positive. Since the change in f is always positive, the f values encountered during the simulation will always increase between layers. As soon as the simulation estimates that a

SIMULATE(*bound, desired, depth, accum, nodes*)
   1. *nodes′* = SIMEXPAND(*depth, nodes*)
   2. *accum′* = *add*(*accum, nodes − nodes′*)
   3. *bound′* = *find_bound*(*accum′, bound, desired*)
   4. if *weight_left_of*(*bound′, nodes − nodes′*) > ϵ
   5.   *depth′* = *depth* + 1
   6.   SIMULATE(*bound′, desired, depth′, accum′, nodes′*)
   7. else return *accum′*

SIMEXPAND(*depth, nodes*)
   8. *nodes′* = new2dhistogramarray
   9. for each *t* and *d* with *weight*(*nodes*[*t, d*]) > 0 do
   10.   *fs* = *nodes*[*t, d*]
   11.   SIMGEN(*depth, t, d, fs, nodes′*)
   12. return *nodes′*

SIMGEN(*depth, t, d, fs, nodes′*)
   13. for each type *t′* and Δd
   14. Δfs = *delta_f_model*[*t′, Δd, d, t*]
   15.   if *weight*(Δfs) > 0 then
   16.   *d′* = *d* + Δd
   17.   *fs′* = CONVOLVE(*fs, Δfs*)
   18.   *nodes′*[*t′, d′*] = *add*(*nodes′*[*t′, d′*], *fs′*)
   19. done

Fig. 2. Pseudo code for the simulation procedure used to estimate the *f* distribution

sufficient number of nodes will be generated to meet our desired count, the maximum *f* value can be fixed as an upper bound since selecting a greater *f* value can only give more nodes than desired. As the simulation proceeds further, we re-evaluate the *f* value that gives our desired number of nodes to account new node generations. This upper bound will continue to decrease and the simulation will estimate fewer and fewer new nodes within the bound at each depth. When the expected number of new nodes is only a fractional value smaller than some ϵ the simulation can stop. In our experiments we use $\epsilon = 10^{-3}$. Additionally, because the *d* value of a node can never be negative, we can prune all nodes that would be generated with $d \leq 0$.

Figure 2 shows the pseudo-code for the procedure that estimates the *f* distribution. The entry point is the SIMULATE function which has the following parameters: the cost bound, desired number of nodes, the current depth, a histogram that contains the accumulated distribution of *f* values so far and a 2-dimensional array of histograms which stores the conditional distribution of *f* values among the nodes at the current depth. SIMULATE begins by simulating the expansion of the nodes at the current depth (line 1). The result of this is the conditional distribution of *f* values for the nodes generated as offspring at the next depth. These *f* values are accumulated into a histogram of all *f* values seen by the simulation thus far (line 2). An upper bound is determined (line 3) and if greater than ϵ new nodes are estimated to be in the next depth then the simulation continues recursively (lines 4–6), otherwise the accumulation of all f values is returned as the final result.

The SIM-EXPAND function is used to build the conditional distribution of the f values for the offspring of the nodes at the current simulation-depth. For each node type t and distance estimate d for which there exist nodes at the current depth, the SIM-GEN function is called to estimate the conditional f distribution of their offspring (lines 9–11). SIM-GEN uses the Δf distribution (line 14) to compute the frequency of f values for nodes generated from parents with the specified combination of type and distance-estimate. Because this distribution is over Δf, t and Δd, we have all of the information that is needed to construct the conditional f distribution for the offspring (lines 16–18).

Warm Starting. As an iterative deepening search progresses, some of the shallower depths become *completely expanded*: no nodes are pruned at that depth or any shallower depth. All of the children of nodes in a completely expanded depth are *completely generated*. When learning the Δf distribution on-line, our incremental model has the exact *depth*, d and f values for all of the layers that have been completely generated. We "warm start" the simulation by seeding it with the perfect information for completed layers and beginning at the first depth that has not been completely generated. This can speed up the computation of the f distribution and can increase accuracy.

4 Empirical Evaluation

In the following sections we show an empirical study of our new model and some of the related previous approaches. We begin by evaluating the accuracy of the incremental model when trained off-line. We then show the accuracy of the incremental model when used on-line to control an IDA* search.

4.1 Off-line Learning

We evaluate the quality of the predictions given by the incremental model when using off-line training by comparing the predictions of the model with the true node expansion counts. For each problem instance the optimal solution cost is used as the cost bound. Because both CDP and the incremental model estimate all of the nodes within a cost bound, the truth values are computed by running a full depth-first search of the tree bounded by the optimal solution cost. This search is equivalent to the final iteration of IDA* assuming that the algorithm finds the goal node after having expanded all other nodes that fall within the cost bound.

Estimation Accuracy. We trained both CDP [14] and an incremental model off-line on ten billion random 15-puzzle states using the Manhattan distance heuristic. We then compared the predictions given by each model to the true number of nodes within the optimal-solution-cost bound for each of the standard

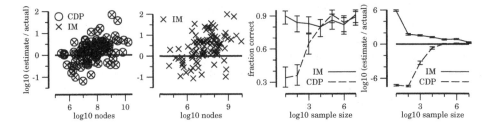

Fig. 3. Accuracy when trained off-line

100 15-puzzle instances due to Korf [3]. The leftmost plot of Figure 3 shows the results of this experiment. The x axis is on a log scale; it shows the actual number of nodes within the cost bound. The y axis is also on a log scale; it shows the ratio of the estimated number of nodes to the actual number of nodes, we call this metric the *estimation factor*. The closer that the estimation factor is to one (recall that $\log_{10}1 = 0$) the more accurate the estimation was. The median estimation factor for the incremental model was 1.435 and the median estimation factor for CDP was 1.465 on this set of instances. From the plot we can see that, on each instance, the incremental model gave estimations that were nearly equivalent to those given by CDP, the current state-of-the-art predictor for this domain.

To demonstrate our incremental model's ability to make predictions in domains with real-valued edge costs and with real-valued heuristic estimates, we created a modified version of the 15-puzzle where each move costs the square root of the tile number that is being moved. We call this problem the square root tiles puzzle and for the heuristic we use a modified version of the Manhattan distance heuristic that takes into account the cost of each individual tile.

As presented by Zahavi et al. [14], CDP is not able to make predictions on this domain because of the real-valued heuristic estimates. The second panel in Figure 3 shows the estimation factor for the predictions given by the incremental model trained off-line on fifty billion random square root tiles states. The same 100 puzzle states were used. Again, both axes are on a log scale. The median estimation factor on this set of puzzles was 2.807.

Small Sample Sizes. Haslum et al. [2] use a technique loosely based on the KRE formula to select between different heuristics for domain independent planning. When given a choice between two heuristic lower bound functions, we would like to select the heuristic that will expand fewer nodes. Using KRE (or CDP) to estimate node expansions requires a very large off-line sample of the heuristic distribution to achieve accurate predictions, which is not achievable in applications such as Haslum et al.'s. Since the incremental model uses Δ values and a backed-off model, however, it is able to make useful predictions with very little training data. To demonstrate this, we created 100 random pairs of instances from Korf's set of 15-puzzles. We used both CDP and the incremental model to estimate the number of expansions required by each instance when given its

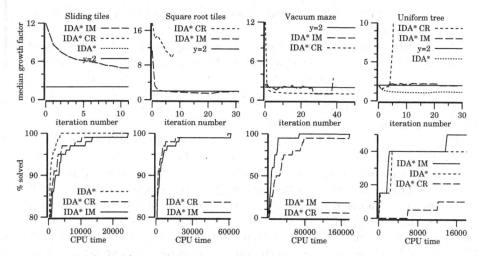

Fig. 4. IDA*, IDA*$_{CR}$ and IDA*$_{IM}$ growth rates and number of instances solved

optimal solution cost. We rated the performance of each model based on the fraction of pairs for which it was able to correctly determine the more difficult of the two instances.

The third plot in Figure 3 shows the fraction of pairs that were ordered correctly by each model for various sample sizes. Error bars represent 95% confidence intervals on the mean. We can see from this plot that the incremental model was able to achieve much higher accuracy when ordering the instances with as few as ten training samples. CDP required 10,000 training samples or more to achieve comparable accuracy. The rightmost plot in this figure shows the $\log_{10}$ estimation factor of the estimates made by each model. While CDP achieved higher quality estimates when given 10,000 or more training instances, the incremental model was able to make much more accurate predictions when trained on only 10, 100 and 1,000 samples.

4.2 On-Line Learning

In this section, we evaluate the incremental model when trained and used on-line during an IDA* search. When it comes time to set the bound for the next iteration, the incremental model is consulted to find a bound that is predicted to double the number of node expansions from that of the previous iteration. We call this algorithm IDA*$_{IM}$. As we will see, because the model is trained on the exact instance for which it will be predicting, the estimations tend to be more accurate than the off-line estimations, even with a much smaller training set. In the following subsections, we evaluate the incremental model by comparing IDA*$_{IM}$ to the original IDA*[3] and IDA*$_{CR}$[10].

Sliding Tiles. The unit-cost sliding tiles puzzle is a domain where standard IDA* search works very well. The minimum cost increase between iterations is two and this leads to a geometric increase in the number of nodes between subsequent iterations.

The top left panel of Figure 4 shows the median *growth factor*, the relative size of one iteration compared to the next, on the y axis, for IDA* , IDA*$_{CR}$ and IDA*$_{IM}$. Ideally, all algorithms would have a median growth factor of two. All three of the lines for the algorithms are drawn directly on top of one another in this plot. While both IDA*$_{CR}$ and IDA*$_{IM}$ attempted to double the work done by subsequent iterations, all algorithms still achieved no less than 5x growth. This is because, due to the coarse granularity of f values in this domain, no threshold can actually achieve the target growth factor. However, the median estimation factor of the incremental model over all iterations in all instances was 1.029. This is very close to the optimal estimation factor of one. So, while granularity of f values made doubling impossible, the incremental model still predicted the amount of work with great accuracy. The bottom panel shows the percentage of instances solved within the time given on the x axis. Because IDA*$_{IM}$ and IDA*$_{CR}$ must use branch-and-bound on the final iteration of search they are unable to outperform IDA* in this domain.

Square Root Tiles. While IDA* works well on the classic sliding tile puzzle, a trivial modification exposes its fragility: changing the edge costs. In this section, we look at the square root cost variant of the sliding tiles. This domain has many distinct f values, so when IDA* increases the bound to the smallest out-of-bound f value, it will visit a very small number of new nodes with the same f in the next iteration. We do not show the results for IDA* on this domain because it gave extremely poor performance. IDA* was unable to solve any instances with a one hour timeout and at least one instance requires more than a week to solve.

The second column of Figure 4 presents the results for IDA*$_{IM}$ and IDA*$_{CR}$. Even with the branch-and-bound requirement, IDA*$_{IM}$ and IDA*$_{CR}$ easily outperform IDA* by increasing the bound more liberally between iterations. While IDA*$_{CR}$ gave slightly better performance with respect to CPU time, its model was not able to provide very accurate predictions. The growth factor between iterations for IDA*$_{CR}$ was no smaller than eight times the size of the previous iteration when the goal was to double. The incremental model, however, was able to keep the growth factor very close to doubling. The median estimation factor was 0.871 for the incremental model which is much closer to the optimal estimation factor of one than when the model was trained off-line. We conjecture that the model was able to learn features that were specific to the instance for which it was predicting.

One reason why IDA*$_{CR}$ was able to achieve competitive performance in this domain is because, by increasing the bound very quickly, it was able to skip many iterations of search that IDA*$_{IM}$ performed. IDA*$_{CR}$ performed no more than 10 iterations on any instance in this set whereas IDA*$_{IM}$ performed up to 33 iterations on a single instance. Although the rapid bound increase was beneficial in the square root tiles domain, in a subsequent section we will see that increasing the bound too quickly can severely hinder performance.

Vacuum Maze. The objective of the vacuum maze domain is to navigate a robot through a maze in order for it to vacuum up spots of dirt. In our experiments, we used 20 instances of 500x500 mazes that were built with a depth-first search. Long hallways with no branching were then collapsed into single edges with a cost equivalent to the hallway length. Each maze contained 10 pieces of dirt and any state in which all dirt had been vacuumed was a goal. The median number of states per instance was 56 million and the median optimal solution cost was 28, 927. The heuristic was the size of the minimum spanning tree of the locations of the dirt and vacuum. The *pathmax* procedure [6] was used to make the f values non-decreasing along a path.

The third column of Figure 4 shows the median growth factor and number of instances solved by each algorithm for a given amount of time. Again, IDA* is not shown due to its very poor performance in this domain. Because there are many dead ends in each maze, the branching factor in this domain is very close to one. The model used by IDA*$_{CR}$ gave very inaccurate predictions and the algorithm often increased the bound by too small of an increment between iterations. IDA*$_{CR}$ performed up to 386 iterations on a single instance. With the exception of a dip near iterations 28–38, the incremental model was able to accurately find a bound that doubled the amount of work between iterations. The dip in the growth factors may be attributed to histogram inaccuracy on the later iterations of the search. The median estimation factor of the incremental model was 0.968, which is very close to the perfect factor of one. Because of the poor predictions given by the IDA*$_{CR}$ model, it was not able to solve instances as quickly as IDA*$_{IM}$ on this domain.

While our results demonstrate that the incremental model gave very accurate predictions in the vacuum maze domain, it should be noted that, due to the small branching factor, iterative searches are not ideal for this domain. A simple implementation of frontier A* [5] was able to solve each instance in this set in no more than 1,887 CPU seconds.

Uniform Trees. We also designed a simple synthetic domain that illustrates the brittleness of IDA*$_{CR}$. We created a set of trees with 3-way branching where each node has outgoing edges of cost 1, 20 and 100. The goal node lies at a depth of 19 along a random path that is a combination of 1- and 20-cost edge and the heuristic $h = 0$ for all nodes. We have found that the model used by IDA*$_{CR}$ will often increase the bound extremely quickly due to the large 100-cost branches. Because of the extremely large searches created by IDA*$_{CR}$ we use a five hour time limit in this domain.

The top right plot in Figure 4 shows the growth factors and number of instances solved in a given amount of time for IDA*$_{IM}$, IDA*$_{CR}$ and IDA*. Again, the incremental model was able to achieve very accurate predictions with a median estimation factor of 0.978. IDA*$_{IM}$ was able to solve ten of twenty instances and IDA* solved eight within the time limit. IDA*$_{IM}$ solved every instance in less time than IDA*. IDA*$_{CR}$ was unable to solve more than two instances within the time limit. It grew the bounds in between iterations extremely quickly, as can be seen in the growth factor plot on the bottom right in Figure 4.

Although IDA* tended to have reasonable CPU time performance in this domain, its growth factors were very close to one. The only reason that IDA* achieved reasonable performance is because expansions in this synthetic tree domain required virtually no computation. This would not be observed in a more realistic domain where expansion required any reasonable computation.

4.3 Summary

When trained off-line, the incremental model was able to make predictions on the 15-puzzle domain that were nearly indistinguishable from CDP, the current state-of-the art. In addition, the incremental model was able to estimate the number of node expansions on a real-valued variant of the sliding tiles puzzle where each move costs the square root of the tile number being moved. When presented with pairs of 15-puzzle instances, the incremental model trained with 10 samples was more accurately able to predict which instance would require fewer expansions than CDP when trained with 10,000 samples.

The incremental model made very accurate predictions across all domains when trained on-line and when used to control the bounds for IDA*, our model made for a robust search. While the alternative approaches occasionally gave extremely poor performance, IDA* controlled by the incremental model achieved the best performance of the IDA* searches in the vacuum maze and uniform tree domains and was competitive with the best search algorithms for both of the sliding tiles domains.

5 Discussion

In search spaces with small branching factors such as the vacuum maze domain, the backed-off model seems to have a greater impact on the accuracy of predictions than in search spaces with larger branching factor such as the sliding tiles domains. Because the branching factor in the vacuum maze domain is small, however, the simulation must extrapolate out to great depths (many of which the model has not been trained on) to accumulate the desired number of expansions. The simple backed-off model used here merely ignored depth. While this tended to give accurate predictions for the vacuum maze domain, a different model may be required for other domains.

6 Conclusion

In this paper, we presented a new incremental model for predicting the distribution of solution cost estimates in a search tree and hence the number of nodes that bounded depth-first search will visit. Our new model is comparable to state-of-the-art methods in domains where those methods apply. The three main advantages of our new model are that it works naturally in domains with real-valued heuristic estimates, it is accurate with few training samples, and it can be trained on-line. We demonstrated that training the model on-line can lead to more accurate predictions. Additionally, we have shown that the incremental model can be used to control an IDA* search, giving a robust algorithm, IDA*$_{IM}$. Given the prevalence of real-valued costs in real-world problems, online incremental models are an important step in broadening the applicability of iterative deepening search.

Acknowledgements. We gratefully acknowledge support from NSF (grant IIS-0812141) and the DARPA CSSG program (grant HR0011-09-1-0021).

References

1. Hart, P.E., Nilsson, N.J., Raphael, B.: A formal basis for the heuristic determination of minimum cost paths. IEEE Transactions of Systems Science and Cybernetics SSC 4(2), 100–107 (1968)
2. Haslum, P., Botea, A., Helmert, M., Bonte, B., Koenig, S.: Domain-independent construction of pattern database heuristics for cost-optimal planning. In: Proceedings of the Twenty-Second Conference on Artificial Intelligence (AAAI 2007) (July 2007)
3. Korf, R.E.: Iterative-deepening-A*: An optimal admissible tree search. In: Proceedings of the International Joint Conference on Artificial Intelligence (IJCAI 1985), pp. 1034–1036 (1985)
4. Korf, R.E., Reid, M., Edelkamp, S.: Time complexity of iterative-deepening-A*. Artificial Intelligence 129, 199–218 (2001)
5. Korf, R.E., Zhang, W., Thayer, I., Hohwald, H.: Frontier search. Journal of the ACM 52(5), 715–748 (2005)
6. Méro, L.: A heuristic search algorithm with modifiable estimate. Artificial Intelligence, 13–27 (1984)
7. Pearl, J.: Heuristics: Intelligent Search Strategies for Computer Problem Solving. Addison-Wesley (1984)
8. Rose, K., Burns, E., Ruml, W.: Best-first search for bounded-depth trees. In: The 2011 International Symposium on Combinatorial Search (SOCS 2011) (2011)
9. Ruml, W.: Adaptive Tree Search. Ph.D. thesis, Harvard University (May 2002)
10. Sarkar, U., Chakrabarti, P., Ghose, S., Sarkar, S.D.: Reducing reexpansions in iterative-deepening search by controlling cutoff bounds. Artificial Intelligence 50, 207–221 (1991)
11. Thayer, J., Ruml, W.: Using distance estimates in heuristic search. In: Proceedings of ICAPS 2009 (2009)
12. Vempaty, N.R., Kumar, V., Korf, R.E.: Depth-first vs best-first search. In: Proceedings of AAAI 1991, pp. 434–440 (1991)
13. Wah, B.W., Shang, Y.: Comparison and evaluation of a class of IDA* algorithms. International Journal on Artificial Intelligence Tools 3(4), 493–523 (1995)
14. Zahavi, U., Felner, A., Burch, N., Holte, R.C.: Predicting the performance of IDA* using conditional distributions. Journal of Artificial Intelligence Research 37, 41–83 (2010)

A Learning Optimization Algorithm in Graph Theory

Versatile Search for Extremal Graphs Using a Learning Algorithm *

Gilles Caporossi and Pierre Hansen

GERAD and HEC Montréal (Canada)
{Gilles.Caporossi,Pierre.Hansen}@gerad.ca

Abstract. Using a heuristic optimization module based upon Variable Neighborhood Search (VNS), the system AutoGraphiX's main feature is to find extremal or near extremal graphs, i.e., graphs that minimize or maximize an invariant. From the so obtained graphs, conjectures are found either automatically or interactively. Most of the features of the system relies on the optimization that must be efficient but the variety of problems handled by the system makes the tuning of the optimizer difficult to achieve. We propose a learning algorithm that is trained during the optimization of the problem and provides better results than all the algorithms previously used for that purpose.

Keywords: extremal graphs, learning algorithm, combinatorial optimization.

1 Introduction

A graph G is defined by a set V of vertices and a set E of edges representing pairs of vertices. A graph invariant is a function $I(G)$ that associates a numerical value to each graph $G = (V, E)$ regardless of the way vertices or edges are labelled. Examples of invariants are the number of vertices $|V| = n$, the number of edges $|E| = m$, the maximum distance between two vertices (diameter), the chromatic number (minimum number of colors needed so that each vertex is colored and two adjacent vertices do not share a color). Some more sophisticated invariants are related to spectral graph theory such as the *index* (largest eigenvalue of the adjacency matrix), the *energy* (sum of the absolute values of the eigenvalues of the adjacency matrix). A graph that minimizes or maximizes an invariant (or a function of invariants, which is also an invariant) is called extremal graph. The system AutoGraphiX (AGX) for computer assisted graph theory was developed at GERAD, Montreal. Since 1997, AGX led to the publication of more than 50 papers. The search for extremal graphs is the first goal of AGX and it is an important tool for graph theorists and mathematical chemists as it may be used to handle the following other goals:

- *Find a graph given some constraints*, achieved by the use of Lagrangian relaxation.
- *Refute or strengthen a conjecture.* Suppose a conjecture says that the invariant I_1 is larger than the invariant I_2 ($I_1 \geq I_2$), minimizing $I_1 - I_2$ could provide a counter-example if a negative value is obtained. Whether a counter-example is found or not,

---

* The authors gratefully acknowledge the support from NSERC (Canada).

Y. Hamadi and M. Schoenauer (Eds.): LION 6, LNCS 7219, pp. 16–30, 2012.
© Springer-Verlag Berlin Heidelberg 2012

looking at the extremal values and the corresponding graphs could help strengthening or correcting the original conjecture.

- *Find conjectures.* Structural or numerical conjectures may be obtained automatically or interactively by analyzing or looking at the extremal graphs obtained.

Based upon the Variable Neighborhood Search metaheuristic (VNS) [16][17], Caporossi and Hansen [9] developped AGX. The extremal graphs obtained by AGX are studied either directly by the researchers or by automated routines that may identify properties of the extremal graphs and deduce conjectures on the problem under study [8][10].

Several graph theorists have used AGX (and the recent AGX2) for study of invariants which most interested them. Applications to mathematics concern spectral graph theory, *i.e.*, the index [11] and the algebraic connectivity [3], as well as several standard graphs invariants [2] and a property of trees [5]. Applications in mathematical chemistry concern the Randić (or connectivity) index [6,7,13,14], the energy [4], indices of irregularity [15] and the HOMO-LUMO gap [12]. This work has led to many extensions by several mathematicians and chemists.

AGX relies on the VNS but also on a large number of transformations used within the search for a local optimum in the Variable Neighborhood Descent (VND) phase of the algorithm. The good performance of the system depends on the user's knowledge to select the correct transformations to use. If the transformations are not appropriate, the optimization will have a poor performance, either because it fails to obtain good solutions or because it takes much too long time. Indeed, choosing a transformation that is too sophisticated will result in a waste of time while a transformation that is too simple will be fast but inefficient. The authors of the system, aware of this problem, proposed in the second version of AGX (called AGX 2) an algorithm that selects automatically its transformations [1], the Adaptive Local Search (ALS). While ALS is a step toward the automation of the selection of the transformations, it cannot as such be considered as a learning algorithm since it is unable to learn on large graphs (its learning is very time consuming on graphs with more than 12 vertices).

In this paper, we propose a new learning algorithm that could replace the original VND used in AGX 1 or the Adaptive Local Search (ALS) of AGX 2. As the ALS, the new Learning Descent (LD) does not require any knowledge in combinatorial optimization and is based upon the concept of transformation matrix. However, its learning capabilities are much more powerful. The next section of the paper describes the various optimization algorithms that have been used to search for extremal graphs. A comparison of the performance of the different algorithms is done in the third section and a short conclusion is drawn at the end of the paper.

2 The Variable Neighborhood Search in AGX

The optimization in AGX is done by VNS which is well suited to handle a wide variety of problems with little tuning, compared to most other methods such as tabu search or genetic algorithms.

Let G be a graph and consider a transformation, for example *move* that consists in removing an edge from G and inserting it in another place on G. This transformation may be used to define $N(G)$, the neighborhood of G, or set of all graphs that may be constructed from G by the transformation *move*. Such neighborhoods could be extended to a succession of transformations. One thus defines the nested neighborhoods $N^k(G)$, the set of graphs that could be constructed by applying k times the chosen transformation to G. This concept of neighborhoods plays an important role in VNS and the definition of a multitude of these neighborhoods is plainly used in the AGX implementation to handle efficiently a wide variety of different problems that would require different neighborhoods (or transformations) for good results.

In AGX, the standard implementation of VNS is used, alternating Local Search and variable magnitude perturbations as described on *figure 1*.

Initialization:
- Select the neighborhood structure N^k and a stopping condition.
- Let G be an initial (usually random) graph.
- Let G^* denote the best graph obtained to date.

Repeat until condition is met:
- Set $k \leftarrow 1$;
- **Until $k = k_{max}$,do:**
  (a) Generate a random graph $G' \in N^k(G)$;
  (b) Apply LS to G'
  Denote G'' the obtained local optimum $G'' = LS(G')$;
  (c) If G'' is better than G,
        Let $G^* \leftarrow G''$ and
        $k \leftarrow 1$
     otherwise,
        set $k \leftarrow k + 1$.

   done

Fig. 1. Rules of Variable Neighborhood Search

2.1   The VND Algorithm in AGX 1

The choice of a good transformation within the local search is a key to success. To add flexibility to the system, different transformations are implemented that could be used one after the other on the same problem. Thus, the Variable Neighborhood Descent is a succession of local searches involving different transformations used for the search. The program performs a local search for each transformation in the list until none of them succeeds. VND could be considered as an extension of local search as it provides a local optimum with respect to each of the transformations used in the search. The VND algorithm is described on figure 2.

While the general VNS parameter k could often be set to the default value, the choice of the list of transformations in VND is much more critical. For instance, if the number of vertices and edges are fixed, any transformation that results in a modification of these numbers will be useless.

A variety of neighborhoods are implemented in AGX to adapt the system to different kinds of problems. Some of these transformations were specially designed to handle special classes of problems (for example, *2-Opt* is well suited for problems with fixed

Initialization:
Select a list of neighborhood structures
$N_l(G)$, $\forall l = 1 \ldots L$, that will be used.
Consider an initial graph G,
set *improved* $\leftarrow$ *true*.
Until *improved* $=$ *false* **do**
    Set *improved* $=$ *false*
    Set $l = 1$
    Until $l = L$ **do**
        (a) Find the best graph $G' \in N_l(G)$.
        (b) If G' is better than G,
            set $G \leftarrow G'$,
            set *improved* $\leftarrow$ *true* and
            return to step *(a)*.
          Otherwise,
            set $l \leftarrow l + 1$;
    done
done

Fig. 2. Rules of Variable Neighborhood Descent

numbers of edges where *simple_move* is very inefficient). The set of transformations used in AGX is described in [9].

To take advantage of the capabilities of AGX, the researcher must have sufficient knowledge in combinatorial optimization, which is not necessarily the case.

2.2 The Adaptive Local Search in AGX 2

The Adaptive Local Search (ALS) may be viewed as meta-transformations that could eventually be used within the VND frame. However, by themselves, ALS replaces most of the transformations available within AGX 1. Each transformation is described as a replacement of an induced subgraph g' of G by another subgraph g''. Considering 4 vertices, at most 6 edges could be present in any graph. It is therefore possible to consider up to $2^6 = 64$ labelled subgraphs on 4 vertices. ALS enumerates all the subgraphs g' with 4 vertices in G. It then considers replacing g' by an alternative subgraph g''. As enumerating and evaluating all the alternative subgraphs g'' to replace g' would be very time consuming, replacing g' by g'' will only be evaluated if there are good reasons to believe it is worthwhile. The implementation of this method encodes each subgraph g' or g'' as a label (number) based upon the 64 patterns as follows.

After relabeling its vertices from 1 to 4 by preserving their order, each subgraph g' is characterized by a unique label from 0 to 63 as follows:
pattern 0 (vector $=$ 000000): empty subgraph
pattern 1 (vector $=$ 000001): $E = \{(1,2)\}$
pattern 2 (vector $=$ 000010): $E = \{(1,3)\}$
:
pattern 13 (vector $=$ 001101): $E = \{(1,2),(1,4),(2,3)\}$
:
pattern 63 (vector $=$ 111111): complete subgraph on 4 vertices.

A 64×64 transformation matrix $T = \{t_{ij}\}$ is used to store information on the performance of each possible transformation from pattern i to pattern j.

In ALS, T is a binary matrix indicating whether a transformation t_{ij} from the pattern i to the pattern j was ever found useful.

Based upon this definition of patterns, the principle of the ALS is to use the selected transformations to try to obtain a better graph. Once the graph could no more be improved by the selected transformations (a local optimum is found with respect to the considered transformations), the algorithm searches for transformations that were not considered but that could improve the current solution. For this search, all potential transformations are considered and those that could improve the graph are added to the set of selected transformations (by setting the corresponding $t_{ij} = true$). This step is very time consuming and is only done for small graphs (12 vertices or less). After selection of new transformations the matrix T is updated to take symmetry into account (the same graph g' may correspond to different patterns according to the labeling of the vertices). A formal description of the algorithm is given on *figure 3*.

When working on large graphs, ALS has to be trained before the optimization as this training will never be modified when optimizing large graphs. The training and optimization phases are thus well separated in ALS (for large graphs, steps 2 and 3 of the algorithm are omitted).

2.3   The Learning Descent

As opposed to the ALS algorithm, the LD algorithm performs the training during the optimization phase and always continues learning. The training and optimization phases occurs at the same time.

The LD algorithm on *figure 5* could be described by the following observations:

1. The pertinence of changing g' into g'' (replacing pattern p' by pattern p'') is memorized in a 64×64 matrix T which is initially set to $T = \{t_{ij} = 0\}$.
2. During the optimization, each induced subgraph g' is considered for replacement by any possible alternative subgraph g'' but this replacement will not necessarily be evaluated.
3. The probability to test the replacement of pattern i (g') by j (g'') is $p = sig(t_{ij}) = \frac{1}{1+e^{-t_{ij}}}$. The initial probability to test a replacement is 50% according to point 1.
4. For any tested transformation, if changing g' (with pattern p') to g'' (with pattern p'') improves the solution, the entry $t_{p',p''}$ of T is increased by δ^+ (and reduced by δ^- otherwise), with $\delta^+ > \delta^-$ because it is more important to use an improving transformation than to avoid a bad one. Also, a good transformation may fail, specially if the graph already has a good performance (here, we use $\delta^+ = 1$ and $\delta^- = 0.1$). The probability to test a transformation increases when it succeeds, but decreases if it does not.

As often used in neural networks, the sigmoid function $sig(x)$ allows the probability to test a transformation to change according to its performance without completely avoiding any transformation (which allows the system to always continue learning). The *figure 4* represents the replacement of *pattern 60* by *pattern 27* on a given graph G for the induced subgraph g' defined by vertices 1, 3, 5 and 6.

Step 1: Initialization
Load the last version of the matrix T for the problem under study if it exists initialize $T = \{t_{ij} = 0\}$ otherwise.

Step 2: Find interesting transformations
Set $f \leftarrow false$ (this flag indicates that no pattern was added to the list at this iteration).
For each subgraph of the current graph with n' vertices do:
    Let p_i be the corresponding pattern
    for each alternative pattern p_j do:
        if replacing the subgraph p_i by the pattern p_j
        would improve the current solution:
            update the matrix T by setting $t_{ij} \leftarrow true$
            set $f \leftarrow true$.
    done
done

Step 3: Update T for symmetry
If $f = true$: Update the matrix T to take symmetry into account:
    for each $t_{ij} = true$ do:
        for each pattern (i', j') obtained from (i, j)
        by relabelling the vertices do:
            set $t_{i'j'} \leftarrow true$.
    done
    done
If $f = false$:
    Stop; a local optimum is found.
    Save the matrix T

Step 4: Apply Local Search
set $improved \leftarrow true$
while $improved = true$ **do:**
    set $improved \leftarrow false$
    For each subgraph g' of G on n' vertices do:
        let p_i be the corresponding pattern
        For each alternative pattern p_j:
            if $t_{ij} = true$ do:
                if replacing p_i by p_j in G improves the
                solution:
                    apply the change
                    set $improved \leftarrow true$
        done
    done
done
Go to Step 2

Fig. 3. Rules of the Adaptive Local Search

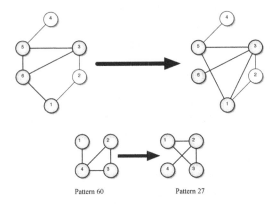

Pattern 60   Pattern 27

Fig. 4. Illustration of the transformation of G (left) to G'(right)

Note that if the algorithm were restricted to Step 2, it would tend to reduce the probability to use any transformation when good solutions are encountered since few transformations would improve such solutions. To avoid this problem, the matrix T is centered after each local search to an average value $\bar{t} = 0$.

3   Performance Comparison

In order to compare the performance of the different versions of the optimization module in AGX-1 and AGX-2, the AGX-1 moves were added to AGX-2. The various difference in the AGX-1 and AGX-2 program are thus not taken into account, which allows a more realistic comparison of both methods.

3.1   Experiments Description

AGX-1 needs an input from the user for good performance. The program was run with two different settings. The novice setting consists in using all the available neighborhoods for the VND, except those that modify the number of vertices. The expert setting consists in properly chosen neighborhoods. To ensure the reliability of a result stating that new methods are better than older ones, the experimental protocol always has a bias in favor of older methods. The choice of the neighborhoods used in the *expert mode* of AGX 1 is done after results obtained by all the combinations of the 4 transformations (add/remove, move, detour/short cut and 2-Opt) are known. The chosen strategy for *AGX-1-e* (expert strategy) was the one providing the best result in success rate as first criterion, using average value and finally cpu time in case of ties. We noticed that the best performance for a given problem but with different number of vertices was not necessarily due to the same scheme. This indicates clearly a bias favorable to *AGX 1-e* since it is difficult that an expert would identify this choice before the tests are done.

Step 1: Initialization
Load the last version of the matrix T for the problem under study if it exists and initialize $T = \{t_{ij} = 0\}$ otherwise.

Step 2: Apply Local Search
set *improved* $\leftarrow$ *true*
while *improved* $=$ *true* **do:**
    set *improved* $\leftarrow$ *false*
    For each subgraph g' **of** G **on** n' **vertices do:**
        let p_i be the corresponding pattern
        For each alternative pattern p_j
        (corresponding to g''**):**
            let x be an uniform 0-1 random number.
            if $x \leq sig(t_{ij})$ **do:**
                if replacing g' by g'' in G improves the
                solution:
                    apply the change
                    set *improved* $\leftarrow$ *true*
                    set $t_{ij} = t_{ij} + \delta^+$.
                otherwise:
                    set $t_{ij} = t_{ij} - \delta^-$.
        done
    done
done
Step 3: Scale the matrix T
    Let $\bar{t}$ be the average value of the terms $t_{ij} \neq 0$.
    For each $t_{ij} \neq 0$:
        set $t_{ij} = t_{ij} - \bar{t}$.
Step 4: Save the matrix T for future usage

Fig. 5. Rules of the Learning Descent

AGX-2 is designed to run without knowledge from the user and was also run in the three following modes : the "complete" mode *AGX 2-c* in which all the possible transformations involving 4 vertices are considered, the adaptive mode *AGX 2-als* in which the software chooses the useful neighborhoods from 5 runs on the same problem restricted to 10 and 12 vertices, and the learning descent mode *AGX 2-ld* in which the probability to use a given transformation is adjusted during the optimization.

To avoid any bias favorable to AGX 2-ld, the training on small graphs for AGX 2-als is done prior to the experiments and the training time (time needed by the system to identify which transformations to use) is not considered. On the opposite, the results of training during the optimization with AGX 2-ld was systematically erased after each optimization, so that the benefits from previous runs is avoided, which is a bias against AGX 2-ld.

The five optimization schemes compared are noted as follows.

- *1 b :* version using the all neighborhoods available in AGX-1,
- *1 e :* version using the best combination of neighborhoods available in AGX-1 (expert mode),

- *2 c :* version using AGX-2 considering all the possible transformations on 4 vertices (with the statistics matrix $T = \{t_{ij} = true\}$),
- *2 als :* version using AGX-2 and the adaptive local search,
- *2 ld :* version using AGX-2 and the learning descent.

The performance of these various algorithms was tested against 12 different and representative problems. The problems used are described in the following section. Each problem was solved 10 times for graphs with 13, 15 and 20 vertices with each of the 5 optimization scheme. In all cases, the total CPU time allowed was 300 seconds and the program was stopped if no improvement was encountered for 60 consecutive seconds. To reduce bias due to the implementation, all the tests were achieved with the same program in which the different versions of the optimizer are available thru parameters. All these tests were achieved on a Sun with 2 Dual Core AMD Opteron(tm) Processor 275 (2.2 GHz) with 4Go RAM memory running Linux CentOS-4 operating system. The performance of each strategy was measured in 3 ways. The first part (Average Z) of each table indicates the average value obtained among the 10 runs; the second (Success) indicates the number of times the best value was obtained and the last (CPU Time) indicates the average CPU time required to reach the best value found. If the best value was never attained by a given strategy, a "-" is displayed.

3.2 Results Analysis

Among the 360 instances tested, *AGX 2-ld* succeeded 275 times (76.4 %), which is the best performance, followed by *AGX 2-als* with 255 successes (70.8 %), *AGX 2-c* with 229 (63.6 %) successes and *AGX 1-e* with 176 (48.8 %) successes, followed by *AGX 1-b* which was only successful 61 times (16.9 %).

Regardless of the problem under study, AGX 1-b (which was often before AGX 2 was developed) shows very poor performance. Even with the experimental bias, the AGX 2 strategies are far better, first because they involve a wide range of transformations that were not implemented in AGX 1, and also because the VND used with AGX 1 spends some time trying to unsuccessfully optimize with a transformation before switching to the next, which is not the case in any of the AGX 2 local search scheme. If one should compare the strategies that do not involve any knowledge of the problem or of the optimization procedure (*1 b, 2 als, 2 c and 2 ld*), which is the most important for the novice point of view, AGX 2 with its stochastic local search seems to be the best choice.

The "Best Z" line on the tables indicates the best obtained value during the whole experiment, which may be (but is not always) the best possible value. The *Min* or *Max* at the top left of each table recalls wether the objective is to be minimized or maximized.

- **Problem 1 :** Minimize the energy E among trees, where $E = \sum_{i=1}^{n} |\lambda_i|$ is the sum of the absolute values of the eigenvalues of the adjacency matrix of the graph. For this problem, the number of edges is fixed by the number of vertices ; not all the transformations are therefore needed. The optimal solution to this problem, is a path and the corresponding value of the energy is $E = 2\sqrt{n-1}$.

- **Problem 2** : Minimize the value of the Randić index [18] among bicyclic connected graphs. The Randić index is defined as $\chi = \Sigma_{(ij)\in E} \frac{1}{\sqrt{d_i d_j}}$ where d_i is the degree of vertex i. The solution to this problem is a star to which are added two edges adjacent to the same vertex. The optimal value is $\chi = \frac{n-4}{\sqrt{n-1}} + \frac{2}{\sqrt{2n-2}} + \frac{1}{\sqrt{3n-3}}$.

- **Problem 3** : Same as *problem 2*, except that the objective function is maximized instead of being minimized. The optimal solution is known to be two cycles sharing an edge or two cycles joined by an edge and the corresponding value is $\chi = \frac{4}{\sqrt{6}} + \frac{3n-10}{6}$.

- **Problem 4** : Minimize the sum of the average degree of the graph $\bar{d}$ and the proximity p, where $\bar{d} = \Sigma_{i=1}^{n} \frac{d_i}{n}$ and $p = \frac{1}{n-1} \min_i(\Sigma_{j=1}^{n} d_{ij})$ where d_{ij} is the distance between the vertices i and j. In this case, the search space is only restricted by the connexity constraint. The optimal solution to the problem is a star and the optimal value is $Z = n + 1 - \frac{2}{n}$.

- **Problem 5** : Maximize the size of the maximum stable set, the maximum number of vertices to select such that no selected vertex is adjacent to another selected vertex, among connected graphs with number of edges equals twice the number of vertices $(m = 2n)$.

- **Problem 6** : Maximize the matching number, the number of edges to be selected such that no vertex is adjacent to two selected edges, among the same set of graphs as *problem 5* $(m = 2n)$. There are lots of graphs maximizing this invariant under the given constraints, but the maximal value cannot exceed $Z = \lfloor \frac{n}{2} \rfloor$, which is attained here.

- **Problem 7** : Maximize the *index*, value of the largest eigenvalue of the adjacency matrix, among the same set of graphs as *problem 5* or *problem 6* $(m = 2n)$.

- **Problem 8** : This problem is the same as *problem 7* except that the objective function is to be minimized instead of being maximized.

- **Problem 9** : Minimize the *index* among trees.

- **Problem 10** : Maximize the *diameter*, maximum distance between two vertices of the graph, among the same set of graphs as *problems 5, 6, 7* and *8*.

- **Problem 11** : Maximize the *diameter* among connected graphs graphs with $m \geq 2n$.

- **Problem 12** : Maximize the size of the maximum stable set among connected graphs graphs with $m \geq 2n$.

Problems 1 to 10 (except problem 4) have a fixed number of edges and of vertices. This corresponds to the problems we encounter most often, particularly for parametric analysis on the order and the size of the graph. In problems 5,6,7,8 and 10, the number of edges is fixed to twice the number of vertices. The number of graphs satisfying this condition is rather large, which makes the combinatorial aspect of the optimization important. Such problems are interesting benchmarks for the optimization routine.

The Problems 5 and 12 are NP-Complete. These two problems provide information on the capability of various strategies to handle problems which are more time consuming. AGX 1-b completely fails, and AGX 2-c is not very efficient either; this is because they are among all the two strategies that perform a large number of useless computations of the objective function.

Table 1. Results for problem 1 (the graphic represents the number of successes)

Min	Average Z			Success			CPU Time		
n	13	15	20	13	15	20	13	15	20
1 b	9.36	12.63	15.69	5	1	0	55.1	50	-
1 e	6.93	7.48	11.60	10	10	1	17.2	47.7	121.8
2 c	6.93	7.48	8.72	10	10	10	0	0	39.2
2 als	6.93	7.48	8.72	10	10	10	0	0	0
2 ld	6.93	7.48	8.72	10	10	10	0.5	0.9	2.4
Best Z	6.93	7.48	8.72						

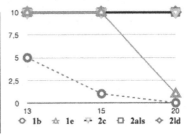

Table 2. Results for problem 2

Min	Average Z			Success			CPU Time		
n	13	15	20	13	15	20	13	15	20
1 b	4.56	5.74	6.95	4	1	0	76.3	99	-
1 e	3.99	4.29	5.53	10	10	2	26.3	78.3	149
2 c	3.99	4.29	4.94	10	10	10	5.4	11.2	71.1
2 als	3.99	4.29	4.94	10	10	10	0.4	0.9	3.7
2 ld	3.99	4.29	4.94	10	10	10	0.5	0.8	2.9
Best Z	3.99	4.29	4.94						

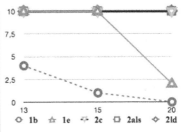

Table 3. Results for problem 3

Max	Average Z			Success			CPU Time		
n	13	15	20	13	15	20	13	15	20
1 b	6.02	6.09	7.67	0	0	0	-	-	-
1 e	6.47	7.46	9.88	10	9	1	22.7	0	0
2 c	6.47	7.47	9.97	10	10	10	4.3	14.1	92.3
2 als	6.47	7.47	9.97	10	10	10	0.3	0.6	3.6
2 ld	6.47	7.47	9.97	10	10	10	0.7	1.3	5.4
Best Z	6.47	7.47	9.97						

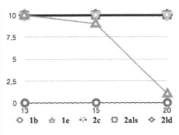

Table 4. Results for problem 4

Min	Average Z			Success			CPU Time		
n	13	15	20	13	15	20	13	15	20
1 b	2.85	2.89	3.26	10	7	0	18	35.9	-
1 e	2.85	2.87	2.95	10	9	4	0.7	25.4	33.4
2 c	2.85	2.87	2.9	10	10	10	4.4	10.7	82.4
2 als	2.85	2.87	2.9	10	10	10	0.6	1.4	6.6
2 ld	2.85	2.87	2.9	10	10	10	0.7	1.2	6.2
Best Z	2.85	2.87	2.9						

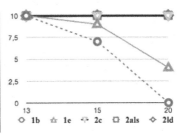

Table 5. Results for problem 5

Max	Average Z			Success			CPU Time		
n	13	15	20	13	15	20	13	15	20
1 b	7.2	8.6	12.1	0	0	0	-	-	-
1 e	8.4	9.4	12.8	5	2	2	24	57.3	5.7
2 c	8.2	9.4	11.5	3	0	0	21.8	-	-
2 als	8.3	9.8	12.7	3	1	2	2.9	10.3	35.4
2 ld	8.7	10	13.1	7	2	4	24.4	39.8	38.9
Best Z	9	11	14						

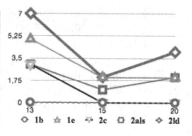

Table 6. Results for problem 6

Max	Average Z			Success			CPU Time		
n	13	15	20	13	15	20	13	15	20
1 b	6	7	8.6	10	10	2	0	0.3	48.5
1 e	6	7	10	10	10	10	0.1	0	0
2 c	6	7	9.8	10	10	8	0	0.5	53.6
2 als	6	7	10	10	10	10	0	0	4.3
2 ld	6	7	10	10	10	10	0	0.1	4.3
Best Z	6	7	10						

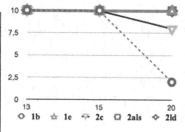

Table 7. Results for problem 7

Max	Average Z			Success			CPU Time		
n	13	15	20	13	15	20	13	15	20
1 b	4.92	5.28	5.91	0	0	0	-	-	-
1 e	5.91	6.2	6.57	1	0	0	100.9	-	-
2 c	5.92	6.18	6.59	10	4	0	27	84.3	-
2 als	5.92	6.29	6.95	10	10	0	3.5	15.8	-
2 ld	5.92	6.28	6.99	10	9	2	6.4	20.9	53.3
Best Z	5.92	6.29	7.25						

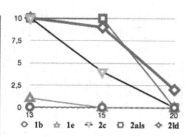

Table 8. Results for problem 8

Min	Average Z			Success			CPU Time		
n	13	15	20	13	15	20	13	15	20
1 b	4.86	5.13	5.65	0	0	0	-	-	-
1 e	4.02	4.03	4.13	4	1	0	0	0	-
2 c	4	4	4.3	10	10	0	11.2	48.3	-
2 als	4	4	4.01	10	10	7	2.7	8.5	131.2
2 ld	4	4	4	10	10	10	3.3	9.6	74.1
Best Z	4	4	4						

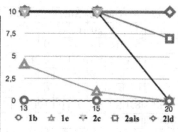

Table 9. Results for problem 9

Min	Average Z			Success			CPU Time		
n	13	15	20	13	15	20	13	15	20
1 b	2.35	2.68	3.26	0	0	0	-	-	-
1 e	1.95	1.96	2.02	10	10	3	0	18.6	52.6
2 c	1.95	1.96	1.98	10	10	10	4.3	12.3	82.3
2 als	1.95	1.96	1.98	10	10	10	0.4	0.7	3.6
2 ld	1.95	1.96	1.98	10	10	10	0.6	1.1	3.6
Best Z	1.95	1.96	1.98						

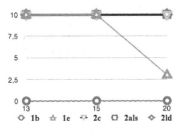

Table 10. Results for problem 10

Max	Average Z			Success			CPU Time		
n	13	15	20	13	15	20	13	15	20
1 b	5.6	5	4.9	1	0	0	84.3	-	-
1 e	6.1	6.7	7	2	0	0	35.4	-	-
2 c	5.9	6.8	10.7	1	0	1	6.9	-	66.6
2 als	6.3	7.5	11.4	3	1	5	19.3	8.6	30.5
2 ld	6.8	7.7	11.1	8	2	4	26.3	45.4	15.6
Best Z	7	9	12						

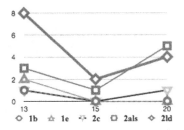

Table 11. Results for problem 11

Max	Average Z			Success			CPU Time		
n	13	15	20	13	15	20	13	15	20
1 b	5.2	3.1	3	0	0	10	-	-	0.5
1 e	6.5	4.3	3	6	1	10	26.3	57.1	0.1
2 c	4.8	3.5	3	0	0	10	-	-	3.4
2 als	4.7	3.6	3	1	0	10	20.7	-	6.7
2 ld	6.1	4.2	3	6	0	10	70.5	-	6.1
Best Z	7	8	3						

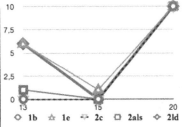

Table 12. Results for problem 12

Max	Average Z			Success			CPU Time		
n	13	15	20	13	15	20	13	15	20
1 b	6.5	6.9	6.4	0	0	0	-	-	-
1 e	8.4	8.6	7.7	1	2	0	81.5	42.7	-
2 c	8.1	8.9	7.9	0	2	0	-	58	-
2 als	8.2	8.9	8.1	0	2	0	-	42.6	-
2 ld	8.4	8.1	8.5	0	0	1	-	-	6.4
Best Z	10	10	10						

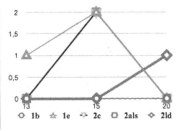

We notice that problems 10 and 11 have a low success rate. The problem 11 on 20 vertices was completely missed by all the strategies as the best found value is only 3 even if the optimal value should be at least as good as that of problem 10. This is an important phenomenon for researchers using AGX; even if the diameter is easy to compute, it has really bad properties from the optimization point of view. Depending on the current graph, changing the value of the diameter by 1 may involve a large number of transformations and no strategy is powerful enough in this case. This phenomena is called *plateau* and some practical ways to handle it are described in [9,1].

4   Conclusion

From these experiments, we first notice that the *VNS-LD* and *VNS-ALS* algorithms are clearly more efficient even if the performance of *VNS-VND e* is overestimated. The good performance of the *VNS-LD* may be due to the wide range of transformations implicitly considered in *VNS-LD* that are not implemented in *VNS-VND*. However, this is not the only reason because some tests were achieved always using all the transformations available in LD (by artificially setting the probability to select any of them to 1) and the best results were only found 269 times. Indeed, one of the forces of LD and ALS are that unlike VND which wastes some time trying to unsuccessfully optimize with a transformation before switching to the next one, LD uses any interesting transformation and concentrates on the most performing ones. Overall, *VNS-LD* performs better than *VNS-ALS* even if this last algorithm's training time is not considered here (training was achieved before the tests on smaller graphs); furthermore, *VNS-LD*'s training was erased between two tests and the system had to learn from scratch at each run.

References

1. Aouchiche, M., Bonnefoy, J.-M., Fidahoussen, A., Caporossi, G., Hansen, P., Lacheré, J., Monhait, A.: Variable neighborhood search for extremal graphs. 14. the autographix 2 system. In: Liberti, L., Maculan, N. (eds.) Global Optimization. From Theory to Implementation. Springer Science, New York (2006)
2. Aouchiche, M., Caporossi, G., Hansen, P.: Variable neighborhood search for extremal graphs 8. variations on graffiti 105. Congressus Numerantium 148, 129–144 (2001)
3. Belhaiza, S., Abreu, N.M.M., Hansen, P., Oliveira, C.S.: Variable neighborhood search for extremal graphs. 11. bounds on algebraic connectivity. In: Hertz, A., Avis, D., Marcotte, O. (eds.) Graph Theory and Combinatorial Optimization. Springer, New York (2005)
4. Caporossi, G., Cvetković, D., Gutman, I., Hansen, P.: Variable neighborhood search for extremal graphs. 2. finding graphs with extremal energy. Journal of Chemical Information and Computer Sciences 39, 984–996 (1999)
5. Caporossi, G., Dobrynin, A.A., Gutman, I., Hansen, P.: Trees with palindromic Hosoya polynomials. Graph Theory Notes of New York 37, 10–16 (1999)
6. Caporossi, G., Gutman, I., Hansen, P.: Variable neighborhood search for extremal graphs: 4. chemical trees with extremal connectivity index. Computers and Chemistry 23, 469–477 (1999)
7. Caporossi, G., Gutman, I., Hansen, P., Pavlović, L.: Graphs with maximum connectivity index. Computational Biology and Chemistry 27, 85–90 (2003)

8. Caporossi, G., Hansen, P.: Finding Relations in Polynomial Time. In: Proceedings of the XVI International Joint Conference on Artificial Intelligence, pp. 780–785 (1999)

9. Caporossi, G., Hansen, P.: Variable neighborhood search for extremal graphs. 1. the autographix system. Discrete Math. 212, 29–44 (2000)

10. Caporossi, G., Hansen, P.: Variable neighborhood search for extremal graphs. v. three ways to automate finding conjectures. Discrete Math. 276, 81–94 (2004)

11. Cvetković, D., Simić, S., Caporossi, G., Hansen, P.: Variable neighborhood search for extremal graphs. iii. on the largest eigenvalue of color-constrained trees. Linear and Multilinear Algebra 49, 143–160 (2001)

12. Fowler, P.W., Hansen, P., Caporossi, G., Soncini, A.: Variable neighborhood search for extremal graph. 7. polyenes with maximum homo-lumo gap. Chemical Physics Letters 342, 105–112 (2001)

13. Gutman, I., Miljković, O., Caporossi, G., Hansen, P.: Alkanes with small and large randić connectivity indices. Chemical Physics Letters 306, 366–372 (1999)

14. Hansen, P., Mélot, H.: Variable neighborhood search for extremal graphs. 6. analyzing bounds for the connectivity index. Journal of Chemical Information and Computer Sciences 43, 1–14 (2003)

15. Hansen, P., Mélot, H.: Variable neighborhood search for extremal graphs. 9. bounding the irregularity of a graph. In: Fajtlowicz, S., Fowler, P., Hansen, P., Janowitz, M., Roberts, F. (eds.) Graphs and Discovery. American Mathematical Society, Providence (2005)

16. Hansen, P., Mladenović, N.: Variable neighborhood search: Principles and applications. European J. Oper. Res. 130, 449–467 (2001)

17. Mladenović, N., Hansen, P.: Variable neighborhood search. Comput. Oper. Res. 24, 1097–1100 (1997)

18. Randić, M.: On characterization of molecular branching. Journal of the American Chemical Society 97, 6609–6615 (1975)

A Math-Heuristic Dantzig-Wolfe Algorithm for the Capacitated Lot Sizing Problem

Marco Caserta[1] and Stefan Voß[2]

[1] IE Business School, Maria de Molina 12, 28006 - Madrid - Spain
mcaserta@faculty.ie.edu
[2] Institute of Information Systems (IWI), University of Hamburg
Von-Melle-Park 5, 20146 Hamburg, Germany
stefan.voss@uni-hamburg.de

Abstract. The multi-item multi-period capacitated lot sizing problem with setups (CLST) is a well known optimization problem with wide applicability in real-world production planning problems. Based on a recently proposed Dantzig-Wolfe approach we present a novel math-heuristic algorithm for the CLST. The major contribution of this paper lies in the presentation of an algorithm that exploits exact techniques (Dantzig-Wolfe) in a metaheuristic fashion, in line with the novel trend of math-heuristic algorithms. To the best of the authors knowledge, it is the first time that such technique is employed within a metaheuristic framework, with the aim of tackling challenging instances in short computational time.

1 Introduction

The Multi-Item Multi-Period Capacitated Lot Sizing Problem with Setups (CLST) is a well known optimization problem that finds a wide variety of real-world applications. The CLST belongs to the class of $\mathcal{NP}$-hard problems [1,15,12]. A mixed-integer formulation of the CLST is:

$$(CLST): \min \quad z = \sum_{j=1}^{n}\sum_{t=1}^{T}(f_{jt}y_{jt} + c_{jt}x_{jt} + h_{jt}s_{jt}) + \sum_{j=1}^{n}h_{j0}s_{j0}$$

$$\text{s.t.} \quad \sum_{j=1}^{n}(a_{jt}x_{jt} + m_{jt}y_{jt}) \leq b_t \quad \forall t$$

$$s_{jt-1} + x_{jt} = d_{jt} + s_{jt} \quad \forall j, t$$

$$x_{jt} \leq My_{jt} \quad \forall j, t$$

$$y_{jt} \in \{0,1\} \quad \forall j, t$$

$$x_{jt}, s_{jt} \geq 0 \quad \forall j, t$$

where items $j = 1, \ldots, n$ should be produced over time periods $t = 1, \ldots, T$. In the CLST formulation, $f_{jt}, c_{jt},$ and h_{jt} indicate the fixed cost, the unitary

Y. Hamadi and M. Schoenauer (Eds.): LION 6, LNCS 7219, pp. 31–41, 2012.
© Springer-Verlag Berlin Heidelberg 2012

production cost and the unitary inventory holding cost for item j in period t, respectively. Parameters m_{jt} and a_{jt} indicate the setup time and the unitary production time, respectively, while b_t stands for the production capacity in period t. Parameter d_{jt} indicates the demand of item j in period t. Finally, in the model, three sets of decision variables are employed, i.e., $y_{jt} \in \{0, 1\}$, which takes value 1 if there is a setup for item j in period t; as well as $x_{jt} \geq 0$ and $s_{jt} \geq 0$, indicating the production volume and the inventory level for item j in period t, respectively. Note that s_{j0} is given as data indicating initial inventory.

Due to its vast industrial applicability, researchers have devoted special attention to the CLST (see, e.g., [5], [13], [6], and [4]). Since the CLST is still difficult to solve to optimality, many researchers have tried to tackle the problem by working on relaxations of the same. A good description of some well studied relaxations of the CLST is provided by [11]. A recent discussion of solution approaches for the CLST can be found in [8].

In recent years, a lot of attention has been devoted to the integration, or hybridization, of metaheuristics with exact methods. This exposition also relates to the term math-heuristics (see, e.g., [9]) which describes works which are, e.g., exploiting mathematical programming techniques in (meta)heuristic frameworks or on granting to mathematical programming approaches the cross-problem robustness and constrained-CPU-time effectiveness which characterize metaheuristics. Discriminating landmark is some form of exploitation of the mathematical formulation of the problems of interest. Here we follow the math-heuristic concept rather than a more general idea of hybridization, where the successful ingredients of various metaheuristics have been combined.

Based on a recently proposed Dantzig-Wolfe approach of [7], in this paper we present a novel math-heuristic algorithm for the CLST. The major contribution of this paper lies in the presentation of an algorithm that exploits exact techniques (Dantzig-Wolfe) in a metaheuristic fashion, in line with the novel trend of math-heuristic algorithms. To the best of the authors' knowledge, it is the first time that such technique is employed within a metaheuristic framework, with the aim of tackling challenging instances in short computational time. In addition, the proposed approach constitutes a clear example of the effectiveness of hybrid approaches, e.g., approaches that intertwine classical mathematical programming techniques with novel metaheuristics.

To measure the effectiveness of the proposed approach, we have tested the algorithm to solve standard benchmark instances from [15], in line with what has been done by a number of authors, e.g., [7], and [3].

The organization of the paper is as follows. In the next section, Section 2, we present a mathematical formulation and a Dantzig-Wolfe decomposition of the CLST; Section 3 illustrates the main ingredients of the proposed math-heuristic algorithm, while Section 4 summarizes the results of the algorithm when tested on a set of benchmark instances. Finally, Section 5 concludes with some final remarks.

2   Mathematical Formulation and Decomposition

Dantzig-Wolfe (DW) decomposition is a well known technique often employed to address mixed integer programs with substructures. This technique has been successfully applied in a number of contexts. (For more details on such technique see, *e.g.*, [16], and [2].) Recently, [7] have presented a DW approach for the CLST, addressing an important structural deficiency of the standard DW approach for the CLST proposed decades ago by [10].

In this section, borrowing ideas from [7], we present a DW reformulation and decomposition for the CLST that is especially suited for the math-heuristic algorithm presented in Section 3. In line with what proposed by [7], we employ a formulation in which setup variables and production variables are dealt with separately. Next, we illustrate how such reformulation leads to a natural implementation of a math-heuristic algorithm aimed at speeding up the column generation phase.

Let us consider the DW decomposition and its associated reformulation. We identify the capacity constraints as the "hard" constraints. Thus, if we eliminate such hard constraints, we obtain a problem that can easily be decomposed into smaller (easy) subproblems, one per item. Let us indicate with:

$$X_j^{\text{sub}} = \left\{ \begin{array}{rcl} s_{jt-1} + x_{jt} & = & d_{jt} + s_{jt} \\ x_{jt} & \leq & My_{jt} \\ y_{jt} & \in & \{0,1\} \\ x_{jt}, s_{jt} & \geq & 0 \end{array} \right\}$$

the set of feasible solutions for the jth subproblem. It is easy to observe that X_j^{sub} defines the feasible region for the single item uncapacitated lot sizing problem.

Let us now indicate with $\{\mathbf{x}_j^k\}$, where $\mathbf{x}_j^k = \left(y_{jt}^k, x_{jt}^k, s_{jt}^k \right)$, with $k = 1, \ldots, K_j$, the set of extreme points of $conv(X_j^{\text{sub}})$. That is, for each item j, the set of extreme points of the corresponding polytope is defined by all the possible feasible schedules and the corresponding dominant production schedules, *i.e.*, production schedules satisfying the Wagner-Whitin condition $s_{jt-1}x_{jt} = 0$ (also known as zero-inventory property). Thus, variables $\mathbf{x}$, $\mathbf{y}$, and $\mathbf{s}$ of the CLST can be rewritten as convex combination of such extreme points. Therefore, we rewrite the CLST as:

$$(M): \min z = \sum_{j=1}^{n} \sum_{k=1}^{K_j} \left[\sum_{t=1}^{T} \left(f_{jt} y_{jt}^k + c_{jt} x_{jt}^k + h_{jt} s_{jt}^k \right) + \sum_{j=1}^{n} h_{j0} s_{j0} \right] \lambda_{jk} \quad (1)$$

$$\text{s.t.} \sum_{j=1}^{n} \sum_{k=1}^{K_j} \left(a_{jt} x_{jt}^k + m_{jt} y_{jt}^k \right) \lambda_{jk} \leq b_t, \quad t = 1, \ldots, T \quad (2)$$

$$\sum_{k=1}^{K_j} \lambda_{jk} = 1, \quad j = 1, \ldots, n \quad (3)$$

$$\lambda_{jt} \in \{0,1\}, \quad j = 1, \ldots, K, \ k = 1, \ldots, K_j \quad (4)$$

The original CLST is rewritten in such a way that every extreme point of the polyhedron of subproblems X_j^{sub} is enumerated and the corresponding weight variable is either set to 1, if the extreme point is selected, or to 0 otherwise. However, it is a well-known fact that, if the capacity constraint is binding for at least one time period, the extreme points of the polytope of the single item uncapacitated lot sizing problem will not necessarily provide an optimal solution to the overall problem [7]. Thus, when imposing the binary constraints on the reformulation variables, i.e., constraint (4), the optimal solution to the original CLST could be missed. In other words, there is not a strict correspondence between the original setup variables y_{jt} and the newly introduced reformulation variables λ_{jk}. As pointed out by [7], the key point here is that the dominant plans are only a subset of the set of extreme points needed to completely define the search space of the original CLST.

What we would need is a convex combination of the setup plans for each single item problem, i.e., we should eliminate constraint (4) from the master problem (M) and, instead of it, we should impose the binary conditions on the original setup variables. This approach is called *convexification approach* to the DW decomposition [16]. Alternatively, one could use a *discretization approach*, in which *all* the integer solutions are enumerated (including interior points of the subsystem polyhedron, i.e., including solutions that make use of non-dominant production plans and yet satisfy the Wagner-Whitin property). The discretization approach would allow to re-introduce constraint (4) into the master problem.

As pointed out by [7], it is also possible to use a third approach, i.e., a discretization approach for the binary variables of the CLST and a convexification approach for the continuous variables. That is, given a feasible setup plan, a convex combination of production plans respecting that setup plan is generated. Therefore, the set of extreme points included in the master program includes non-dominant plans satisfying the Wagner-Whitin property, i.e., there will be solutions for which, in some periods, it is possible that there exists a setup but no production.

Consequently, in the sequel, we will introduce two sets of reformulation variables, i.e., (i) λ_{jk}, to select exactly one setup plan among the ones included in the master problem, and (ii) μ_{jkw}, to select a convex combination of production schedules (dominant and non-dominant) arising from a given setup schedule.

An Example. Let us consider a single-item lot sizing problem with five time periods. Let us fix $d_t = 10$, for $t = 1, \ldots, 5$. Let us assume we are given a setup schedule, in which there exists a setup in the first, third, and fifth period (as presented in Table 1). In the sequel, for the sake of clarity, inventory levels s_t are omitted. In the table, the row associated to solution x_1 corresponds to the dominant production plan associated to the given setup plan. However, in the reformulation, we also want to consider non-dominant production plans that are still feasible with respect to the given setup plan, e.g., production plans x_2 to x_4. Every production plan presented in Table 1 still satisfies the Wagner-Whitin property. However, except for the production schedule indicated by x_1, all the production schedules are non-dominant plans, i.e., there exist some periods in which a setup is not accompanied by production.

Table 1. An example of the convexification-discretization approach

t	1	2	3	4	5
d_t	10	10	10	10	10
y_t^k	1	0	1	0	1
x_1^k	20	0	20	0	10
x_2^k	50	0	0	0	0
x_3^k	20	0	30	0	0
x_4^k	40	0	0	0	10

In line with what is mentioned above, the *convexification-discretization* approach defines two sets of dual variables, one for the setup plan and one for the production plan. Therefore, with respect to Table 1, we will use:

- A binary variable $\lambda_k \in \{0,1\}$ to determine whether the setup plan described by $\mathbf{y}^k$ should be selected; and
- A set of continuous variables μ_w^k, with $w = 1, \ldots, 4$, to define a convex combination of production plans arising from the same setup plan.

To formalize, and in line with what is presented in [7], let us assume we are considering a single-item lot sizing problem, *e.g.*, item $j \in [1, n]$. Let us also assume we indicate with

$$\mathcal{Y} = \left\{ \mathbf{y}^1, \ldots, \mathbf{y}^k, \ldots, \mathbf{y}^{K_j} \right\}$$

the set of all feasible setup plans for item j. Let us now consider a setup schedule $\mathbf{y}^k = \left(y_1^k, \ldots, y_T^k \right) \in \mathcal{Y}$ and let us define the set of induced setup schedules as:

$$\mathcal{Y}(k) = \left\{ (y_1, \ldots, y_T) : y_t \leq y_t^k, y_t \in \{0,1\}, t = 1, \ldots, T \right\}$$

That is, given a setup schedule $\mathbf{y}^k$, the set of induced setup schedules, indicated with $\mathcal{Y}(k)$, is built by taking all the possible schedules dominated by $\mathbf{y}^k$, *i.e.*, assuming that $\mathbf{y}^k$ defines s setups, the set $\mathcal{Y}(k)$ contains 2^s setup schedules. Obviously, for each setup schedule in $\mathcal{Y}(k)$ there exists a unique dominant Wagner-Whitin production plan. Thus, the set of extreme points defined by the setup schedule $\mathbf{y}^k$ is defined by all the combinations of the Wagner-Whitin production plans arising from the induced set $\mathcal{Y}(k)$ with the original setup plan $\mathbf{y}^k$.

With respect to the example provided in Table 1, we have that $\mathbf{y}^k = (1, 0, 1, 0, 1)$. Thus, the set of induced setup schedules is:

$$\mathcal{Y}(k) = \left\{ (1,0,1,0,1), (1,0,0,0,0), (1,0,1,0,0), (1,0,0,0,1) \right\}.$$

As previously mentioned, since the setup schedule $\mathbf{y}^k$ defines three setups, and considering the setup in the first period as fixed, the setup schedule $\mathbf{y}^k$ gives rise to $2^{3-1} = 4$ induced setup plans. Each setup plan in $\mathcal{Y}(k)$, in turn, defines a unique Wagner-Whitin schedule, as indicated in Table 1. Thus, we now have four extreme points that could be added to the master plan, *i.e.*, $(\mathbf{y}^k, \mathbf{x}_1^k)$, $(\mathbf{y}^k, \mathbf{x}_2^k)$, $(\mathbf{y}^k, \mathbf{x}_3^k)$, and $(\mathbf{y}^k, \mathbf{x}_4^k)$.

Finally, with respect to the dual variables of the master problem, *i.e.*, the new columns of program (M), we generate the following 2^{s-1} new columns, *i.e.*, (λ_k, μ_1^k), (λ_k, μ_2^k), (λ_k, μ_3^k), and (λ_k, μ_4^k).

Finally, once we generated a set of (dominant and non-dominant) production plans associated to a given setup plan, the relation between these two is established using the following constraints (considering a single-item problem):

$$\sum_{k=1}^{K_j} \lambda_k = 1 \tag{5}$$

$$\sum_{w=1}^{|\mathcal{Y}(k)|} \mu_w^k = \lambda_k, \quad k = 1, \ldots, K_j \tag{6}$$

$$\lambda_k \in \{0, 1\}, \quad k = 1, \ldots, K_j \tag{7}$$

$$\mu_w^k \geq 0, \quad w = 1, \ldots, |\mathcal{Y}(k)|, \ k = 1, \ldots, K_j \tag{8}$$

Thus, via Equations (5) and (7) we enforce the discretization mechanism for the (binary) setup variables, *i.e.*, one setup plan must be selected; on the other hand, with Equations (6) and (8), we enforce the convexification mechanism, *i.e.*, once a setup plan is selected (*e.g.*, $\lambda_k = 1$ for any $k \in [1, K_j]$), we allow for a convex combination of (dominant and non-dominant) induced production plans to be selected.

Let us now present the Dantzig-Wolfe master reformulation and the derived subproblems.

$$\min z = \sum_{j=1}^{n} \sum_{k=1}^{K_j} \left[\sum_{t=1}^{T} \left(f_{jt} y_{jt}^k \lambda_{jk} + \sum_{w=1}^{|\mathcal{Y}(k)|} \left(c_{jt} x_{jt}^w + h_{jt} s_{jt}^w \right) \mu_{jk}^w \right) \right] \tag{9}$$

$$\text{s.t.} \ \sum_{j=1}^{n} \sum_{k=1}^{K_j} \left(m_{jt} y_{jt}^k \lambda_{jk} + \sum_{w=1}^{|\mathcal{Y}(k)|} a_{jt} x_{jt}^w \mu_{jk}^w \right) \leq b_t, \quad t = 1, \ldots, T \tag{10}$$

$$\sum_{k=1}^{K_j} \lambda_{jk} = 1, \quad j = 1, \ldots, n \tag{11}$$

$$\sum_{w=1}^{|\mathcal{Y}(k)|} \mu_{jk}^w = \lambda_{jk}, \quad j = 1, \ldots, n, \ k = 1, \ldots, K_j \tag{12}$$

$$\lambda_{jt} \in \{0, 1\}, \quad j = 1, \ldots, K, \ k = 1, \ldots, K_j \tag{13}$$

$$\mu_{jk}^w \geq 0, \quad j = 1, \ldots, K, \ k = 1, \ldots, K_j, \ w = 1, \ldots, |\mathcal{Y}(k)| \tag{14}$$

Due to the large number of variables of the master problem, the linear programming relaxation of the master is solved using column generation. The subproblem used to price in new columns is (separable over single items j):

$$\min \ r_j = \sum_{t=1}^{T} \left((f_{jt} - u_t m_{jt}) \, y_{jt} + (c_{jt} - u_t a_{jt}) \, x_{jt} + h_{jt} s_{jt} \right) - \alpha_j \quad (15)$$

$$\text{s.t.} \ \ x_{jt} + s_{jt-1} = d_{jt} + s_{jt}, \quad t = 1, \ldots, T \quad\quad\quad (16)$$

$$x_{jt} \le M y_{jt}, \quad t = 1, \ldots, T \quad\quad\quad\quad\quad (17)$$

$$y_{jt} \in \{0, 1\}, \quad t = 1, \ldots, T \quad\quad\quad\quad\quad (18)$$

$$x_{jt}, s_{jt} \ge 0, \quad t = 1, \ldots, T \quad\quad\quad\quad\quad (19)$$

where $\mathbf{u}$ are the dual variables associated to the constraint (10) and α are the dual variables associated to constraint (11). As long as the pricing problem returns negative reduced costs for at least one item j, with $j = 1, \ldots, n$, these columns are added to the master, and the linear programming relaxation of the master is re-optimized. However, it is worth noting that, due to the discretization-convexification approach, the number of new columns generated every time the subproblem is solved can become large. Thus, in the next section, we propose a metaheuristic approach, based on the corridor method, aimed at finding a "good" set of new columns to be added to the master problem within a prespecified amount of computational time.

3 A Math-Heuristic Approach

Let us now present in detail the math-heuristic column generation approach used to solve the subproblems.

The proposed approach belongs to the class of math-heuristic algorithms, since we use mathematical programming techniques in a heuristic fashion [9]. The basic steps of a column generation approach are here briefly highlighted:

1. Populate the master problem with an initial set of columns.
2. Solve the LP relaxation of the master.
3. Solve the subproblems defined using the current dual values obtained from the master.
4. Price in new columns: If there exists at least one new column with negative reduced cost, add such column(s) to the master. Otherwise, stop.

The main contribution of this paper is related to the use of a corridor method-inspired scheme [14] to address step 3. As presented in Section 2, we use a discretization-convexification approach for the master problem, i.e., we employ discretization for the selection of the setup variables and convexification for the selection of a combination of induced production plans. Therefore, in phase 3 of the column generation approach, we need to devise an efficient method aimed at obtaining a "good" setup plan and a sufficiently large set of induced production plans. Given a current setup plan $\mathbf{y}^k$, there exist 2^{s-1} induced production plans, i.e., $|\mathcal{Y}(k)| = 2^{s-1}$, where s is the total number of setups in $\mathbf{y}^k$, i.e. $\sum_{t=1}^{T} y_t^k = s$. Therefore, a complete enumeration of all the induced production plans for a given setup plan $\mathbf{y}^k$ is infeasible for practical, real-world size instances.

Due to the aforementioned drawback, we propose here a math-heuristic approach to generate a sufficiently large set $\mathcal{Y}(k)$ of induced production plans in a controlled amount of computational time. Let us present here the steps of the corridor method-inspired scheme used to generate new columns (the scheme is presented for a single item). In Figure 1, we present in details the steps of the third phase of the algorithm. Basically, we first generate a Wagner-Whitin solution for the current subproblem. Then, we apply a corridor method algorithm to collect a set of induced solutions. Such solutions are stored in a pool of solutions Ω. The corridor method phase stops when either the optimal solution to the constraint problem or a maximum running time have been reached. Subsequently, the solutions in the pool Ω are priced in and, whenever a solution with negative reduced cost is found, such solution is added to the master.

The overall algorithm begins by creating an initial set of columns for the master problem. We first add to the master a set of Wagner-Whitin solutions obtained fixing all the dual values equal to zero, along with the solutions in which all the demand is satisfied using initial inventory s_{j0}. Once these columns are added to the master, the linear programming relaxation of problem (9)-(14) is solved to optimality using a standard LP solver. The dual values obtained after solving problem (M) are then used to define the subproblems (15)-(19) (one per item). Each subproblem is then solved with the proposed algorithm of Figure 1 and new columns are priced into the master problem (9)-(14). The master-subproblem cycle is repeated until there exist no new columns with negative reduced cost. In that case, the algorithm stops and problem (9)-(14) is finally solved to optimality and the best feasible solution for the original CLST is returned.

4   Computational Results

In this section we present the computational results of the algorithm on a set of well-known benchmark instances. We report results on six instances taken from the test set used by [15], as reported by [3]. The same instances have been tackled by [7] and, therefore, constitute an interesting test bed for the preliminary evaluation of the proposed algorithm. As reported by [7], these instances are hard to solve to optimality. However, as presented in Table 2, we could solve all these instances to optimality in a reasonable amount of computational time using IBM CPLEX. Thus, we could also measure how far our heuristic solution was from an optimal solution. Optimal values for these instances were also reported by [11].

The algorithm proposed in this paper was coded in C++ and compiled using the GNU g++ 4.5.2 compiler on a dual core Pentium 1.8GHz Linux workstation with 4Gb of RAM. Throughout the computational experiment phase, the maximum running time for each "constrained" subproblem $r_j(\mathbf{u}, \alpha, \mathbf{y}^k)$ was kept fixed to one second, while the number of columns generated in each iteration δ was fixed to 100.

In Table 2, the first column reports the instance name, columns two and three report the optimal value and the running time obtained using IBM CPLEX 12.1

S1. Initialization

1. Solve problem $r_j(\mathbf{u}, \alpha)$ using Wagner-Whitin $\Rightarrow (\mathbf{y}^k, \mathbf{x}_1^k)$
2. If $r_j(\mathbf{u}, \alpha) \geq 0$, STOP.

S2. Corridor Method$(\mathbf{y}^k, \delta)$

1. Define a neighborhood around the Wagner-Whitin solution as:

$$\mathcal{N}(\mathbf{y}^k) = \left\{ y \in \{0,1\}^T : y_{jt} \leq y_{jt}^k \right\} \tag{20}$$

2. Add the following *corridor constraint* to the subproblem $r_j(\mathbf{u}, \alpha)$:

$$y_{jt} \leq y_{jt}^k, \quad t = 1, \ldots, T \tag{21}$$

and solve the resulting "constrained" subproblem $r_j(\mathbf{u}, \alpha, \mathbf{y}^k)$.

3. While solving to optimality $r_j(\mathbf{u}, \alpha, \mathbf{y}^k)$, collect the best $\delta \geq 1$ feasible solutions and store them in a pool:

$$\Omega = \left\{ (\mathbf{y}^{kw}, \mathbf{x}^{kw}) : \mathbf{y}^{kw} \in \mathcal{N}(\mathbf{y}^k) \right\} \tag{22}$$

where $\mathbf{y}^{kw}$ is a setup plan satisfying constraint (21), and $\mathbf{x}^{kw}$ is the dominant Wagner-Whitin solution associated to the setup plan $\mathbf{y}^{kw}$.

4. Stop the corridor method when one of the two criteria has been reached:
   a. maximum running time, or
   b. optimal solution.

S3. Pricing$(\mathbf{y}^k, \Omega)$

1. Price in new columns until all the solutions in Ω have been examined and Ω is empty:
   (a) Select a solution from the pool Ω and compute the reduced cost of the composed solution $(\mathbf{y}^k, \mathbf{x}^{kw})$, with $w = 1, \ldots, |\Omega|$.
   (b) If the reduced cost of the current solution is negative, *i.e.*, $r_j < 0$, add the column $(\mathbf{y}^k, \mathbf{x}^{kw})$ to the master. Otherwise, discard the column, eliminate it from Ω, and go back to step 1a.

Fig. 1. Outline of the proposed corridor method-inspired algorithm for the generation of new columns

as MIP solver. Columns four and five provide the best value and the running time of [3], while columns six and seven provide the same information as presented in [7]. Finally, the last two columns provide the best result and the running time of the proposed algorithm. With respect to running times, [3] has a limit of 900 seconds, while [7] stopped the algorithm after vising 2000 nodes.

From the table, we can observe that the proposed algorithm is competitive in terms of both solution quality and running time, especially for larger instances.

Table 2. Results on six instances from [15]. The optimal values and corresponding running times reported here have been obtained using CPLEX 12.1.

Instance	Optimal		BW		DJ		CM	
	z^*	$T^\natural$	z	$T^\dagger$	z	$T^\ddagger$	z	$T^\natural$
Tr6-15	37,721	1.24	37,721	38.4	38,162	29	37,721	1.62
Tr6-30	61,746	126.2	61,806	900	62,644	359	62,885	6.45
Tr12-15	74,634	2.67	74,799	900	75,035	66	74,727	29.65
Tr12-30	130,596	154.41	132,650	900	131,234	215	131,185	19.91
Tr24-15	136,509	23.70	136,872	900	136,860	44	136,556	9.34
Tr24-30	287,929	110.99	288,424	900	288,383	306	287,974	40.73

$\natural$: time on a dual core pentium 1.8GHz Linux workstation.
$\dagger$: time on a 200MHz Windows workstation.
$\ddagger$: time on a pentium III 750MHz Windows workstation.

Running times among the three algorithms cannot be meaningfully compared, due to the huge differences in machines speed. However, we can conclude that the proposed approach is faster than CPLEX alone in solving these instances, of course at a price of delivering a near-optimal solution.

While it is true that no robust conclusions can be drawn on a such a limited benchmark test, the results presented in this paper do show that the proposed approach is promising.

5   Conclusions

In this paper, we have presented a novel math-heuristic for a well known optimization problem, the lot sizing problem with setup times and setup costs. The main contribution of the work lies in the introduction of a math-heuristic approach for the column generation phase of a Dantzig-Wolfe algorithm.

Starting from an observation of [7], we presented a Dantzig-Wolfe reformulation in which the setup variables and the production variables are dealt with separately. More specifically, for any given setup plan, we generate columns in which the production plan needs not be a dominant plan, *i.e.*, we introduce into the master problem columns corresponding to solutions in which, for certain periods, there might be a setup without having a production. Thus, a single setup plan induces a number of non-dominant production plans. However, due to the large size of the set of non-dominant plans induced by each setup plan, we designed a mechanism inspired in the corridor method to bound the search of non-dominant production plans in the neighborhood of the Wagner-Whitin dominant plan associated to the current setup plan. By adding an exogenous constraint to the pricing problem, we collect a set of new columns and, subsequently, we price into the master problem those columns with negative reduced costs. Finally, the column generation approach is repeated until no new columns with negative reduced costs are found.

The proposed algorithm has been tested on a well-known set of benchmark instances and the results obtained have been compared with two approaches from the literature, as well as with the optimal solutions obtained by IBM CPLEX 12.1. While additional results on other problems are still needed, the results presented in the computational section allow to conclude that the proposed approach is promising, both in terms of solution quality and running time, and leaves various options for future applicability in other types of problems.

References

1. Aras, O.A.: A Heuristic Solution Procedure for a Single-Facility Multi-Item Dynamic Lot Sizing and Sequencing Problem. PhD thesis, Syracuse University (1981)
2. Barnhart, C., Johnson, E.L., Nemhauser, G.L., Savelsbergh, M.W.P., Vance, P.H.: Branch and Price: Column Generation for Huge Integer Programs. Operations Research 46(3), 316–329 (1998)
3. Belvaux, G., Wolsey, L.A.: BC-PROD: A Specialized Branch-and-Cut System for Lot-Sizing Problems. Management Science 46(5), 724–738 (2000)
4. Constantino, M.: A Cutting Plane Approach to Capacitated Lot-Sizing with Start-up Costs. Mathematical Programming B 75(3), 353–376 (1996)
5. Dzielinski, M.A.H., Gomory, R.E.: Optimal Programming of Lot-Sizing Inventory and Labor Allocation. Management Science 11(9), 874–890 (1965)
6. Eppen, G.D., Martin, R.K.: Solving Multi-Item Lot-Sizing Problems Using Variable Redefinition. Operations Research 35(6), 832–848 (1987)
7. Jans, R., Degraeve, Z.: A New Dantzig-Wolfe Reformulation and Branch-and-Price Algorithm for the Capacitated Lot Sizing Problem with Setup Times. Operations Research 55(5), 909–920 (2007)
8. Jans, R., Degraeve, Z.: Meta-heuristics for Dynamic Lot Sizing: A Review and Comparison of Solution Approaches. European Journal of Operational Research 177(3), 1855–1875 (2007)
9. Maniezzo, V., Stützle, T., Voß, S. (eds.): Matheuristics – Hybridizing Metaheuristics and Mathematical Programming. Annals of Information Systems, vol. 10. Springer (2010)
10. Manne, A.S.: Programming of Economic Lot Sizes. Management Science 4(2), 115–135 (1958)
11. Miller, A.J., Nemhauser, G.L., Savelsbergh, M.W.: Solving Multi-Item Capacitated Lot-Sizing Problems with Setup Times by Branch and Cut. Core dp 2000/39, Université Catholique de Louvain, Louvain-la-Neuve, Belgium (August 2000)
12. Nemhauser, G.L., Wolsey, L.A.: Integer and Combinatorial Optimization. Wiley-Interscience Series in Discrete Mathematics and Optimization. Wiley (1999)
13. Pochet, Y., Wolsey, L.A.: Polyhedra for Lot Sizing with Wagner-Whitin Costs. Mathematical Programming B 67(1-3), 297–323 (1994)
14. Sniedovich, M., Voß, S.: The Corridor Method: A Dynamic Programming Inspired Metaheuristic. Control and Cybernetics 35(3), 551–578 (2006)
15. Trigeiro, W.W., Thomas, L.J., McClain, J.O.: Capacitated Lot-sizing With Setup Times. Management Science 35(3), 353–366 (1989)
16. Vanderbeck, F., Savelsbergh, M.W.P.: A Generic View of Dantzig-Wolfe Decomposition in Mixed Integer Programming. Operations Research Letters 34(3), 296–306 (2006)

Application of the Nested Rollout Policy Adaptation Algorithm to the Traveling Salesman Problem with Time Windows

Tristan Cazenave[1] and Fabien Teytaud[1,2]

[1] LAMSADE, Université Paris Dauphine, France
[2] HEC Paris, CNRS, 1 rue de la Libération 78351 Jouy-en-Josas, France

Abstract. In this paper, we are interested in the minimization of the travel cost of the traveling salesman problem with time windows. In order to do this minimization we use a Nested Rollout Policy Adaptation (NRPA) algorithm. NRPA has multiple levels and maintains the best tour at each level. It consists in learning a rollout policy at each level. We also show how to improve the original algorithm with a modified rollout policy that helps NRPA to avoid time windows violations.

Keywords: Nested Monte-Carlo, Nested Rollout Policy Adaptation, Traveling Salesman Problem with Time Windows.

1 Introduction

In this paper we are interested in the minimization of the travel cost of the Traveling Salesman Problem with Time Windows. Recently, the use of a Nested Monte-Carlo algorithm (combined with expert knowledge and an evolutionary algorithm) gave good results on a set of state of the art problems [13]. However, as it has been pointed out by the authors, the effectiveness of the Nested Monte-Carlo algorithm decreases as the number of cities increases. When the number of cities is too large (greater than 30 for this set of problems), the algorithm is not able to find the state of the art solutions.

A natural extension to the work presented in [13], which consists in the application of the Nested Monte-Carlo algorithm on a set of Traveling Salesman Problems with Time Windows, is to study the efficiency of the Nested Rollout Policy Adaptation algorithm on the same set of problems.

In this work we study the use of a Nested Rollout Policy Adaptation algorithm on the Traveling Salesman Problem with Time Windows. The Nested Rollout Policy Adaptation algorithm has recently been introduced in [15], and provides good results, including records in Morpion Solitaire and crossword puzzles.

We improve this algorithm by replacing the standard random policy used in the rollouts with a domain-specific one, defined as a mixture of heuristics. These domain-specific heuristics are presented in Section 4.2.

The paper is organized as follows. Section 2 describes the Traveling Salesman Problem with Time Windows, Section 3 presents the Nested Monte-Carlo

Y. Hamadi and M. Schoenauer (Eds.): LION 6, LNCS 7219, pp. 42–54, 2012.
© Springer-Verlag Berlin Heidelberg 2012

algorithm (Section 3.1) and its application to the Traveling Salesman Problem with Time Windows (Section 3.2). Section 4 presents the Nested Rollout Policy Adaptation algorithm (Section 4.1) and its application to the Traveling Salesman Problem with Time Windows (Section 4.2). Section 5 presents a set of experiments concerning the application of the Nested Rollout Policy Adaptation algorithm on the Traveling Salesman Problem with Time Windows.

2   The Traveling Salesman Problem with Time Windows

The Traveling Salesman Problem (TSP) is a famous logistic problem. Given a list of cities and their pairwise distances, the goal of the problem is to find the shortest possible path that visits each city only once. The path has to start and finish at a given depot. The TSP problem is NP-hard [8]. In this work, we are interested in a similar problem but more difficult, the Traveling Salesman Problem with Time Windows (TSPTW). In this version, a difficulty is added. Each city has to be visited within a given period of time.

A survey of efficient methods for solving the TSPTW can be found in [9]. Existing methods for solving the TSPTW are numerous. First, branch and bound methods were used [1,3]. Later, dynamic programing based methods [5] and heuristics based algorithms [17,7] have been proposed. More recently, methods based on constraint programming have been published [6,10].

An algorithm based on the Nested Monte-Carlo Search algorithm has been proposed [13] and is summarized in Section 3.2.

The TSPTW can be defined as follow. Let G be an undirected complete graph. $G = (N, A)$, where $N = 0, 1, \ldots, n$ corresponds to a set of nodes and $A = N \times N$ corresponds to the set of edges between the nodes. The node 0 corresponds to the depot. Each city is represented by the n other nodes. A cost function $c : A \to \mathbb{R}$ is given and represents the distance between two cities. A solution to this problem is a sequence of nodes $P = (p_0, p_1, \ldots, p_n)$ where $p_0 = 0$ and $(p_1, \ldots, p_n)$ is a permutation of $[1, N]$. Set $p_{n+1} = 0$ (the path must finish at the depot), then the goal is to minimize the function defined in Equation 1.

$$cost(P) = \sum_{k=0}^{n} c(a_{p_k}, a_{p_{k+1}}) \tag{1}$$

As said previously, the TSPTW version is more difficult because each city i has to be visited in a time interval $[e_i, l_i]$. This means that a city i has to be visited before l_i. It is possible to visit a cite before e_i, but in that case, the new departure time becomes e_i. Consequently, this case may be dangerous as it generates a penalty. Formally, if r_{p_k} is the real arrival time at node p_k, then the departure time d_{p_k} from this node is $d_{p_k} = max(r_{p_k}, e_{p_k})$.

In the TSPTW, the function to minimize is the same as for the TSP (Equation 1), but a set of constraint is added and must be satisfied. Let us define $\Omega(P)$ as the number of violated windows constraints by tour (P).

Two constraints are defined. The first constraint is to check that the arrival time is lower than the fixed time. Formally,

$$\forall p_k, r_{p_k} < l_{p_k}.$$

The second constraint is the minimization of the time lost by waiting at a city. Formally,

$$r_{p_{k+1}} = \max(r_{p_k}, e_{p_k}) + c(a_{p_k, p_{k+1}}).$$

With the algorithm used in this work, paths with violated constraints can be generated. As presented in [13] , a new score $Tcost(p)$ of a path p can be defined as follow:

$$Tcost(p) = cost(p) + 10^6 * \Omega(p),$$

with, as defined previously, $cost(p)$ the cost of the path p and $\Omega(p)$ the number of violated constraints. 10^6 is a constant chosen high enough so that the algorithm first optimizes the constraints.

The TSPTW is much harder than the TSP, consequently new algorithms have to be used for solving this problem.

In the next sections, we define two algorithms for solving the TSPTW. The first one, in Section 3, is the Nested Monte-Carlo algorithm from [13], and the second one, in Section 4, is the Nested Rollout Policy Adaptation algorithm, which is used in this work to solve the TSPTW. We eventually present results in Section 5.

3   The Nested Monte-Carlo Search Algorithm

First, in Section 3.1 we present the Nested Monte-Carlo Search, and then in Section 3.2 the application done in [13] to the Traveling Salesman Problem with Time Windows.

3.1   Presentation of the Algorithm

The basic idea of Nested Monte-Carlo Search (NMC) is to find a solution path of cities with the particularity that each city choice is based on the results of a lower level of the algorithm [2]. At level 1, the lower level search is simply a playout (i.e. each city is chosen randomly).

Figure 1 illustrates a level 1 Nested Monte-Carlo search. Three selections of cities at level 1 are shown. The leftmost tree shows that, at the root, all possible cities are tried and that for each possible decision a playout follows it. Among the three possible cities at the root, the rightmost city has the best result of 30, therefore this is the first decision played at level 1. This brings us to the middle tree. After this first city choice, playouts are performed again for each possible city following the first choice. One of the cities has result 20 which is the best playout result among his siblings. So the algorithm continues with this decision as shown in the rightmost tree. This algorithm is presented in Algorithm 1.

Algorithm 1. Nested Monte-Carlo search

```
nested (level,node)
if level==0 then
  ply ← 0
  seq ← {}
  while num_children(node) > 0 do
    CHOOSE seq[ply] ← child i with probability 1/num_children(node)
    node ← child(node,seq[ply])
    ply ← ply+1
  end while
  RETURN (score(node),seq)
else
  ply ← 0
  seq ←{}
  best_score ← ∞
  while num_children(node) > 0 do
    for children i of node do
      temp ← child(node,i)
      (results,new) ← nested(level-1,temp)
      if results<best_score then
        best_score ← results
        seq[ply]=i
        seq[ply+1...]=new
      end if
    end for
    node=child(node,seq[ply])
    ply←ply+1
  end while
  RETURN (best_score,seq)
end if
```

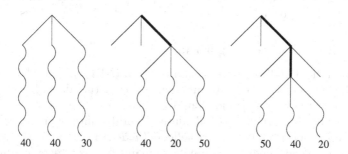

Fig. 1. This figure explains three steps of a level 1 search. At each step of the playout of level 1 shown here with a bold line, an NMC of level 1 performs a playout (shown with wavy lines) for each available decision and selects the best one.

At each choice of a playout of level 1 it chooses the city that gives the best score when followed by a single random playout. Similarly for a playout of level n it chooses the city that gives the best score when followed by a playout of level $n - 1$.

3.2   Application to the TSPTW

In [13], a NMC algorithm is used in order to solve a set of TSPTW. With the intention of having a competitive algorithm they add expert-knowledge to the NMC algorithm. Three heuristics are added and used to bias the Monte-Carlo simulations thanks to a Boltzmann softmax policy. The principle is to define a new policy for the rollout phase. This policy is defined by the probability $\pi_\theta(p, a)$ of choosing the action a in a position p:

$$\pi_\theta(p, a) = \frac{e^{\phi(p,a)^T \theta}}{\sum_b e^{\phi(p,b)^T \theta}},$$

where $\phi(p, a)$ is a vector of heuristics and θ is a vector of heuristic weights.

The three heuristics used are the same as the ones defined in [17], and are summarised as follows:

- The distance to the last city.
- The amount of wasted time because a city is visited too-early.
- The amount of time left until the end of the time window of a city.

For the tuning of the weights of each heuristic, they used an evolution strategy [12], more precisely a Self-Adaptive Evolution Strategy [12,16], known for its robutsness.

4   The Nested Rollout Policy Adaptation Algorithm

First, in Section 4.1, we present the Nested Rollout Policy Adaptation algorithm. In Section 4.2 we present some modifications done on this algorithm in order to improve it on the Traveling Salesman Problem with Time Windows.

4.1   Presentation of the Algorithm

The Nested Rollout Policy Adaptation algorithm (NRPA) is an algorithm that learns a playout policy. There are different levels in the algorithm. Each level is associated to the best sequence found at that level. The playout policy is a vector of weights that are used to calculate the probability of choosing a city. A city is chosen proportionally to $\exp(pol[\text{code}(node,\text{i})])$. pol(x) is the adaptable weight on code x. code($node$,i) is a unique domain-specific integer leading from a situation $node$ to its ith child. This is comparable with previous known learning of Monte-Carlo simulations [4,14].

Learning the playout policy consists in increasing the weights associated to the best cities and decreasing the weights associated to the other cities. The algorithm is given in algorithm 2.

Algorithm 2. Nested Rollout Policy Adaptation

NRPA (*level,pol*)
if *level* = 0 then
   node ← root
   ply ← 0
   seq ← {}
   while there are possible moves do
      CHOOSE *seq[ply]* ← child i the with probability proportional to exp(pol[code($node$,i)])
      node ← child($node$, *seq [ply]*)
      ply ←*ply* + 1
   end while
   return (score ($node$), *seq*)
else
   bestScore ← ∞
   for N iterations do
      (result,new) ← NRPA (*level* − 1, *pol*)
      if result ≤ *bestScore* then
         bestScore ← result
         seq ← new
      end if
      pol ← Adapt(*pol,seq*)
   end for
end if
return (*bestScore*,seq)

Adapt (*pol,seq*)
node ← root
pol' ← *pol*
for *ply* ← 0 to length(*seq*) - 1 do
   pol'[code($node$,seq[ply])] += Alpha
   z ← SUM exp(pol[code($node$,i)]) over node's children i
   for children i of *node* do
      pol'[code($node$,i)] -= Alpha × exp(pol[code($node$,i)]) / z
   end for
   node ← child($node$, *seq [ply]*)
end for
return *pol'*

4.2   Application to the TSPTW

As for the NMC algorithm, adding expert-knowledge is possible in order to improve this generic algorithm. Consequently, the generality of the resulting algorithm is lower. We implement a NRPA algorithm with a specific Monte-Carlo policy.

The idea of this algorithm is first to force to visit cities as soon as they go after their window end. The reason is that cities that are after their window end should have been visited earlier and that must be taken into account for the continuation of the playout. If we force to visit them, the algorithm will try more to visit them in time.

The second idea of the algorithm is to avoid visiting a city if it makes another city go after its window end. It considers all the moves that do not make any city go after its window end.

Algorithm 3. Playout policy for NRPA_EK

---

possibleMoves ()
$s \leftarrow \{\}$
for all not yet visited cities c **do**
   if going to the city c arrives after the window end of the city **then**
      add the city c to the set s
   end if
end for
if $s = \{\}$ **then**
   for all not yet visited cities c **do**
      $tooLate \leftarrow$ false
      for all not yet visited cities d different from c **do**
         if going to the city d arrives before the window end of the city d and going
         to the city c arrives after the window end of the city d **then**
            $tooLate \leftarrow$ true
         end if
      end for
      if not $tooLate$ **then**
         add the city c to the set s
      end if
   end for
end if
if $s = \{\}$ **then**
   for all not yet visited cities c **do**
      add the city c to the set s
   end for
end if
return s

These moves that avoid some cities can never be moves that force the algorithm into a suboptimal answer. These moves always imply a violation of the time window. Therefore they only change invalid solutions of the problem. An optimal move will not be pruned by our expert knowledge since it does not violate a constraint.

If there are still no possible moves after these two tests, the algorithm considers all the possible moves.

The resulting algorithm is labeled NRPA_EK and is given in algorithm 3.

5   Experiments

First, in Section 5.1, we study the behaviour of the NRPA algorithm. Second, in Section 5.2, we compare it with the version defined in Section 4.2 on two problems among the set of problems from [11]. Finally, in Section 5.3, we provide a comparison of the two algorithms studied in this work, the NMC algorithm in [13] and the state of the art results found in [9].

In all our experiments we set $\alpha = 1$ for the NRPA and the NRPA_EK algorithms.

5.1   The Behaviour of the NRPA Algorithm

It has been found for the NMC algorithm and the NRPA algorithm (both on Morpion Solitaire) that a plateau is reached for each level of the algorithms, and then consequently, that increasing the level improves the results of the algorithms. In figure 2, similar results are shown for a TSPTW (on the problem rc204.1 from the set of problems from [11]). We measure the score of a NRPA algorithm as a function of the time T for different levels. The time T represents the number of evaluations done for each level. Formally, $T = N^{level}$. We recall that N is the number of iterations done for each level (> 0) of the NRPA algorithm. Each point is the average of 30 runs. Plateaus are here well represented. For the level 1, $T = N$, and we can note that increasing N beyond 1000 does not improve the algorithm. The level 2 of the NRPA algorithm is quickly better than a level 1. It is better to use the level 3 of the algorithm than the level 2 around $T = 30000$, this means, approximately for $N = 170$ for the level 2 and $N = 30$ for the level 3. The level 4 of the algorithm becomes better than the level 3 around $T = 330000$, so it corresponds to $N = 575$ for the level 3 and $N = 70$ for the level 4.

5.2   NRPA against NRPA_EK

In this experiment we compare the NRPA algorithm (as defined in Algorithm 2) and the version of NRPA with expert knowledge, presented in Section 4.2. This last algorithm is labeled NRPA_EK in all our experiments.

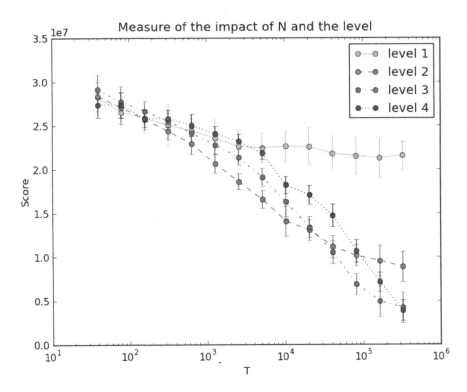

Fig. 2. Score as a function of T. Average on 30 runs. Plateaus are reached for the first three levels, and increasing the level of the algorithm improves the results.

We test these two algorithms on two problems from the set of problems from [11], the problem rc204.3 and the problem rc204.1. This last problem is the hardest one among all the problems of the set, and has 46 cities. In all this experiment, N is fixed at 50. Results are presented in Figure 3 for the problem rc204.3 and in Figure 4 for the problem rc204.1.

Results of the rc204.3 problem (Figure 3) are close, because this problem is simple enough for both algorithms and they are able to quickly find good solutions. However, the NRPA_EK is slightly better for both the levels 3 and 4. On the hardest problem (Figure 4), results are much more significant. The level 3 of both the algorithms are not able to find a path with respect to all the time constraints. In average, the NRPA_EK version is able to solve one more constraint than the NRPA algorithm. For the level 4 comparison, the NRPA_EK is by far better than the classic version of the algorithm and is able to find a path without any violated constraint in all the runs.

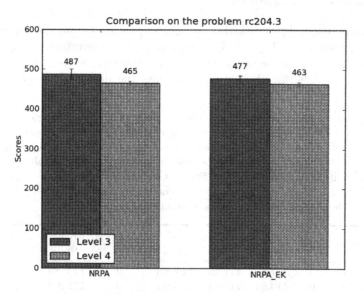

Fig. 3. Comparison between the NRPA algorithm and the NRPA_EK algorithm of the problem rc204.3. For both the algorithms $N = 50$.

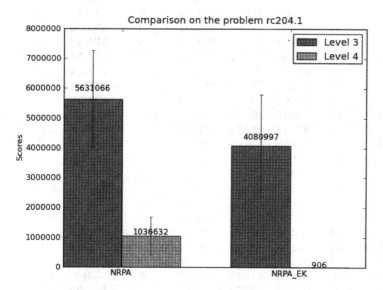

Fig. 4. Comparison between the NRPA algorithm and the NRPA_EK algorithm of the problem rc204.1. For both the algorithms $N = 50$.

5.3   State of the Art Problems

We experiment the NRPA algorithm on all the problems from the set of [11]. Results are presented in Table 1. Problems are sorted according to n (i.e. the number of cities). The state of the art results are found by the ant colony algorithm from [9]. The fourth column represents the best score found in [13].

Table 1. Results on all problems from the set from Potvin and Bengio [11]. First Column is the problem, second column the number of nodes, third column the best score found in [9], fourth column the best score found by the NMC algorithm with heuristics from [13], fifth column is the best score found by the NRPA algorithm and sixth column is its corresponding RPD. Seventh and eighth columns are respectively the score of the NRPA_EK algorithm and the corresponding RPD. The problems for which we find the state of the art solutions are in bold.

Problem	n	State of the art	NMC score	NRPA score	RPD	NRPA_EK	
rc206.1	4	117.85	117.85	**117.85**	0	**117.85**	0
rc207.4	6	119.64	119.64	**119.64**	0	**119.64**	0
rc202.2	14	304.14	304.14	**304.14**	0	**304.14**	0
rc205.1	14	343.21	343.21	**343.21**	0	**343.21**	0
rc203.4	15	314.29	314.29	**314.29**	0	**314.29**	0
rc203.1	19	453.48	453.48	**453.48**	0	**453.48**	0
rc201.1	20	444.54	444.54	**444.54**	0	**444.54**	0
rc204.3	24	455.03	455.03	**455.03**	0	**455.03**	0
rc206.3	25	574.42	574.42	**574.42**	0	**574.42**	0
rc201.2	26	711.54	711.54	**711.54**	0	**711.54**	0
rc201.4	26	793.64	793.64	**793.64**	0	**793.64**	0
rc205.2	27	755.93	755.93	**755.93**	0	**755.93**	0
rc202.4	28	793.03	793.03	800.18	0.90	**793.03**	0
rc205.4	28	760.47	760.47	765.38	0.65	**760.47**	0
rc202.3	29	837.72	837.72	839.58	0.22	839.58	0.22
rc208.2	29	533.78	536.04	537.74	0.74	**533.78**	0
rc207.2	31	701.25	707.74	702.17	0.13	**701.25**	0
rc201.3	32	790.61	790.61	796.98	0.81	**790.61**	0
rc204.2	33	662.16	675.33	673.89	1.77	664.38	0.34
rc202.1	33	771.78	776.47	775.59	0.49	772.18	0.05
rc203.2	33	784.16	784.16	**784.16**	0	**784.16**	0
rc207.3	33	682.40	687.58	688.50	0.83	**682.40**	0
rc207.1	34	732.68	743.29	743.72	1.51	738.74	0.83
rc205.3	35	825.06	828.27	828.36	0.40	**825.06**	0
rc208.3	36	634.44	641.17	656.40	3.46	650.49	2.53
rc203.3	37	817.53	837.72	820.93	0.42	**817.53**	0
rc206.2	37	828.06	839.18	829.07	0.12	**828.06**	0
rc206.4	38	831.67	859.07	831.72	0.01	**831.67**	0
rc208.1	38	789.25	797.89	799.24	1.27	793.60	0.55
rc204.1	46	868.76	899.79	883.85	1.74	880.89	1.40

[13] uses an evolutionary algorithm for tuning heuristics used to bias the level 0 of a Nested Monte-Carlo algorithm. The NRPA algorithm is somehow close to the NMC algorithm, and a comparison between these two algorithms on a set of TSPTW is interesting. The classic version of the NRPA algorithm does not use any expert knowledge, but is still able to achieve good results on a lot of problems (Table 1, fifth column). We provide the Relative Percentage Deviation (RPD) for both NRPA (column 6) and NRPA_EK (column 8). The RPD is $100 \times \frac{value - record}{record}$. Here again, we show that adding expert knowledge is useful and improves the algorithm. The NRPA_EK version is able to find most state of the art results (76.66%), as shown in column 7. For difficult problems, the best results were obtained after 2 to 4 runs of the algorithm.

6   Conclusion

In this paper we study the generality of a nested rollout policy adaptation algorithm by applying it to traveling salesman problems with time windows. Even with no expert knowledge at all, the NRPA algorithm is able to find state of the art results for problems for which the number of nodes is not too large. We also experiment the addition of expert knowledge in this algorithm. With the good results of the algorithm NRPA_EK, we show the feasability of guiding the rollout policy using domain-specific knowledge.

We show that adding expert knowledge significantly improves the results. It has been shown in our experiments that the NRPA_EK algorithm (the expert knowledge version of the NRPA algorithm) is able to find most state of the art results (76.66%), and has good results on other problems.

An extension of this work is the use of a pool of policies (instead of just having one), in order to have an algorithm more robust in front of local optima. Adapting the parameters of the NRPA algorithm according to the problem difficulty is also an interesting work.

Acknowledgement. This work has been supported by French National Research Agency (ANR) through COSINUS program (project EXPLORA ANR-08-COSI-004).

References

1. Baker, E.: An exact algorithm for the time-constrained traveling salesman problem. Operations Research 31(5), 938–945 (1983)
2. Cazenave, T.: Nested Monte-Carlo search. In: IJCAI, pp. 456–461 (2009)
3. Christofides, N., Mingozzi, A., Toth, P.: State-space relaxation procedures for the computation of bounds to routing problems. Networks 11(2), 145–164 (1981)
4. Drake, P.: The last-good-reply policy for monte-carlo go. ICGA Journal 32(4), 221–227 (2009)
5. Dumas, Y., Desrosiers, J., Gelinas, E., Solomon, M.: An optimal algorithm for the traveling salesman problem with time windows. Operations Research 43(2), 367–371 (1995)

6. Focacci, F., Lodi, A., Milano, M.: A hybrid exact algorithm for the tsptw. Informs Journal on Computing 14(4), 403–417 (2002)
7. Gendreau, M., Hertz, A., Laporte, G., Stan, M.: A generalized insertion heuristic for the traveling salesman problem with time windows. Operations Research 46(3), 330–335 (1998)
8. Johnson, D., Papadimitriou, C.: Computational complexity and the traveling salesman problem. Mass. Inst. of Technology, Laboratory for Computer Science (1981)
9. López-Ibáñez, M., Blum, C.: Beam-ACO for the travelling salesman problem with time windows. Computers & OR 37(9), 1570–1583 (2010)
10. Pesant, G., Gendreau, M., Potvin, J., Rousseau, J.: An exact constraint logic programming algorithm for the traveling salesman problem with time windows. Transportation Science 32(1), 12–29 (1998)
11. Potvin, J., Bengio, S.: The vehicle routing problem with time windows part II: genetic search. Informs Journal on Computing 8(2), 165 (1996)
12. Rechenberg, I.: Evolutionsstrategie: Optimierung technischer Systeme nach Prinzipien der biologischen Evolution, Fromman-Holzboog, Stuttgart, German (1973)
13. Rimmel, A., Teytaud, F., Cazenave, T.: Optimization of the Nested Monte-Carlo Algorithm on the Traveling Salesman Problem with Time Windows. In: Di Chio, C., Brabazon, A., Di Caro, G.A., Drechsler, R., Farooq, M., Grahl, J., Greenfield, G., Prins, C., Romero, J., Squillero, G., Tarantino, E., Tettamanzi, A.G.B., Urquhart, N., Uyar, A.Ş. (eds.) EvoApplications 2011, Part II. LNCS, vol. 6625, pp. 501–510. Springer, Heidelberg (2011)
14. Rimmel, A., Teytaud, F., Teytaud, O.: Biasing Monte-Carlo Simulations through RAVE Values. In: van den Herik, H.J., Iida, H., Plaat, A. (eds.) CG 2010. LNCS, vol. 6515, pp. 59–68. Springer, Heidelberg (2011)
15. Rosin, C.D.: Nested rollout policy adaptation for monte carlo tree search. In: Walsh, T. (ed.) IJCAI, pp. 649–654. IJCAI/AAAI (2011)
16. Schwefel, H.: Adaptive Mechanismen in der biologischen Evolution und ihr Einfluß auf die Evolutionsgeschwindigkeit. Interner Bericht der Arbeitsgruppe Bionik und Evolutionstechnik am Institut für Mess-und Regelungstechnik Re 215(3) (1974)
17. Solomon, M.: Algorithms for the vehicle routing and scheduling problems with time window constraints. Operations Research 35(2), 254–265 (1987)

Parallel Algorithm Configuration

Frank Hutter, Holger H. Hoos, and Kevin Leyton-Brown

University of British Columbia, 2366 Main Mall, Vancouver BC, V6T 1Z4, Canada
{hutter,hoos,kevinlb}@cs.ubc.ca

Abstract. State-of-the-art algorithms for solving hard computational problems often expose many parameters whose settings critically affect empirical performance. Manually exploring the resulting combinatorial space of parameter settings is often tedious and unsatisfactory. Automated approaches for finding good parameter settings are becoming increasingly prominent and have recently lead to substantial improvements in the state of the art for solving a variety of computationally challenging problems. However, running such automated algorithm configuration procedures is typically very costly, involving many thousands of invocations of the algorithm to be configured. Here, we study the extent to which parallel computing can come to the rescue. We compare straightforward parallelization by multiple independent runs with a more sophisticated method of parallelizing the model-based configuration procedure SMAC. Empirical results for configuring the MIP solver CPLEX demonstrate that near-optimal speedups can be obtained with up to 16 parallel workers, and that 64 workers can still accomplish challenging configuration tasks that previously took 2 days in 1–2 hours. Overall, we show that our methods make effective use of large-scale parallel resources and thus substantially expand the practical applicability of algorithm configuration methods.

1   Introduction

Heuristic algorithms are often surprisingly effective at solving hard combinatorial problems. However, heuristics that work well for one family of problem instances can perform poorly on another. Recognizing this, algorithm designers tend to parameterize some of these design choices so that an end user can select heuristics that work well in the solver's eventual domain of application. The way these parameters are chosen often makes the difference between whether a problem ends up being "easy" or "hard", with differences in runtime frequently spanning multiple orders of magnitude.

Traditionally, the problem of finding good parameter settings was left entirely to the end user (and/or the algorithm designer) and was solved through manual experiments. Lately, automated tools have become available to solve this *algorithm configuration (AC) problem*, making it possible to explore large and complex parameter spaces (involving up to more than seventy parameters with numerical, categorical, ordinal and conditional values; *e.g.*, [1, 2, 3]). Procedures for automated algorithm configuration have advanced steadily over the past few years, and now can usually outperform default settings determined by human experts even in very large and challenging configuration scenarios. However, automated algorithm configuration is computationally expensive: AC procedures involve repeatedly running the algorithm to be configured (the so-called

Y. Hamadi and M. Schoenauer (Eds.): LION 6, LNCS 7219, pp. 55–70, 2012.
© Springer-Verlag Berlin Heidelberg 2012

target algorithm) with different parameter settings, and hence consume orders of magnitude more computing time than a single run of the target algorithm.

Substantial amounts of computing time are now readily available, in the form of powerful, multi-core consumer machines, large compute clusters and on-demand commodity computing resources, such as Amazon's Elastic Compute Cloud (EC2). Nevertheless, the computational cost of AC procedures remains a challenge—particularly because the wall-clock time required for automated algorithm configuration can quickly become a limiting factor in real-world applications and academic studies. As ongoing increases in the amount of computing power per cost unit are now almost exclusively achieved by means of parallelization, the effective use of parallel computing resources becomes an important issue in the use and design of AC procedures.

Most AC procedures are inherently sequential in that they iteratively perform target algorithm runs, learn something about which parameters work well, and then perform new runs taking this information into account. Nevertheless, there are significant opportunities for parallelization, in two different senses. First, because all state-of-the-art AC procedures are randomized, the entire procedure can be run multiple times in parallel, followed by selection of the best configuration thus obtained. Second, a finer-grained form of parallelism can be used, with a centralized process distributing sets of target algorithm runs over different processor cores. Indeed, the literature on algorithm configuration already contains examples of both the first [1, 4, 2, 3] and second [4, 5] forms of parallelization.

Here, we present a thorough investigation of parallelizing automated algorithm configuration. We explore the efficacy of parallelization by means of multiple independent runs for two state-of-the-art algorithm configuration procedures, PARAMILS [4] and SMAC [6], investigating both how to make the best use of wall-clock time (what if parallel resources were free?) and CPU time (what if one had to pay for each CPU hour?). We present a method for introducing fine-grained parallelism into model-based AC procedures and apply it to SMAC. We evaluate the performance of the resulting distributed AC procedure, D-SMAC, as in contrast to and in combination with parallelization by means of multiple independent runs. Overall, we found that D-SMAC outperformed independent parallel runs and achieved near-perfect speedups when using up to 16 parallel worker processes. Using 64 parallel workers, we still obtained 21-fold to 52-fold speedups and were thus able to solve AC problems that previously took 2 days in 1-2 hours.

2   Methods for Parallelizing Algorithm Configuration

In this section, we describe two methods for parallelizing algorithm configuration: performing multiple independent runs and distributing target algorithms runs in the model-based configuration procedure SMAC.

2.1   Multiple Independent Runs

Any randomized algorithm with sufficiently large variation in runtime can usefully be parallelized by a simple and surprisingly powerful method: performing multiple independent runs (see, *e.g.*, [7, 8, 9]). In particular, it has been shown that the runtime of certain classes of local search procedures closely follows an exponential distribution [10], implying that optimal speedups can be achieved by using additional processor cores. Likewise, some complete search procedures have been shown to exhibit heavy-tailed runtime distributions, in which case multiple independent runs (or, equivalently, random restarts) can yield parallelization speedups greater than the number of parallel processes [8].

Multiple independent runs have also been used routinely in algorithm configuration (although we are not aware of any existing study that characterizes the runtime distributions of these procedures). In our research on PARAMILS, we have adopted the policy of performing 10 to 25 parallel independent runs, returning the configuration found by the run with best training performance [1, 4, 2, 3], which can be formalized as follows:

Definition 1 (*k*-**fold independent parallel version of configurator** *C*). *The k-fold independent parallel version of configurator C, denoted $k \times C$, is the configurator that executes k runs of C in parallel, and whose incumbent at each time t is the incumbent of the run with the best training performance at time t.*

For a given time budget, PARAMILS may not be able to evaluate target algorithm performance on all given training instances. In previous work, in such cases, we would sometimes measure training performance on the entire training set at the end of each of the independent configuration runs, selecting the final configuration based on those data. To keep computational costs manageable, we did not do this in the work described here. Also in previous work, we have observed that PARAMILS runs occasionally stagnate at rather poor configurations, and that in such cases $k \times$ PARAMILS can dramatically improve performance. However, to the best of our knowledge, this effect has never been quantified for PARAMILS, nor for any other AC procedure.

2.2   D-SMAC: SMAC with Distributed Target Algorithm Runs

Any state-of-the-art AC procedure could in principle be parallelized by distributing target algorithm runs over multiple cores. However, we are only aware of two examples from the literature describing AC solvers that implement such fine-grained parallelism. The first is the genetic algorithm GGA [5]; however, GGA always uses eight local workers, regardless of machine architecture, and is unable to distribute runs on a cluster. Second, in our own PARAMILS variant BASICILS [4], target algorithm runs were distributed over a cluster with 110 CPUs [11]. This, however, took advantage of the fact that BASIC-ILS performs a large number of runs for *every* configuration considered, and the same fact explains why our standard PARAMILS variant FOCUSEDILS typically outperforms BASICILS. [1] To the best of our knowledge, the effect of the number of parallel processes on overall performance has not been studied for any of these configurators.

Here, we present a general and principled method for adding fine-grained parallelization to SMAC, a recent model-based AC procedure [6]. SMAC is the focus of our current work, because (1) it achieves state-of-the-art performance for AC [6] and (2) its explicit

---

[1] The latest implementation of iterated F-Race [12] also supports parallelization of target algorithm runs, but this feature has not (yet) been described in the literature.

Algorithm 1. Sequential Model-Based Algorithm Configuration (SMAC)
R keeps track of all performed target algorithm runs and their performances (*i.e.*, SMAC's training data); $\mathcal{M}$ is SMAC's model; and Θ_{new} is a list of promising configurations.

> **Input** : Target algorithm with parameter configuration space Θ; instance set Π; cost metric $\hat{c}$
> **Output** : Optimized (incumbent) parameter configuration, θ_{inc}
> 1 $[\mathbf{R}, \theta_{inc}] \leftarrow Initialize(\Theta, \Pi)$;
> 2 **repeat**
> 3 $\mathcal{M} \leftarrow FitModel(\mathbf{R})$;
> 4 $\Theta_{new} \leftarrow SelectConfigurations(\mathcal{M}, \theta_{inc}, \Theta)$;
> 5 $[\mathbf{R}, \theta_{inc}] \leftarrow Intensify(\Theta_{new}, \theta_{inc}, \mathbf{R}, \Pi, \hat{c})$;
> 6 **until** *total time budget for configuration exhausted*;
> 7 **return** θ_{inc};

---

model of algorithm performance promises to be useful beyond merely finding good configurations (*e.g.*, for selecting informative problem instances or for gaining deeper insights into the impact of parameter settings on target algorithm performance).

SMAC operates in 4 phases (see Algorithm 1). First, it initializes its data and incumbent configuration θ_{inc}—the best configuration seen thus far—using algorithm runs from an initial design. Then it iterates between learning a new model, selecting new configurations based on that model and performing additional runs to compare these selected configurations against the incumbent.

The selection of new configurations is performed by optimizing a desirability function $d(\theta)$ defined in terms of the model's predictive distribution for θ. This desirability function serves to address the exploration/exploitation tradeoff between learning about new, unknown parts of the parameter space and intensifying the search locally in the best known region. Having found an incumbent with training performance f_{min}, SMAC uses a classic desirability function measuring the expected positive improvement over f_{min}, $\mathbb{E}[I(\theta)] = \mathbb{E}[\max\{0, f_{min} - f(\theta)\}]$. Many other desirability functions have been defined, such as the probability of improvement $\mathbb{P}[I(\theta) > 0]$ [13], generalizations of expected improvement $\mathbb{E}[I^g(\theta)]$ for $g > 1$ [14], and the optimistic confidence bound $(-\mu_\theta + \lambda\sigma_\theta)$ for $\lambda > 0$ [13, 15].[2] High values of all of these desirability functions reward low predictive mean (to encourage minimization of the performance metric) and high predictive variance (to encourage exploration of new regions).

Several methods have been proposed for identifying multiple desirable inputs to be evaluated in parallel. Ginsbourger et al. [16] introduced the multipoints expected improvement criterion, as well as the "constant liar approach": greedily select one new input θ, using expected improvement, hallucinate that its response equals the current model's predictive mean μ_θ, refit the model, and iterate. Jones [13] demonstrated that maximizing the optimistic confidence bound $(-\mu_\theta + \lambda\sigma_\theta)$ with different values of λ yields a diverse set of points whose parallel evaluation is useful.

---

[2] For maximization problems, this desirability function is $(\mu_\theta + \lambda\sigma_\theta)$ and is called the *upper confidence bound (UCB)*.

In our distributed version of SMAC we follow this latter approach (slightly deviating from it by sampling λ uniformly at random from an exponential distribution with mean 1 instead of using a fixed set of values for λ), since it also allows for the selection step to be parallelized: each of k workers can sample a value for λ and then optimize $(-\mu_\theta + \lambda\sigma_\theta)$ independently. Surprisingly, although we originally chose it in order to facilitate parallelization, in our experiments we found that this modified desirability function sometimes substantially improved SMAC's performance and never substantially degraded it compared to the expected improvement criterion we used previously (see Table 2 in Section 4).

The simplest way to parallelize SMAC would be to maintain the structure of Algorithm 1, but to execute each major component in parallel, synchronizing afterwards. The initialization can be parallelized easily, as it consists of randomly chosen target algorithm runs. Model fitting also parallelizes, as SMAC uses random forest models: each tree can be learned independently, and even subtrees of a single tree are independent. Gathering k desirable and diverse configurations can be parallelized as described above, and one can also parallelize the comparison of these configurations against the incumbent. We experimented with this approach, but found that when running on a compute cluster, it suffered from high communication overhead: learning the model requires the full input data, and optimizing the desirability function requires the model. Furthermore, the model learning phase is unlikely to parallelize perfectly, since a constant fraction of the time for building a regression tree is spent at the root of the tree. While we can still parallelize perfectly *across* trees, we typically only use 10 trees in practice and are interested in scaling to much larger numbers of parallel processes.

The parallelized version of SMAC we present here (dubbed D-SMAC) is therefore based on a different approach, slightly changing the structure of SMAC to bypass the chokepoint wherein more workers are available than can meaningfully be used to learn the model. Algorithm 2 illustrates the new control flow. The important difference is that D-SMAC maintains a queue of algorithm runs that is replenished whenever its current state drops below the number of runs than can be handled in one iteration by the parallel processes available. The intensification step—which compares challengers to the incumbent—now merely queues up runs rather than executing them. The benefit is that a master process can learn the model and select desirable new configurations *while* worker processes are performing target algorithm runs (typically the most expensive operation in SMAC). The master could also execute target runs or parallelize model learning and selecting configurations as necessary, to further balance load with the workers. In our current implementation, we simply use a separate processor for SMAC's master thread; since the model overhead was low in our experiments, these master threads spent most of their time idling, and we started several master processes on a single CPU.[3]

We employed a lightweight solution for distributing runs on compute clusters. The D-SMAC master writes command line call strings for target algorithm runs to designated files on a shared file system. Worker jobs submitted via the respective cluster's

---

[3] The model overhead grows with the number of data points, meaning that for long enough configuration runs it could become a chokepoint. In such settings, the master could delegate model learning to a slave, and update its model whenever the slave completed this work.

Algorithm 2. Distributed Sequential Model-Based Algorithm Configuration (D-SMAC)

Q is a queue of target algoritm runs to be executed; **A** is a set of runs currently assigned to workers; **R** keeps track of all executed runs and their performances (*i.e.*, SMAC's training data); $\mathcal{M}$ is SMAC's model, and Θ_{new} is a list of promising configurations. *Initialize* performs $\sqrt{k}$ runs for the default configuration and one run each for other configurations from a Latin Hypercube Design.

Input : Target algorithm with parameter configuration space Θ; instance set Π; cost metric $\hat{c}$; number of workers, k

Output : Optimized (incumbent) parameter configuration, θ_{inc}

1 $\mathbf{Q} \leftarrow$ *Initialize*($\Theta, \Pi, 2k$);
2 $\mathbf{A} \leftarrow \emptyset$; Move first k runs from **Q** to **A** and start the workers;
3 **repeat**
4 $\quad$ Wait for workers to finish, move the finished runs from **A** to **R**;
5 $\quad$ $\Theta_{new} \leftarrow \{\theta \mid \mathbf{R}$ received at least one new run with $\theta\}$;
6 $\quad$ $[\mathbf{Q}, \theta_{inc}] \leftarrow$ *Intensify*($\Theta_{new}, \theta_{inc}, \mathbf{Q}, \mathbf{R}, \Pi, \hat{c}$);
7 $\quad$ Move first k runs from **Q** to **A** and start the workers;
8 $\quad$ **if** $|Q| < k$ **then**
9 $\quad\quad$ $\mathcal{M} \leftarrow$ *FitModel*(**R**);
10 $\quad\quad$ $\Theta_{new} \leftarrow$ *SelectConfigurations*($\mathcal{M}, \theta_{inc}, \Theta, k - |Q|$);
11 $\quad\quad$ $[\mathbf{Q}, \theta_{inc}] \leftarrow$ *Intensify*($\Theta_{new}, \theta_{inc}, \mathbf{Q}, \mathbf{R}, \Pi, \hat{c}$);
12 **until** *total time budget for configuration exhausted*;
13 **return** θ_{inc};

---

queueing software (in our case, Torque) listen on the designated files, carry out the requested target run, and write the resulting performance to designated output files to be read by the master. On the cluster we used, we found the overhead for this job dispatch mechanism to be comparable to starting jobs on the local machine. Larger-scale deployment of D-SMAC would benefit from the use of an experimental framework such as HAL [17] or EDACC [18].

3   Experimental Setup

Our configuration experiments in this paper focus on the optimization of the solution quality that the mixed integer solver CPLEX can achieve in a fixed runtime. Specifically, we employed the five solution quality AC scenarios introduced in [2], as well as one additional scenario described below. All of these AC scenarios use a lexicographic objective function that first minimizes the number of instances for which no feasible solution was found, and then breaks ties by the average optimality gap. To use this objective function in SMAC and D-SMAC (whose modelling step requires scalar objective functions), we counted the "optimality gap" of runs that did not find a feasible solution as $10^{10}\%$. For a configuration scenario with test instance set S and fixed time limit per CPLEX run L, we defined the *test performance* of a configuration run R as the average optimality gap CPLEX achieved on S in runs with time limit L when using the incumbent parameter configuration of R.

Table 1. Overview of CPLEX parameters and MIP benchmark sets used

Parameter type	# parameters of this type	# values considered	Total # configurations
Boolean	6 (7)	2	
Categorical	45 (43)	3–7	$1.90 \cdot 10^{47}$
Integer	18	5–7	
Continuous	7	5–8	

Benchmark	Description	# instances		Default performance	
		training	test	% infeasible	mean opt. gap when feasible
MIK	Mixed integer knapsack [19]	60	60	0%	0.142%
CLS	Capacitated lot-sizing [20]	50	50	0%	0.273%
REGIONS200	Combinatorial winner determination [21]	1000	1000	0%	1.87%
CORLAT	Wildlife corridor [22]	1000	1000	28%	4.43%
MASS	Multi-activity shift scheduling [23]	50	50	64%	1.91%
RCW	Spread of red-cockaded woodpecker [24]	1000	1000	0%	49%

Throughout our experiments, in order to study the test performance of $k \times \mathcal{C}$, the k-fold independent parallel version of AC procedure $\mathcal{C}$, we employed a bootstrap analysis. Given a large population $\mathcal{P}$ of independent runs of $\mathcal{C}$, we evaluated $k \times \mathcal{C}$ by repeatedly drawing k runs of $\mathcal{C}$ from $\mathcal{P}$ (with repetitions) and computing the test performance of the best of the k runs (best in terms of training performance). This process yielded a bootstrap distribution of test performance; we plot the median of this distribution at each time step, show boxplots for the final state and carry out a Mann-Whitney U-test for differences across different configurators (or different parallelization options).

To carry out a robust bootstrap analysis of $k \times \mathcal{C}$, a population of roughly $3 \cdot k$ runs of $\mathcal{C}$ is required. Since we wanted to evaluate the benefit of up to 64 independent runs, we had to run each configurator 200 times on each configuration scenario. As a result, we carried out over 5000 configuration runs, more than in all of our previously published works on AC combined. Note that each individual configuration run involved thousands of target algorithm runs. In total, the experiments for this paper (including offline validation) took roughly 20 CPU years.

To fit within this time budget, we kept the original, relatively small configuration budget for the five AC scenarios taken from [2]: five hours per AC run and ten seconds per CPLEX run. Since the machines we used[4] are a (surprisingly constant) factor of just above 2 times faster than the machines used in [2], we divided both the runtime for configuration runs and for individual CPLEX runs by 2 to keep the characteristics of the AC scenarios as similar as possible to previously published work.

For the same reason, we used exactly the same parameter configuration space of CPLEX 12.1, and the same mixed integer problems (MIPs) as in the original scenarios from [2]. Briefly, we considered 76 parameters that directly affect the performance of CPLEX. We carefully kept all parameters fixed that change the problem formulation (e.g., numerical precision parameters). The 76 parameters we selected affect all aspects of CPLEX. They include 12 preprocessing parameters; 17 MIP strategy parameters; 11

---

[4] All of our experiments were carried out on the Westgrid Orcinus cluster (http://www.westgrid.ca/), comprising 384 nodes with two Intel X5650 six-core 2.66 GHz processors each.

cut parameters; 9 MIP limits parameters; 10 simplex parameters; 6 barrier optimization parameters; and 11 further parameters. Table 1 gives an overview of these parameters and of the MIP benchmarks we used; full details can be found in [2].

To study whether our findings for the short configuration runs above translate to longer runs of the most recent CPLEX version (12.3) on more challenging benchmark sets, we also carried out experiments on a new configuration scenario. The MIP instances in this scenario come from the domain of computational sustainability; they model the spread of the endangered red-cockaded woodpecker (RCW), conditional on decisions about certain parcels of land to be protected. We generated 2000 instances using the generator from [24] (using the five hardest of their eleven maps). CPLEX 12.3's default configuration could solve 7% of these instances in two minutes and 75% in one hour. The objective in our RCW configuration scenario was to minimize the optimality gap CPLEX could achieve within two minutes, and the AC budget was two days.

Throughout our experiments, we accounted for the inherent runtime overheads for building and using models, but we did not count the constant overhead of starting jobs (either as part of the per-run budget or of the configuration budget), since this can be reduced to almost zero in a production system. We computed the wall clock time for each iteration of D-SMAC as the maximum of the master's model learning time and the maximum of the CPU times of the parallel algorithm runs it executed in parallel.

4   Experiments

We studied the parallelization speedups obtained by using multiple independent runs and by using fine-grained parallelism in D-SMAC. As a side result, we were able to show for the first time that SMAC (in its sequential version) achieves state-of-the-art performance for optimizing a measure of solution quality that can be obtained in a fixed time (rather than minimizing the runtime required to solve a problem).

4.1   Multiple Independent Runs

First, we assessed the baseline performance of the three sequential AC procedures we used: PARAMILS, SMAC, and D-SMAC(1). PARAMILS has been shown to achieve state-of-the-art performance for the five configuration scenarios we study here [2], and Table 2 demonstrates that SMAC and D-SMAC(1) perform competitively, making all procedures natural candidates for parallelization. The right part of Table 2 compares the performance of the multiple independent run versions 25×PARAMILS, 25×SMAC, and 25×D-SMAC, showing that PARAMILS benefitted more from multiple runs than the two SMAC versions. The raw data (not shown) explains this: the variance of PARAMILS's performance was higher than for either SMAC version.

Table 3 quantifies the speedups gained by multiple independent runs of the AC procedures. For the two versions of SMAC, speedups were consistent and sometimes near-perfect with up to 4 independent runs. Due to larger performance variation

Table 2. Statistics for baseline comparison of configuration procedures. We show median test performances achieved by the base AC procedures (left), and their k-fold parallel independent run versions with $k = 25$ (recall that test performance is the average optimality gap across test instances, counting runs with infeasible solutions as a gap of $10^{10}\%$). We bold-faced entries for configurators that are not significantly worse than the best configurator for the respective configuration space, based on a Mann-Whitney U test.

Scenario	Unit	Median test performance			Median test performance		
		PILS	SMAC	d-SMAC(1)	25×PILS	25×SMAC	25×d-SMAC(1)
CLS	[0.1%]	2.36	2.43	**2.00**	1.38	1.41	**1.35**
CORLAT	[10^8%]	17.6	3.17	**2.95**	4.20	0.82	**0.72**
MIK	[0.01%]	6.56	6.59	**2.78**	**0.44**	2.08	0.73
Regions200	[1%]	**1.69**	1.8	1.83	**0.85**	1.16	1.14
MASS	[10^9%]	6.40	3.68	**3.47**	4.00	2.36	**2.29**

Table 3. Speedups achieved by using independent parallel runs of various AC procedures C. We give the speedups of $4 \times C$ over C, $16 \times C$ over $4 \times C$, and $64 \times C$ over $16 \times C$. The speedup of procedure C_1 over procedure C_2 is defined as the time allocated to C_2 divided by the time C_1 required to reach (at least) C_2's final solution quality. We do not report speedups of $16 \times C$ and $64 \times C$ over C directly since C often found very poor results in the small configuration budget allowed, the time to find which is not indicative of a procedure's ultimate performance.

Scenario	PARAMILS			SMAC			D-SMAC(1)		
	1→4×	4→16×	16→64×	1→4×	4→16×	16→64×	1→4×	4→16×	16→64×
CLS	5.02	2.87	1.66	5.72	2.33	1.50	1.92	2.09	1.75
CORLAT	12.1	4.75	4.22	2.45	2.10	1.15	3.93	2.31	1.00
MIK	8.29	3.10	2.29	3.22	3.45	4.01	2.37	2.91	1.02
Regions200	5.59	3.65	2.94	3.04	1.49	1.76	3.14	3.08	2.39
MASS	4.00	5.78	1.00	1.62	1.44	1.36	2.24	1.49	1.00

between independent runs, the parallelization speedups obtained for PARAMILS were more pronounced: perfect or higher-than-perfect speedups were observed for all scenarios with up to 4 independent runs, and the speedup factor obtained when moving from 4 to 16 independent parallel runs was still almost 4.

Figures 1 and 2 visualize the speedups achieved for two representative configuration scenarios. As the left column of Figure 1 shows, for benchmark set Regions200 additional independent runs yielded consistent speedups in wall clock time for all configurators. The right column shows that, as the runtime spent in a PARAMILS or D-SMAC(1) run increases for this benchmark set, $k×$PARAMILS and $k×$D-SMAC(1) tend to use their combined CPU time about as well as their respective sequential versions with a k-fold larger time budget. Figure 2 visualizes the results for benchmark set CORLAT, showing an interesting effect by which $k×$PARAMILS and $k×$SMAC can actually perform *worse* in their early phases as k increases. This effect is due to the fact that training

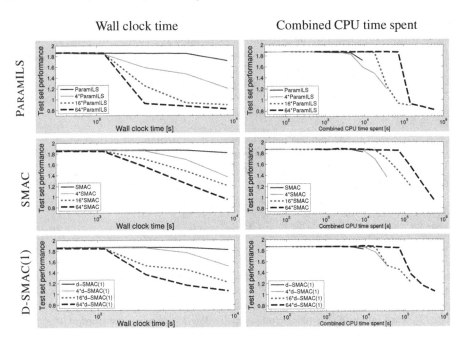

Fig. 1. Evaluation of k-fold parallel independent run versions of PARAMILS, SMAC, and D-SMAC(1) on benchmark set Regions200. For each configurator C, $k*C$ denotes $k\times C$.

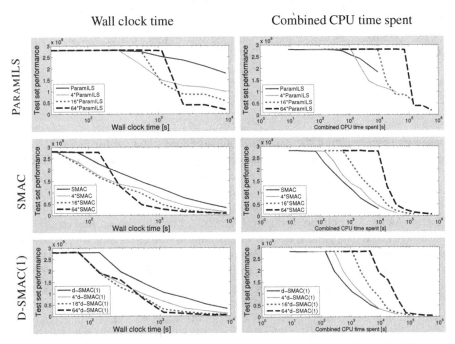

Fig. 2. Evaluation of k-fold parallel independent run versions of PARAMILS, SMAC, and D-SMAC(1) on benchmark set CORLAT

Table 4. Wall clock speedups over D-SMAC(1) with different numbers of distributed workers in SMAC. The speedup of procedure C_1 over procedure C_2 is defined as the time allocated to C_2 divided by the time C_1 required to reach (at least) C_2's final solution quality. For consistency with Table 3, we give the speedups of $4 \times C$ over C, $16 \times C$ over $4 \times C$, and $64 \times C$ over $16 \times C$. We also report the speedups of $16 \times C$ over C, and of $64 \times C$ over C.

Scenario	$1 \to 4\times$	$4 \to 16\times$	$16 \to 64\times$	$1 \to 4\times$	$1 \to 16\times$	$1 \to 64\times$
CLS	6.22	2.87	2.55	6.22	16.2	41.2
CORLAT	4.04	3.35	1.95	4.04	13.7	27.3
MIK	2.40	4.70	1.56	2.40	11.9	21.3
Regions200	2.61	10.9	1.25	2.61	41.3	52.3
MASS	2.21	3.44	2.41	2.21	9.76	21.5

performance can be *negatively* correlated with progress in the early phase of the search;[5] it is clearly visible as the crossing of lines in Figure 2 (left column). Another interesting effect for this scenario is that $4\times$PARAMILS achieved higher-than-perfect speedups (visible as the crossing of lines for PARAMILS in the right column of Figure 2).

4.2   Distributed SMAC

We now evaluate the parallelization speedups obtained by D-SMAC with a varying number of parallel worker processes. As shown in Table 4, these speedups were greater than those for multiple independent runs, with near-perfect speedups up to 16 workers and speedup factors between 1.2 and 2.6 for increasing the number of workers by another factor of 4 to 64. Overall, D-SMAC(64)'s speedups in the time required to find configurations of the same quality as D-SMAC(1) were between 21 and 52. Figure 3 visualizes the results for three configuration scenarios. The left side of this figure demonstrates that the substantial speedups D-SMAC achieved with additional workers were consistent across scenarios and across D-SMAC's trajectory.[6] In particular, speedups for early phases of the search were much more robust than for parallelization by multiple independent runs. The right side of Figure 3 demonstrates that D-SMAC(p) used its combined CPU time almost as well as D-SMAC(1) would, but required a factor p less wall clock time.

---

[5] PARAMILS starts from the default configuration, which finds a feasible solution for 72% of the instances. The configuration scenario's objective function heavily penalizes target algorithm runs that do not find a feasible solution and no configuration is found that finds a feasible solution for *all* training instances. Thus, any configuration run that has made enough progress will have a worse training performance than configuration runs that are still stuck having done only a few successful runs on the default. The larger we grow k in $k\times$PARAMILS, the more likely it is that one of the runs will be stuck at the default up to any given time (having seen only successful runs for the default), making $k\times$PARAMILS's incumbent the default configuration.

[6] The only exception is that D-SMAC(4) performed better than D-SMAC(16) early in the search for scenario CORLAT. Here, several of the D-SMAC(4) runs started out an order magnitude faster than D-SMAC(1); however, after about 60 seconds of search time D-SMAC(16) dominated D-SMAC(4) as expected.

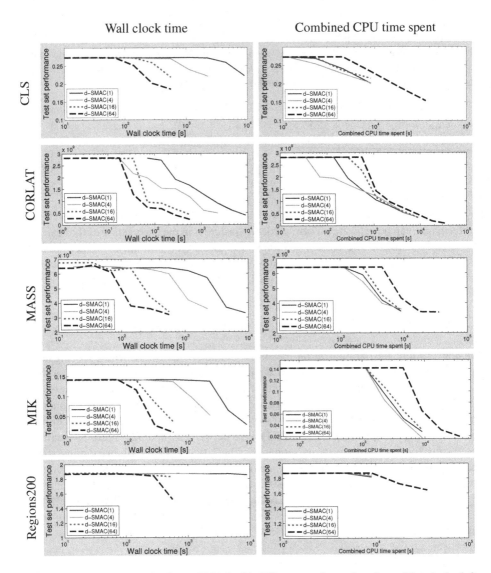

Fig. 3. Parallelization benefits for D-SMAC with different numbers of workers. (Plots in the left and right columns are based on different bootstrap samples.)

4.3 Multiple Independent Runs of Distributed SMAC

Next, we studied various allocations of a fixed number of $N = 64$ CPUs to independent runs of D-SMAC (that is, different variants of $k\times$ D-SMAC(p) with constant $k \times p = 64$). Figure 4 shows that $1\times$ D-SMAC(64) tended to perform better than the other combinations across all domains and time budgets. Table 5 shows that, given the same time budget of 10 CPU hours (or 562.5 wall seconds on 64 processors), $1\times$ D-SMAC(64) statistically significantly outperformed all other combinations we tested in

Table 5. Performance comparison of various possibilities of allocating 64 cores for a wall clock time of 560 seconds in D-SMAC. For each combination of independent runs and number of workers in D-SMAC, we show median test performance; we bold-faced entries for configurators that were not significantly worse than the best configurator for the respective configuration space, based on a Mann-Whitney U test.

Scenario	Unit	Bootstrap median of average test set performance			
		$64 \times$ d-SMAC(1)	$16 \times$ d-SMAC(4)	$4 \times$ d-SMAC(16)	d-SMAC(64)
CLS	$[0.1\%]$	2.37	1.96	**1.76**	**1.81**
CORLAT	$[10^8\%]$	10.9	3.41	**1.96**	**2.26**
MIK	$[0.01\%]$	8.68	**1.2**	2.03	2.46
Regions200	$[1\%]$	1.91	1.77	1.58	**1.52**
MASS	$[10^9\%]$	3.88	4.00	3.39	**3.2**

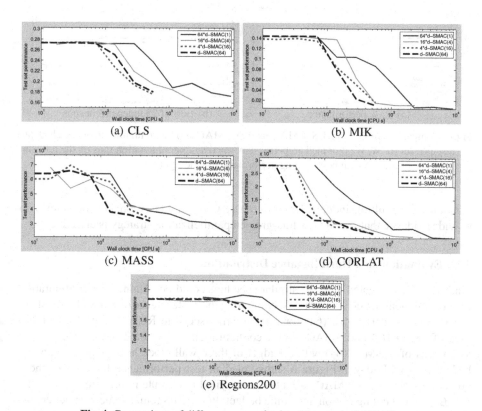

(a) CLS   (b) MIK

(c) MASS   (d) CORLAT

(e) Regions200

Fig. 4. Comparison of different ways of using 64 cores in d-SMAC

2 of 5 cases, and tied for best on the remaining 3 cases. These results demonstrate that performing a small number of parallel independent runs of D-SMAC(p) can be useful, but that using all available processors in D-SMAC(p) tends to yield the best performance. While these results do not preclude the possibility that for even higher degrees of

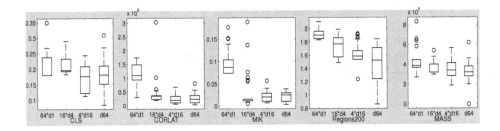

Fig. 5. Performance comparison of various possibilities of allocating 64 cores in d-SMAC; $k*\mathrm{d}p$ denotes k independent runs of D-SMAC(p), with $k * p = 64$. For each combination of k and p, we show boxplots of test set performance, using the same data as underlying Figure 4 .

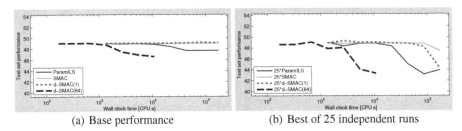

(a) Base performance   (b) Best of 25 independent runs

Fig. 6. Comparison of PARAMILS, SMAC, and D-SMAC(64) for configuration on the challenging instance set RCW. We plot median performance across 25 configuration runs on the left, and performance of the run with best training performance on the right.

parallelization multiple independent runs of D-SMAC might be more beneficial, they do provide evidence that D-SMAC's fine-grained parallelization strategy is effective.

4.4 Evaluation on a Hard Instance Distribution

Finally, we investigated whether similar speedups could be obtained for configuration on more challenging benchmark sets, comparing the performance of PARAMILS, SMAC, D-SMAC(1), and D-SMAC(64) for configuration scenario RCW. We performed 200 runs of PARAMILS and SMAC with a configuration budget of 2 days each, as well as 25 runs of D-SMAC(64) with a budget of three wall clock hours (for a combined budget of 8 CPU days). Figure 6 shows median test performance for each of these procedures. While the SMAC variants did not yield noticeable improvements over the CPLEX default configuration in its time budget, PARAMILS found somewhat better configurations. D-SMAC(64) already improved over the default configuration after roughly 20 wall clock minutes, required less than one wall clock hour to find a configuration as good as the one PARAMILS found after 2 days, and consistently improved afterwards. We also studied the performance of 25×PARAMILS, 25×SMAC, 25×D-SMAC(1), and 25×D-SMAC(64) for this benchmark set. Multiple independent runs improved the performance of all configurators. 25×D-SMAC(64) performed best, requiring roughly two hours to achieve the performance 25×PARAMILS and 25×D-SMAC(1) achieved in two days. It also matched D-SMAC(64)'s final performance in roughly a quarter of the time and found substantially better configurations afterwards. While a single run of D-SMAC (1600) might have yielded even better performance (we did not try, for lack of computing resources), this result shows that even AC procedures that implement large-scale fine-grained parallelism can benefit from performing multiple independent runs.

5 Conclusion

Parallel computing is key to reducing the substantial amount of time required by automatic algorithm configuration methods. Here, we presented the first comparative study of the two fundamental approaches for parallelizing automated configuration procedures—multiple independent runs and fine-grained parallelization—investigating how effective each of them is in isolation and to which extent they complement each other. We showed that the generic multiple independent runs parallelization approach is suprisingly effective when applied to the state-of-the-art configuration procedures PARAMILS and SMAC. We also introduced D-SMAC, a fine-grained parallelization of the state-of-the-art model-based algorithm configuration procedure SMAC, and showed that it achieves even better parallelization efficiencies, with speedups up to around 50 when using 64 parallel worker processes on a cluster of standard quad-core machines. Overall, we showed that using 64 parallel workers can reduce the wall clock time necessary for a range of challenging algorithm configuration tasks from 2 days to 1-2 hours. We believe that reductions of this magnitude substantially expand the practical applicability of existing algorithm configuration procedures and further facilitate their integration into the algorithm design process.

References

[1] Hutter, F., Babić, D., Hoos, H.H., Hu, A.J.: Boosting verification by automatic tuning of decision procedures. In: Proc. of FMCAD 2007, pp. 27–34. IEEE Computer Society (2007)

[2] Hutter, F., Hoos, H.H., Leyton-Brown, K.: Automated Configuration of Mixed Integer Programming Solvers. In: Lodi, A., Milano, M., Toth, P. (eds.) CPAIOR 2010. LNCS, vol. 6140, pp. 186–202. Springer, Heidelberg (2010)

[3] Fawcett, C., Helmert, M., Hoos, H.H., Karpas, E., Röger, G., Seipp, J.: FD-Autotune: Domain-specific configuration using fast-downward. In: Proc. of ICAPS-PAL 2011, 8 p. (2011)

[4] Hutter, F., Hoos, H.H., Leyton-Brown, K., Stützle, T.: ParamILS: an automatic algorithm configuration framework. Journal of Artificial Intelligence Research 36, 267–306 (2009)

[5] Ansótegui, C., Sellmann, M., Tierney, K.: A Gender-Based Genetic Algorithm for the Automatic Configuration of Algorithms. In: Gent, I.P. (ed.) CP 2009. LNCS, vol. 5732, pp. 142–157. Springer, Heidelberg (2009)

[6] Hutter, F., Hoos, H.H., Leyton-Brown, K.: Sequential Model-Based Optimization for General Algorithm Configuration. In: Coello, C.A.C. (ed.) LION 2011. LNCS, vol. 6683, pp. 507–523. Springer, Heidelberg (2011)

[7] Hoos, H.H., Stützle, T.: Local search algorithms for SAT: An empirical evaluation. Journal of Automated Reasoning 24(4), 421–481 (2000)

[8] Gomes, C.P., Selman, B., Crato, N., Kautz, H.: Heavy-tailed phenomena in satisfiability and constraint satisfaction problems. Journal of Algorithms 24(1) (2000)

[9] Ribeiro, C.C., Rosseti, I., Vallejos, R.: On the Use of Run Time Distributions to Evaluate and Compare Stochastic Local Search Algorithms. In: Stützle, T., Birattari, M., Hoos, H.H. (eds.) SLS 2009. LNCS, vol. 5752, pp. 16–30. Springer, Heidelberg (2009)

[10] Hoos, H.H., Stützle, T.: Towards a characterisation of the behaviour of stochastic local search algorithms for SAT. Artificial Intelligence 112(1-2), 213–232 (1999)

[11] Hutter, F.: Automated Configuration of Algorithms for Solving Hard Computational Problems. PhD thesis, University Of British Columbia, Department of Computer Science, Vancouver, Canada (October 2009)

[12] López-Ibáñez, M., Dubois-Lacoste, J., Stützle, T., Birattari, M.: The irace package, iterated race for automatic algorithm configuration. Technical Report TR/IRIDIA/2011-004, IRIDIA, Université Libre de Bruxelles, Belgium (2011)

[13] Jones, D.R.: A taxonomy of global optimization methods based on response surfaces. Journal of Global Optimization 21(4), 345–383 (2001)

[14] Schonlau, M., Welch, W.J., Jones, D.R.: Global versus local search in constrained optimization of computer models. In: Flournoy, N., Rosenberger, W.F., Wong, W.K. (eds.) New Developments and Applications in Experimental Design, vol. 34, pp. 11–25. Institute of Mathematical Statistics, Hayward (1998)

[15] Srinivas, N., Krause, A., Kakade, S., Seeger, M.: Gaussian process optimization in the bandit setting: No regret and experimental design. In: Proc. of ICML 2010 (2010)

[16] Ginsbourger, D., Le Riche, R., Carraro, L.: Kriging Is Well-Suited to Parallelize Optimization. In: Tenne, Y., Goh, C.-K. (eds.) Computational Intel. in Expensive Opti. Prob. ALO, vol. 2, pp. 131–162. Springer, Heidelberg (2010)

[17] Nell, C., Fawcett, C., Hoos, H.H., Leyton-Brown, K.: HAL: A Framework for the Automated Analysis and Design of High-Performance Algorithms. In: Coello, C.A.C. (ed.) LION 2011. LNCS, vol. 6683, pp. 600–615. Springer, Heidelberg (2011)

[18] Balint, A., Diepold, D., Gall, D., Gerber, S., Kapler, G., Retz, R.: EDACC - An Advanced Platform for the Experiment Design, Administration and Analysis of Empirical Algorithms. In: Coello, C.A.C. (ed.) LION 2011. LNCS, vol. 6683, pp. 586–599. Springer, Heidelberg (2011)

[19] Atamtürk, A.: On the facets of the mixed–integer knapsack polyhedron. Mathematical Programming 98, 145–175 (2003)

[20] Atamtürk, A., Muñoz, J.C.: A study of the lot-sizing polytope. Mathematical Programming 99, 443–465 (2004)

[21] Leyton-Brown, K., Pearson, M., Shoham, Y.: Towards a universal test suite for combinatorial auction algorithms. In: Proc. of EC 2000, pp. 66–76 (2000)

[22] Gomes, C.P., van Hoeve, W.-J., Sabharwal, A.: Connections in Networks: A Hybrid Approach. In: Trick, M.A. (ed.) CPAIOR 2008. LNCS, vol. 5015, pp. 303–307. Springer, Heidelberg (2008)

[23] Cote, M., Gendron, B., Rousseau, L.: Grammar-based integer programing models for multiactivity shift scheduling. Technical Report CIRRELT-2010-01, Centre interuniversitaire de recherche sur les réseaux d'entreprise, la logistique et le transport (2010)

[24] Ahmadizadeh, K., Dilkina, B., Gomes, C.P., Sabharwal, A.: An Empirical Study of Optimization for Maximizing Diffusion in Networks. In: Cohen, D. (ed.) CP 2010. LNCS, vol. 6308, pp. 514–521. Springer, Heidelberg (2010)

Community Detection in Social and Biological Networks Using Differential Evolution

Guanbo Jia[1], Zixing Cai[1], Mirco Musolesi[4], Yong Wang[1], Dan A. Tennant[3],
Ralf J.M. Weber[2], John K. Heath[2], and Shan He[2,4,*]

[1] School of Information Science and Engineering, Central South University,
Changsha 410083, China
[2] Center for Systems Biology, School of Biological Sciences
[3] School of Cancer Sciences
[4] School of Computer Science, University of Birmingham,
Birmingham, B15 2TT, United Kingdom

Abstract. The community detection in complex networks is an important problem in many scientific fields, from biology to sociology. This paper proposes a new algorithm, Differential Evolution based Community Detection (DECD), which employs a novel optimization algorithm, differential evolution (DE) for detecting communities in complex networks. DE uses network modularity as the fitness function to search for an optimal partition of a network. Based on the standard DE crossover operator, we design a modified binomial crossover to effectively transmit some important information about the community structure in evolution. Moreover, a biased initialization process and a clean-up operation are employed in DECD to improve the quality of individuals in the population. One of the distinct merits of DECD is that, unlike many other community detection algorithms, DECD does not require any prior knowledge about the community structure, which is particularly useful for its application to real-world complex networks where prior knowledge is usually not available. We evaluate DECD on several artificial and real-world social and biological networks. Experimental results show that DECD has very competitive performance compared with other state-of-the-art community detection algorithms.

Keywords: Community structure, graph partitioning, evolutionary computation, Differential Evolution.

1 Introduction

In the fields of science and engineering, there exist various kinds of complex systems which can be represented as complex networks naturally, such as social networks [26] and the Internet [7]. A complex network consists of nodes (or vertices) and edges (or links) which respectively represent the individual members and their relationships in systems [5]. In recent years, the study of complex networks has attracted more and more attention [1,13,16,30].

---

* Corresponding author.

Y. Hamadi and M. Schoenauer (Eds.): LION 6, LNCS 7219, pp. 71–85, 2012.
© Springer-Verlag Berlin Heidelberg 2012

Complex networks possess many distinctive properties [12], of which community structure [4] is one of the most studied. The community structure is usually considered as the division of networks into subsets of vertices within which intra-connections are dense while between which inter-connections are sparse [4,12]. Identifying the community structure is very helpful to obtain some important information about the relationship and interaction among nodes.

To detect the underlying community structure in complex networks, many successful algorithms have been proposed so far [4,12]. However, the community detection in networks is a nondeterministic polynomial (NP) hard problem. Most of current community detection algorithms based on greedy algorithms perform poorly on large complex networks. Moreover, many algorithms for community detection also require some prior knowledge about the community structure, e.g., the number of the communities, which is very difficult to be obtained in real-world networks.

To overcome these drawbacks, this paper proposes a new community detection algorithm based on Differential Evolution (DE), named DECD. To the best of our knowledge, it is the first time DE is introduced for community detection. In DECD, DE is used to evolve a population of potential solutions for network partitions to maximize the network modularity [20]. It is worth mentioning that DECD does not require any prior knowledge about the community structure when detecting communities in networks, which is is beneficial for its applications to real-world problems where prior knowledge is usually not available.

Apart from introducing DE for community detection, other key contributions of this paper include: 1) the design of an improved version of the standard binomial crossover in DE to transmit some important information about the community structure during evolution in DECD; 2) a biased process and a clean-up operation similar to [31] is introduced to DECD to improve the quality of the individuals in the population; 3) a thorough evaluation of the performance of DECD on artificial and two real-world social networks, which achieved better results than other state-of-the-art community detection algorithms. 4) the application of DECD to a Yeast interacting protein dataset [10], which achieve the best results in the literature.

The remainder of this paper is organized as follows. Section 1.1 introduces some basic ideas of DE. In Section 1.2, some of the most popular algorithms for community detection are briefly reviewed. Section 2 presents a detailed description of DECD. In Section 3, the performance of DECD is tested on artificial and real-world networks and then the experimental results are discussed. Finally, Section 4 concludes this paper.

1.1 Differential Evolution

Differential evolution (DE) is a very simple yet efficient evolutionary algorithm proposed by Storn and Price in 1995 [29]. DE starts the search with an initial population containing NP individuals randomly sampled from the search space. Then, one individual called the target vector in the population is used to generate a mutant vector by the mutation operation. The most popular mutation strategy [17,18] which is also employed in DECD is the "rand/1" strategy as follows:

$$v_i = x_{r1} + F \times (x_{r2} - x_{r3}), \tag{1}$$

where $i \in \{1, 2, \ldots, NP\}$, $r1$, $r2$ and $r3$ are integers randomly selected from $1, 2, \ldots, NP$ and satisfy $r1 \neq r2 \neq r3 \neq i$, the scaling factor F is usually a real number between 0 and 1, the decision vector $x_i = (x_{i,1}, x_{i,2}, \ldots, x_{i,n})$ with n decision variables is the individual in the population and also called the target vector, and $v_i = (v_{i,1}, v_{i,2}, \ldots, v_{i,n})$ is the mutant vector.

After mutation, all the components of the mutant vector are checked whether they violate the boundary constraints. If the jth component $v_{i,j}$ of the mutant vector v_i violates the boundary constraint, $v_{i,j}$ is reflected back from the violated boundary constraint as follows [14]:

$$
v_{i,j} = \begin{cases} 2LB_j - v_{i,j}, & \text{if } v_{i,j} < LB_j \\ 2UB_j - v_{i,j}, & \text{if } v_{i,j} > UB_j \\ v_{i,j} & \text{otherwise}, \end{cases} \tag{2}
$$

where LB_j and UB_j are the lower and upper bounds of the ith decision variable x_i, respectively.

Subsequently, the crossover operation is implemented on the mutant vector v_i and the target vector x_i to generate a trial vector u_i. A commonly used crossover operation is the binomial crossover which is executed as follows:

$$
u_{i,j} = \begin{cases} v_{i,j}, & \text{if } rand \leq CR \text{ or } j = j_{rand} \\ x_{i,j}, & \text{otherwise}, \end{cases} \tag{3}
$$

where $i \in 1, 2, \ldots, NP, j \in 1, 2, \ldots, n$, $rand$ is a uniformly distributed random number between 0 and 1, j_{rand} is a randomly selected integer from 1 to n, CR is the crossover control parameter, and $u_{i,j}$ is the jth component of the trial vector u_i.

Finally, the target vector x_i is compared with the trial vector in terms of the objective function value and the better one survives into the next generation:

$$
x_i = \begin{cases} u_i, & \text{if } f(u_i) \leq f(x_i) \\ x_i, & \text{otherwise}. \end{cases} \tag{4}
$$

1.2   Related Work

During the past decade, the research on analyzing the community structure in complex networks has drawn a great deal of attention. Meanwhile, various kinds of algorithms have been proposed. Some of the most known algorithms are reviewed as follows.

Girvan and Newman [12] proposed the Girvan-Newman (GN) algorithm which is one of the most known algorithms proposed so far. This algorithm is a divisive method and iteratively removes the edges with the greatest betweenness value based on betweenness centrality [9]. Newman [19] presented an agglomerative hierarchical clustering method based on the greedy optimization of the network modularity. This method iteratively joins communities of nodes in pairs and chooses the join with the greatest increase in the network modularity at each step. Moreover, based on the original strategies, its faster version [4] was proposed by using some shortcuts and some sophisticated data structures. Radicchi et al. [24] presented the definitions of communities in both a strong sense and a weak sense. Moreover, in their paper a division algorithm [24] was proposed to detect communities by removing edges with the smallest value of edge cluster coefficient. Duch and Arenas [6] presented a division method which uses a heuristic search based on the extremal optimization to optimize the network modularity to detect communities in networks. Rosvall and Bergstrom [25] developed an algorithm based on an information-theoretic framework which identifies the communities by finding an optimal compression of the topology and capitalizing on regularities in the structure of networks.

However, some of the above community detection algorithms have large computational complexity and are unsuitable for very large networks. Moreover, a priori knowledge about the community structure (e.g., the number of communities) which is not easy or impossible to obtain in real-world networks is also required in most of the above algorithms [31]. To overcome the drawbacks, algorithms based on evolutionary algorithms have been proposed. These algorithms are very effective for community detection especially in very large complex networks. Tasgin and Bingol [31] presented an approach based on a genetic algorithm to optimize the network modularity in order to find community structures in networks. Pizzuti [21] proposed a method based on a genetic algorithm to discover communities in networks. This method defines the community score to measure the quality of a partitioning in communities of networks and uses a genetic algorithm to optimize the community score. Chen et al. [3] presented an algorithm based on the immune clone selection algorithm which is employed to optimize the modularity density [15] to identify communities in networks.

2 The Proposed Algorithm

In this paper, a new algorithm based on DE called DECD is proposed for community detection in complex networks. DECD uses DE as the search engine and employs the network modularity as the fitness function to evolve the population. Next, DECD is described in detail.

2.1 Individual Representation

DECD uses the community identifier-based representation proposed in [31] to represent individuals in the population for the community detection problem. For a graph $G = (V, E)$ with n nodes modeling a network, the kth individual in the population is constituted of n genes $\boldsymbol{x}_k = \{x_1, x_2, \ldots, x_n\}$ in which each gene x_i can be assigned an allele value j in the range $\{1, 2, \ldots, n\}$. The gene and allele represent the node and the community identifier (commID) of communities in G respectively. Thus, $x_i = j$ denotes that the node i belongs to the community whose commID is j, and nodes i and d belong to the same community if $x_i = x_d$. Since DECD puts nodes in communities randomly when initializing, at most n communities exist in G and then the maximum value of commID is n.

In the above representation, all the communities in G and all the nodes belonging to each community can be identified straightforwardly from individuals in the population. The community identifier-based representation is very simple and effective. Moreover, the number of communities is automatically determined by the individuals and no de-coding process is required in this representation.

For example, Figure 1.(a) shows a network containing 12 nodes numbered from 1 to 12. According to the definition of the community structure, the network is divided into three communities visualized by different colors of nodes. Figure 1.(b) is the genotype of the optimal solution for the community structure of the network, while the graph structure of the genotype is given in Figure 1.(c).

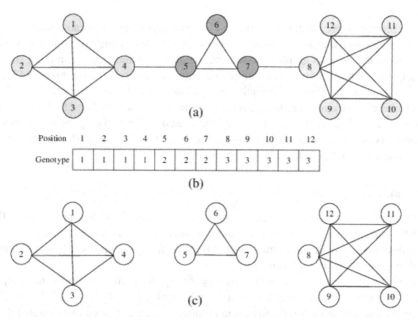

Fig. 1. (a) a graph model of a network; (b) the community identifier-based representation of a genotype; (c) the graph structure of the genotype

2.2 Fitness Function

Newman and Girvan [20] proposed the network modularity to measure the strength of the community structure found by algorithms. The network modularity is a very efficient quality metric for estimating the best partition of a network into communities. It has been used by many community detection algorithms recently [4,19,31].

DECD also employs the network modularity which is maximized as the fitness function to evaluate individuals in the population. The network modularity is defined as follows [31].

$$Q = \sum_{j=1}^{m} \left[\frac{l_j}{L} - \left(\frac{d_j}{2L} \right)^2 \right],$$

(5)

where j is the commID, m is the total number of communities, l_j is the number of links in module j, L is the total number of edges in the network and d_j is the degree of all nodes in module j.

2.3 Initialization

At the beginning of the initialization process, DECD places each node into a random community by assigning a random commID and generates individuals in the initial population.

However, the above random generation of individuals is likely to cause some unreasonable results such that a community contains some nodes having no connectivity with each other in the original graph. Considering that nodes in the same community should connect with each other and in the simple case are neighbors, a biased process [31] is used to overcome the above drawbacks, that is, once an individual is generated, some nodes represented by genes in the individual are randomly selected and their commIDs are assigned to all of their neighbors. By the biased process, the space of the possible solutions is restricted and the convergence of DECD is improved. Through these operations, the initial population P_0 is generated.

2.4 Mutation

DECD employs the "rand/1" strategy to mutate individuals in the population. The "rand/1" strategy is a very efficient mutation strategy. It has no bias to any special search directions and chooses new search directions in a random manner by randomly selecting individuals for mutation [33].

When implementing the "rand/1" strategy, firstly three different individuals x_{r1} , x_{r2} and x_{r3} are randomly selected from P_t, where $r1, r2, r3 \in \{1, \ldots, NP\}$, NP is the population size and t is the generation number. Then these three individuals follow the equation (1) and generate a mutant vector v which is put into the mutant population V_t. The above two steps are executed iteratively until the population size of V_t is NP.

Subsequently, all the components of each mutant vector in V_t are checked whether they violate the boundary constraints. If violating the boundary constraint, the component is reflected back from the violated boundary by following the equation (2). Then, a mutant population V_t satisfying all the boundary constraints is obtained.

2.5 Crossover

Since DECD randomly assigns an integer in the range $\{1, 2, \ldots, n\}$ to each individual in the population as its commID, no one-to-one absolute corresponding relationship exists between the communities and commIDs. That is, for different individuals the same community may have different commIDs while the same commID is likely to represent different communities. For example, with respect to two individuals $(1, 2, 1, 3)$ and $(2, 1, 2, 3)$, the community with the commID equals to 1 for the first individual and the one with the commID equals to 2 for the second individual are the same community. Meanwhile, the communities with the commID equals to 1 for these two individuals are different although they have the same commID.

For individuals represented by the community identifier-based representation, the traditional crossover operators (e.g., the binomial crossover) do not work well. This is because they just simply change the commIDs and never consider nodes in their communities. As a result, the offspring individuals fail to inherit good genes from the parent individuals and the search ability is heavily impaired.

Considering the above reasons, DECD designs a modified binomial crossover based on the binomial crossover to enhance the search ability. Inspired by the modified crossover operation [31], the modified binomial crossover assigns the commIDs of some nodes in an individual to those corresponding nodes in another individual. The implementation of the modified binomial crossover is as follows.

Firstly, the trial vector $u_i = x_i$ is set for every $i \in 1, 2, \ldots, NP$. Subsequently, the jth components in u_i and the mutant vector v_i are considered for every $j \in \{1, 2, \ldots, n\}$ and every $i \in \{1, 2, \ldots, NP\}$. If $rand \leq CR$ or $j = j_{rand}$, all the nodes in the community whose commID is $v_{i,j}$ of v are found, and then the commIDs of all those corresponding nodes of u_i are assigned the value $v_{i,j}$ which means all those corresponding nodes of u_i are put into the community whose commID is $v_{i,j}$; otherwise, no operation will be performed on u_i . Therein, $rand$ is a uniformly distributed random number between 0 and 1, j_{rand} is a randomly selected integer from 1 to n, and CR is the crossover control parameter. Finally, the trial vectors $U_t = \{u_1, \ldots, u_{NP}\}$ are obtained.

From the above process, it can be seen that the trial vectors are able to obtain some useful information about the community structure from both the target vectors and the trial vectors. Therefore, the modified binomial crossover is very helpful for identifying communities in networks.

2.6 Clean-Up Step

Since DE is a stochastic optimization algorithm, the solutions for the community division are likely to have some mistakes in evolution, that is, some nodes may be put into wrong communities. These mistakes impair the search ability of DECD and make it get stuck in a local optimum, ultimately leading to community divisions with inferior quality.

To solve the above problem, DECD adopts the clean-up operation proposed by Tasgin and Bingol [31], which effectively corrects the mistakes of putting nodes into wrong communities in both mutant and trial vectors and improves the search ability of. The clean-up operation is based on the community variance $CV(i)$, which is defined as the fraction of the number of different communities among the node i and its neighbors to the degree of the node i as follows:

$$CV(i) = \frac{\sum_{(i,j) \in E} \text{neq}(i,j)}{\deg(i)}, \tag{6}$$

where $\text{neq}(i,j) = \begin{cases} 1, & \text{if commID}(i) \neq \text{commID}(j) \\ 0, & \text{otherwise} \end{cases}$, $\deg(i)$ is the degree of the ith node, E is the set of edges, and commID is the community containing ith node.

According to the classic definition of the community structure, a community should contain more internal edges among nodes inside the community than external edges with other communities. Thus, a node and all its neighbors should be in the same community with a high probability, and then the community variance of this node should be low in a good community division. Based on the above analysis, the clean-up operation is performed as follows. Firstly, some nodes are randomly selected. Then, for each of these nodes, its community variance is computed and compared with a threshold value η which is a predefined constant obtained after some experiments. If the community variance of this node is larger than this threshold value η, which indicates that the node has been put into a wrong community, then this node and all its neighbors are placed into the same community containing the highest number of nodes in the neighborhood of this node. Otherwise, no operation is executed for this node.

2.7 DECD Algorithm Framework

Finally, the framework of DECD is described as follows:

Step 1) Set $t = 0$ where t denotes the generation number.

Step 2) Generate the initial population $P_0 = \{x_1, \ldots, x_{NP}\}$ by uniformly and randomly sampling NP points from the search space S.

Step 3) Compute the network modularity value $Q(x_i)$ of each individual x_i in P_0.

Step 4) Perform the mutation operation (see Section 2.4 for details) on each individual x_i in P_t and obtain the mutant vectors $V_t = \{v_1, \ldots, v_{NP}\}$.

Step 5) Correct the mistakes in each mutant vector v_i in V_t by executing the clean-up operation (see Section 2.6 for details).

Step 6) Execute the modified binomial crossover (see Section 2.5 for details) on each mutant vector v_i in V_t and generate the trial vectors $U_t = \{u_1, \ldots, u_{NP}\}$.

Step 7) Correct the mistakes in each trail vector u_i in U_t by executing the clean-up operation (see Section 2.6 for details).

Step 8) Calculate the network modularity value $Q(u_i)$ of each trial vector u_i in U_t.

Step 9) Compare x_i with u_i $(i = 1, \ldots, NP)$ in terms of the network modularity value by following the equation (4), and put the winner into the next population P_{t+1}.

Step 10) Set $t = t + 1$.

Step 11) If the termination criterion is not satisfied, go to **Step 4**; otherwise, stop and output the best individual x_{best} in P_t.

3   Experiments and Results

In this section, the performance of DECD is tested on a class of widely used artificial networks and some well studied real world social and biological networks. DECD is implemented in MATLAB and all the experiments are performed on Windows XP SP2 with Pentium Dual-Core 2.5GHz processor and 2.0GB RAM. The parameters in DECD are set as follows: the population size $NP = 200$, the scaling factor $F = 0.9$, the control crossover parameter $CR = 0.3$, the threshold value $\eta = 0.35$ and the maximum number of generations is 200. In each run, DECD is stopped when the maximum number of generations is reached.

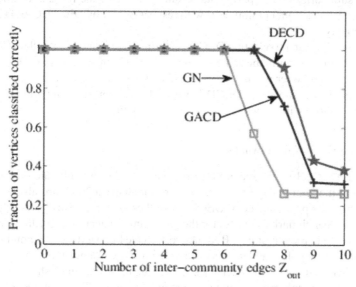

Fig. 2. The fraction of vertices correctly classified by DECD, GACD and GN as the average z_{out} of inter-community edges per vertex is varied for the computer-generated networks

For comparison, we implement another community detection algorithm based on GA, named GACD. We adopt the MATLAB Genetic Algorithm Optimization Toolbox (GAOT) to optimize the network modularity to detect communities in networks. The GA we use is real encoded GA with heuristic crossover and uniform mutation. Moreover, for the sake of fairness, the same biased process and the clean-up operation in DECD are employed. The values of all the parameters use in the experiments are the default parameters in GAOT. We also adopt MATLAB implementations of Girvan-Newman (GN) algorithm from Matlab Tools for Network Analysis (http://www.mit.edu/~gerganaa) for comparison.

3.1 Artificial Networks

To evaluate the performance of DECD of detecting network community structure, artificial computer-generated networks are employed. The computer-generated networks were proposed by Girvan and Newman and have been widely used to benchmark the community detection algorithms [12]. Each network has 128 nodes which are divided into 4 communities each with 32 nodes. Each node has an average z_{in} edges connecting it to members of the same community and z_{out} edges to members of other communities. Moreover, z_{in} and z_{out} are chosen to satisfy the total expected degree of a node $z_{in} + z_{out} = 16$. The community detection algorithms with good performance should discover all the communities in the network with $z_{in} > z_{out}$ which indicates that the neighbors of a node inside its community are more than the neighbors contained by the other three communities. According to the definition of the community structure, the community structure in the network becomes vaguer with the z_{out} increases.

Figure 2 summarizes the experimental results and shows the fraction of nodes correctly divided into the four communities with respect to z_{out} by DECD, GACD and GN [12], respectively.

From Figure 2, it can be seen that DECD performs significantly better than GN when $z_{out} > 6$ while they perform the same when $z_{out} \leq 6$. Comparing with GACD, DECD also performs much better when $z_{out} > 7$ and they have the same performance when $z_{out} \leq 7$. The results show that DECD is very effective for detecting communities in networks, even those with very vague community structures.

3.2   Real-World Social Networks

In this paper, two real-world social networks, i.e., the Zachary's karate club network [36] and the American college football network cite8Girvan2002, are also employed to further verify the performance of DECD. Both these two real-world social networks are well-known benchmark examples for the community detection algorithms and have been well studied in the literatures. Both networks have known (true) community structures, which provide gold-standard for validating our DECD algorithm.

The first social network is the Zachary's karate club network, which shows the friendships between the members of a karate club at an University. The network contains 34 nodes and 78 edges. The most interesting feature of the network is that the club split into two as a result of an internal dispute, which provides a ground-true community division of the network.

The American college football network [12] has a known community structure. The network is a representation of the schedule of Division I games for the 2000 games. In the football network, there are 115 nodes and 616 edges divided into 12 communities, where nodes and edges represent the teams (identified by their college names) and the regular season games between the two teams they connect, respectively. The teams are divided into "conferences" and each conference contains around 8 to 12 teams. The teams play an average of about 4 inter-conference games and 7 intra-conference games which indicates that games are more frequent between members of the same conference than between members of different conferences.

As pointed out in [28] and [34], performance metrics based on network modularity Q is not reliable. Therefore, apart from Q, we also adopt accuracy as a quantitative measure for validating DECD as used in [28]:

$$\text{Accuracy} = \frac{\sum_{k=1}^{n} \text{equal}(t_k, p_k)}{n},\tag{7}$$

where

$$\text{equal}(x, y) = \begin{cases} 1, & \text{if commID}(x) = \text{commID}(y) \\ 0, & \text{otherwise} \end{cases},$$

and t_k is the kth node in the true (known) network structure, and p_k is the kth node in the predicted network structure.

Since DECD and GACD are stochastic optimization algorithms, we perform the experiments 30 times on these two networks. The average values Q_{avg} and Acc_{avg} of Q

and accuracy and their best values Q_{bst} and Acc_{bst}, are compared with that obtained by GN (a deterministic algorithm) from one run of experiment in Table 1. In order to run the GN algorithm, we need to specify the number of communities. In our experiment, we use the number of known communities in the two networks, e.g., 2 for the karate network and 12 for the football network as number of communities for the GN algorithm.

Table 1. Experimental results of the Zachary's karate club network and the American college football network. N_{pr} is the average number of communities; Q_{avg} and Q_{bst} are the average and best values of modularity Q, respectively; and Acc_{avg} and Acc_{bst} are the average and best accuracy, respectively.

Network	Algorithm	N_{pr}	Q_{avg}	Q_{bst}	Acc_{avg}	Acc_{bst}
Karate	DECD	3.467 ± 0.730	$\mathbf{0.385 \pm 0.013}$	**0.416**	$\mathbf{0.972 \pm 0.009}$	1
	GACD	2.900 ± 1.094	0.369 ± 0.022	0.402	0.959 ± 0.020	0.971
	GN	2	0.360	0.360	0.971	0.971
Football	DECD	11.233 ± 0.504	$\mathbf{0.596 \pm 0.003}$	**0.605**	$\mathbf{0.930 \pm 0.004}$	**0.939**
	GACD	8.767 ± 1.073	0.587 ± 0.015	**0.605**	0.913 ± 0.010	0.930
	GN	12	0.455	0.455	0.904	0.904

From Table 1, it can be seen that DECD perform better than the two competitors, i.e., GACD and GN on both the karate and football networks. It is interesting to see that, for the smaller scale karate network, the performance of DECD is just slightly better than GN. However, for the larger football network, the performance of DECD is significantly better than GN. Such results indicate that DECD is more effective in detecting communities in larger complex networks than GN.

In [28], the authors tested several state-of-the-art community detection algorithms. In the paper, for the karate network, the best algorithm in terms of Q was Fast GN (denoted as FastQ in their paper) [4], which generated Q value of 0.381 from one run of experiment, which is slightly better than that from the GN algorithm used in our paper but still worse than that from our DECD. In terms of accuracy, their best algorithm (Random walks) achieved 1, which is as same as Acc_{bst} found by our DECD in 30 runs. In [28], for football network, the best result in terms of Q was 0.604 (WalkTrap [22]), which is slightly worse than Q_{bst} generated by our DECD. In terms of accuracy, DECD also generated the same $Acc_{bst} = 0.939$ as the best result in [28], which was generated by Random walks and MCL [32].

3.3 Yeast Protein-Protein Interaction Network

We apply our DECD algorithm to a Yeast Protein-Protein Interaction (PPI) Network [11], which contains 1430 proteins and 6535 interactions. We use CYC2008 [23], a complete and up-to-date set of yeast protein complexes (also called modules or communities) as reference set to evaluate the predicted modules by DECD (In the following text, we will use module and complex instead of community, which is a less popular term in bioinformatics). We compute precision, recall and F-measure to measure the

Table 2. Experimental results of the Yeast Protein-Protein Interaction Network

ω	Algorithm	#. pred. complex	Precision	Recall	F-measure
0.2	DECD	115	**0.6696**	**0.4976**	**0.5709**
	GACD	106	0.6132	0.4902	0.5488
	GECSS	157	0.5063	0.2802	0.3608
	MCODE	39	0.2278	0.4871	0.3104
	MCL	200	0.5063	0.2050	0.2918
0.5	DECD	115	0.4696	**0.2390**	**0.3168**
	GACD	106	0.4340	0.2220	0.2937
	GN	65	0.5231	0.0568	0.1025
	CMCD	65	**0.6154**	0.0691	0.1242

performance of DECD. The performance of DECD is compared with GACD and GN. We also adopt results from recent literature, e.g., [35,27] for comparison.

Similar to the experiments in [35,27], we use affinity score to decide whether a predicted module is matched with a reference complex:

$$\text{Affinity}(A, B) = \frac{|A \bigcup B|^2}{|A| \times |B|}, \tag{8}$$

where A and B are two modules of proteins, e.g., one of predicted module or reference complexes. We assume a module A matches module B if and only if $\text{Affinity}(A, B)$ is above a predefined threshold ω. Then we can define $Hit(\mathcal{A}, \mathcal{B})$ which contains all the matched modules:

$$Hit(\mathcal{A}, \mathcal{B}) = \{A_i \in \mathcal{A} | \text{Affinity}(A_i, B_j) > \omega, \exists B_j \in \mathcal{B}\}. \tag{9}$$

We define precision, recall and F-measure as follows:

$$\text{Recall} = \frac{|Hit(\mathcal{R}, \mathcal{P})|}{|\mathcal{R}|}, \tag{10}$$

$$\text{Precision} = \frac{|Hit(\mathcal{P}, \mathcal{R})|}{|\mathcal{P}|}, \tag{11}$$

$$\text{F-measure} = \frac{2 \times \text{Recall} \times \text{Precision}}{\text{Recall} + \text{Precision}}, \tag{12}$$

where $\mathcal{P}$ is the predicted module set and $\mathcal{R}$ is the reference complex set.

Following the experimental settings in [35,27], we set $\omega = 0.2$ and 0.5 in order to compare with their algorithms fairly. We adopt the results of algorithms tested in [35,27], which include Critical Module Community Detection algorithm (CMCD) [27], Gene Expression Condition Set Similarity (GECSS) algorithm [35], Molecular Complex Detection (MCODE) algorithm [2] and Markov clustering (MCL) algorithm [32] . It is worth mentioning that, due to the large size of the PPI network, the GN algorithm in Matlab Tools for Network Analysis (http://www.mit.edu/~gerganaa) failed to produce results in reasonable time. Therefore, we adopt the results of the GN algorithm from [27] for comparison.

From Table 2, we can see that compared with GACD and other algorithms tested in [35,27], DECD has better performance. It is interesting to see that, the performance gap between DECD and GACD is not as significant as those between DECD and other non-population-based algorithms, e.g., MCL. Such results indicate that, at least for medium size networks, population-based algorithms are preferred because of their better predictive performance.

4 Conclusion

In this paper, we have introduced differential evolution (DE) to detect community structure in complex networks. To the best of our knowledge, it is the first time DE has been applied to community detection problems. The proposed algorithm DECD uses DE to search the best network partition of a complex network that can achieve an optimal network modularity value. Based on the standard binomial crossover of DE, we designed a modified binomial crossover to transmit some important information about the community structure during evolution. We have also introduced a biased process and a clean-up operation similar to [31] to improve the quality of the individuals in the population.

We have tested our DECD on the artificial networks and the real-world social and biological networks in comparison with GACD and GN algorithms. Apart from the modularity value, for the real-world networks, we have also employed accuracy based on true community structure as a performance metric [28], which provided reliable performance information. The experimental results have demonstrated that DECD is very effective for community detection in complex networks, including those with very vague community structures, e.g., the artificial networks with larger values of z_{out}. In addition to its excellent performance, another merit of DECD is that it does not require any prior knowledge about the community structure when detecting communities in networks.

The limitation of this work is that we only used modularity as the objective function to find the optimal community structure of a network. However, it has been recently pointed out that such approach might suffer from the so-called resolution limit problem, that is, some modules smaller than a specific scale will not be detected by the algorithms that only optimize modularity [8]. Although we have achieved better community detection results than many other algorithms on 4 complex networks, we do not anticipate DECD can avoid this resolution limit problem. We will further investigate this problem in our future work.

Acknowledgment. This work is sponsored by ESPRC (EP/J501414/1). We acknowledge the financial support of the Systems Science for Health initiative of the University of Birmingham.

References

1. Albert, R., Jeong, H., Barabasi, A.: Error and attack tolerance of complex networks. Nature 406, 378–382 (2000)
2. Bader, G.D., Hogue, C.W.V.: An automated method for finding molecular complexes in large protein interaction networks. BMC Bioinformatics 4 (2003)

3. Chen, G., Wang, Y., Yang, Y.: Community detection in complex networks using immune clone selection algorithm. International Journal of Digital Content Technology and its Applications 5, 182–189 (2011)
4. Clauset, A., Newman, M.E.J., Moore, C.: Finding community structure in very large networks. Physical Review E 70, 066111 (2004)
5. Dorogovtsev, S.N., Mendes, J.F.F.: Evolution of networks. Adv. Phys. 51, 1079 (2001)
6. Duch, J., Arenas, A.: Community detection in complex networks using extremal optimization. Physical Review E 72, 027104 (2005)
7. Faloutsos, M., Faloutsos, P., Faloutsos, C.: On power-law relationships of the internet topology. ACM SIGCOMM Computer Communications Review 29, 251–262 (1999)
8. Fortunato, S., Barthelemy, M.: Resolution limit in community detection. Proceedings of the National Academy of Sciences 104, 36–41 (2007)
9. Freeman, L.: A set of measures of centrality based upon betweenness. Sociometry 40, 35–41 (1977)
10. Gavin, A.C., et al.: Proteome survey reveals modularity of the yeast cell machinery. Na 440, 631–636 (2006)
11. Gavin, A.C., et al.: Proteome survey reveals modularity of the yeast cell machinery. Nature 440, 631–636 (2006)
12. Girvan, M., Newman, M.E.J.: Community structure in social and biological networks. Proceedings of the National Academy of Sciences 99, 7821–7826 (2002)
13. Guimera, R., Amaral, L.: Functional cartography of complex metabolic networks. Nature 433, 895–900 (2005)
14. Kukkonen, S., Lampinen, J.: Constrained real-parameter optimization with generalized differential evolution. In: Proceedings of the Congress on Evolutionary Computation (CEC 2006). IEEE Press, Sheraton Vancouver Wall Centre Hotel, Vancouver (2006)
15. Li, Z., Zhang, S., Wang, R., Zhang, X., Chen, L.: Quantitative function for community detection. Physical Review E 77, 036109 (2008)
16. Liu, Y., Slotine, J., Barabasi, A.: Controllability of complex networks. Nature 473, 167–173 (2011)
17. Mezura-Montes, E., Miranda-Varela, M., Gómez-Ramón, R.: Differential evolution in constrained numerical optimization: An empirical study. Information Sciences 180, 4223–4262 (2010)
18. Neri, F., Tirronen, V.: Recent advances in differential evolution: a survey and experimental analysis. Artificial Intelligence Review 33, 61–106 (2010)
19. Newman, M.E.J.: Fast algorithm for detecting community structure in networks. Physical Review E 69, 026113 (2004)
20. Newman, M.E.J., Girvan, M.: Finding and evaluating community structure in networks. Physical Review E 69, 026113 (2004)
21. Pizzuti, C.: GA-Net: A Genetic Algorithm for Community Detection in Social Networks. In: Rudolph, G., Jansen, T., Lucas, S., Poloni, C., Beume, N. (eds.) PPSN 2008. LNCS, vol. 5199, pp. 1081–1090. Springer, Heidelberg (2008)
22. Pons, P., Latapy, M.: Computing communities in large networks using random walks. J. of Graph Alg. and App. Bf 10, 284–293 (2004)
23. Pu, S., Wong, J., Turner, B., Cho, E., Wodak, S.J.: Up-to-date catalogues of yeast protein complexes. Nucleic Acids Res. 37, 825–831 (2009)
24. Radicchi, F., Castellano, C., Cecconi, F., Loreto, V., Parisi, D.: Defining and identifying communities in networks. Proceedings of the National Academy of Sciences 101, 2658–2663 (2004)
25. Rosvall, M., Bergstrom, C.: An information-theoretic framework for resolving community structure in complex networks. Proceedings of the National Academy of Sciences 104, 7327–7331 (2007)

26. Scott, J.: Social network analysis: A Handbook. Sage Publications, London (2000)
27. Sohaee, N., Forst, C.V.: Modular clustering of protein-protein interaction networks. In: 2010 IEEE Symposium on Computational Intelligence in Bioinformatics and Computational Biology, CIBCB (2010)
28. Steinhaeuser, K., Chawla, N.V.: Identifying and evaluating community structure in complex networks. Pattern Recognition Letters 31, 413–421 (2009)
29. Storn, R., Price, K.: Differential evolution a simple and efficient adaptive scheme for global optimization over continuous spaces. Journal of Global Optimization 11, 341–359 (1997)
30. Strogatz, S.H.: Exploring complex networks. Nature 410, 268–276 (2001)
31. Tasgin, M., Bingol, H.: Community detection in complex networks using genetic algorithm. In: Proceedings of the European Conference on Complex Systems (2006)
32. van Dongen, S.: Graph Clustering by Flow Simulation. PhD thesis, University of Utrecht (2000)
33. Wang, Y., Cai, Z., Zhang, Q.: Differential evolution with composite trial vector generation strategies and control parameters. IEEE Transactions on Evolutionary Computation 15, 55–66 (2011)
34. Yang, Y., Sun, Y., Pandit, S., Chawla, N.V., Han, J.: Is objective function the silver bullet? a case study of community detection algorithms on social networks. In: International Conference on Advances in Social Network Analysis and Mining, pp. 394–397 (2011)
35. Yeu, Y., Ahn, J., Yoon, Y., Park, S.: Protein complex discovery from protein interaction network with high false-positive rate. In: Evolutionary Computation, Machine Learning and Data Mining in Bioinformatics 2011, EvoBio 2011 (2011)
36. Zachary, W.W.: An information flow model for conflict and fission in small groups. Journal of Anthropological Research 33, 452–473 (1977)

A Study on Large Population MOEA Using Adaptive ε-Box Dominance and Neighborhood Recombination for Many-Objective Optimization

Naoya Kowatari[1], Akira Oyama[2], Hernán E. Aguirre[1], and Kiyoshi Tanaka[1]

[1] Faculty of Engineering, Shinshu University
4-17-1 Wakasato, Nagano, 380-8553 JAPAN
[2] Institute of Space and Astronautical Science, Japan Aerospace Exploration Agency
{kowatari@iplab,ahernan,ktanaka}@shinshu-u.ac.jp, oyama@flab.isas.jaxa.jp

Abstract. Multi-objective evolutionary algorithms are increasingly being investigated to solve many-objective optimization problems. However, most algorithms recently proposed for many-objective optimization cannot find Pareto optimal solutions with good properties on convergence, spread, and distribution. Often, the algorithms favor one property at the expense of the other. In addition, in some applications it takes a very long time to evaluate solutions, which prohibits running the algorithm for a large number of generations. In this work to obtain good representations of the Pareto optimal set we investigate a large population MOEA, which employs adaptive ε-box dominance for selection and neighborhood recombination for variation, using a very short number of generations to evolve the population. We study the performance of the algorithm on some functions of the DTLZ family, showing the importance of using larger populations to search on many-objective problems and the effectiveness of employing adaptive ε-box dominance with neighborhood recombination that take into account the characteristics of many-objective landscapes.

1 Introduction

Recently, there is a growing interest on applying multi-objective evolutionary algorithms (MOEAs) to solve many-objective optimization problems, where the number of objective functions to optimize simultaneously is considered to be more than three. Historically, most applications of MOEAs have dealt with two and three objective problems leading to the development of several evolutionary approaches that work successfully in these low dimensional objective spaces. However, it is well known that conventional MOEAs [1, 2] scale up poorly with the number of objectives of the problem. The poor performance of conventional MOEAs is attributed to an increased complexity inherent to high dimensional spaces and to the use of inappropriate selection and variation operators that fail to take into account the characteristics of many-objective landscapes [3–5].

Y. Hamadi and M. Schoenauer (Eds.): LION 6, LNCS 7219, pp. 86–100, 2012.
© Springer-Verlag Berlin Heidelberg 2012

MOEAs seek to find trade-off solutions with good properties of convergence to the Pareto front, well spread, and well distributed along the front. These three properties are especially difficult to achieve in many-objective problems and most search strategies for many-objective optimization proposed recently compromise one in favor of the other [6]. In several application domains, such as multidisciplinary multi-objective design optimization, a large number of Pareto optimal solutions that give a good representation of the true Pareto front in terms of convergence, spread, and distribution of solutions are essential to extract relevant knowledge about the problem in order to provide useful guidelines to designers during the implementation of preferred solutions. Moreover, in some applications it takes a very long time to evaluate solutions, which prohibits running the evolutionary algorithm for a large number of generations. Thus, in addition to the difficulties imposed by high-dimensional spaces, we are usually constrained by time.

From this point of view, in this work to obtain good representations of the Pareto optimal set we investigate a large population MOEA, which employs adaptive ε-box dominance for selection and neighborhood recombination for variation, using a very short number of generations to evolve the population. The motivation to use large populations is twofold. One is that we need many more solutions to properly approximate the Pareto optimal set of many-objective problems. The other one is that large populations may support better the evolutionary search on high dimensional spaces. That is, large populations may be more suitable to deal with the increased complexity inherent to high dimensional spaces. We assume that all solutions in the population can be evaluated simultaneously and in parallel, i.e. the time to evaluate one generation equals the time required to evaluate one solution, independently of the number of solutions we use in the population. So, our limitations on time are directly related to the number of generations rather than to the total number of fitness function evaluations. The motivation to use adaptive ε-box dominance for selection and neighborhood recombination for variation is to enhance the design of the algorithm incorporating selection and recombination operators that interpret better the characteristics of many-objective landscapes.

We study the performance of the algorithm using some test functions of the DTLZ family [7]. Our experiments reveal the importance of using a large population to search in many-objective problems and the effectiveness of employing ε-Box dominance and neighborhood recombination.

2   Proposed Method

2.1   Concept

Two important characteristics of many-objective optimization problems are that the number of non-dominated solutions increases exponentially [3, 4] with the number of objectives of the problem, and that these solutions become spread over broader regions in variable space [5]. These characteristics of many-objective landscapes must be considered when we design the major components of the algorithm, namely ranking, density estimators, mating, and variation operators.

In this work, selection is improved by incorporating adaptive ε-box dominance during the process of ranking and selecting solutions for the next generation. The effectiveness of the recombination operator is improved by incorporating a neighborhood to mate and cross individuals that are close in objective space. In the following we describe adaptive ε-box dominance and neighborhood recombination.

2.2   Adaptive ε-Box Dominance

In the proposed method we use adaptive ε-box dominance to rank solutions and select a ε-Pareto set [8] of non-dominated solutions for the next generation. In [8] an archiving strategy that updates a ε-Pareto set with a newly generated individual was proposed to guarantee convergence and diversity properties of the solutions found. The principles of the above archiving strategy [8] is applied to modify non-domination sorting used in NSGA-II [9]. The main steps of ε-box non-domination sorting are as follow.

Step 1. Similar to [8], ε-box non-domination sorting implicitly divides the objective space into non-overlapping boxes, where each solution is uniquely mapped to one box. That is, the box index vector $\boldsymbol{b}^{(i)} = (b_1^{(i)}, \cdots, b_m^{(i)})$ of the i-th solution in the combined population of parents P and offspring Q is calculated by

$$b_k^{(i)}(\boldsymbol{x}) = \left\lfloor \frac{\log_{10} f_k^{(i)}(\boldsymbol{x})}{\log_{10}(1+\varepsilon)} \right\rfloor \ (k = 1, 2, \cdots m), \tag{1}$$

where $f_k^{(i)}$ is the fitness value in the k-th objective of the i-th solution $i = 1, 2, \cdots, |P| + |Q|$, m the number of objectives, and ε a parameter that controls the size of the box.

Step 2. Pareto dominance is calculated using the box indexes $\boldsymbol{b}^{(i)}$ of solution to get a set of non-dominated ε-boxes.

Step 3. Form a front of solutions by picking one solution from each non-dominated ε-box. If there is more than one solution in a box, we calculate Pareto dominance among solutions within the box. Thus, in each ε-box there could be either a dominating solution or several non-dominated solutions, and possibly one or more dominated solutions. To form the front we chose from each box the dominating solution, or select randomly one of the non-dominating solutions.

Step 4. Go to Step 2 to form subsequent fronts, excluding solutions already included in a previous front. Solutions located in a non-dominated ε-box but not selected as part of a previous front are considered to form the next front. Thus, compared to conventional non-domination sorting based on Pareto dominance, the proposed method reduces the ranking of some non-dominated solutions, namely those located in the same ε-box but not

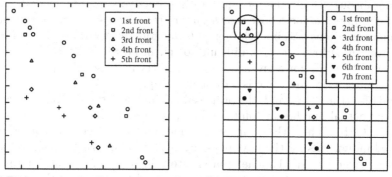

(a) Front sorting by Pareto dominance   (b) Front sorting by ε-Box Dominance

Fig. 1. Solution ranking by ε-Box Dominance

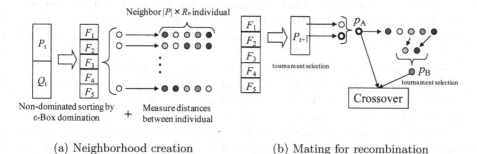

(a) Neighborhood creation   (b) Mating for recombination

Fig. 2. Neighborhood Recombination

chosen to form the front. Note that in the archiving strategy proposed in [8], dominated solutions and not-selected non-dominated solutions within a ε-box are eliminated. Here, we keep those solutions but reduce their ranking.

Fig. 1 illustrates front non-domination sorting by Pareto dominance and by ε-box dominance. In our illustration we assume a two-objective maximization problem. Note that non-dominated solutions that fall within the same ε-box are given different rank by ε-box non-domination sorting.

The ε-box non-domination sorting groups solutions in fronts of ε-box non-dominated solutions, denoted as F_j^ε, where j indicates the front number. Then, solutions are assigned a primary rank equal to the front number j it belongs to.

In many-objective problems, the number of Pareto non-dominated solutions $|F_1|$ obtained from the combined population of parents P and offspring Q is expected to surpass the size of the parent population since early generations, i.e. $|F_1| > |P|$. Since only one solution is selected from each ε-box to form a front, the number of solutions in the first front after applying ε-box non-domination sorting is expected to be smaller than the number of Pareto optimal solutions, $|F_1^\varepsilon| < |F_1|$, and its exact number depends on the value set to $\varepsilon > 0$ and on the instantaneous distribution of solutions in objective space. In general, larger values of ε imply that the ε-boxes cover larger areas, increasing the likelihood of having more solutions within each box and therefore less solutions in the finally formed front F_1^ε. However, it is difficult to tell in advance exactly how many solutions will be in F_1^ε for a given value of ε and trying to set this parameter by hand to achieve a desired value is a difficult and problem depending task.

Instead of setting ε by hand, the proposed method adapts ε at each generation so that the actual number of solutions in F_1^ε is close to the size of the parent population P [10]. Thus, the adaptive method aims to select a sample of non-dominated individuals for the next population that are distributed according to the spacing given by Eq. (1). The appropriate value of ε that renders a number of solutions close to the desired number is expected to change as the evolution process proceeds and it is affected by the stochastic nature of the search that alters the instantaneous distributions of solutions in objective space. Thus, in addition to adapting ε, the step of adaptation Δ is also important to properly follow the dynamics of the evolutionary process on a given problem. For this reason, the proposed adaptive procedure adapts ε and its step of adaptation Δ as well.

The method to adapt ε before it is used in Eq. (1) is as follows. First, before start searching solutions, we set initial values for ε and the step of adaptation Δ, set ε's lower bound $\varepsilon_{min} > 0.0$ and Δ's lower and upper bound, Δ_{min} and Δ_{max}, such that $0.0 < \Delta_{min} < \Delta_{max}$. Next, at every generation we count the number of solutions obtained in the first front F_1^ε by ε-box non-domination sorting and compare with the size of the population P. If $|F_1^\varepsilon| < |P|$ the step of adaptation is updated to $\Delta = \Delta/2$ and $\varepsilon = \varepsilon - \Delta$, to make the grid fine grained. Otherwise, $\Delta = \Delta \times 2$ and $\varepsilon = \varepsilon + \Delta$, to make the grid coarser. If after updating ε or Δ their values go above or below their established bounds, they are reset to their corresponding bound. In this work, the initial value set to ε is 0.01, its lower bound ε_{min} is 10^{-8}, the initial step of adaptation Δ is 0.01, its upper bound Δ_{max} is 1, and its lower bound Δ_{min} is 0.0001.

2.3   Neighborhood Recombination

As mentioned above, it has been shown that in many-objective problems the number of non-dominated solutions grows exponentially with the number of objectives of the problem. A side effect of this is that non-dominated solutions in problems with a large number of objectives tend to cover a larger portion of objective and variable space compared to problems with less number of objectives. Thus, in many objective problems, the difference between individuals in the instantaneous population is expected to be larger. This could affect the effectiveness of recombination because recombining two very different individuals could be too disruptive.

In this work, we encourage mating between individuals located close to each other, aiming to improve the effectiveness of recombination in high dimensional objective spaces. To accomplish that, the proposed method calculates the distance between individuals in objective space and keeps a record of the $|P| \times R_n$

closest neighbors of each individual during the ε-dominance process. However, note that when the ranked population of size $|P| + |Q|$ is truncated to form the new population of size $|P|$, some individuals would be deleted from the neighborhood of each individual. During mating for recombination, the first parent p_A is chosen from the parent population P using a binary tournament selection, while the second parent p_B is chosen from the neighborhood of p_A using another binary tournament. Then, recombination is performed between p_A and p_B. That is, between p_A and one of its neighbors in objective space. If all neighbors of individual p_A were eliminated during truncation, the second parent p_B is selected from the population P similar to p_A. Fig.2 illustrates the neighborhood creation and mating for recombination. In this work, we set the parameter R_n that defines the size of the neighborhood of each individual to 2%C5%Cand 10% of the parent population P.

3   Test Problems and Performance Indicators

3.1   Test Problems

In this work, we study the performance of the algorithms in continuous functions DTLZ2, DTLZ3, and DTLZ4 of the DTLZ test functions family [7]. These functions are scalable in the number of objectives and variables and thus allow for a many-objective study. In our experiments, we vary the number of objectives from $m = 4$ to 6 and set the total number of variables to $n = m+9$. DTLZ2 has a non-convex Pareto-optimal surface that lies inside the first quadrant of the unit hyper-sphere. DTLZ3 and DTLZ4 are variations of DTLZ2. DTLZ3 introduces a large number of local Pareto-optimal fronts in order to test the convergence ability of the algorithm. DTLZ4 introduces biases on the density of solutions to some of the objective-space planes in order to test the ability of the algorithms to maintain a good distribution of solutions. For a detailed description of these problems the reader is referred to [7].

3.2   Performance Indicators

In this work we evaluate the Pareto optimal solutions obtained by the algorithms using the quality indicators described below.

Proximity Indicator(I_p) [11]: Measures the convergence of solutions using equation 2, where P denotes the population and x a solution in the population. Smaller values of I_p indicate that the population P is closer to the Pareto front. That is, smaller values of I_p mean high convergence of solutions.

$$I_p = \underset{x \in P}{median} \left\{ \left[\sum_{i=1}^{m} (f_i(x))^2 \right]^{\frac{1}{2}} - 1 \right\} \tag{2}$$

C-metric [12]: Let us denote A and B the sets of non-dominated solutions found by two algorithms. $C(A, B)$ gives the fraction of solutions in B that are dominated at least by one solution in A. More formally,

$$C(A, B) = \frac{\mid \{b \in B | \exists a \in A : f(a) \succeq f(b)\} \mid}{\mid B \mid}. \tag{3}$$

$C(A, B) = 1.0$ indicates that all solutions in B are dominated by solutions in A, whereas $C(A, B) = 0.0$ indicates that no solution in B is dominated by solutions in A. Since usually $C(A, B) + C(B, A) \neq 1.0$, both $C(A, B)$ and $C(B, A)$ are required to understand the degree to which solutions of one set dominate solutions of the other set.

Hypervolume (HV) [12]: HV calculates the volume of the m-dimensional region in objective space enclosed by a set of non-dominated solutions and a dominated reference point r, giving a measure of convergence and diversity of solutions. In general, larger values of HV indicate better convergence and/or diversity of solutions. Thus, MOEAs that find Pareto optimal solutions that lead to larger values of HV are consider as algorithms with better search ability. We use Fonseca et al. [13] algorithm to calculate the hypervolume.

4   Simulation Results and Discussion

4.1   Preparation

In this work we use NSGA-II, a well known representative of the class of dominance based MOEAs. In this framework we include Adaptive ε-Box dominance and Neighborhood Recombination. In the following we call for short $A\varepsilon B$ and $A\varepsilon BNR$ the MOEAs that include Adaptive ε-Box dominance and Adaptive ε-Box dominance & Neighborhood Recombination, respectively. As genetic operators we use SBX crossover and Polynomial Mutation, setting their distribution exponents to $\eta_c = 15$ and $\eta_m = 20$, respectively. The parameter for the operators are crossover rate $p_c = 1.0$, crossover rate per variable $p_v = 0.5$, and mutation rate $p_m = 1/n$, where n is the number of variables. The number of generations is fixed to $T = 100$ and the population sizes varies from $|P| = 100$ to 5000 solutions. Result reported here are average results obtained running the algorithms 30 times.

4.2   Effects of Increasing Population Size

First, we focus our analysis on DTLZ2. Fig.3 shows I_p obtained at the final generation varying the number of solutions in the population P from 100 to 5000. It could be seen that I_p reduces when a larger population size $|P|$ is used, regardless of the number of objectives. That is, a larger population size improves convergence of the algorithm. Especially note that for $m = 4, 5$ the reduction of I_p is remarkable and that the values of $|P|$ that lead to a pronounced decline on I_p are different, depending on the number of objectives. In the case of $m = 5$ objectives, it can be seen that a larger reduction occurs when the population

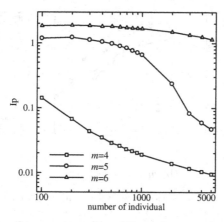

Fig. 3. Effect of increasing population size in NSGA-II (DTLZ2)

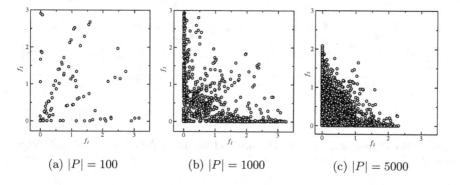

(a) $|P| = 100$ (b) $|P| = 1000$ (c) $|P| = 5000$

Fig. 4. Distribution of solutions by NSGA-II using various population sizes $|P|$, DTLZ2 $m = 5$ objectives

size increases from $|P| = 1000$ to 2000Cthan from 2000 to 3000. In the case of $m = 6$ objectives, a sharp reduction on Ip is not observed for a population size of up to 5000. It would be interesting to verify the effects of population sizes larger than 5000 on $m = 6$ objectives.

Next, Fig.4 shows for $m = 5$ objectives the distribution of solutions in objective space projected to the $f_1 - f_2$ plane. In the DTLZ2 problem, keeping the population fixed and increasing the number of objectives, it is common to see that solutions tend to concentrate along the axis. These solutions are known as dominance-resistance solutions [14], which are favored by selection due to its inability to discriminate based on dominance while actively promoting diversity, and may cause the algorithm to diverge from the true Pareto front. From the figure, it can be seen that when the population increases the population tends to cluster towards the central regions of objective space, helping to control the presence of dominance-resistance solutions and their negative effect on convergence.

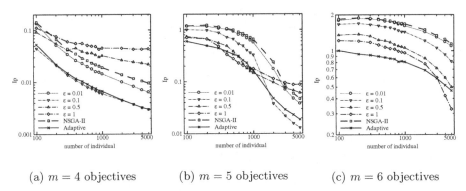

(a) $m = 4$ objectives   (b) $m = 5$ objectives   (c) $m = 6$ objectives

Fig. 5. Effect of increasing population size and including ε-Box Dominance (DTLZ2)

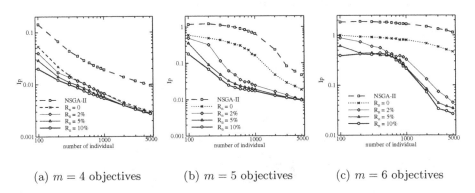

(a) $m = 4$ objectives   (b) $m = 5$ objectives   (c) $m = 6$ objectives

Fig. 6. Effect of increasing population size and including Adaptive ε-Box and Neighborhood Recombination (DTLZ2)

4.3　Effects of Adaptive ε-Box Dominance

In this section we compare results by conventional NSGA-II, NSGA-II enhanced with ε-box dominance setting a fixed value of $\varepsilon = \{0.01, 0.1, 0.5, 1\}$, and the Adaptive ε-Box dominance algorithm $A\varepsilon B$. The I_p by these methods is shown in Fig.5. Note that ε-Box dominance with fixed ε achieves better or worse I_p than conventional NSGA-II in $m = 4$ objectives, depending on the value set to ε. However, ε-Box dominance improves remarkably I_p in $m = 5, 6$, showing a bigger effect for larger population sizes $|P|$. On the other hand, using $A\varepsilon B$ stable and satisfactory performance is achieved, independently of the number of objectives. Especially, in the case of $m = 6$ objectives, $A\varepsilon B$ achieves best performance, with better effectiveness observed for larger population size.

4.4　Effects of Neighborhood Recombination

Next, we study the effects of incorporating Neighborhood Recombination in addition to Adaptive ε-Box dominance. Fig.6 shows results by NSGA-II, $A\varepsilon B$, and $A\varepsilon BNR$ for comparison. From the figure we can see that increases in population size $|P|$ and the inclusion of Neighborhood Recombination improves further convergence. Note that the effect of Neighborhood Recombination becomes larger with the number of objectives. Especially, in $m = 6$ objectives, comparing with $A\varepsilon B$ ($R_n = 0$), note that the inclusion of Neighborhood Recombination and increasing population size $|P|$ in $A\varepsilon BNR$ ($R_n > 0$) improves convergence remarkably.

Summarizing, increasing population size $|P|$ improves convergence of MOEAs in multi-objective problems. In addition, the inclusion of Adaptive ε-Box further improves convergence, with larger improvements observed as we increase the number of objectives. Furthermore, the inclusion of Neighborhood Recombination leads to an additional remarkable improvement in convergence, which also gets larger as we increase the number of objectives. For example in the case of $m = 6$ objectives, a drastic reduction in I_p from 1.84 to 0.026 is observed if we compare the performance of conventional NSGA-II using population size $|P| = 100$ and $A\varepsilon BNR$ using a population size $|P| = 5000$.

4.5 Comparison Using the C-Metric and Distribution of Solutions

In this section, we compare the search ability of conventional NSGA-II, $A\varepsilon B$ ($R_n = 0\%$) and $A\varepsilon BNR$ ($R_n = 10\%$) using the C-metric performance indicator and analyze the distribution of solutions rendered by these algorithms.

Fig.7 shows results of a pairwise comparison between algorithms using the C-metric. First, from Fig.7 (a) it can be seen that a significant fraction of Pareto optimal solutions (POS) by the enhanced algorithm with $A\varepsilon BNR$ dominate POS by conventional NSGA-II; whereas no solution by conventional NSGA-II dominates solutions by $A\varepsilon BNR$. Also note that the fraction of dominated solutions gets larger as we increase the number of objectives. Second, from Fig.7 (b), comparing conventional NSGA-II and the enhanced $A\varepsilon BNR$, a similar tendency as the one describe above can be observed, but with better fractions of dominated solutions in favor of the enhanced algorithm. Third, from Fig.7 (c), comparing the enhanced algorithms, note that $A\varepsilon BNR$ that includes both Adaptive ε-Box dominance and Neighborhood Recombination dominates a fraction of solutions found by the algorithm $A\varepsilon B$ that only includes Adaptive ε-Box dominance; whereas the opposite is not true. Note that this effect becomes remarkable when the number of objectives increases. From these results on the C-metric we conclude that the inclusion of Adaptive ε-Box dominance and Neighborhood Recombination leads to remarkable increase on the number of solutions with better convergence in the POS found by the algorithm.

Next, Fig.8 shows the distribution of solutions in objective space projected to the $f_1 - f_2$ plane by these three algorithms, for $m = 6$ objectives and population size $|P| = 5000$. Note that solutions by conventional NSGA-II are broadly spread, however they lack convergence to the true Pareto front as shown in Fig.8 (a). When Adaptive ε-Box is introduced, convergence of solutions improves but their distribution is biased towards the extreme regions of the Pareto front as seen in Fig.8 (b). This is because Adaptive ε-Box strengthen the trend to favor solutions towards the axis of the multi-objective space. On the other hand, when

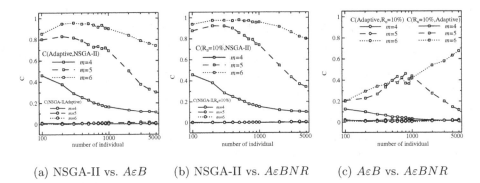

(a) NSGA-II vs. $A\varepsilon B$ (b) NSGA-II vs. $A\varepsilon BNR$ (c) $A\varepsilon B$ vs. $A\varepsilon BNR$

Fig. 7. Comparison among NSGA-II, $A\varepsilon B$, and $A\varepsilon BNR$ using the C-metric (DTLZ2)

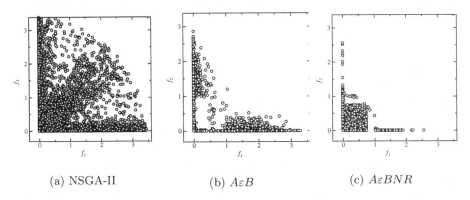

(a) NSGA-II   (b) $A\varepsilon B$ (c) $A\varepsilon BNR$

Fig. 8. Distribution of solutions found by NSGA-II, $A\varepsilon B$, and $A\varepsilon BNR$ (DTLZ2, $|P| = 5000, m = 6$)

Neighborhood Recombination is added convergence of the population of solutions improves further, solutions get compactly distributed around the Pareto optimal region, and the bias towards the axis of objective space becomes almost unnoticed as shown in Fig.8 (c). This is because recombination of solutions that are neighbor in objective space allows a better exploitation of the search, especially in problems where objective and variable space are not strongly uncorrelated. However, the different density of solutions in objective space produced by the uneven granularity of the grid used by Adaptive ε-Box should be investigated with more detail in the future.

4.6   Comparison Using HV

The HV measures both convergence and diversity (spread) of solutions. However, we can emphasize one over the other depending on the reference point r used to calculate HV. When the reference point is close to the Pareto front, convergence of solutions is emphasized. On the other hand, when the reference point is far away from the Pareto from, the diversity of solutions is emphasized.

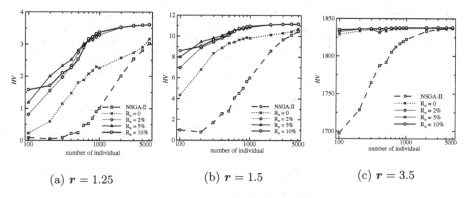

Fig. 9. HV on DTLZ2 $m = 6$ objectives

Fig.9 shows the HV for DTLZ2 with $m = 6$ objectives, varying population size $|P|$ from 100 to 5000, and varying the reference point to r=1.25C1.5C3.5. From this figure, note that when the reference point is set to $r = 3.5$, which emphasizes the estimation of diversity of solutions, similar values of HV are observed by the improved algorithms for all population sizes. This is because spread of solutions by the improved algorithms is similar. The HV by NSGA-II appears very low for small populations, but approaches the HV achieved by the improved algorithm for very large population sizes. These values reflect the fact that a considerable number of solutions by NSGA-II with small populations are past the reference point, and thus do not contribute to the hypervolume calculation. In the cases of $r = 1.5$ and $r = 1.25$, where convergence of solutions is emphasized, differences on HV between the improved algorithms and NSGA-II can be clearly seen even for very large populations. These results on the hypervolume are in accordance with our analysis discussed on previous sections.

4.7   Results on DTLZ3 and DTLZ4 Functions

In previous sections we focused our analysis on the DTLZ2 function. Here, we include and analyze results on DTLZ3 and DTLZ4 functions. Results for DTLZ3 are show in Fig.10, whereas results for DTLZ4 are shown in Fig.11. Results for DTLZ2 using similar configurations are shown in Fig.6.

From Fig.11 it can be seen that results on DTLZ4 are similar to those observed on DTLZ2, but note that in DTLZ4 convergence improvement due to bigger population sizes becomes more significant. This is because DTLZ4 is a problem that favors diversity of solutions, where an algorithm that selects solutions based on crowding distance[1] is expected to improve convergence of solutions specially in extreme regions of the Pareto front. On the other hand, from Fig.10 note that in DTLZ3 although convergence improves by using $A\varepsilon BNR$, compared to the I_p values achieved on DTLZ2 and DTLZ4 it can be seen that convergence is still insufficient.

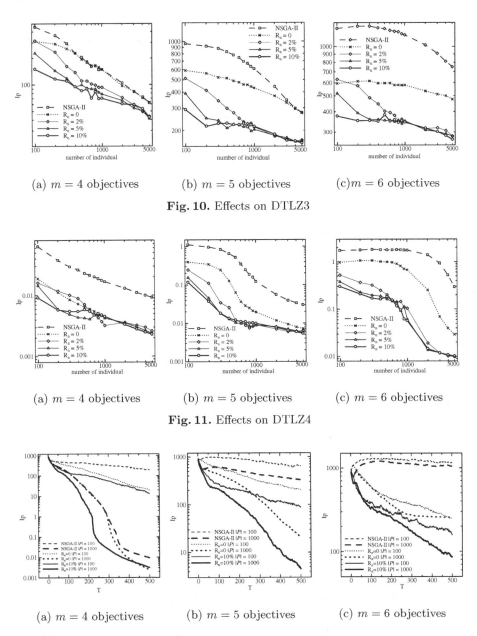

(a) $m = 4$ objectives   (b) $m = 5$ objectives   (c) $m = 6$ objectives

Fig. 10. Effects on DTLZ3

(a) $m = 4$ objectives   (b) $m = 5$ objectives   (c) $m = 6$ objectives

Fig. 11. Effects on DTLZ4

(a) $m = 4$ objectives   (b) $m = 5$ objectives   (c) $m = 6$ objectives

Fig. 12. I_p convergence on DTLZ3 increasing the number of generations to $T = 500$

To study with more detail the insufficient convergence in DTLZ3, Fig.12 show I_p by the three algorithms on $m = 4, 5, 6$ objectives, setting the population size $|P|$ to 100 and 1000 individuals, and extending the number of generations from 100 to $T = 500$. From the figure note that independently of the number of objectives, increasing the number of generations is effective to improve convergence of

$A\varepsilon B$ and $A\varepsilon BNR$. Similarly, better results are observed using a population size of 1000 individuals than a population size of 100. However, note that the reduction speed of I_p and its achieved value differ greatly according to the number of objectives. That is, the algorithms find it more difficult to reduce I_p conform the number of objectives increases. In the future, it is necessary to investigate ways to achieve good convergence within a minimum number of generations in this kind of problem.

5 Conclusions

In this work, we have studied the optimization of $m = 4, 5, 6$ many-objective problems under a fixed, restricted, small number of generations. First, we investigated the effect of population size on a conventional NSGA-II, verifying that convergence of solutions improves when we increase the population size. Next, we investigated the effects of including ε-Box dominance and Adaptive ε-Box dominance into NSGA-II, verifying that Adaptive ε-Box dominance improves convergence of solutions without the need to set the parameter ϵ by hand. Moreover, we investigated the effects of Adaptive ε-Box Dominance and Neighborhood Recombination, verifying that convergence of solutions improves substantially, especially when bigger populations are used and the number of the objectives of the problem becomes larger.

In the future, we would like to study the effects of using larger populations, explore parallelization, and develop MOEAs that can effectively evolve solutions independently of the characteristics of problem under a restricted number of generations.

References

1. Deb, K.: Multi-Objective Optimization using Evolutionary Algorithms. John Wiley & Sons, Chichester (2001)
2. Coello, C., Van Veldhuizen, D., Lamont, G.: Evolutionary Algorithms for Solving Multi-Objective Problems. Kluwer Academic Publishers, Boston (2002)
3. Aguirre, H., Tanaka, K.: Insights on Properties of Multi-objective MNK-Landscapes. In: Proc. 2004 IEEE Congress on Evolutionary Computation, pp. 196–203. IEEE Service Center (2004)
4. Aguirre, H., Tanaka, K.: Working Principles, Behavior, and Performance of MOEAs on MNK-Landscapes. European Journal of Operational Research 181(3), 1670–1690 (2007)
5. Sato, H., Aguirre, H.E., Tanaka, K.: Genetic Diversity and Effective Crossover in Evolutionary Many-objective Optimization. In: Coello, C.A.C. (ed.) LION 2011. LNCS, vol. 6683, pp. 91–105. Springer, Heidelberg (2011)
6. Ishibuchi, H., Tsukamoto, N., Nojima, Y.: Evolutionary Many-Objective Optimization: A Short Review. In: In Proc. IEEE Congress on Evolutionary Computation (CEC 2008), pp. 2424–2431. IEEE Press (2008)
7. Deb, K., Thiele, L., Laumanns, M., Zitzler, E.: Scalable Multi-Objective Optimization Test Problems. In: Proc. 2002 Congress on Evolutionary Computation, pp. 825–830. IEEE Service Center (2002)

8. Laumanns, M., Thiele, L., Deb, K., Zitzler, E.: Combining Convergence and Diversity in Evolutionary Multi-objective Optimization. Evolutionary Computation 10(3), 263–282 (Fall 2002)
9. Deb, K., Agrawal, S., Pratap, A., Meyarivan, T.: A Fast Elitist Non-Dominated Sorting Genetic Algorithm for Multi-Objective Optimization: NSGA-II, KanGAL report 200001 (2000)
10. Aguirre, H., Tanaka, K.: Adaptive ε-Ranking on Many-Objective Problems. Evolutionary Intelligence 2(4), 183–206 (2009)
11. Purshouse, R.C., Fleming, P.J.: Evolutionary Many-Objective Optimization: An Exploratory Analysis. In: Proc. IEEE CEC 2003, pp. 2066–2073 (2003)
12. Zitzler, E.: Evolutionary Algorithms for Multiobjective Optimization:Methods and Applications. PhD thesis, Swiss Federal Institute of Technology, Zurich (1999)
13. Fonseca, C., Paquete, L., López-Ibáñez, M.: An Improved Dimension-sweep Algorithm for the Hypervolume Indicator. In: Proc. 2006 IEEE Congress on Evolutionary Computation, pp. 1157–1163. IEEE Service Center (2006)
14. Ikeda, K., Kita, H., Kobayashi, S.: Failure of Pareto-based MOEAs: does non-dominated really mean near to optimal? In: Proc. of the 2001 Congress on Evolutionary Computation, vol. 2, pp. 957–962. IEEE Service Center (2001)

A Non-adaptive Stochastic Local Search Algorithm for the CHeSC 2011 Competition

Franco Mascia and Thomas Stützle

IRIDIA - Université Libre de Bruxelles
{fmascia,stuetzle}@ulb.ac.be

Abstract. In this work, we present our submission for the Cross-domain Heuristic Search Challenge 2011. We implemented a stochastic local search algorithm that consists of several algorithm schemata that have been offline-tuned on four sample problem domains. The schemata are based on all families of low-level heuristics available in the framework used in the competition with the exception of crossover heuristics. Our algorithm goes through an initial phase that filters dominated low-level heuristics, followed by an algorithm schemata selection implemented in a race. The winning schema is run for the remaining computation time. Our algorithm ranked seventh in the competition results. In this paper, we present the results obtained after a more careful tuning, and a different combination of algorithm schemata included in the final algorithm design. This improved version would rank fourth in the competition.

1 Introduction

Recent years have seen progressive abstractions in the design of stochastic local search (SLS) algorithms. From the first heuristics designed to solve instances of specific hard combinatorial optimisation problems, the community of researchers moved towards the engineering of more generic algorithm schemata that could be applied across different problems. Among these SLS methods, also referred to in the literature as meta-heuristics, there are Tabu Search [6, 7], Memetic Algorithms [17], Iterated Local Search [15], Iterated Greedy [19], and Variable Neighbourhood Search [8, 16].

Hyper-heuristics were introduced in 2001 by Cowling et al. [3], and, as meta-heuristics, they aim at selecting, combining, or adapting low-level heuristics to solve optimisation problems. Hyper-heuristics are usually classified in two families: the first one is composed of algorithms that select the best low-level heuristics for the problem being optimised; the second family is composed of the algorithms that generate or adapt low-level heuristics for the problem at hand. For a recent survey on the subject, see [2].

The first Cross-domain Heuristic Search Challenge (CHeSC 2011) is a competition devised by Burke et al. [18], which aims at encouraging the design of generic heuristic algorithms that can be successfully applied across different problem domains. In this competition, the distinction between the problem domain and the hyper-heuristic is very clear-cut; in fact, the contestants were asked

Y. Hamadi and M. Schoenauer (Eds.): LION 6, LNCS 7219, pp. 101–114, 2012.
© Springer-Verlag Berlin Heidelberg 2012

to implement a hyper-heuristic using HyFlex [18], a Java framework that has some design decisions that emphasise the separation between problem domains and hyper-heuristics.

The framework makes available a series of low-level heuristics for different problem domains. These low-level heuristics are categorised in four different families: (i) mutation heuristics, which perturb the current solution without any guarantee of improving the solution quality; (ii) local search heuristics, which are hill-climbers that apply mutation heuristics accepting only non-deteriorating solutions; (iii) ruin-and-recreate heuristics, which remove solutions components from the current solution and reconstruct it with problem specific constructive heuristics; (iv) crossover heuristics, which combine different solutions to obtain a new solution. The primitives exposed by the framework allow the developers to know how many heuristics of each specific family are available, to set the intensity of mutation of ruin-and-recreate and of mutation heuristics and the depth of search of local search heuristics, to apply a low-level heuristic to the current solution, to know the quality of the current solution, to store different solutions in memory, and to know how many CPU-seconds are available before the end of the allocated time.

Before the competition, four sample problem domains and ten instances for each domain had been made available to the contestants to test their implementations. The problem domains were: boolean satisfiability (MAX-SAT), one dimensional bin packing, permutation flow shop scheduling (PFSP), and personnel scheduling. During the competition, three instances of the ten sample instances of each problem domain were selected, and other two instances for each problem domain were added. Moreover, two hidden problem domains were revealed with five instances each. The two hidden problem domains were the traveling salesman problem (TSP) and the vehicle routing problem (VRP).

The competition rules imposed a specific setting in which the submitted SLS algorithms had no possibility of recognising the problem or the instance being optimised, and no information about a solution could be extracted, except for the solution quality. Some limitations imposed by the rules are relatively unrealistic, and hardly encountered in real-world scenarios. For example, one way to enforce the separation between the hyper-heuristic and the problem domain is to randomly shuffle the low-level heuristics identifier. We decided to participate to the competition with an off-line tuned algorithm even though competition rules do not favour such techniques. Our algorithm is based on the rationale that with an appropriately chosen fixed algorithm schema, a problem can be solved competitively even when not adapting the heuristic selection mechanisms at runtime. Obviously, because heuristic identifiers are randomly shuffled in the framework, we cannot offline tune the sequence of the heuristics to be applied. As a work-around to this missing information, we applied a heuristic selection phase, in which we identify dominated heuristics that are eliminated from further consideration when running our algorithm schemata. The algorithm remains in the sense non-adaptive, in that there is no adaptation of the parameters and no selection of the low level heuristics throughout the search. Moreover, experiments we performed after the submission showed a rather limited impact of this initial phase of filtering low level heuristics. To emphasise the aspect that the heuristics selection is not adapted during the main execution phase of the algorithm, we call it a Non-Adaptive SLS Algorithm (NA-SLS). In the end, our NA-SLS ranked seventh in the competition. This paper describes an improved version of our submission, which shares the same code but uses a more thorough off-line tuning, and adds some more analysis on the algorithm. The version presented in this paper would rank fourth among the algorithms that participated to the competition.

The rest of the paper is organised as follows. Section 2 presents the analysis on the low-level heuristics; Section 3 describes the algorithmic schemata we implemented as building blocks for NA-SLS; Section 4 presents the results of the tuning of these schemata on the four sample problem domains; Section 5 describes the design of NA-SLS; and Section 6 presents the results that would have been obtained at the competition. Eventually, Section 7 draws the conclusion.

2   Analysis of the Low-Level Heuristics

Before actually running NA-SLS, the available low-level heuristics are tested to identify whether any of the heuristics are dominated. The criterion of dominance takes into account the computation time and the solution quality. This first phase of the analysis of the low-level heuristics is run for at most 7.5% of the total allocated runtime or for at most 25 runs of each heuristic. During each run a random solution is constructed and heuristics belonging to the local search, and to ruin-and-recreate families are applied to it. For each low-level heuristic A, the median run-time t_A and the median solution quality q_A are computed. A heuristic A is dominated by heuristic B if $t_B < t_A$ and $q_B < q_A$. The aim of this phase, is to enforce that only non-dominated heuristics will be available for the remaining runtime. Nevertheless, in order to avoid that a single heuristic that dominates all others is the only one available for the following phases, we make sure that the two heuristics having the lowest median solution quality are never discarded. Figure 1 shows an example of the analysis performed on an instance of the personnel scheduling problem. The plot on the left shows all local search heuristics with different values of the parameter that set the intensity of mutation. The plot on the right shows the non-dominated heuristics after the filtering.

3   Design and Implementation of Algorithmic Schemata

We implemented a series of algorithmic schemata that use all low-level heuristics (except crossovers) as basic building blocks. Among the large number of implemented schemata there are several algorithms that are well established in the literature. In the following we list the most relevant, with their variants and the parameters defined for the tuning:

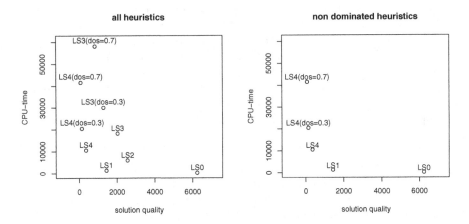

Fig. 1. Example for the analysis of the low-level heuristics on an instance of the personnel scheduling domain. LS4(dos=0.7) corresponds to local search heuristic number 4, with the depth of search parameter set to 0.7. Where not specified, the depth of search is set to the default 0.1. LS4(dos=0.7), LS3(dos=0.3), LS3, and LS2 are the dominated heuristics.

- Randomized Iterative Improvement (RII) [9]: the parameters defined for the tuning are two continuos parameters representing the probability of selecting a mutation or a ruin-and-recreate heuristic, respectively, and a continuous parameter representing the probability of accepting worsening solutions.
- Probabilistic Iterative Improvement (PII) [9]; this algorithm is the same as the previous, but it uses a Metropolis condition for computing the probability of accepting worsening solutions and has continuous parameter for the temperature.
- Variable Neighbourhood Descent (VND) [8]; for this algorithm we implemented two versions. The first one uses local search heuristics, and the second one uses ruin-and-recreate heuristics. In both cases, the heuristics are applied in the order of decreasing median computation time, which is computed by running the low-level heuristics several times from an initial solution. The parameters defined for this schema are a categorical parameter that allows to select among the two variants, and a continuos parameter representing the probability of accepting worsening solutions as in RII.
- Iterated Local Search (ILS) [15]; few variations of this algorithm have been implemented. We implemented the ILS described in [1], an ILS that uses the VND variants described above as subsidiary local search, and a Hierarchical ILS [10]. Hierarchical ILS uses an ILS as subsidiary local search and applies a strong perturbation in the outer ILS and a small perturbation in the inner ILS. We defined a categorical parameter that allows to select among the variants, a continuous parameter for the probability of accepting a worsening solution, and an integer parameter for determining the perturbation size,

which corresponds to the number of times a random mutation heuristic is applied to the current solution. For the Hierarchical ILS we also used the same parameters for the inner ILS, and a further continuos parameter for the time allocated to the inner ILS as a fraction of the total run time.

- Simulated Annealing (SA) [11, 20]; this algorithm estimates the initial temperature by measuring the average solution quality improvement after applying a fixed number of steps of first improvement. The initial acceptance probability and the number of steps for this initial phase are a continuos and an integer parameter respectively. The algorithm continues with a cooling schedule that is defined by a constant multiplicative factor (continuos parameter) applied regularly after a specified number of steps (integer parameter) until a final temperature is reached (continuous parameter). When the final temperature is reached, the algorithm restarts from the initial estimated temperature.

- Iterated Greedy (IG) [19]; we implemented IG by means of repeated applications of a random ruin-and-recreate heuristic with a probability of accepting worsening solutions, which is specified by a continuous parameter. We also implemented few variations of this algorithmic schema, which differ for a probabilistic acceptance criterion, and different selection strategies of low-level heuristics.

For all schemata we defined two continuos parameters that set the intensity of mutation for the ruin-and-recreate and mutation heuristics, and the depth of search for local search heuristics. The parameters range from 0 to 1, with 1 representing the maximum intensity of mutation and the maximum depth of search respectively. If the application of a specific local search heuristic takes more than 10 CPU-seconds, the depth of search parameter for that specific heuristic is fixed to the default value defined in the framework, which is 0.1. For RII, PII, and the ILS variants, we implemented also a fixed restart policy. We defined a categorical parameter that could select the restart, and a continuous parameter representing the fraction of the allocated runtime between restarts.

Besides the algorithms that are well-established in the literature, we implemented a further algorithm we called tuneable SLS algorithm (TSA), which is a juxtaposition of blocks of low-level heuristics. TSA has been designed with the aim to see how good could perform an algorithm with less rationale in the design and a larger parameter space for automatic tuning. Listing 1.1 shows the pseudo-code of TSA with the parameters defined for the tuning. The algorithm starts from an initial random solution s, and until the allocated time has not expired, it goes through a series of phases characterised by the application of heuristics belonging to a specific family. In the first phase, the algorithm applies sequentially n_{rr} randomly selected ruin-and-recreate heuristics which, if defined in the framework, exploit constructive heuristics for the problem at hand. Non improving solutions are accepted with a probability $p_{a_{rr}}$. In the following phase, the solution that is constructed by the ruin-and-recreate heuristics is (possibly) improved by the application of n_{ls} randomly selected local search heuristics. Non-improving solutions in this phase are accepted with probability $p_{a_{ls}}$. After

Listing 1.1. Tuneable SLS algorithm pseudo-code.

```
1   procedure TSA(n_rr, p_a_rr, n_ls, p_a_ls, p_a, p_m, p_restart, dos, iom, iom_rr)
2       s' ← s ← random initial solution
3       set depth of search of local search heuristics to dos
4       set intensity of mutation of mutation heuristics to iom
5       set intensity of mutation of ruin−and−recreate heuristics to iom_rr
6       while time has not expired:
7           for n_rr times:
8               apply randomly selected heuristic of type ruin−and−recreate
9               accept non−improving solution with probability p_a_rr
10          for n_ls times:
11              apply randomly selected heuristic of type local search
12              accept non−improving solution with probability p_a_ls
13          if f(s) < f(s') or rand(0, 1) < p_a:
14              s' ← s
15          else
16              s ← s'
17          with probability p_m:
18              apply randomly selected heuristic of type mutation
19          with probability p_restart:
20              s ← random initial solution
```

the first two phases, an improving solution is always accepted and stored in s', while worsening solutions are accepted with probability p_a. Before going again through the series of ruin-and-recreate heuristics of the first phase, the current solution is perturbed by the application of a random mutation heuristic with probability p_m. Finally, with probability $p_{restart}$ the algorithm restarts from a new random solution.

4 Off-line Tuning of the Algorithmic Schemata

After the implementation of the algorithmic schemata, we tuned them on the four problem domains and the ten instances that were available before the competition, and selected a small subset that would be part of the NA-SLS algorithm.

For the off-line tuning, we used irace [13], a software package that implements an iterated racing procedure for the off-line tuning of categorical, integer, and continuous parameters. Moreover, irace allows to specify conditional parameters and therefore tune the parameters regarding a specific algorithm schemata, only when the specific algorithm schemata has been selected by a categorical parameter. These characteristics and a number of successful applications [4, 5, 12–14], made it the natural choice for tuning the parameters of our SLS algorithm.

The off-line tuning is divided in two phases. In the first phase, we tuned the parameters of each algorithmic schema on the single problem domains. In the second phase, for each problem domain, we selected the best tuned schemata. The off-line tuning of both phases was performed with irace on cluster nodes with

Table 1. Selected algorithmic schemata for the problem domains

Problem domain	Algorithmic Schema
MAX-SAT	TSA, parameter setting MAX-SAT
Bin packing	TSA, parameter setting bin packing
Personnel scheduling	TSA, parameter setting personnel scheduling
PFSP	IG with probabilistic acceptance criterion

Table 2. Parameters of the tuned TSA schemata on three problem domains

Problem domain	Parameters									
	n_{rr}	$p_{a_{rr}}$	n_{ls}	$p_{a_{ls}}$	p_a	p_m	$p_{restart}$	dos	iom	iom_{rr}
MAX-SAT	1	0.4966	50	0.9265	0.2341	0.0959	0.0008	0.6	0.1063	0.2
Bin packing	4	0.057	4	0.43	0.0001	0.29	0.014	0.46	0.11	0.69
Personnel scheduling	11	0.54	15	0.72	0.54	0.2	0.67	0.85	0.43	0.21

16GB of RAM and 2 AMD Opteron 6128 CPUs, each with eight cores running at 2GHz. During the two tuning phases, the number of different configurations tested amounted to 136,276 for a total of 3 years of CPU-time.

Table 1 shows the best performing algorithms for the single domains. Surprisingly, three differently tuned TSA were selected for three out of four problem domains, namely MAX-SAT, bin packing and personnel scheduling. The parameter settings for the TSAs are shown in Table 2. For PFSP, the best schema is an IG with a probabilistic acceptance criterion. The tuned parameters lead to an IG that executes at each iteration one or two random ruin-and-recreate heuristics. If no such heuristics are available for the problem domain, the algorithm executes one or two mutation heuristics. After the perturbation, one local search heuristic is executed. Worsening solutions are accepted with probability $\exp\{-\Delta f/0.87\}$.

All algorithms in Table 1 were included in the algorithm we submitted for the competition. In the implementation described in this paper we decided to extend the pool of available algorithms by looking at the second ranking algorithm schemata for the four problem domains.

- For MAX-SAT, the second ranking algorithm was an ILS with VND as subsidiary local search (ILS+VND); the probability of accepting a worsening solution was 0.99 and the perturbation size amounted to the application of 2 random mutation heuristics. The intensity of mutation was 0.3 and the depth of search was 0.41.
- For personnel scheduling, it was also an ILS+VND; the tuned probability of accepting a worsening solution was 0.23 and the perturbation size amounted to the application of 6 random mutation heuristics. The intensity of mutation was 0.08 and the depth of search was 0.25.
- For PFSP again an ILS+VND ranked second, with a low probability of accepting worsening solutions, i.e. 0.39, and large perturbations amounting

to 11 applications of random mutation heuristics. The intensity of mutation was 0.59 and the depth of search was 0.022.

– For bin packing an ILS with acceptance probability equal to 0.037 ranked second; for this schema the amount of perturbation selected was of 4 random mutation heuristics. The intensity of mutation was 0.033 and the depth of search was 0.31.

We decided to include in the pool of available schemata three of the four first ranking algorithms, i.e., TSA for bin packing, TSA for personnel scheduling and IG for PFSP. We did not include TSA for MAX-SAT since the performances were pretty close to the second ranking and less tailored ILS+VND. In three domains out of four, ILS+VND ranked second, therefore we added two different ILS+VND schemata, namely the ILS+VND tuned for MAX-SAT and the ILS+VND tuned for personnel scheduling. The rationale behind this choice is to add some well-known meta-heuristics that even if tuned on specific problem domains could be effective on the two hidden problem domains.

5   Final SLS Algorithm

NA-SLS is composed of the following three phases.

Phase 1: Analysis of the low-level heuristics
This phase, which lasts at most 7.5% of the total runtime, is devoted to the analysis of the low-level heuristics available for the problem being optimised. As described in Section 2, only non dominated low-level heuristics will be available for the five selected algorithm schemata in the following phases.

Phase 2: Algorithm selection
In this phase, a fraction of the remaining time is allocated for selecting the best performing schema for the problem at hand among the schemata with fixed parameters described in Section 4. The selection is as in a race, where at each step the worst candidate schema is eliminated. Starting from the best solution found during the first phase, the algorithmic schemata are run in an interleaved manner each for a fraction of CPU-seconds that amounts to 2.5% of the time remaining after the first phase. After each race, the worst performing schema is discarded, and a new race is performed on the remaining algorithms. The phase terminates when only one schema remains. In order to keep this phase as short as possible, if this phase lasts 25% of the total computation time, the phase is automatically terminated and the best of the remaining algorithms is kept as the winner of the race. This is a very simplistic algorithm selection, which could be seen as a workaround to the competition rules. By using a more sophisticated schema as in [21], we probably would make a faster and better selection.

Phase 3: Run
The best-performing algorithm is executed for the remaining allocated time.

6   Experimental Results

The competition organisers made available an updated version of the HyFlex framework with all instances and all domains used during the competition, i.e., the four sample ones and the two hidden ones, TSP and VRP. In order to facilitate the comparison with the algorithms submitted to the competition, the organisers published also all detailed results obtained by all algorithms on all instances, and the random seed used to select the five random instances for each problem domain.

During the competition, algorithms have been run 31 times for 600 seconds on each domain and each instance. The median values have been selected and the algorithms have been ranked. The ranking system is mutuated from the Formula-1 point system. The first eight best heuristics receive 10, 8, 6, 5, 4, 3, 2, and 1 point respectively. In case of ties, the points of the concerned positions are summed and equally divided by the algorithms having the same median solution quality.

We run NA-SLS on one node of the cluster for 1176 CPU-seconds which correspond to 600 CPU-seconds on the competition reference machine. The speedup between the machines has been computed with a benchmark tool supplied by the organisers of the competition. We verified that the amount of seconds computed by the benchmark tool was correct by running on the same node of the cluster

Table 3. Comparison with CHeSC 2011 contestants on MAX-SAT, bin packing, and PFSP

MAX-SAT			Bin packing			PFSP		
Rank	Algorithm	Score	Rank	Algorithm	Score	Rank	Algorithm	Score
1	**NA-SLS**	44.2	1	AdapHH	45	1	ML	38
2	AdapHH	29.28	2	ISEA	30	2	AdapHH	36
3	VNS-TW	28.08	3	**NA-SLS**	23	3	VNS-TW	32
4	HAHA	27.28	4	ACO-HH	19	4	**NA-SLS**	26
5	KSATS-HH	19.5	5	GenHive	14	5	EPH	21
6	ML	12.30	6	XCJ	12	6	HAEA	10
7	AVEG-Nep	12.1	6	DynILS	12	7	PHUNTER	9
8	PHUNTER	9.3	6	ML	12	7	ACO-HH	9
9	ISEA	4.10	9	KSATS-HH	11	9	GenHive	7
10	MCHH-S	3.75	10	EPH	9	10	ISEA	3.5
11	XCJ	3.60	11	PHUNTER	3	10	HAHA	3.5
12	GISS	0.75	11	VNS-TW	3	12	AVEG-Nep	0
12	SA-ILS	0.75	13	HAEA	2	12	GISS	0
14	ACO-HH	0	14	AVEG-Nep	0	12	SA-ILS	0
14	GenHive	0	14	GISS	0	12	SelfSearch	0
14	SelfSearch	0	14	SA-ILS	0	12	Ant-Q	0
14	Ant-Q	0	14	HAHA	0	12	XCJ	0
14	EPH	0	14	SelfSearch	0	12	DynILS	0
14	DynILS	0	14	Ant-Q	0	12	KSATS-HH	0
14	HAEA	0	14	MCHH-S	0	12	MCHH-S	0

Table 4. Comparison with CHeSC 2011 contestants on personnel scheduling, TSP, and VRP

Personnel scheduling			TSP			VRP		
Rank	Algorithm	Score	Rank	Algorithm	Score	Rank	Algorithm	Score
1	VNS-TW	39.5	1	AdapHH	41.25	1	PHUNTER	33
2	ML	31	2	EPH	36.25	2	HAEA	28
3	HAHA	25	3	PHUNTER	26.25	3	KSATS-HH	23
4	SA-ILS	18.5	4	VNS-TW	17.25	4	ML	22
5	ISEA	14.5	5	DynILS	13	5	AdapHH	16
6	PHUNTER	12.5	5	ML	13	5	HAHA	16
7	GISS	10	7	**NA-SLS**	12	7	EPH	12
7	EPH	10	8	ISEA	11	8	AVEG-Nep	10
9	AdapHH	9	8	HAEA	11	9	GISS	6
9	KSATS-HH	9	10	ACO-HH	8	9	GenHive	6
11	GenHive	6.5	11	GenHive	3	9	VNS-TW	6
12	SelfSearch	5	11	SelfSearch	3	12	ISEA	5
13	**NA-SLS**	2.5	13	AVEG-Nep	0	12	XCJ	5
14	HAEA	2	13	GISS	0	14	SA-ILS	4
15	AVEG-Nep	0	13	SA-ILS	0	15	ACO-HH	2
15	ACO-HH	0	13	HAHA	0	16	DynILS	1
15	Ant-Q	0	13	Ant-Q	0	17	**NA-SLS**	0
15	XCJ	0	13	XCJ	0	17	SelfSearch	0
15	DynILS	0	13	KSATS-HH	· 0	17	Ant-Q	0
15	MCHH-S	0	13	MCHH-S	0	17	MCHH-S	0

our original submission and verifying (with the same random seed used in the competition) that the results obtained in 1176 CPU-seconds corresponded to the results obtained in the competition.

Table 3 and Table 4 show the results on the different problem domains. On three out of four of the sample domains we would score high achieving the first, the third, and the fourth position. For personnel scheduling, we already knew before the final competition that our results would not be very competitive; this is confirmed by the thirteenth position achieved by our algorithm with only 2.5 points. On the two hidden domains, we scored relatively good on TSP and for us surprisingly bad on VRP, where we did not score any point. Future work will be devoted to understand the reasons behind the poor performances on VRP. Overall, with a more careful tuning, and a different combination of schemata, our algorithm would have ranked fourth at the competition.

In order to better understand the results, after the competition, we run an analysis of the impact of the heuristics selection phase. We ran an identical copy of NA-SLS where we allowed all low-level heuristics to be used. The results showed that the impact was more limited than expected. In fact, by allowing dominated heuristics, the final algorithm would still score fourth in the final ranking with 108.2 points, which is more than 107.7 points achieved by the version with the heuristic selection phase. For what concerns the single domains, the ranking and points would not change on MAX-SAT, bin packing, TSP, and

Table 5. Comparison with CHeSC 2011 contestants on all problem domains

All problem domains

Rank	Algorithm	Score
1	AdapHH	176.53
2	ML	128.3
3	VNS-TW	125.83
4	**NA-SLS**	107.7
5	PHUNTER	93.05
6	EPH	88.25
7	HAHA	71.78
8	ISEA	68.1
9	KSATS-HH	62.5
10	HAEA	53
11	ACO-HH	38
12	GenHive	36.5
13	DynILS	26
14	SA-ILS	23.25
15	AVEG-Nep	22.1
16	XCJ	20.6
17	GISS	16.75
18	SelfSearch	8
19	MCHH-S	3.75
20	Ant-Q	0

VRP. On PFSP the algorithm would actually gain 3 points and still score fourth with 29 points; and for personnel scheduling it would lose the 2.5 points and rank fourteenth.

In order to analyse the impact of the algorithm selection, we implemented two alternative versions of NA-SLS. In the first one, a candidate schema is selected at random from the pool of the available ones. In the second one, an oracle always chooses the best a posteriori candidate schema for the problem domain being optimised. These two variants represent a lower and an upper bound to the desired behaviour of the algorithm selection we implemented. In the case of random selection, NA-SLS would rank seventeenth at the competition with only 8 points; while the version implementing the oracle, would rank second with 138.8 points. Table 6 breaks down the results for the single problems domains. On the test instances used in the competition, the best algorithmic schema for MAX-SAT, bin packing, and PFSP correspond to the schemata that had been tuned for the same problem domains. For personnel scheduling the best schema would have been an ILS+VND tuned for MAX-SAT, but also the ILS+VND tuned for personnel scheduling would have scored close and rank fifth with 10.5 points. For the hidden domain TSP the best schema would have been the TSA tuned for bin packing, and for the hidden domain VRP, both ILS+VND tuned for MAX-SAT and ILS+VND tuned for personnel scheduling would have ranked seventeenth with one point.

Table 6. Results of NA-SLS, where the algorithmic schemata selection is performed by an oracle that knows which schema will be the best choice for the problem being optimised

Problem domain	Rank	Score	Algorithmic Schema
MAX-SAT	1	40.8	ILS+VND, parameter setting MAX-SAT
Bin packing	3	24	TSA, parameter setting bin packing
Personnel scheduling	6	12	ILS+VND, parameter setting MAX-SAT
PFSP	1	39	IG with probabilistic acceptance criterion
TSP	4	22	TSA, parameter setting bin packing
VRP	17	1	ILS+VND, parameter setting MAX-SAT
All problem domains	2	138.8	-

The bounds obtained show that our simplistic algorithm selection is already close to the upper bound, and that there is still a gap for further improvement. Nevertheless, this gap could be small since the alternative versions of NA-SLS used to compute the bounds do not spend time for the for the algorithm selection phase, and have more time available for the search.

7   Conclusions

In this paper we presented in detail a further development of our submission for the CHeSC 2011 challenge. Our algorithm is composed by different schemata we tuned on four sample problem domains supplied by the competition organisers. After tuning each algorithmic schema on each problem domain, we selected a pool of five schemata that would be part of the final algorithm and that would be selected at runtime with a simplistic algorithm selection mechanism. The experimental results shows that our algorithm would have ranked fourth at the competition. Even if in the end the algorithm did not rank first, the testing of such a large number of algorithm configurations would have been unfeasible without the automatic tuning.

There are several ad-hoc choices that were done without much analysis, for example the algorithm selection schema sounded a reasonable strategy to try, but it could be replaced with more sophisticated schemes that would probably allow for a faster and better selection. Heuristics and parameter adaptation schemes considering the results of the other algorithms could be another step to apply. Eventually, it would also be interesting to test a different version of the competition, in which the low level heuristics are not reshuffled and therefore their choice or the sequence of their execution could be directly tuned.

Acknowledgments. This work was supported by the Meta-X project funded by the Scientific Research Directorate of the French Community of Belgium. Thomas Stützle acknowledges support from the Belgian F.R.S.-FNRS, of which he is a Research Associate.

References

1. Burke, E., Curtois, T., Hyde, M., Kendall, G., Ochoa, G., Petrovic, S., Vazquez-Rodriguez, J.A., Gendreau, M.: Iterated local search vs. hyper-heuristics: Towards general-purpose search algorithms. In: IEEE Congress on Evolutionary Computation, pp. 1–8. IEEE Press, Piscataway (2010)
2. Chakhlevitch, K., Cowling, P.: Hyper-heuristics: Recent developments. In: Cotta, C., Sevaux, M., Sörensen, K. (eds.) Adaptive and Multilevel Metaheuristics. SCI, vol. 136, pp. 3–29. Springer (2008)
3. Cowling, P.I., Kendall, G., Soubeiga, E.: A Hyperheuristic Approach to Scheduling a Sales Summit. In: Burke, E., Erben, W. (eds.) PATAT 2000. LNCS, vol. 2079, pp. 176–190. Springer, Heidelberg (2001)
4. Dubois-Lacoste, J., López-Ibáñez, M., Stützle, T.: Automatic configuration of state-of-the-art multi-objective optimizers using the TP+PLS framework. In: Krasnogor, N., et al. (eds.) GECCO 2011, pp. 2019–2026. ACM Press, New York (2011)
5. Dubois-Lacoste, J., López-Ibáñez, M., Stützle, T.: A hybrid TP+PLS algorithm for bi-objective flow-shop scheduling problems. Computers & Operations Research 38(8), 1219–1236 (2011)
6. Glover, F.: Future paths for integer programming and links to artificial intelligence. Computers & Operations Research 13(5), 533–549 (1986)
7. Hansen, P., Jaumard, B.: Algorithms for the maximum satisfiability problem. Computing 44, 279–303 (1990)
8. Hansen, P., Mladenovic, N.: Variable neighborhood search: Principles and applications. European Journal of Operational Research 130(3), 449–467 (2001)
9. Hoos, H.H., Stützle, T.: Stochastic Local Search: Foundations and Applications. Elsevier, Amsterdam (2004)
10. Hussin, M.S., Stützle, T.: Hierarchical Iterated Local Search for the Quadratic Assignment Problem. In: Blesa, M.J., Blum, C., Di Gaspero, L., Roli, A., Sampels, M., Schaerf, A. (eds.) HM 2009. LNCS, vol. 5818, pp. 115–129. Springer, Heidelberg (2009)
11. Kirkpatrick, S., Gelatt, C.D., Vecchi, M.P.: Optimization by simulated annealing. Science 220(4598), 671–680 (1983)
12. Liao, T., Montes de Oca, M.A., Aydin, D., Stützle, T., Dorigo, M.: An incremental ant colony algorithm with local search for continuous optimization. In: Krasnogor, N., et al. (eds.) GECCO 2011, pp. 125–132. ACM Press, New York (2011)
13. López-Ibáñez, M., Dubois-Lacoste, J., Stützle, T., Birattari, M.: The irace package, iterated race for automatic algorithm configuration. Tech. Rep. TR/IRIDIA/2011-004, IRIDIA, Université Libre de Bruxelles, Belgium (2011),
http://iridia.ulb.ac.be/IridiaTrSeries/IridiaTr2011-004.pdf
14. López-Ibáñez, M., Stützle, T.: The automatic design of multi-objective ant colony optimization algorithms. IEEE Transactions on Evolutionary Computation (accepted, 2012)
15. Lourenço, H.R., Martin, O., Stüttzle, T.: Iterated local search: Framework and applications. In: Gendreau, M., Potvin, J.Y. (eds.) Handbook of Metaheuristics, 2nd edn. International Series in Operations Research & Management Science, vol. 146, ch. 9, pp. 363–397. Springer, New York (2010)
16. Mladenovic, N., Hansen, P.: Variable neighbourhood search. Computers and Operations Research 24(11), 71–86 (1997)

17. Moscato, P.: Memetic algorithms: a short introduction, pp. 219–234. McGraw-Hill Ltd., UK (1999)
18. Ochoa, G., Hyde, M., Curtois, T., Vazquez-Rodriguez, J.A., Walker, J., Gendreau, M., Kendall, G., McCollum, B., Parkes, A.J., Petrovic, S., Burke, E.K.: HyFlex: A Benchmark Framework for Cross-Domain Heuristic Search. In: Hao, J.-K., Middendorf, M. (eds.) EvoCOP 2012. LNCS, vol. 7245, pp. 136–147. Springer, Heidelberg (2012)
19. Ruiz, R., Stützle, T.: A simple and effective iterated greedy algorithm for the permutation flowshop scheduling problem. European Journal of Operational Research 177(3), 2033–2049 (2007)
20. Černý, V.: Thermodynamical approach to the traveling salesman problem: An efficient simulation algorithm. Journal of Optimization Theory and Applications 45, 41–51 (1985)
21. Xu, L., Hutter, F., Hoos, H.H., Leyton-Brown, K.: SATzilla: Portfolio-based algorithm selection for SAT. Journal of Artificial Intelligence Research 32(1), 565–606 (2008)

Local Search and the Traveling Salesman Problem: A Feature-Based Characterization of Problem Hardness

Olaf Mersmann[1], Bernd Bischl[1], Jakob Bossek[1], Heike Trautmann[1], Markus Wagner[2], and Frank Neumann[2]

[1] Statistics Faculty, TU Dortmund University, Germany
{olafm,bischl,bossek,trautmann}@statistik.tu-dortmund.de
[2] School of Computer Science, The University of Adelaide, Australia
{markus.wagner,frank.neumann}@adelaide.edu.au

Abstract. With this paper we contribute to the understanding of the success of 2-opt based local search algorithms for solving the traveling salesman problem (TSP). Although 2-opt is widely used in practice, it is hard to understand its success from a theoretical perspective. We take a statistical approach and examine the features of TSP instances that make the problem either hard or easy to solve. As a measure of problem difficulty for 2-opt we use the approximation ratio that it achieves on a given instance. Our investigations point out important features that make TSP instances hard or easy to be approximated by 2-opt.

Keywords: TSP, 2-opt, Classification, Feature Selection, MARS.

1 Introduction

Metaheuristic algorithms such as local search, simulated annealing, evolutionary algorithms, and ant colony optimization have produced good results for a wide range of NP-hard combinatorial optimization problems. One of the most famous NP-hard combinatorial optimization problems is the traveling salesman problem (TSP). Given a set of N cities and positive distances d_{ij} to travel from city i to city j, $1 \leq i, j \leq N$ and $i \neq j$, the task is to compute a tour of minimal traveled distance that visits each city exactly once and returns to the origin.

The perhaps simplest NP-hard subclass of TSP is the Euclidean TSP where the cities are points in the Euclidean plane and the distances are the Euclidean distances between them. We will focus on the Euclidean TSP. It is well known that there is a polynomial time approximation scheme (PTAS) for this problem [2]. However, this algorithm is very complicated and hard to implement.

Many heuristic approaches have been proposed for the TSP. Often local search methods are the preferred methods used in practice. The most successful algorithms rely on the well-known 2-opt operator, which removes two edges from a current tour and connects the resulting two parts by two other edges such that a different tour is obtained [9]. Despite the success of these algorithms for a wide

Y. Hamadi and M. Schoenauer (Eds.): LION 6, LNCS 7219, pp. 115–129, 2012.
© Springer-Verlag Berlin Heidelberg 2012

range of TSP instances, it is still hard to understand 2-opt from a theoretical point of view.

Theoretical studies regarding 2-opt have investigated the approximation behavior as well as the time to reach a local optimum. Chandra et al [4] have studied the worst-case approximation ratio that 2-opt achieves for different classes of TSP instances. Furthermore, they investigated the time that a local search algorithm based on 2-opt needs to reach a locally optimal solution. Englert et al. [6] have shown that there are even instances for the Euclidean TSP where a deterministic local search algorithm based on 2-opt would take exponential time to find a local optimal solution. Furthermore, they have shown polynomial bounds on the expected number of steps until 2-opt reaches a local optimum for random Euclidean instances and proved that such a local optimum gives a good approximation for the Euclidean TSP. These results also transfer to simple ant colony optimization algorithms as shown in [12]. Most previously mentioned investigations have in common that they either investigate the worst local optimum and compare it to a global optimal solution or investigate the worst case time that such an algorithm needs to reach a local optimal solution. Although these studies provide interesting insights into the structure of TSP instances they do not provide much insights into what is actually going on in the application of 2-opt based algorithms. In almost all cases the results obtained by 2-opt are much better than the actual worst-case guarantees given in these papers. These motivates the studies carried out in this paper, which aim to get further insights into the search behavior of 2-opt and to characterize hard and easy TSP instances for 2-opt.

We take a statistical meta-learning approach to gain new insights into which properties of a TSP instance make it difficult or easy to solve for 2-opt. Analyzing different features of TSP instances and their correlation we point out how they influence the search behavior of local search algorithms based on 2-opt. To generate hard or easy instances for the TSP we use an evolutionary algorithm approach similar to the one of [21]. However, instead of defining hardness by the number of 2-opt steps to reach a local optimum, we define hardness by the approximation ratio that such an algorithm achieves for a given TSP instance compared to the optimum solution. This is motivated by classical algorithmic studies for the TSP problem in the field of approximation algorithms. Having generated instances that lead to a bad or good approximation ratio, the features of these instances are analyzed and classification rules are derived, which predict the type of an instance (easy, hard) based on its feature levels. In addition, instances of moderate difficulty in between the two extreme classes are generated by transferring hard into easy instances based on convex combinations of both instances, denoted as morphing. Systematic changes of the feature levels along this "path" are identified and used for a feature based prediction of the difficulty of a TSP instance for 2-opt-based local search algorithms.

The structure of the rest of this paper is as follows. In Section 2, we give an overview about different TSP solvers, features to characterize TSP instances and indicators that reflect the difficulty of an instance for a given solver. Section 3 introduces an evolutionary algorithm for evolving TSP instances that are hard or easy to approximate and carries out a feature based analysis of the hardness of TSP instances. Finally, we finish with concluding remarks and an outlook on further research perspectives in Section 4.

2   Local Search and the Traveling Salesman Problem

Local search algorithms are frequently used to tackle the TSP problem. They iteratively improve the current solution by searching for a better one in its pre-defined neighborhood. The algorithm stops when there is no better solution in the given neighborhood or if a certain number of iterations has been reached.

Historically, 2-opt [5] was one of the first successful algorithms to solve larger TSP instances. It is a local search algorithm whose neighbourhood is defined by the removal of two edges from the current tour. The resulting two parts of the tour are reconnected by two other edges to obtain a new solution. Later on, this idea has been extended to 3-opt [14] where three connections in a tour are first deleted, and then the best possible reconnection of the network is taken as a new solution. Lin and Kernighan [13] extended the idea to more complex neigh-bourhoods by making the number of performed 2-opt and 3-opt steps adaptive. Nowadays, variants of these seminal algorithms represent the state-of-the-art in heuristic TSP optimizers.

Among others, memetic algorithms and subpath ejection chain procedures have shown to be competitive alternatives, with hybrid approaches still being investigated today. In the bio-inspired memetic algorithms for the TSP problem (see [16] for an overview) information about subtours is combined to form new tours via 'crossover operators'. Additionally, tours are modified via 'mutation operators', to introduce new subtours. The idea behind the subpath ejection chain procedures is that in a first step a dislocation is created that requires further change. In subsequent steps, the task is to restore the system. It has been shown that the neighbourhoods investigated by the ejection chain procedures form supersets of those generated by the Lin-Kernighan heuristic [8].

In contrast to the above-mentioned iterative and heuristic algorithms, Concorde [1] is an exact algorithm that has been successfully applied to TSP instances with up to 85,900 vertices. It follows a branch-and-cut scheme [17], embedding the cutting-plane algorithm within a branch-and-bound search. The branching steps create a search tree, with the original problem as the root node. By traversing the tree it is possible to establish that the leafs correspond to a set of subproblems that include every tour for our TSP.

2.1   Characterization of TSP Instances

The theoretical assessment of problem difficulty of a TSP instance at hand a-priori to optimization is usually hard if not impossible. Thus, research has fo-cussed on deriving and extracting problem properties, which characterize and relate to the hardness of TSP instances (e.g. [21,10,20]). We refer to these prop-erties as features in the following and provide an overview subsequently. Features that are based on knowledge of the optimal tour [22,11] cannot be used to characterize an instance a priori to optimization. They are not relevant in the context of this paper and thus are not discussed in detail.

An intuitive and considered feature is the number of nodes N of the TSP instance [21,10,20]). Kanda et al. [10] assume the number of edges to be important as well and introduce a set of features that are based on summary statistics of the edge cost distribution. We will use edge cost or edge weight as a synonym of distance between nodes in the following. The lowest, highest, mean and median edge cost are considered as well as the respective standard deviation and the sum of N edges with lowest edge cost values. Furthermore, the quantity of edges with costs lower than the mean or median edge cost is taken into account. Additional features are the number of modes of the edge cost distribution and related features such as the frequency of the modes and the mean of the modal values.

Smith-Miles et al. [20,21] list features that assume that the existence and number of node clusters affect the performance of TSP solvers. Derived features are the cluster ratio, i.e. the number of clusters divided by N, and the mean distances to the cluster centers. Uniformity of an instance is further reflected by the minimum, maximum, standard deviation and the coefficient of variation of the normalized nearest-neighbor distances (nnd) of each node. The outlier ratio, i.e. the number of outliers divided by N, and the number of nodes near the edge of the plane are additionally considered. The centroid together with the mean distance from the nodes to the centroid and the bounding box of the nodes reflect the 'spread' of the instance on the plane. The feature list is completed by the fraction of distinct distances, i.e. different distance levels, and the standard deviation of the distance matrix.

Note that in order to allow for a fair comparison of features across instances of different sizes N the features have to be normalized appropriately. This means that all distances and edge costs have to be divided by their total sum. Analogously, all quantities have to be expressed relatively to the corresponding maximum quantity. Ideally, all instances should be normalized to the domain $[0, 1]^2$ to get rid of scaling issues.

We will use the approximation ratio that an algorithm achieves for a given instance as the optimization accuracy. The approximation ratio is given by the relative error of the tour length resulting from 2-opt compared to the optimal tour length and is a classical measure in the field of approximation algorithms [23]. Based on the approximation ratio that the 2-opt algorithm achieves, we will classify TSP instances either as easy or hard. Afterwards, we will analyze the features of hard and easy instances.

3   Analysis of TSP Problem Difficulty

In this section, we analyze easy and hard instances for the TSP. We start by describing an evolutionary algorithm that we used to generate easy and hard instances. Later on, we characterize these instances by the different features

---

Algorithm 1. Generate a random TSP instance.

---

function RANDOMINSTANCE(*size*)
    for $i = 1 \rightarrow size$ **do**
        $instance[i, 1] \leftarrow \mathcal{U}(0, 1)$ ▷ Uniform random number between 0 and 1
        $instance[i, 2] \leftarrow \mathcal{U}(0, 1)$ ▷ Uniform random number between 0 and 1
    end for
    return *instance*
end function

Algorithm 2. EA for evolving problem easy and hard TSP instances

---

function EA(*popSize, instSize, generations, time_limit, digits, repetitions, type*)
    $poolSize \leftarrow \lfloor popSize/2 \rfloor$
    for $i = 1 \rightarrow popSize$ **do**
        $population[i] \leftarrow$ RANDOMINSTANCE(*instSize*)
    end for
    for $generation = 1 \rightarrow generations$ **do**
        for $k = 1 \rightarrow popSize$ **do**
            $fitness[k] \leftarrow$ COMPUTEFITNESS(*population[k], repetitions*)
        end for
        $matingPool \leftarrow$ CREATEMATINGPOOL(*poolSize, population, fitness*)
        $nextPopulation[1] \leftarrow population[$BESTOF(*fitness*)$]$ ▷ 1-elitism
        for $k = 2 \rightarrow popSize$ **do**
            $parent1 \leftarrow$ RANDOMELEMENT(*population*)
            $parent2 \leftarrow$ RANDOMELEMENT(*population*)
            $offspring \leftarrow$ UNIFORMMUTATION(UNIFORMCROSSOVER(*parent1, parent2*))
            $nextPopulation[k] \leftarrow$ ROUND(NORMALMUTATION(*offspring*))
        end for
        $population \leftarrow nextPopulation$
        if over time limit *time_limit* **then**
            return *population*
        end if
    end for
end function

---

that we analyzed and point out which features make a TSP instance difficult to be solved by 2-opt.

3.1   EA-Based Generation of Easy and Hard TSP Instances

As the aim is to identify the features that are crucial for predicting the hardness of instances for the 2-opt heuristic, a representative set of instances is required which consists of a wide range of difficulties. It turned out that the construction of such a set is not an easy task. The generation of instances in a random manner did not provide a sufficient spread with respect to the instance hardness. The same is true for instances contained in the TSPLIB [18] of moderate size, i.e. lower than 1000 nodes, for which, in addition, the number of instances is not high enough to provide an adequate data basis. Higher instance sizes were excluded

Algorithm 3. Compute Fitness

---

function COMPUTEFITNESS($instance, repetitions$)
    $optimalTourLength \leftarrow$ CONCORDE($instance$)
    for $j \leftarrow 1, repetitions$ **do**
        $twoOptTourLengths[j] \leftarrow$ TWOOPT($instance$)   ▷ Two Opt Tour length
    end for
    return $\frac{\text{MEAN}(twoOptTourLengths)}{optimalTourLenght}$
end function

Algorithm 4. Mating pool creation

---

function CREATEMATINGPOOL(poolSize, population, fitness)
    for $i = 1 \rightarrow poolSize$ **do**
        $matingPool[i]$
        $\leftarrow$ BETTEROF(RANDOMELEMENT($population$), RANDOMELEMENT($population$))
    end for
    return $matingPool$
end function

---

due to the large computational effort required for their analysis, especially the computation of the optimal tours.

Therefore, two sets of instances are constructed in the $[0, 1]$-plane, which focus on reflecting the extreme levels of difficulty. An evolutionary algorithm (EA) is used for this purpose (see Algs. 1 - 4 for a description), which can be parameterized such that its aim is to evolve instances that are either as easy or as hard as possible for a given instance size. The approach is conceptually similar to [21] but focusses on approximation quality rather than on the number of swaps. Since some features depend on equal distances between the cities, we opted to implement a rounding scheme in the mutation step to force all cities to lie on a predefined grid. Initial studies also showed, that a second mutation strategy was necesarry. "Local mutation" was achieved by adding a small normal pertubation to the location, "global mutation" was performed by replacing each coordinate of the city with a new uniform random value. This later step was performed with a very low probability. All parameters are given at the end of this section.

The fitness function to be optimized is chosen as the approximation quality of 2-opt, estimated by the arithmetic mean of the tour lengths of a fixed number of 2-opt runs, on a given instance divided by the optimal tour length which is calculated using Concorde [1]. In general other summary statistics instead of the arithmetic mean could be used as well such as the maximum or minimum approximation quality achieved. Note that randomness is only induced by varying the initial tour whereas the 2-opt algorithm is deterministic in always choosing the edge replacement resulting in the highest reduction of the current tour length. Depending on the type of instance that is desired, the BETTEROF and BESTOF operators are either chosen to minimze or maximize the approximation quality.

We use a 1-elitism strategy such that only the individual with the current best fitness value survives and will be contained in the next population. The population is completed by iteratively choosing two parents from the mating pool, applying uniform crossover, uniform and normal mutation and adding the offspring to the population. This procedure is repeated until the population size is reached. Two sequential mutation strategies enable small local as well as global structural changes of the offspring resulting from the crossover operation.

In the experiments 100 instances each for the two instance classes (easy, hard) with a fixed instance size of 100 are generated. The remaining parameters are set as follows: $popSize = 30$, $generations = 1500$, $time\_limit = 22h$, $uniformMutationRate = 0.001$, $normalMutationRate = 0.01$, $digits = 2$, and the standard deviation of the normal distribution used in the $normalMutation$ step equals $normalMutationSd = 0.025$. The parameter levels were chosen based on initial experiments. However, a matter of future research will be a systematic tuning of the EA parameters in order to check if the results can be significantly improved. The number of 2-opt repetitions for calculating the approximation quality is set to 500.

3.2   Characterization of the Generated Instances

The average approximation qualities and respective standard deviations of the evolved easy and hard instances are (1.032 ± 0.0041) and (1.177 ± 0.0044), i.e. for the easy instances the average tour length of the 2-opt is about three percent higher than the optimal tour. The corresponding value for the hard instances is 18 percent, which results in a sufficiently high performance discrepancy between the two evolved sets.

In Figure 1 three EA generated instances of both classes are shown together with the corresponding optimal tours computed by Concorde. The main visual observations can be summarized as follows:

- The distances of the cities on the optimal tour appear to be more uniform for the hard instances than it is the case for the easy ones. This is supported by Figure 2 that shows boxplots of the standard deviations of the edge weights on the optimal tour. There we see that respective standard deviations of the easy instances are roughly twice as high than for the hard instances.
- The optimal tours of the hard instances are more similar to a "U-shape" whereas the optimal tours of the easy instances rather match an "X-shape".
- It seems that the easy instances consist of many small clusters of cities whereas this is not the case for the hard instances up to the same extent.

3.3   Feature-Based Prediction of TSP Problem Hardness

A decision tree [3] is used to differentiate between the two instance classes. This leads to the following classification rule, which is based on the coefficient of variation of the nearest neighbor distances (CVND) and the highest edge cost value (HEC):

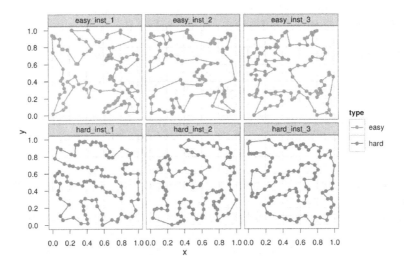

Fig. 1. Examples of the evolved instances of both types (easy, hard) including the optimal tours computed by Concorde

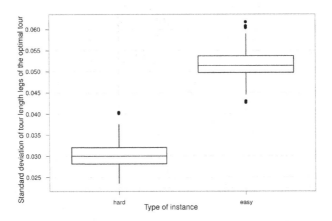

Fig. 2. Boxplots of the standard deviations of the tour length legs of the optimal tour, both for the evolved easy and hard instances

```
coefficient_of_variation_of_nnds>=0.5167739 → easy
coefficient_of_variation_of_nnds< 0.5167739
    ↪ highest_edge_cost>=0.000485 → easy
    ↪ highest_edge_cost< 0.000485 → hard
```

The ten-fold cross-validated error rate is very low, i.e. equals 3.02% so that an almost perfect classification of instances into the two classes based on only two features is possible. Basically, the classification relies on the single feature

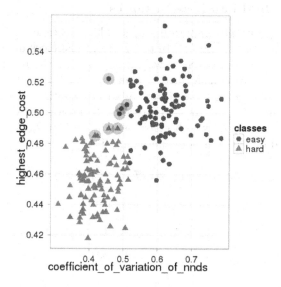

Fig. 3. Scatterplot of the features CVND and HEC for all evolved instances. Feature classes predicted by the decision tree are marked by different colors. Incorrectly classified instances during cross-validation are labeled by a grey circle.

CVND, which is shown in Figure 3. The two-dimensional feature space of the 200 instances is visualized using the features selected by the decision tree and labels the instances based on the presented classification rule. Incorrectly classified instances are marked by a grey circle.

It can be seen that the key feature for classifying the evolved instances into the two classes is CVND, which is perfectly in line with the exploratory analysis in Section 3.2. As the nearest neighbor of a node is the most likely candidate to be chosen in the course of the construction of the optimal tour, the CVND is highly correlated with the standard deviation of the distances on the optimal tour. In addition, the interpretation of the subrule regarding the feature HEC allows the same interpretation such that with increasing HEC value the less likely a uniform distribution of the edge weights becomes.

Classification rules generated for classifying easy and hard instances w.r.t. the Chained Lin-Kernighan (CLK) and Lin-Kernighan with Cluster Compensation (LKCC) algorithms in [21] also incorporate the feature CVND but the rule points into the opposite direction for CLK. Low CVND values characterize instances that are easy to solve for CLK. Although in [21] the approximation quality is measured by the number of swaps rather than by the resulting tour length relative to the optimal one, this is an interesting observation. In contrast, the results for LKCC are similar to the 2-opt rule, i.e. instances are classified as easy for high CVND values in combination with a low number of distinct distances.

3.4   Morphing Hard into Easy Instances

We are now in the position to separate easy and hard instances with the classification rule presented in Section 3.3. In this section, instances in between, i.e. of moderate difficulty, are considered as well. Starting from the result in [6] that a hard TSP instance can be transformed into an easy one by slight variation of the node locations, we studied the 'transition' of hard to easy instances by morphing a single hard instance I_H into an easy instance I_E by a convex combination of the points of both instances, which generates an instance I_{new} in between the original ones based on random point matching, i.e.

$$I_{new} = \alpha \cdot I_H + (1 - \alpha) \cdot I_E \quad \text{with} \quad \alpha \in [0, 1].$$

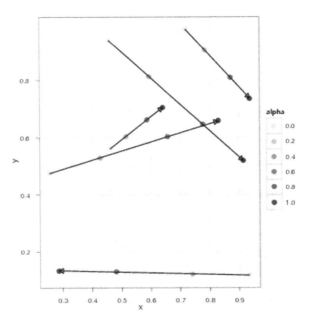

Fig. 4. Example: Morphing of one instance into a different instance for different α-levels of the convex combination. Nodes of the same color belong to the same instance.

An example of morphing is shown in Figure 4. The morphing strategy is applied to all possible combinations of single hard and easy instances of the two evolved instance sets using 51 levels of α ($\alpha \in \{0, 0.02, 0.04, ..., 1\}$). Each generated instance is characterized by the levels of the features discussed in Section 2.1. Thus, the changes of the feature levels with increasing α can be studied which is of interest as it should lead to an understanding of the influence of the different features on the approximation quality.

Figure 5 shows the approximation quality for the instances of all 10 000 morphing sequences for the various α levels in the bottom subfigure. Starting from a hard instance on the left side of the plot ($\alpha = 0$) the findings of [6] are confirmed. The approximation quality of 2-opt quickly increases, i.e. the value of *approx* decreases, with slight increases of α. Interestingly, roughly for $\approx 0.1 < \alpha < 0.9$ the approximation quality is quite stable, whereas it nonlinearly increases for $\alpha >= 0.9$. Additionally, the feature levels of the generated instances are visualized.

On the one hand features exist that do not show any systematic relationship with the approximation quality, e.g. all features related to the modes, lowest_edge_cost or cities_near_edge_ratio_one_percent. Other features exhibit a convex or concave shape obtaining similar values for the extreme α-levels and minimum resp. maximum value at $\alpha \approx 0.5$, e.g. mean_distance_to_cluster _centroids_0.1, cities_near_edge_ratio_five_percent or edges_lower_than_average _cost. The ratio of distinct_distances is only low for $\alpha \in \{0, 1\}$ and rather constant on a much higher level in between. A systematic nonlinear relationship can be detected for the features on which the classification rule is based, i.e. the CVND and HEC as well as for the mean_of_normalized_nnds and sum_of_ _lowest_edge_values.

In order to get a more accurate picture of the relationship between the approximation quality and the features a Multivariate Adaptive Regression Splines (MARS) [7] model is constructed in order to directly predict the expected approximation quality of 2-opt on a given instance based on the candidate features. Only a subset of the data is considered for the analysis (all morphed instances with $\alpha \in \{0, 0.2, ..., 1\}$), in order to adequately balance the data w.r.t. the various levels of 2-opt approximation quality. Had all 51 α-levels been used, the moderately difficult instances would have been massively overrepresented compared to the easy and hard instances.

We used a MARS model with interaction effects up to the second degree. Although this model class consists of an internal variable selection strategy a forward selection process together with a threefold cross-validation is applied in order to systematically focus on model validation and minimizing the root mean squared error (RMSE) of the prediction. Starting from an empty model, successively the feature that maximally reduces the RMSE is added to the existing model until the RMSE improvement falls below the threshold $t = 0.0005$. The results of the modeling procedure are shown in Table 1. The final root mean squared error is 0.0113.

The main and interaction effects of the model are visualized in Figure 6. Analogously to the classification rule the CVND is a key feature in predicting the

Table 1. Results of the MARS model

Feature list	RMSE
empty model	0.0484
+ coefficient of variation of nnds (CVND)	0.0246
+ distinct distances (DD)	0.0163
+ highest edge cost (HEC)	0.0119
+ sum of lowest edge values (SLEV)	0.0113

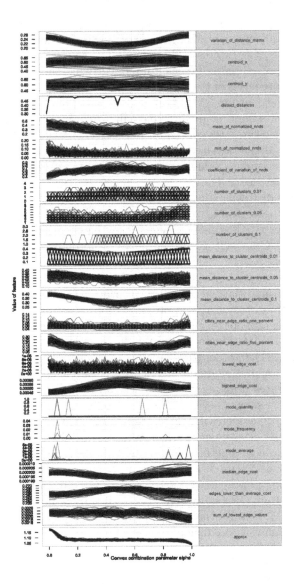

Fig. 5. Approximation quality and feature values for different α levels of all conducted morphing experiments. The annotations "_0.01","_0.05" and "_0.1" identify different levels of the reachability distance as a parameter of GDBSCAN [19] used for clustering.

approximation quality. From the plots, it is obvious that high values of the approximation ratio only occur for very low CVND values. However, the main effect in this case does not reflect the classification rule generated before. The HEC is only part of an interaction with the remaining features. The problem hardness tends to be higher for low HEC values combined with high DD values and low CVND values. In addition, problem hardness decreases with lower SLEV values.

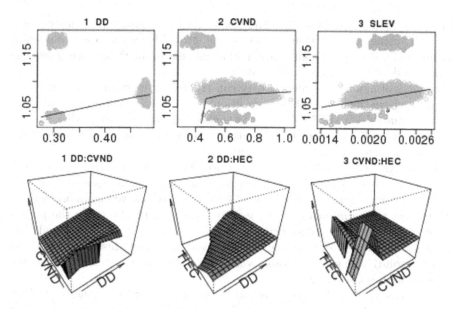

Fig. 6. Effect plots resulting from the MARS model. Main effects are plotted (top) together with the feature interactions (bottom). Coefficient of variation of nnds (CVND), highest edge cost (HEC), distinct distances (DD) and sum of lowest edge values (SLEV) were selected. In all plots the last axis reflects the approximation quality of 2-opt. The plots are generated by setting the non-displayed variables to their median values.

The selection of the DD feature seems to be somewhat arbitrary in that the slope of the effect line could just as well have been negative from visual analysis. Summarizing, the interpretation of the model results is not straightforward but nevertheless a quite accurate prediction of 2-opt approximation quality on the considered instances is achieved.

4   Summary and Outlook

In this paper we investigated concepts to predict TSP problem hardness for 2-opt based local search strategies on the basis of experimental features that characterize the properties of a TSP instance. A crucial aspect was the generation of a representative instance set as a basis for the analysis. This turned out to be far from straightforward. Therefore it was only possible to generate very hard and very easy instances using sophisticated (evolutionary) strategies. Summarizing, we managed to generate classes of easy and hard instances for which we are able to predict the correct instance class based on the corresponding feature levels with only marginal errors. The coefficient of variation of nearest neighbor distances was identified as the key feature for differentiating between hard and easy instances, and the results are supported by exploratory analysis of the evolved instances and the respective optimal tours. However, it should be noted that most probably not the whole space of possible hard instances is covered by using our evolutionary method, i.e. probably only a subset of possible characteristics or feature combinations that make a problem hard for 2-opt can be identified by the applied methodology.

Instances of moderate difficulty were constructed by morphing hard into easy instances where the effects of the transition on the corresponding feature levels could be studied. A MARS model was successfully applied to predict the approximation quality of 2-opt based on the features of an adequate subset of the generated instances with very high accuracy.

The analysis offers promising perspectives for further research. A generalization of the results to other instance sizes should be addressed as well as a systematic comparison to other local and global search as well as hybrid solvers with respect to the influence of the feature levels of an instance on the performance of the respective algorithms. However, it has to be kept in mind that the computational effort intensely increases with increasing instance size as the optimum solution, e.g. computable via Concorde, is required to calculate the approximation quality of 2-opt.

The extension of the feature set is another relevant topic that could be studied. For example, in the context of benchmarking algorithms on continuous black-box optimization problems the extraction of problem properties that might influence algorithm performance is an important and current focus of research, denoted as exploratory landscape analysis (ELA, [15]). Although the TSP search space is not continuous, e.g. the ELA feature that tries to capture the number of modes of an empirical density function, could be transferred to the problem at hand. Furthermore, possible advantages of sophisticated point matching strategies during the morphing of hard into easy instances can be investigated. Finally, it is open how representative the generated instances are for real-world TSP instances. It is therefore very desirable to create a much larger pool of small to medium sized, real-world, TSP instances for comparison experiments.

Acknowledgements. This work was partly supported by the Collaborative Research Center SFB 823, the Graduate School of Energy Efficient Production and Logistics and the Research Training Group "Statistical Modelling" of the German Research Foundation.

References

1. Applegate, D., Cook, W.J., Dash, S., Rohe, A.: Solution of a min-max vehicle routing problem. Informs Journal on Computing 14(2), 132–143 (2002)
2. Arora, S.: Polynomial time approximation schemes for euclidean traveling salesman and other geometric problems. J. ACM 45(5), 753–782 (1998)
3. Breiman, L., Friedman, J.H., Olshen, R.A., Stone, C.J.: Classification and Regression Trees. Wadsworth, Belmont (1984)
4. Chandra, B., Karloff, H.J., Tovey, C.A.: New results on the old k-Opt algorithm for the traveling salesman problem. SIAM J. Comput. 28(6), 1998–2029 (1999)
5. Croes, G.A.: A method for solving traveling-salesman problems. Operations Research 6(6), 791–812 (1958)

6. Englert, M., Röglin, H., Vöcking, B.: Worst case and probabilistic analysis of the 2-opt algorithm for the tsp: extended abstract. In: Bansal, N., Pruhs, K., Stein, C. (eds.) SODA, pp. 1295–1304. SIAM (2007)

7. Friedman, J.H.: Multivariate adaptive regression splines. Annals of Statistics 19(1), 1–67 (1991)

8. Glover, F.: Ejection chains, reference structures and alternating path methods for traveling salesman problems. Discrete Applied Mathematics 65(1-3), 223–253 (1996)

9. Johnson, D.S., McGeoch, L.A.: The traveling salesman problem: A case study in local optimization. In: Aarts, E.H.L., Lenstra, J.K. (eds.) Local Search in Combinatorial Optimization. Wiley (1997)

10. Kanda, J., Carvalho, A., Hruschka, E., Soares, C.: Selection of algorithms to solve traveling salesman problems using meta-learning. Hybrid Intelligent Systems 8, 117–128 (2011)

11. Kilby, P., Slaney, J., Walsh, T.: The backbone of the travelling salesperson. In: Proc, of the 19th International Joint Conference on Artificial intelligence, IJCAI 2005, pp. 175–180. Morgan Kaufmann Publishers Inc., San Francisco (2005)

12. Kötzing, T., Neumann, F., Röglin, H., Witt, C.: Theoretical Properties of Two ACO Approaches for the Traveling Salesman Problem. In: Dorigo, M., Birattari, M., Di Caro, G.A., Doursat, R., Engelbrecht, A.P., Floreano, D., Gambardella, L.M., Groß, R., Şahin, E., Sayama, H., Stützle, T. (eds.) ANTS 2010. LNCS, vol. 6234, pp. 324–335. Springer, Heidelberg (2010)

13. Lin, S., Kernighan, B.: An effective heuristic algorithm for the traveling salesman problem. Operations Research 21, 498–516 (1973)

14. Lin, S.: Computer solutions of the travelling salesman problem. Bell Systems Technical Journal 44(10), 2245–2269 (1965)

15. Mersmann, O., Bischl, B., Trautmann, H., Preuss, M., Weihs, C., Rudolph, G.: Exploratory landscape analysis. In: Proc. of the 13th Annual Conference on Genetic and Evolutionary Computation, GECCO 2011, pp. 829–836. ACM, New York (2011)

16. Merz, P., Freisleben, B.: Memetic algorithms for the traveling salesman problem. Complex Systems 13(4), 297–345 (2001)

17. Padberg, M., Rinaldi, G.: A branch-and-cut algorithm for the resolution of large-scale symmetric traveling salesman problems. SIAMR 33(1), 60–100 (1991)

18. Reinelt, G.: Tsplib - a traveling salesman problem library. ORSA Journal on Computing 3(4), 376–384 (1991)

19. Sander, J., Ester, M., Kriegel, H., Xu, X.: Density-based clustering in spatial databases: The algorithm gdbscan and its applications. Data Mining and Knowledge Discovery 2(2), 169–194 (1998)

20. Smith-Miles, K., van Hemert, J.: Discovering the suitability of optimisation algorithms by learning from evolved instances. Annals of Mathematics and Artificial Intelligence (2011) (forthcoming)

21. Smith-Miles, K., van Hemert, J., Lim, X.Y.: Understanding TSP Difficulty by Learning from Evolved Instances. In: Blum, C., Battiti, R. (eds.) LION 4. LNCS, vol. 6073, pp. 266–280. Springer, Heidelberg (2010)

22. Stadler, P.F., Schnabl, W.: The Landscape of the Traveling Salesman Problem. Physics Letters A 161, 337–344 (1992)

23. Vazirani, V.V.: Approximation algorithms. Springer (2001)

Evaluating Tree-Decomposition Based Algorithms for Answer Set Programming

Michael Morak, Nysret Musliu, Reinhard Pichler,
Stefan Rümmele, and Stefan Woltran

Institute of Information Systems, Vienna University of Technology
{surname}@dbai.tuwien.ac.at

Abstract. A promising approach to tackle intractable problems is given by a combination of decomposition methods with dynamic algorithms. One such decomposition concept is tree decomposition. However, several heuristics for obtaining a tree decomposition exist and, moreover, also the subsequent dynamic algorithm can be laid out differently. In this paper, we provide an experimental evaluation of this combined approach when applied to reasoning problems in propositional answer set programming. More specifically, we analyze the performance of three different heuristics and two different dynamic algorithms, an existing standard version and a recently proposed algorithm based on a more involved data structure, but which provides better theoretical runtime. The results suggest that a suitable combination of the tree decomposition heuristics and the dynamic algorithm has to be chosen carefully. In particular, we observed that the performance of the dynamic algorithm highly depends on certain features (besides treewidth) of the provided tree decomposition. Based on this observation we apply supervised machine learning techniques to automatically select the dynamic algorithm depending on the features of the input tree decomposition.

1 Introduction

Many instances of constraint satisfaction problems and other NP-hard problems can be solved in polynomial time if their treewidth is bounded by a constant. This suggests two-phased implementations where first a tree decomposition [25] of the given problem is obtained which is then used in the second phase to solve the problem under consideration by a (usually, dynamic) algorithm traversing the tree decomposition. The running time of the dynamic algorithm[1] mainly depends on the width of the provided tree decomposition. Hence, the overall process performs well on instances of small treewidth (formal definitions of tree decompositions and treewidth are given in Section 2), but can also be used in general in case the running time for finding a tree decomposition remains low. Thus, instead of complete methods for finding a tree decomposition, heuristic methods are often employed. In other words, to gain a good performance for this combined tree-decomposition dynamic-algorithm (TDDA, in the following) approach we require efficient tree decomposition techniques which still provide results for which the running time of the dynamic algorithm is feasible.

---

[1] We use – throughout the paper – the term "dynamic algorithm" as a synonym for "dynamic *programming* algorithm" to avoid confusion with the concept of Answer-Set *programming*.

Y. Hamadi and M. Schoenauer (Eds.): LION 6, LNCS 7219, pp. 130–144, 2012.
© Springer-Verlag Berlin Heidelberg 2012

Tree-decomposition based algorithms have been used in several applications including probabilistic networks [18] or constraint satisfaction problems such as MAX-SAT [17]. The application area we shall focus on here is propositional Answer-Set Programming (ASP, for short) [20,23] which is nowadays a well acknowledged paradigm for declarative problem solving with many successful applications in the areas of AI and KR.[2] The problem of deciding ASP consistency (i.e. whether a logic program has at least one answer set) is Σ_2^P-complete in general but has been shown tractable [12] for programs of bounded treewidth. In this paper, we consider a certain subclass of programs, namely head-cycle free programs (for more formal definitions, we again refer to Section 2); for such programs the consistency problem is NP-complete.

Let us illustrate here the functioning of ASP on a typical example. Consider the problem of 3-colorability of an (undirected) graph and suppose the vertices of a graph are given via the predicate vertex($\cdot$) and its edges via the predicate edge($\cdot, \cdot$). We employ a disjunctive rule to guess a color for each node in the graph, and then check in the remaining three rules whether adjacent vertices have indeed different colors:

$$r(X) \vee g(X) \vee b(X) \leftarrow \text{vertex}(X);$$
$$\bot \leftarrow r(X), r(Y), \text{edge}(X, Y);$$
$$\bot \leftarrow g(X), g(Y), \text{edge}(X, Y);$$
$$\bot \leftarrow b(X), b(Y), \text{edge}(X, Y);$$

Assume a simple input database with facts vertex(a), vertex(b) and edge(a, b). The above program (together with the input database) yields six answer sets. In fact, the above program is head-cycle free. Many NP-complete problems can be succinctly represented using head-cycle free programs (in particular, the disjunction allows for a direct representation of the guess; in our example the guess of a coloring); see [19] (Section 3) for a collection of problems which can be represented with head-cycle free programs as opposed to problems which require the full power of ASP. However, the above program contains variables and thus has to be grounded yet. So-called grounders turn such programs into variable-free (i.e., propositional) ones which are then fed into ASP-solvers. The algorithms discussed in this paper work on variable-free programs. We emphasize at this point a valuable side-effect. For our example above, it turns out that if the input graph has small treewidth, then the grounded variable-free program has small treewidth as well (see Section 2 for a continuation of the example). This not only holds for the encoding of the 3-colorability problem, but for many other ASP programs (in particular, programs without recursive rules). Thus the class of propositional programs with low treewidth is indeed important also in the context of ASP with variables.

A dynamic algorithm for general propositional ASP has already been presented in [15]. Recently, a new algorithm was proposed for the fragment of head-cycle free programs [21]. Their main differences are as follows: the algorithm from [15] is based on ideas from dynamic SAT algorithms [26] and explicitly takes care of the minimality checks following the standard definition of answer sets; thus it requires double-exponential time in the width of the provided tree decomposition. The algorithm

---

[2] See http://www.cs.uni-potsdam.de/~torsten/asp/ for a collection.

proposed in [21] follows a more involved characterization [5] which applies to head-cycle free programs and thus calls for a more complex data structure and operations. However, it runs in single-exponential time wrt. the width of the provided tree decomposition. Both algorithms have been integrated into a novel TDDA system for ASP, which we call dynASP[3]. For the tree-decomposition phase, dynASP offers three different heuristics, namely Maximum Cardinality Search (MCS) [29], Min-Fill and Minimum Degree (see [7] for a survey on such heuristics). According to [11], the min-fill heuristic usually produces tree decompositions of lower width than the other heuristics.

By the above considerations, one would naturally expect that computing a tree decomposition with the min-fill heuristic (which usually yields the lowest width) and applying the dedicated dynamic algorithm from [21] for head-cycle free ASPs (which is single-exponential wrt. to the width of the tree decomposition) yields the best two-phased algorithm for head-cycle free ASPs. Surprisingly, extensive testing with our dynASP system has by no means confirmed these expectations: First, the TDDA algorithm is not always most efficient when the best heuristic for tree decomposition is used. Second, the specialized algorithm for head-cycle free programs does not always perform better than the general algorithm, although the worst-case running time of the latter is double-exponential in the treewidth while the running time of the former is only single-exponential.

The goal of this paper is to get a deeper understanding of the interplay between tree decompositions and dynamic algorithms and to arrive at an optimal configuration of the two-phased dynamic algorithm. The above mentioned experimental results suggest that the width of the tree decomposition is not the only significant parameter for efficiency of our dynamic algorithms. Therefore, we identify other important features of tree decompositions that influence the running time of the dynamic algorithms. Based on these observations, we propose the application of machine learning techniques to automatically select the best dynamic algorithm for the given input instance. We successfully apply classification techniques for algorithm selection in this domain. Additionally, we exploit regression techniques that are used to predict the runtime of our dynamic algorithms based on input instance features.

Note that the proposed features of tree decompositions are independent of the application domain of ASP. We therefore expect that our insights into the influence of various characteristics of tree decompositions on the performance of TDDAs are generally applicable to tree-decomposition based algorithms and that they are by no means restricted to ASPs. The same holds true for the methodology developed here in order to arrive at an optimal algorithm configuration of such two-phased algorithms.

2 Preliminaries

Answer Set Programming. A (propositional) disjunctive logic program (program, for short) is a pair $\Pi = (\mathcal{A}, \mathcal{R})$, where $\mathcal{A}$ is a set of propositional atoms and $\mathcal{R}$ is a set of rules of the form:

---

[3] A preliminary version of this system has been presented in [22], see
http://dbai.tuwien.ac.at/proj/dynasp.

$$a_1 \vee \cdots \vee a_l \leftarrow a_{l+1}, \ldots, a_m, \neg a_{m+1}, \ldots, \neg a_n \qquad (1)$$

where "$\neg$" is default negation[4] $n \geq 1$, $n \geq m \geq l$ and $a_i \in \mathcal{A}$ for all $1 \leq i \leq n$. A rule $r \in \mathcal{R}$ of the form (1) consists of a head $H(r) = \{a_1, \ldots, a_l\}$ and a body $B(r) = B^+(r) \cup B^-(r)$, given by $B^+(r) = \{a_{l+1}, \ldots, a_m\}$ and $B^-(r) = \{a_{m+1}, \ldots, a_n\}$. A set $M \subseteq \mathcal{A}$ is a called a model of r, if $B^+(r) \subseteq M \wedge B^-(r) \cap M = \emptyset$ implies that $H(r) \cap M \neq \emptyset$. We denote the set of models of r by $Mod(r)$ and the models of a program $\Pi = (\mathcal{A}, \mathcal{R})$ are given by $Mod(\Pi) = \bigcap_{r \in \mathcal{R}} Mod(r)$.

The reduct Π^I of a program Π w.r.t. an interpretation $I \subseteq \mathcal{A}$ is given by $(\mathcal{A}, \{r^I : r \in \mathcal{R}, B^-(r) \cap I = \emptyset)\})$, where r^I is r without the negative body, i.e., $H(r^I) = H(r)$, $B^+(r^I) = B^+(r)$, and $B^-(r^I) = \emptyset$. Following [10], $M \subseteq \mathcal{A}$ is an *answer set* of a program $\Pi = (\mathcal{A}, \mathcal{R})$ if $M \in' Mod(\Pi)$ and for no $N \subset M$, $N \in Mod(\Pi^M)$.

We consider here the class of *head-cycle free programs* (HCFPs) as introduced in [5]. We first recall the concept of *(positive) dependency graphs*. A dependency graph of a program $\Pi = (\mathcal{A}, \mathcal{R})$ is given by $\mathcal{G} = (V, E)$, where $V = \mathcal{A}$ and $E = \{(p, q) \mid r \in \mathcal{R}, p \in B^+(r), q \in H(r)\}$. A program $\Pi = (\mathcal{A}, \mathcal{R})$ is called head-cycle free if its dependency graph does not contain a directed cycle going through two different atoms which jointly occur in the head of a rule in $\mathcal{R}$.

Example 1. We provide the fully instantiated (i.e. ground) version of our introductory example from Section 1, which solves the 3-colorability for the given input database vertex(a), vertex(b) and edge(a, b), yielding five rules (taking straight forward simplifications as performed by state-of-the-art grounders into account):

$r1 : r(a) \vee g(a) \vee b(a) \leftarrow \top;$ $r2 : r(b) \vee g(b) \vee b(b) \leftarrow \top;$
$r3 : \bot \leftarrow r(a), r(b);$ $r4 : \bot \leftarrow g(a), g(b);$
$r5 : \bot \leftarrow b(a), b(b);$

Tree Decomposition and Treewidth. A *tree decomposition* of a graph $\mathcal{G} = (V, E)$ is a pair $\mathcal{T} = (T, \chi)$, where T is a tree and χ maps each node t of T (we use $t \in T$ as a shorthand below) to a *bag* $\chi(t) \subseteq V$, such that (1) for each $v \in V$, there is a $t \in T$, s.t. $v \in \chi(t)$; (2) for each $(v, w) \in E$, there is a $t \in T$, s.t. $\{v, w\} \subseteq \chi(t)$; (3) for each $r, s, t \in T$, s.t. s lies on the path from r to t, $\chi(r) \cap \chi(t) \subseteq \chi(s)$.

A tree decomposition (T, χ) is called *normalized* (or *nice*) [16], if (1) each $t \in T$ has ≤ 2 children; (2) for each $t \in T$ with two children r and s, $\chi(t) = \chi(r) = \chi(s)$; and (3) for each $t \in T$ with one child s, $\chi(t)$ and $\chi(s)$ differ in exactly one element, i.e. $|\chi(t) \Delta \chi(s)| = 1$.

The *width* of a tree decomposition is defined as the cardinality of its largest bag minus one. Every tree decomposition can be normalized in linear time without increasing the width [16]. The *treewidth* of a graph $\mathcal{G}$, denoted by $tw(\mathcal{G})$, is the minimum width over all tree decompositions of $\mathcal{G}$.

For a given graph and integer k, deciding whether the graph has treewidth at most k is NP-complete [2]. For computing tree decompositions, different complete [27,11,3] and heuristic methods have been proposed in the literature. Heuristic techniques are mainly

---

[4] We omit strong negation as considered in [5]; our results easily extend to programs with strong negation.

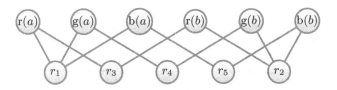

Fig. 1. The incidence graph of the ground program of Example 1

based on searching for a good elimination ordering of graph nodes. Several heuristics that run in polynomial time have been proposed for finding a good elimination ordering of nodes. These heuristics select the ordering of nodes based on different criteria, such as the degree of the nodes, the number of edges to be added to make the node simplicial (a node is simplicial if its neighbors form a clique) etc. We briefly mention three of them: (i) Maximum Cardinality Search (MCS) [29] initially selects a random vertex of the graph to be the first vertex in the elimination ordering (the elimination ordering is constructed from right to left). The next vertex will be picked such that it has the highest connectivity with the vertices previously selected in the elimination ordering. The ties are broken randomly. MCS repeats this process iteratively until all vertices are selected. (ii) The min-fill heuristic first picks the vertex which adds the smallest number of edges when eliminated (the ties are broken randomly). The selected vertex is made simplicial and it is eliminated from the graph. The next vertex in the ordering will be any vertex that adds the minimum number of edges when eliminated from the graph. This process is repeated iteratively until the whole elimination ordering is constructed. (iii) The minimum degree heuristic picks first the vertex with the minimum degree. The selected vertex is made simplicial and it is removed from the graph. Further, the vertex that has the minimum number of unselected neighbors will be chosen as the next node in the elimination ordering. This process is repeated iteratively. MCS, min-fill, and min-degree heuristics run in polynomial time and usually produce a tree decomposition of reasonable width. For other types of heuristics and metaheuristic techniques based on the elimination ordering of nodes, see [7].

Tree Decompositions of Logic Programs. To build tree decompositions for programs, we use incidence graphs.[5] Thus, for program $\Pi = (\mathcal{A}, \mathcal{R})$, such a graph is given by $\mathcal{G} = (V, E)$, where $V = \mathcal{A} \cup \mathcal{R}$ and E is the set of all pairs (a, r) with an atom $a \in \mathcal{A}$ appearing in a rule $r \in \mathcal{R}$. Thus the resulting graphs are bipartite.

For normalized tree decompositions of programs, we thus distinguish between six types of nodes: *leaf* (L), *join* or *branch* (B), *atom introduction* (AI), *atom removal* (AR), *rule introduction* (RI), and *rule removal* (RR) node. The last four types will be often augmented with the element e (either an atom or a rule) which is removed or added compared to the bag of the child node.

Figures 1 and 2 show the incidence graph of Example 1 and a corresponding tree decomposition.

---

[5] See [26] for justifications why incidence graphs are favorable over other types of graphs.

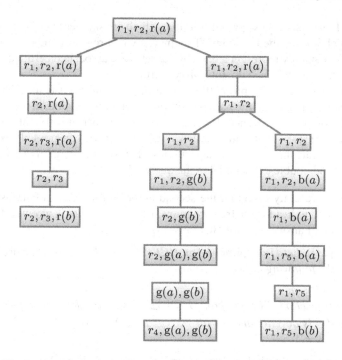

Fig. 2. A normalized tree decomposition of the graph shown in Figure 1

3   Dynamic Algorithms for ASP

Tree-decomposition based dynamic algorithms start at the leaf nodes and traverse the tree to the root. Thereby, at each node a set of partial solutions is generated by taking those solutions into account that have been computed for the child nodes. The most difficult part in constructing such an algorithm is to identify an appropriate data structure to represent the partial solutions at each node: on the one hand, this data structure must contain sufficient information so as to compute the representation of the partial solutions at each node from the corresponding representation at the child node(s). On the other hand, the size of the data structure must only depend on the size of the bag (and not on the size of the entire answer set program).

In this section we review two completely different realizations of this data structure, leading to algorithms which we will call Dyn-ASP1 and Dyn-ASP2.

Dyn-ASP1. The first algorithm was presented in [15]. It was proposed for propositional disjunctive programs Π which are not necessarily head-cycle free. Its data structure, called tree interpretation, follows very closely the characterization of answer sets presented in Section 2. A tree interpretation for tree decomposition $\mathcal{T}$ is a tuple $(t, M, \mathcal{C})$, where t is a node of $\mathcal{T}$, $M \subseteq \chi(t)$ is called assignment, and $\mathcal{C} \subseteq 2^{\chi(t)}$ is called certificate. The idea is that M represents a partial solution limited to what is visible in the bag $\chi(t)$. That means it contains parts of a final answer set as well as all those rules which are already satisfied. The certificate $\mathcal{C}$ takes care of the minimality criteria for answer sets. It is a list of those partial solutions which are smaller than M together with the rules which are satisfied by them. This means when reaching the root node of $\mathcal{T}$, assignment M can only represent a real answer set if the associated certificate is empty or contains only entries which do not satisfy all rules.

It turns out that due to the properties of tree decompositions it is indeed enough to store only the information of the partial solution which is still visible in the current bag of the tree decomposition. Hence, for each node the number of different assignments M is limited single exponential in the treewidth. Together with the possible exponential size of the certificate this leads to an algorithm with a worst case running time linear in the input size and double exponential in the treewidth.

Dyn-ASP2. We recently proposed the second algorithm in [21]. In contrast to Dyn-ASP1 it is limited to head-cycle free programs. Its data structure is motivated by a new characterization of answer sets for HCFPs:

Theorem 1 ([21]). *Let $\Pi = (\mathcal{A}, \mathcal{R})$ be an HCFP. Then, $M \subseteq \mathcal{A}$ is an answer set of Π if and only if the following holds:*

- *$M \in Mod(\Pi)$, and*
- *there exists a set $\rho \subseteq \mathcal{R}$ such that, $M \subseteq \bigcup_{r \in \rho} H(r)$; the derivation graph induced by M and ρ is acyclic; and for all $r \in \rho$: $B^+(r) \subseteq M$, $B^-(r) \cap M = \emptyset$, and $|H(r) \cap M| = 1$.*

Here the derivation graph induced by M and ρ is given by $V = M \cup \rho$ and E is the transitive closure of the edge set $E' = \{(b, r) : r \in \rho, b \in B^+(r) \cap M\} \cup \{(r, a) : r \in \rho, a \in H(r) \cap M\}$.

Hence, the data structure used in Dyn-ASP2 is a tuple (G, S), where G is a derivation graph (extended by a special node due to technical reasons) and S is the set of satisfied rules used to test the first condition in Theorem 1. Again it is enough to limit G and S to the elements of the current bag $\chi(t)$. Therefore the number of possible tuples (G, S) in each node is at most single exponential in the treewidth. This leads to an algorithm with a worst case running time linear in the input size and single exponential in the treewidth.

4   Evaluation of Tree Decompositions for ASP

In this section we give an extensive evaluation of dynamic algorithms based on tree decompositions for solving benchmark problems in answer set programming. In Figure 3 our solver based on tree decompositions and dynamic algorithms is presented, where Dyn-ASP1 and Dyn-ASP2 refers to the two algorithms described Section 3. Moreover, note that tree decompositions have to be normalized to be amenable to the two dynamic algorithms. The efficiency of our solver depends on the tree decomposition module and the applied dynamic algorithm. Regarding the tree decomposition we evaluated three heuristics which produce different tree decompositions. Furthermore, we analyzed the impact of tree decomposition features on the efficiency of the dynamic algorithms. Observing that neither dynamic algorithm dominates the other on all instances, we propose an automated selection of a dynamic algorithm during the solving process based on the features of the produced tree decomposition.

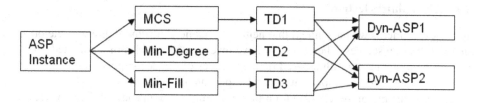

Fig. 3. Architecture of the TDDA-based ASP solver

Benchmark Description: To identify tree decomposition features that impact the runtime of our Dyn-ASP1 and Dyn-ASP2, different logic programs were generated and different tree decompositions were computed for these programs.

Programs were generated in two ways: Firstly, by generating a random SAT instance using MKCNF[6]. These CNF formulas were then encoded as a logic program and passed to the dynASP program. MKCNF was called with the following parameters: Number of clauses ranging from 150 to 300, clause-size ranging from 3 to 13 and number of variables calculated by 10 × number of clauses × clause-size.

The second method used for program generation closely follows the one described in [31]. For rule-length n, from a set $\mathcal{A}$ of atoms, a head atom and $n - 1$ body atoms are randomly selected. Each of the body atoms is negated with a probability of 0.5. Here the rule-length ranges from 3 to 7 and the number of rules ranges from 20 to 50. The number of atoms is always $\frac{1}{5}$ of the number of rules, which is, according to [31], a hard region for current logic program solvers.

For each of these programs, three different tree decompositions are computed using the three heuristics described below. Each of these tree decompositions is then normalized, as both algorithms currently only handle "nice" tree compositions.

Applied Tree-Decomposition Algorithms: As we described in Section 2 different methods have been proposed in the literature for constructing of tree decompositions with small width. Although complete methods give the exact treewidth, they can be used only for small graphs, and were not applicable for our problems which contains up to 20000 nodes. Therefore, we selected three heuristic methods (MCS, min-fill, and min-degree) which give a reasonable width in a very short amount of time. We have also considered using and developing new metaheuristic techniques. Although such an approach slightly improves the treewidth produced by the previous three heuristics, they are far less efficient compared to the original variants. In our experiments we have observed that a slightly improved treewidth does not have a significant impact on the efficiency of the dynamic algorithm for our problem domain and therefore we decided to use the three heuristics directly. We initially used an implementation of these heuristics available in a state-of-the-art libraries [8] for tree/hypertree decomposition. Further, we implemented new data structures that store additional information about vertices, their adjacent edges and neighbors to find the next node in the ordering faster. With these new data structures the performance of Min-fill and MCS heuristics was improved by factor 2–3.

---

[6] ftp://dimacs.rutgers.edu/pub/challenge/satisfiability/
   contributed/UCSC

4.1 Algorithm Selection

In our experiments we have noted that neither dynamic algorithm dominates the other in all problem instances. Therefore, we have investigated the idea of automated selection of the dynamic algorithm based on the features of the decomposition. Automated algorithm selection is an important research topic and has been investigated by many researchers in the literature (c.f. [28] for a survey). However, to the best of our knowledge, algorithm selection has not yet been investigated for tree decompositions.

To achieve our goal we identified important features of tree decompositions and applied supervised machine learning techniques to select the algorithm that should be used on the particular tree decomposition. We have provided training sets to the machine learning algorithms and analyzed the performance of different variants of these algorithms on the testing set. The detailed performance results of the machine learning algorithm are presented in the next section.

Structural Properties of Tree Decompositions: For every tree decomposition, a number of features are calculated to identify the properties that make them particularly suitable for one of the algorithms (or conversely, particularly unsuitable). The following features (besides treewidth) were used:

– Percentage of join nodes in the normalized tree decomposition (*jpct*)
– Percentage of join nodes in the non-normalized decomposition (*tdbranchpct*)
– Percentage of leaf nodes in the non-normalized decomposition (*tdleafpct*)
– Average distance between two join nodes in the decomposition (*jjdist*)
– Relative size increase of the decomposition during normalization (*nsizeinc*)
– Average bag size of join nodes (*jwidth*)
– Relative size of the tree decomposition (i.e. number of tree nodes) compared to the size (vertices + edges) of the incidence graph (*reltdsize*)

We note that our data set also includes features of the graph from which the tree decomposition is constructed. These features include number of edges of the graph, number of vertices, minimum degree, maximum degree etc. Because the graph features had a minor impact on the machine learning algorithms, the discussion in this paper is concentrated on tree decomposition features.

Experiments: All experiments were performed on a 64bit Gentoo Linux machine with an Intel Core2Duo P9500 2.53GHz processor and 4GB of system RAM. For each generated head-cycle free logic program, 50 tree decompositions were computed with each of the three heuristics available. For each of these 150 decompositions, the two algorithms described in Section 3 were run in order to determine which one works best on the given tree decomposition. Thus, a tuple in the benchmark dataset consists of the generated program and a tree decomposition, and for each tuple it is stored which algorithm performed better and its corresponding runtime.

Based on this generated dataset, using the WEKA toolkit [13], a machine learning approach was used to try to automatically select the best dynamic algorithm for an already computed tree decomposition. Trying to select the best combination of both tree decomposition heuristic and dynamic algorithm unfortunately seems impractical, as the

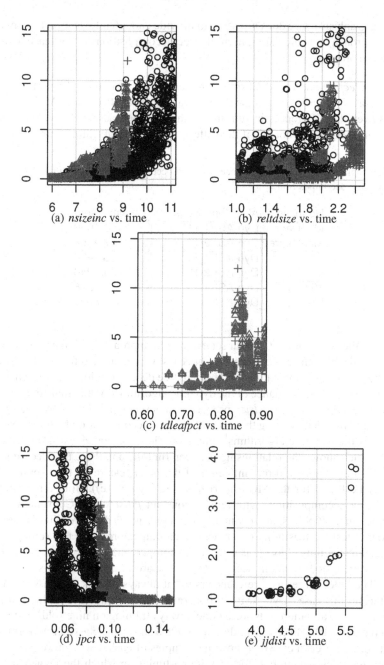

Fig. 4. Every benchmark instance (i.e. each calculated tree decomposition) contributes one data-point to the plots above. Usage of the MCS, Min-Degree and Min-Fill heuristics are represented by black circles, grey triangles and light grey crosses respectively. Note that the latter two almost always overlap. The Y scale measures overall running time of the best algorithm in seconds. Plots (a)–(d) use the full benchmark set, (e) uses MKCNF 21000 300 7.

underlying graph structure does not seem to provide enough information for a machine learning approach and calculating multiple tree decompositions is not feasible, as it is an expensive process.

Algorithm selection based on classification techniques: Based on the performance of the two algorithms, each tuple in the dataset was either labelled "Dyn-ASP1" or "Dyn-ASP2". Given the differences in runtime as shown in extracts in Table 1, the overall runtime can be improved notably if the better-performing algorithm is run.

Table 1. Exemplary performance differences that can occur in our two algorithms when working on the same tree decomposition

Heuristic	Algorithm	TD width	Runtime (sec)
Min-Degree	Dyn-ASP1	11	53.1629
Min-Degree	Dyn-ASP2	11	7.4058
MCS	Dyn-ASP1	10	6.2420
MCS	Dyn-ASP2	10	268.2940
Min-Fill	Dyn-ASP1	10	9.8325
Min-Fill	Dyn-ASP2	10	2.6030

By using the well-known CFS subset evaluation approach implemented in WEKA (see [14] for details), the *jjdist* and *jpct* properties were identified to correlate strongly with the best algorithm, indicating that they are tree decomposition features which have a high impact on the performance of the dynamic algorithms. When ranked by information gain (see Table 2), the *reltdsize* property ranks second, followed by *tdleafpct, td-branchpct* and *jwidth* indicating that all of these tree decomposition features bear some influence on the dynamic algorithms' runtimes. These outcomes can also be seen in Figure 4, which shows the relationship between runtime and these tree decomposition properties. Interestingly, a direct influence of the *jjdist* feature on the overall running could only be found for the MCS heuristics (see Figure 4(e)). Both other heuristics produced tree decompositions with almost constant *jjdist* value. Conversely, for the *tdleafpct* feature, MCS was the only heuristic not producing direct results (Figure 4(c)).

In order to test the feasibility of a machine learning approach in this setting, a number of machine learning algorithms were run to compare their performance. Three such classifiers were tested: Random decision trees, k-nearest neighbor and a single rule algorithm. The latter serves as a reference point, it always returns the class that occurs most often (in this case "Dyn-ASP2"). For training, the dataset was split tenfold and ten training- and validation runs were done, always training on nine folds and validating with the 10th (10-fold cross-validation). Table 3 shows the classifier performance in detail. It shows for each classifier, how many tuples of each class (the "correct" class) were incorrectly classified, e.g. for all training-tuples on which the Dyn-ASP1 algorithm performed better, the k-NN classifier (wrongly) chose the Dyn-ASP2 algorithm in only 10.8% of the cases.

Algorithm selection based on regression techniques: The second approach that we applied for selection of the best dynamic algorithm on the particular tree decomposition is

Table 2. Feature ranking based on Information Gain, using 10-fold cross-validation

Average merit	Average rank	Attribute
0.436 ± 0.002	1 ± 0	jpct
0.422 ± 0.004	2 ± 0	reltdsize
0.386 ± 0.004	3.2 ± 0.4	jjdist
0.372 ± 0.012	4.2 ± 0.87	tdleafpct
0.357 ± 0.006	5.2 ± 0.6	tdbranchpct
0.354 ± 0.01	5.4 ± 0.8	jwidth

Table 3. Different classifiers and percentages of incorrectly classified instances

Classifier	Correct class	Incorrectly classified
Single-rule	Dyn-ASP1	23.1%
Single-rule	Dyn-ASP2	18.4%
k-NN, k=10	Dyn-ASP1	10.8%
k-NN, k=10	Dyn-ASP2	18.5%
Random forest	Dyn-ASP1	10.6%
Random forest	Dyn-ASP2	18.4%

based on regression techniques. The main idea is to use machine learning algorithms to first predict the runtime of each dynamic algorithm in a particular instance, and then select the algorithm that has better predicted runtime. To learn the model for runtime prediction we provide for each dynamic algorithm a training set that consists of instances that include features of input tree decomposition (and the input graph). Additionally, for each example are given the information for the time needed to construct the tree decomposition and the running time of the particular dynamic algorithm.

We experimented with several machine learning algorithms for regression available in WEKA, and compared their performance regarding the selection accuracy of the fastest dynamic algorithm for the given input instance. For each machine learning algorithm we provided a training set consisting of 6090 examples. The testing set contained 3045 examples.

The algorithm k-NN (k=5) gave best results among these machine learning algorithms regarding runtime prediction for both dynamic algorithms. To illustrate the performance of k-NN algorithm regarding the runtime prediction we present the actual runtime and the predicted runtime for both dynamic algorithms in Figure 5. Results for the first 30 examples in the testing set are given.

Regarding the algorithm selection based on the runtime prediction, we present in Table 4 the best current results that we could obtain with two machine learning algorithms k-NN (k-nearest neighbors, see [1]) and M5P (Pruned regression tree, see [24] and [30]). As we can see the accuracy of selecting the right (fastest) dynamic algorithm for a particular instance is good. In particular, the k-NN algorithm selects the best algorithm for the 88% of the test instances.

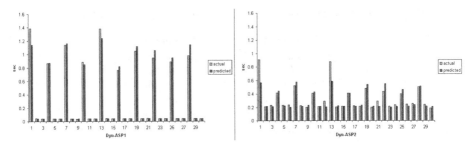

Fig. 5. Actual and predicted time with k-NN for first 30 test examples

Table 4. Two regression algorithms and their accuracy regarding the selection of better dynamic algorithm

Algorithm	Dynamic algorithm selection accuracy
M5P	80.2%
k-NN, k=5	88.1%

5   Discussion

In our experiments, we have identified several important tree decomposition features. As these features can have a high impact on the performance of a subsequent dynamic algorithm, heuristics should try to create "good" decompositions also with respect to these features and not only with respect to the width. It has become apparent that a higher width can be compensated by such a decomposition, e.g. in our benchmarks, the MCS heuristic always produced the worst width, but actually speeds up our dynamic algorithms. Moreover, these features have turned out to be well suitable for classification and regression methods.

The good results that were obtained by our machine learning approach clearly suggest that two-phased algorithms like our dynASP system significantly profit from an automatic selection of the dynamic algorithm in the second phase – based on the tree decomposition features identified here. Given the effectiveness of the single-rule classifier, by simply implementing this rule (which is equivalent to a simple if-statement), the dynamic algorithm can be effectively selected once the tree decomposition has been computed. By utilizing a k-nearest neighbor or random decision tree approach, on average more than 85% of the decisions made are correct, yielding further improvements. Machine learning approaches (like portfolio solvers) are already in use for ASP (see e.g. [4,9]), however these are specific to ASP, whereas our approach, using tree decomposition features for decisions, can generally be used for all TDDA approaches.

6   Conclusion

In this paper we have studied the interplay between three heuristics for the computation of tree decompositions and two different dynamic algorithms for head-cycle free programs, an important subclass of disjunctive logic programs. We have identified features beside the width of tree decompositions that influence the running time of our dynamic algorithms. Based on these observations, we have proposed and evaluated algorithm selection via different machine learning techniques. This will help to improve our prototypical TDDA system dynASP.

For future work, we plan to study the possibilities to not only perform algorithm selection for the dynamic algorithm but also for the heuristic to compute the tree decomposition. Furthermore, our results suggest that heuristic methods for tree decompositions should not only focus on minimizing the width but should also take some other features as objectives into account. Finally, we expect that our observations are independent of the domain of answer set programming. We therefore plan to evaluate tree-decomposition based algorithms for further problems from various other areas [6].

Acknowledgments. The work was supported by the Austrian Science Fund (FWF): P20704-N18, and by the Vienna University of Technology program "Innovative Ideas".

References

1. Aha, D.W., Kibler, D.F., Albert, M.K.: Instance-based learning algorithms. Machine Learning 6, 37–66 (1991)
2. Arnborg, S., Corneil, D.G., Proskurowski, A.: Complexity of finding embeddings in a k-tree. SIAM J. Alg. Disc. Meth. 8, 277–284 (1987)
3. Bachoore, E.H., Bodlaender, H.L.: A Branch and Bound Algorithm for Exact, Upper, and Lower Bounds on Treewidth. In: Cheng, S.-W., Poon, C.K. (eds.) AAIM 2006. LNCS, vol. 4041, pp. 255–266. Springer, Heidelberg (2006)
4. Balduccini, M.: Learning and using domain-specific heuristics in ASP solvers. AI Commun. 24(2), 147–164 (2011)
5. Ben-Eliyahu, R., Dechter, R.: Propositional semantics for disjunctive logic programs. Ann. Math. Artif. Intell. 12, 53–87 (1994)
6. Bodlaender, H.L.: A tourist guide through treewidth. Acta Cybern. 11(1-2), 1–22 (1993)
7. Bodlaender, H.L., Koster, A.M.C.A.: Treewidth computations I. Upper Bounds. Inf. Comput. 208(3), 259–275 (2010)
8. Dermaku, A., Ganzow, T., Gottlob, G., McMahan, B., Musliu, N., Samer, M.: Heuristic Methods for Hypertree Decomposition. In: Gelbukh, A., Morales, E.F. (eds.) MICAI 2008. LNCS (LNAI), vol. 5317, pp. 1–11. Springer, Heidelberg (2008)
9. Gebser, M., Kaminski, R., Kaufmann, B., Schaub, T., Schneider, M.T., Ziller, S.: A Portfolio Solver for Answer Set Programming: Preliminary Report. In: Delgrande, J.P., Faber, W. (eds.) LPNMR 2011. LNCS, vol. 6645, pp. 352–357. Springer, Heidelberg (2011)
10. Gelfond, M., Lifschitz, V.: Classical negation in logic programs and disjunctive databases. New Generation Comput. 9(3/4), 365–386 (1991)
11. Gogate, V., Dechter, R.: A complete anytime algorithm for treewidth. In: Proc. UAI 2004, pp. 201–208. AUAI Press (2004)
12. Gottlob, G., Pichler, R., Wei, F.: Bounded treewidth as a key to tractability of knowledge representation and reasoning. In: Proc. AAAI 2006, pp. 250–256. AAAI Press (2006)
13. Hall, M., Frank, E., Holmes, G., Pfahringer, B., Reutemann, P., Witten, I.H.: The WEKA data mining software: an update. SIGKDD Explorations 11(1), 10–18 (2009)
14. Hall, M.A., Smith, L.A.: Practical feature subset selection for machine learning. In: Proc. ACSC 1998, pp. 181–191. Springer (1998)

15. Jakl, M., Pichler, R., Woltran, S.: Answer-set programming with bounded treewidth. In: Proc. IJCAI 2009, pp. 816–822. AAAI Press (2009)
16. Kloks, T.: Treewidth, computations and approximations. LNCS, vol. 842. Springer, Heidelberg (1994)
17. Koster, A., van Hoesel, S., Kolen, A.: Solving partial constraint satisfaction problems with tree-decomposition. Networks 40(3), 170–180 (2002)
18. Lauritzen, S., Spiegelhalter, D.: Local computations with probabilities on graphical structures and their application to expert systems. Journal of the Royal Statistical Society, Series B 50, 157–224 (1988)
19. Leone, N., Pfeifer, G., Faber, W., Eiter, T., Gottlob, G., Perri, S., Scarcello, F.: The DLV system for knowledge representation and reasoning. ACM Trans. Comput. Log. 7(3), 499–562 (2006)
20. Marek, V.W., Truszczyński, M.: Stable Models and an Alternative Logic Programming Paradigm. In: The Logic Programming Paradigm – A 25-Year Perspective, pp. 375–398. Springer (1999)
21. Morak, M., Musliu, N., Pichler, R., Rümmele, S., Woltran, S.: A new tree-decomposition based algorithm for answer set programming. In: Proc. ICTAI, pp. 916–918 (2011)
22. Morak, M., Pichler, R., Rümmele, S., Woltran, S.: A Dynamic-Programming Based ASP-Solver. In: Janhunen, T., Niemelä, I. (eds.) JELIA 2010. LNCS, vol. 6341, pp. 369–372. Springer, Heidelberg (2010)
23. Niemelä, I.: Logic programming with stable model semantics as a constraint programming paradigm. Ann. Math. Artif. Intell. 25(3-4), 241–273 (1999)
24. Quinlan, R.J.: Learning with continuous classes. In: 5th Australian Joint Conference on Artificial Intelligence, Singapore, pp. 343–348 (1992)
25. Robertson, N., Seymour, P.D.: Graph minors II: Algorithmic aspects of tree-width. Journal Algorithms 7, 309–322 (1986)
26. Samer, M., Szeider, S.: Algorithms for propositional model counting. J. Discrete Algorithms 8(1), 50–64 (2010)
27. Shoikhet, K., Geiger, D.: A practical algorithm for finding optimal triangulations. In: Proc. AAAI 1997, pp. 185–190. AAAI Press/The MIT Press (1997)
28. Smith-Miles, K.: Cross-disciplinary perspectives on meta-learning for algorithm selection. ACM Comput. Surv. 41(1) (2008)
29. Tarjan, R., Yannakakis, M.: Simple linear-time algorithm to test chordality of graphs, test acyclicity of hypergraphs, and selectively reduce acyclic hypergraphs. SIAM J. Comput. 13, 566–579 (1984)
30. Wang, Y., Witten, I.H.: Induction of model trees for predicting continuous classes. In: Poster Papers of the 9th European Conference on Machine Learning (1997)
31. Zhao, Y., Lin, F.: Answer Set Programming Phase Transition: A Study on Randomly Generated Programs. In: Palamidessi, C. (ed.) ICLP 2003. LNCS, vol. 2916, pp. 239–253. Springer, Heidelberg (2003)

High-Dimensional Model-Based Optimization Based on Noisy Evaluations of Computer Games

Mike Preuss[1], Tobias Wagner[2], and David Ginsbourger[3]

[1] Technische Universität Dortmund, Chair of Algorithm Engineering (LS 11)
Otto-Hahn-Str. 14, Dortmund, D-44227, Germany
mike.preuss@tu-dortmund.de
[2] Technische Universität Dortmund, Institute of Machining Technology (ISF)
Baroper Str. 301, Dortmund, D-44227, Germany
wagner@isf.de
[3] University of Bern, Institute of Mathematical Statistics and Actuarial Science
Sidlerstr. 5, Bern, CH-3012, Switzerland
david.ginsbourger@stat.unibe.ch

Abstract. Most publications on surrogate models have focused either on the prediction quality or on the optimization performance. It is still unclear whether the prediction quality is indeed related to the suitability for optimization. Moreover, most of these studies only employ low-dimensional test cases. There are no results for popular surrogate models, such as kriging, for high-dimensional ($n > 10$) noisy problems. In this paper, we analyze both aspects by comparing different surrogate models on the noisy 22-dimensional car setup optimization problem, based on both, prediction quality and optimization performance. In order not to favor specific properties of the model, we run two conceptually different modern optimization methods on the surrogate models, CMA-ES and BOBYQA. It appears that kriging and random forests are very good modeling techniques with respect to both, prediction quality and suitability for optimization algorithms.

Keywords: Computer Games, Design and Analysis of Computer Experiments, Kriging, Model-Based Optimization, Sequential Parameter Optimization, The Open Racing Car Simulator.

1 Introduction

Over the last 15 years, the use of surrogate-model-assisted optimization approaches has obtained a high popularity in almost all application areas [12, 13, 17, 23]. Within this period, the research on model-based optimization has mainly focused on low-dimensional problems and noise-free evaluations. In particular, kriging has been shown to be well-suited for modeling deterministic data of computer experiments (design and analysis of computer experiments, DACE [21]) with low or moderate input dimension $n \in [1, 10]$. In the modeling and optimization of practical problems, however, e.g., in the computer games community,

Y. Hamadi and M. Schoenauer (Eds.): LION 6, LNCS 7219, pp. 145–159, 2012.
© Springer-Verlag Berlin Heidelberg 2012

high-dimensional parameter spaces and noisy responses have to be considered. Consequently, the modern kriging models of DACE have been enhanced to cope with noisy data in recent years [5, 6, 11]. For high-dimensional data, however, almost no results of kriging-based modeling approaches have been reported.

In this paper we thus investigate how these kriging variants and other popular surrogate modeling techniques can assist in optimizing a 22-dimensional problem from the domain of computer games - the car setup optimization problem based on the open racing car simulator (TORCS). The response to be modeled is the distance obtained by a racing car with a specific car setup encoded by the input parameters. Based on a short evaluation time on TORCS with a quasi-random starting point on the track, this response is very noisy. The analysis and evaluation of the surrogate models is two-fold. First, their global prediction qualities based on the initial design are evaluated. Then, the capability of the models to guide and tune the optimization [19] is assessed by performing a global optimization on the model and compare the predicted optimum with the quality of the evaluation on TORCS (one-step approach). Almost all previous studies using a one-step approach have focused only on one of these aspects – the prediction quality or the results of a model-based optimization. Based on the combined analysis, some important questions can be addressed:

1. Is the prediction quality a good indicator for the optimization capability of a surrogate model?
2. Are certain surrogate models particularly well suited for high-dimensional noisy problems?
3. Can the successful results of kriging-based optimization approaches be transferred to higher dimensions and noisy data?

In the following section, the basic principles of the considered surrogate models are described. The car setup optimization problem and TORCS are briefly summarized in section 3. The two main sections of the papers address the prediction quality and the optimization results obtained by the different surrogate models. In the final section 6, the results are summarized, conclusions are drawn, and an outlook on future research topics is given.

2   Surrogate Models

For almost all real-world applications, the evaluation of parameter vectors is time-consuming and/or expensive, e. g., because a finite-element analysis, a computational fluid dynamics calculation or a real-world experiment have to be performed. In these cases, a model-based approach is often used. Here, we focus on one-step approaches. Based on an initial design of the problem parameters, a model is fitted which is then used as a surrogate for the actual experiment, e. g., the parameter vector resulting in the optimal model prediction is directly used as a solution or the model is used as a surrogate for tuning optimization algorithms in order to use the tuned variant on the actual problem [19]. For both kinds of applications, the surrogate model should

1. be as close to the true response as possible (prediction quality), and
2. reflect the characteristics of the optima of the true response surface (model optimization).

In the following subsections, some popular surrogate models are described and discussed. Due to the extremely high number of approaches, we restrict our description to the models considered in our experiments. More methods and additional details to the presented approaches can be found in Hastie et al. [9].

2.1 First Order Response Surface

In the first order (linear) response surface model (LM), the relationship between the control variables $\mathbf{x}_i$ and the corresponding observations y_i is described by

$$y_i = \mathbf{x}_i \beta + \varepsilon_i. \tag{1}$$

Equation 1 is set up for each pair of parameter vector $\mathbf{x}_i$ and observation y_i ($i = 1, \ldots, N$) in the initial design. The least-squares estimate $\hat{\beta}$ of the coefficients β is then calculated as the solution to the corresponding system of linear equations [10, p. 11]. With this $\hat{\beta}$, equation 1 can be used for prediction of unknown parameter vectors $\mathbf{x}$.

2.2 Generalized Additive Model

The Generalized Additive Model (GAM) [8] replaces the linear form of equation 1 by a sum of smoothing functions for single parameters $\beta + \sum s_j(x_j)$ ($j = 1, \ldots, k$), where an iterative algorithm is employed to decide about the important variables x_j and the corresponding smooth functions s_j. Contrary to the first order response surface (LM), the GAM also allows nonlinear smoothing functions to be specified. The employed R package GAM[1] supports local polynomial regression and smoothing splines.

2.3 Random Forest

Random forests [2] consist of huge ensembles (typically 500 or more) of decision trees, whereby each of them is trained on a randomly chosen subset of the available observations. The prediction of the random forest is then computed as the average of the predictions of the individuals trees. Random forests are usually used for classification, but also regression can be realized by implementing regressing decision trees, as done in the R package randomForest[2].

---

[1] http://cran.r-project.org/web/packages/gam/index.html
[2] http://cran.r-project.org/web/packages/randomForest/index.html

2.4 Kriging

Kriging is a surrogate model originated from geosciences [4] which has become popular in the DACE [21] and machine learning [20] communities. In ordinary kriging, the response of interest can be considered as one realization of a random variable $Y(\mathbf{x}) = \mu + Z(\mathbf{x})$, where $\mu \in \mathbb{R}$ is an intercept used for centering the stationary zero-mean Gaussian process (GP) Z. Z depends on a covariance kernel of the form $(\mathbf{x}, \mathbf{x}') \to D^2 : k(\mathbf{x}, \mathbf{x}') \mapsto \sigma^2 r(\mathbf{x} - \mathbf{x}'; \psi)$ for a correlation function r with parameters ψ.

The predictions of the kriging model can be obtained by taking the conditional expectation $m(\mathbf{x}) = \mathbb{E}[Y(\mathbf{x})|Y(\mathbf{x}_i) = y_i]$ of Y based on the N current pairs of parameter vectors $\mathbf{x}_i$ and observations y_i. Consequently, $m(\mathbf{x})$ is also denoted as the kriging mean. It provides a prediction for each observation $\mathbf{x}$ by enhancing the constant trend using the correlation to the existing observations. It thus explicitly uses the information of each observation. For an efficient evaluation, the kriging mean can be computed in closed form

$$m(\mathbf{x}) = \widehat{\mu} + \mathbf{k}(\mathbf{x})^T \mathbf{K}^{-1} (\mathbf{y} - \widehat{\mu}\mathbf{1}), \tag{2}$$

using the observations $\mathbf{y} = (y_1, \ldots, y_n)^T$, the covariance matrix of the experiments $\mathbf{K} = (k(\mathbf{x}_{i_1}, \mathbf{x}_{i_2}))$, the covariance vector $\mathbf{k}_n(\mathbf{x}) = (k(\mathbf{x}, \mathbf{x}_1), \ldots, k(\mathbf{x}, \mathbf{x}_n))^T$ of $\mathbf{x}$ and the existing design points, and the maximum likelihood estimation of the trend

$$\widehat{\mu} = \frac{\mathbf{1}^T \mathbf{K}^{-1} \mathbf{y}}{\mathbf{1}^T \mathbf{K}^{-1} \mathbf{1}}.$$

For noisy evaluations $\widetilde{Y}_i := Y(\mathbf{x}_i) + \varepsilon_i$, the GP is conditioned based on a sum of random variables – one following a GP and one for the noise. Assuming independence between the random variables as well as between different realizations of the noise, the kriging mean can still be computed using equation 2, only the intercovariance matrix $\mathbf{K}$ is replaced by $\bar{\mathbf{K}} = \mathbf{K} + \tau^2 \mathbf{I}$ at every occurrence. The additional term τ^2 denotes the noise variance which is only added for identical observations. In the case of heterogeneous noise variances, i.e., $var(\varepsilon_1) = \tau_1^2 \neq \ldots \neq var(\varepsilon_N) = \tau_N^2$, $\mathbf{K}$ is replaced by $\bar{\mathbf{K}} = \mathbf{K} + diag([\tau_1^2 \ldots \tau_n^2])$. Contrarily to the noiseless case, these models do not interpolate the noisy observations.

The choice of the covariance kernel and its parameters determines the shape (smoothness, modality) and the flexibility of the response surfaces predicted by the kriging model. In this paper, two popular kernels implemented in the R package DiceKriging[3] are considered:

1. the Gaussian kernel:

$$k(\mathbf{x}, \mathbf{x}') = \sigma^2 \exp\left[-\sum_{j=1}^{n} \left(\frac{x_j - x_j'}{\theta_j} \right)^2 \right] \tag{3}$$

---

[3] cran.r-project.org/web/packages/DiceKriging/index.html

2. the Matérn kernel with $\nu = 5/2$:

$$k(\mathbf{x}, \mathbf{x}') = \sigma^2 \prod_{j=1}^{n} \left[1 + \sqrt{5}D_j + \frac{5}{3}D_j^2\right] \exp\left[-\sqrt{5}D_j\right], \; D_j = \frac{|x_j - x_j'|}{\theta_j} \quad (4)$$

Both kernels depend on a set of parameters, σ^2 and $\{\theta_1, \ldots, \theta_d\}$, which are often referred to respectively as *process variance* and *ranges*. They have to be fitted based on the available evaluations, for which we use maximum-likelihood estimation in the experiments.

3 Car Setup Optimization Problem

The car setup optimization problem originates from a competition held at the EvoStar 2010 conference[4]. It is based on the open source car racing simulator (TORCS)[5] which is used as simulation engine for the evaluations. The task in this competition is to find a near optimal setting for the 22 car parameters listed in Table 1. Performance is measured by the track distance covered within this time frame. In order to avoid handling different parameter ranges within the optimization, all parameters are scaled to the interval $[0, 1]$ by the interface.

Table 1. The 22 car setup optimization parameters of the EvoStar 2010 competition and their original ranges, taken from [3]

parameter	section	name	unit	min	max
1	gearbox/gears/2	ratio	SI	0	5
2	gearbox/gears/3	ratio	SI	0	5
3	gearbox/gears/4	ratio	SI	0	5
4	gearbox/gears/5	ratio	SI	0	5
5	gearbox/gears/6	ratio	SI	0	5
6	rear wing	angle	deg	0	18
7	front wing	angle	deg	0	12
8	brake system	front-rear brake repartition	SI	0.3	0.7
9	brake system	max pressure	kPa	100	150000
10	front anti-roll bar	spring	lbs/in	0	5000
11	rear anti-roll bar	spring	lbs/in	0	5000
12	front left-right wheel	ride height	mm	100	300
13	front left-right wheel	toe	deg	-5	5
14	front left-right wheel	camber	deg	-5	-3
15	rear left-right wheel	ride height	mm	100	300
16	rear left-right wheel	camber	deg	-5	-2
17	front left-right suspension	spring	lbs/in	0	10000
18	front left-right suspension	suspension course	m	0	0.2
19	rear left-right suspension	spring	lbs/in	0	10000
20	rear left-right suspension	suspension course	m	0	0.2
21	front left-right brake	disk diameter	mm	100	380
22	rear left-right brake	disk diameter	mm	100	380

In the competition, a time frame of one million *tics* of 20 *ms* each was allowed as budget for the optimization algorithm. The algorithm can distribute the available time arbitrarily between different settings, i.e., each evaluation can take as

---

[4] http://cig.ws.dei.polimi.it/?page_id=103
[5] http://torcs.sourceforge.net/

long as desired. For comparable results, however, the evaluation time should be fixed. The evaluations are made in a row while the game is running, where short breaks are required in order to brake down the car to a standstill. Therefore, different parts of the track are used for measuring the performance, which is a major source of the noise in the evaluations. Recent parameter studies [16] have shown that below a certain limit of around 250 tics (5 s), measured values become so noisy that they are unsuitable for optimization. Longer evaluations spent the budget more quickly so that Kemmerling recommends evaluations of 2000 tics (40 s), resulting in only 500 evaluations of the simulator. In this time, around one third of the Suzuka F1 track (wheel-2 in TORCS) can be covered. This track is shown in Fig. 1. It combines many challenges, such as high speed parts and different curve types, and is, thus, used for evaluations in this paper.

Fig. 1. Screenshot of TORCS, in which the solution obtained from the model-based optimization on the (Matérn covariance) kriging model is driving the reference track (Suzuka F1). A minimap of the track is shown in the top right corner.

Summarizing, the car setup optimization problem can be regarded as a high-dimensional noisy practical problem with a very limited budget of evaluations. Consequently, it is hard to solve[6]. In the computer games context, such problems appear whenever an implemented, parameterizable component (as a car driving bot) must be adapted to other components, be they provided by the user or procedurally generated [16]. As known from the formula 1 races, the optimal setup will change whenever one of the components (car, driver, track) is modified. Thus, the results of the EvoStar2010 competition where other tracks then the Suzuka F1 were considered cannot be directly compared to our setup.

---

[6] See also the results of the EvoStar 2010 competition at
http://www.slideshare.net/dloiacono/
car-setup-oprimization-competition-evostar-2010

4 Prediction Quality of the Models

In the first part of the experimental analysis, the prediction quality of the surrogate models on the car setup problem is evaluated. The results build the basis for the combined analysis in the next section.

Setup. In order to evaluate the prediction quality of the models, sets for the training and the validation of the surrogate model had to be prepared. The budget for the training set was chosen according to the specification of the car setup competition, where 20000 s of running the simulator had been allowed. We chose an evaluation time of 40 s which resulted in a size of 500 design points in the training set. As mentioned in the previous section, the evaluations of a car setup were noisy. In order to allow the effect of the accuracy of an evaluation to be considered, two different training sets were used, one having 125 mutual design points with 4 replications $\mathbf{X}_{125,4}$ and one having 500 mutual design points without replications $\mathbf{X}_{500,1}$. For validation, a larger and more accurate set of $N = 440$ randomly distributed design points with 20 replications was employed. The root mean squared error $RMSE = \sqrt{\sum_{i=1}^{N}(\hat{y}_i - \bar{y}_i)^2}$ was used as performance measure, where $\hat{y}_i$ denotes the prediction of the model for the i-th design point and $\bar{y}_i$ is the mean of the 20 observations in the validation set. For some parameter vectors, the damage of the car exceeds a specified threshold. In these cases, the output of the simulator is not the distance reached by the car, but a high penalty encoding the damage. In order to not deteriorate the quality of the models by integrating discontinuities and different scales, these values were removed from the training and validation sets. If a subset of the repeats of a parameter vector is penalized, only the remaining results were used for the computation of the mean performance.

For kriging, we analyzed the effect of the covariance kernel and the use of the estimated variances of set $\mathbf{X}_{125,4}$ in a heterogeneous kriging model. More specifically, we consider the Gaussian and the Matérn kernel with $\nu = 5/2$ and the homogeneous (single τ^2) and heterogeneous (vector of τ_i^2) formulation of kriging for noisy observations, as presented in section 2.4. The other surrogate models have been run with their standard parameters. For LM and GAM, the set $\mathbf{X}_{125,4}$ was tested with and without using variance-depending weights $w_i = 1/var(y_i)$ of the observations, where the latter approach results in a weighted least squares fit.

Pre-experimental Planning. The 22 design variables of the car setup problem directly result in 23 model parameters (θ_j and σ) in the likelihood optimization. In order to avoid a deterioration of the kriging results based on a bad fit of the internal parameters, the optimization of the kriging parameters was analyzed before the experiments. Accounting for the multimodality of the log likelihood, the global optimization strategy of the DICEKriging package based on the Genetic Optimizer Using Derivatives (RGenOUD)[7] was considered. It was observed that

---

[7] http://cran.r-project.org/web/packages/rgenoud/index.html

the standard parameters, population size $P = 20$, wait generations $W = 2$, maximum generation limit $L = 5$, and the generation in which the gradient-based refinement is performed for the first time (BFGS burnin) $B = 0$ did not robustly find the maximum of the log likelihood for the large data set $\mathbf{X}_{500,1}$. Thus, a small experiment was conducted in preparation for the actual study. Based on a 43 point Latin hypercube design (LHD) [14] of the four RGenOUD parameters, the performance of log likelihood optimization was analyzed. The results of the set $\mathbf{X}_{500,1}$ were defined as the test case. The experiment was conducted for the Gaussian and the Matérn kernel with $\nu = 5/2$. Based on the results and allowing a slightly larger computational budget for ensuring robust results, we recommend to use $(P, W, L, B) = (10, 1, 120, 40)$. The computation time for $\mathbf{X}_{500,1}$ is still below 4 minutes on a 3 GHz PC.

The focus of the experiments reported in the paper is clearly on surrogate models based on or related to kriging. This originates from our experience in using these models. Other popular surrogate models, such as support vector regression and neural networks, were also used in the beginning of the study, but since the results of standard R implementations (e1071 package[8] based on the libsvm[9] and the nnet package[10]) were much worse and we did not manage to appropriately adjust these models, we excluded them from the paper. Nevertheless, our results are based on open R packages and the test sets can be downloaded online[11], which allows a comparison of experts in these areas with our results to be realized.

Task. According to the scope of the experiment, four hypotheses were established:

1. More inaccurate points are better than a few with higher accuracy.
2. The use of the estimated noise variances can improve the results on training set $\mathbf{X}_{125,4}$.
3. The Matérn kernel with $\nu = 5/2$ is superior to the standard Gaussian kernel.
4. Kriging is the superior model with respect to prediction quality.

The first hypothesis was based on former results of kriging on noisy data sets [1] and is generally related to the bias-variance tradeoff in machine learning [9]. The second one was straightforward, but surprises with respect to a bad estimation of the variances or an increase of model complexity might occur. The third one was based on the weaker assumptions of the Matérn kernel compared to the Gaussian with respect to the differentiability of the response surface – twice compared to infinitely often. The fourth one was driven by the hope to transfer the results of kriging in lower dimensions to higher ones. However, this was questionable due to former results in the literature [22].

Results/Visualization. The results of the experiments with regard to the prediction quality are summarized in Table 2.

---

[8] http://cran.r-project.org/web/packages/e1071/index.html
[9] http://www.csie.ntu.edu.tw/~cjlin/libsvm/
[10] http://cran.r-project.org/web/packages/nnet/index.html
[11] http://ls11-www.cs.uni-dortmund.de/rudolph/kriging/
    applications?&#video_game_data

Table 2. Summary of the results with respect to the prediction quality of the models

method	data set	var. used?	repeats	mean RMSE	std RMSE
LM	$X_{125,4}$	no	1	0.1835478	-
	$X_{125,4}$	yes	1	0.2428571	-
	$X_{500,1}$	no var.	1	0.1699225	-
GAM	$X_{125,4}$	no	1	0.1346388	-
	$X_{125,4}$	yes	1	0.1803923	-
	$X_{500,1}$	no var.	1	0.0976035	-
Random Forest	$X_{125,4}$	no	10	0.1259052	0.0010350
	$X_{500,1}$	no	10	0.1066116	0.0007794
	$X_{125,4}$	no	10	0.1447966	0.0329848
Kriging, Gauss	$X_{125,4}$	yes	10	0.1283082	0.0000040
	$X_{500,1}$	no var.	10	0.0935839	0.0000007
	$X_{125,4}$	no	10	0.1202643	0.0000065
Kriging, Matérn	$X_{125,4}$	yes	10	0.1283162	0.0273566
	$X_{500,1}$	no var.	10	0.0937369	0.0000006

Observations. Based on the results of Table 2, the hypotheses can be tested[12]:

1. More inaccurate points are indeed better than a few with higher accuracy. The results of the training set $X_{500,1}$ are significantly improving the prediction quality for all considered surrogate models.

2. The results concerning this hypothesis show no clear trend. For the LM and the GAM, the results of the weighted least squares fit are worse compared to the standard one. For the Gaussian kernel, the mean prediction quality is improved by using the estimated variances while still being robust with respect to the model fitting. For the Matérn kernel, the mean prediction quality and the robustness of the model fitting decrease with estimated variances. This result, however, is based on the fact that the best model (RMSE ≈ 0.115) is only found in 8 of the 10 repeats. In the two remaining cases, bad models (RMSE > 0.178) are returned which result in the observed decrease in mean performance.

3. The only situation in which the Matérn kernel with $\nu = 5/2$ is indeed superior to the usually applied Gaussian kernel is the one for the set $X_{125,4}$ with unknown variances. In all other scenarios, no significant results can be obtained with respect to the covariance kernel which is of course also based on the high standard deviation of the Matérn kernel on set $X_{125,4}$ with estimated variances.

4. Kriging is indeed the superior model with respect to prediction quality. Although the superiority is significant, the improvement over Random Forests (set $X_{125,4}$) and the GAM (set $X_{500,1}$) is only small.

---

[12] Due to the partly deterministic results and very low variances of the stochastic approaches, no additional statistical tests have been performed.

Discussion. The most important effect with respect to the prediction quality is the one of the training set. A diverse set of many inaccurate solutions results in a higher prediction quality for all considered models. This effect may be caused by the diminishing gain of information obtained by replications [11]. In addition, four repeats are not enough for a sensible approximation of the variance corresponding to an observation. This conjecture is also based on the bad results of the weighted least squares approaches.

5   Model-Based Optimization

Related to the questions formulated in the introduction, we now analyze whether the established models are useful for optimization, and if there is a relation to the prediction quality we can exploit to predict this suitability.

Pre-experimental Planning. Based on the results of the previous section, we only focused on models based on the set $\mathbf{X}_{500,1}$. We used a stationary approach in which the surrogate model is not refined. Our focus is on the capability of the model to reflect the characteristics of the true response surface based on the large initial design, e. g., for a one-step optimization approach or for tuning optimization algorithms [19]. In order not to bias the results by the choice of the optimizer, two different optimization strategies were used. The *covariance-matrix-adaptation evolution strategy* (CMA-ES)[13] [7] is a powerful evolutionary algorithm for global optimization, whereas *boundary optimization by quadratic approximation* (BOBYQA)[14] [18] is a modern gradient-free local search strategy for box-constrained optimization.

Task. We want to decide if model quality may be employed as a guideline for choosing a model for optimization, namely by judging these two hypotheses:

1. The prediction quality can be used as an indicator for the suitability for a one-step optimization.
2. The Kriging models offering a high prediction quality are particularly suited for reflecting the characteristics of nonlinear problems and finding its optima.

The first hypothesis expresses the common belief of just considering one of these indicators for assessing surrogate models. The last one was based on the huge number of publications on kriging metamodeling [23].

Setup. In order to evaluate the suitability for optimization, one model of each type was chosen as representative. This choice was made based on the internal quality criterion of the model – log likelihood for kriging, out-of-bag error for random forest. For the other approaches, no variation in the model fitting exists. On each representative, 20 runs of BOBYQA and the CMA-ES were conducted. The obtained local optima were then evaluated on TORCS for 10 times.

---

[13] http://cran.r-project.org/web/packages/cmaes/
[14] http://cran.r-project.org/web/packages/minqa/

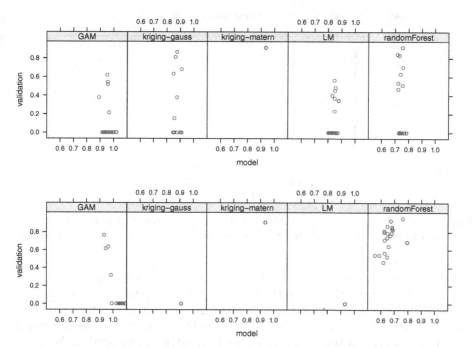

Fig. 2. Comparison of the predicted ('model', x-axis) and actual performance ('validation', y-axis) of the local optima (top: CMA-ES, bottom: BOBYQA.). The potentially global optimum on the model is colored in red.

The resulting values provided the basis for the analysis. If the predicted values of the potential optima also result in locally or globally optimal values on the simulator, the characteristics of the model are well reflected. The assessment of the correlation between the prediction quality and the optimization performance is performed by comparing the RMSE of the representative model and the actual quality of the approximated optima.

Results/Visualization. The results of the optimization on the representative models are shown in Fig. 2. The correlation between the prediction quality and the optimization performance can be assessed based on Fig. 3. It looks conspicuous that a lot of the potential optima on the models result in a validated value of zero. These values are caused by parameter vectors that could not be evaluated properly on TORCS because the damage threshold of the car is exceeded (cf. section 3), making this parameter vector infeasible.

Observations. The first hypothesis can clearly be rejected. Based on Fig. 3, the random forest provides a better optimization performance than Gaussian kriging and GAM, whereas its RMSE is worse. The RMSE can only successfully distinguish between the worst (LM) and the best (Matérn kriging) approach with respect to both indicators.

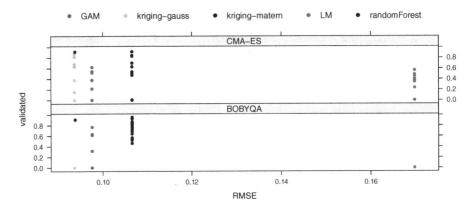

Fig. 3. Correlation between the RMSE of the representative model and the actual averaged distance achieved in TORCS by the approximated optimum parameter setting

With respect to the second hypothesis, no objective results can be seen from Fig. 2. Whereas the CMA-ES on the Gaussian kriging model returned many local optima with an almost equal performance on the model, but high variation on the actual problem, the optimization on the model using the Matérn kernel always resulted in the same optimum which is indeed a good parameter setting. All other nonlinear surrogate models also resulted in a multimodal response surface with different local optima, where the performance variation on the actual problem is huge.

Discussion. The high number of infeasible solutions proposed by LM, GAM, and Gaussian kriging is based on the extrapolation properties of these approaches. Since the parameter vectors of the infeasible solutions are not used for fitting the model, no points for interpolation are provided in these parameter regions. Nevertheless, the assumption of an underlying model (LM and GAM) or the strong differentiability assumed by Gaussian kriging result in local optima within these areas. This is highly undesired for the focused application setting, where only one iteration of model-based optimization is performed, because it may result in an infeasible solution after optimization. In contrast, the Matérn kernel seems to result in a more data-dependent prediction without extrapolation effects.

The difference between the RMSE and the optimization performance may consequently be caused by the different extrapolation properties of the approaches. Because the infeasible values are also removed from the validation set, no predictions in these areas are considered. In addition, the predictions of the random forest seem to be more conservative. Whereas all other modeling approaches predict their local optimal values between 0.8 and 1, the optimal values of the random forest are between 0.6 and 0.8. It seems like the characteristics of the problem are well covered, but the variation of the response values is decreased which results in an increased RMSE.

In order to judge whether the response surfaces reflect the characteristics of the true problem, the true modality of the problem is of interest. In figure 4, we

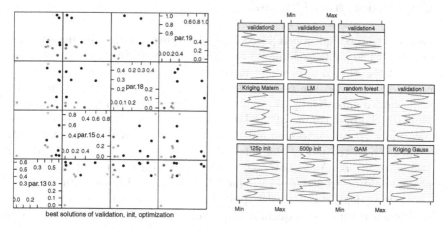

Fig. 4. Visualizations of the 4 best of the 440 solutions of our validation set (blue) together with the best initial solutions from the 125 point design (gray) and the 500 point design (black) and the best solutions obtained from the CMA-ES runs, GAM (red), Kriging Gauss (orange), Kriging Matern (green), LM (violet), random forest (indianred). Left: scatter plot of the 4 most important variables as indicated in [15]. Note the different range for parameter 13 due to infeasible solutions outside this range. Right: parallel coordinate plot over all 22 parameters.

depict the best 4 of the 440 solutions in the validation set (together with the best initials solutions and the best from the optimization runs). The plots clearly show that there are several distinct local optima – even if only the four most important variables according to importance estimations of the GAM model and further studies [15] are considered. There is seemingly not one dominant basin of attraction. Based on this fact, employing the Matérn kriging model that always leads the optimization algorithm to one search space region may become risky. Obtaining different local optima – if they indeed exist – is surely preferable in order to have alternatives for the optimal solution on the model, in case it is infeasible on the actual problem. In particular since the optimization on the surrogate model is very fast compared to an evaluation on the original problem. However, the validated result of the Matérn kriging is extremely good (0.912). The best of the initial points only obtains a validated score of 0.900. The model thus manages to find better solutions. Only four of the space-filling 440 points of the validation set (0.938, 0.928, 0.918, 0.915) are better. For the best validated solution of the random forest (0.917), the situation is similar. However, this solution is not identified as the global optimum of the model (cf. Figure 2).

Summarizing, the second hypothesis cannot be completely accepted. The kriging models either simplify the characteristics of the true problem (Matérn) or result in undesired solutions (Gauss), whereby the Matérn model is surely the better choice. It may just have the empirical information to detect only one of the basins. The multimodality predicted by the Gaussian kernel is also based on undesired extrapolation effects.

6   Conclusion and Outlook

With respect to three questions posed in the introduction, the first question has to be negated. The prediction quality can only roughly distinguish the suitability of a model for optimization. As an answer to the second question, the Matérn kriging model obtained the best results for both, prediction quality and optimization performance. The performance of the approximated optimum is competitive with the best results of the large validation set which required around 20 times more resources to compute. This achievement, however, comes with a simplification of the model characteristics, maybe caused by a smoothing of true response surface. As a consequence, the successful results of kriging-based optimizers could be transferred to higher dimensions and noisy data, although there is a strong dependence on the applied covariance kernel. For a more robust optimization performance, sequential approaches, which refine the model based on the additional evaluations, can be considered [5, 11, 13, 14].

The results obtained in this study are a first step towards a combined analysis of prediction quality and optimization capability on complex practical problems. Their generality is of course questionable due to the restriction to a single problem instance. In the future, automatically generated instances of the car optimization problem with different tracks, cars, and bots and also instances from other related problems can be used to perform a much broader simulation study in order to improve on this drawback. Results of other surrogate modeling approaches on the open test cases defined in this paper can assist in providing guidelines for applications. In any case we would like to emphasize that modeling difficult noisy high-dimensional problems is obviously useful and possible.

Acknowledgments. This paper is based on investigations of project D5 *Synthesis and multi-objective model-based optimisation of process chains for manufacturing parts with functionally graded properties* as part of the collaborative research center SFB/TR TRR 30, which is kindly supported by the Deutsche Forschungsgemeinschaft (DFG). We would like to thank Markus Kemmerling for providing the evaluator code for the car setup optimization competition and Olaf Mersmann for consulting with respect to the R implementation of the CMA-ES.

References

1. Biermann, D., Weinert, K., Wagner, T.: Model-based optimization revisited: Towards real-world processes. In: Michalewicz, Z., Reynolds, R.G. (eds.) Proc. 2008 IEEE Congress on Evolutionary Computation (CEC 2008), pp. 2980–2987. IEEE Press, Piscataway (2008)
2. Breiman, L.: Random forests. Machine Learning 45(1), 5–32 (2001)
3. Cardamone, L., Loiacono, D., Lanzi, P.L.: Car Setup Optimization. Competition Software Manual. Tech. Rep. 2010.1, Dipartimento di Elettronica e Informazione, Politecnico di Milano, Italy (2010)
4. Cressie, N.: The origins of kriging. Mathematical Geology 22(3), 239–252 (1990)

5. Forrester, A.I.J., Keane, A.J., Bressloff, N.W.: Design and analysis of 'noisy' computer experiments. AIAA Journal 44(10), 2331–2339 (2006)
6. Ginsbourger, D., Picheny, V., Roustant, O., Richet, Y.: Kriging with heterogeneous nugget effect for the approximation of noisy simulators with tunable fidelity. In: Proc. Joint Meeting of the Statistical Society of Canada and the Société Francaise de Statistique (2008)
7. Hansen, N., Ostermeier, A.: Completely derandomized self-adaptation in evolution strategies. Evolutionary Computation 9(2), 159–195 (2001)
8. Hastie, T., Tibshirani, R.: Generalized additive models. Statistical Science 1(3), 297–318 (1986)
9. Hastie, T., Tibshirani, R., Friedman, J.: The Elements of Statistical Learning: Data Mining, Inference, and Prediction, 2nd edn. Springer (2009)
10. Horton, R.L.: The General Linear Model. McGraw-Hill, New York (1978)
11. Huang, D., Allen, T.T., Notz, W.I., Zheng, N.: Global optimization of stochastic black-box systems via sequential kriging meta-models. Journal on Global Optimization 34(4), 441–466 (2006)
12. Jin, Y.: A comprehensive survey of fitness approximation in evolutionary computation. Soft Computing 9(1), 3–12 (2005)
13. Jones, D.R.: A taxonomy of global optimization methods based on response surfaces. Journal of Global Optimization 21(4), 345–383 (2001)
14. Jones, D.R., Schonlau, M., Welch, W.J.: Efficient global optimization of expensive black-box functions. Journal of Global Optimization 13(4), 455–492 (1998)
15. Kemmerling, M.: Optimierung der Fahrzeugabstimmung auf Basis verrauschter Daten einer Autorennsimulation. Diplomarbeit, TU Dortmund (2010) (in German)
16. Kemmerling, M., Preuss, M.: Automatic adaptation to generated content via car setup optimization in torcs. In: Proceedings of the 2010 IEEE Conference on Computational Intelligence and Games, CIG 2010, pp. 131–138. IEEE (2010)
17. Kleijnen, J.P.C.: Kriging metamodeling in simulation: A review. European Journal of Operational Research 192(3), 707–716 (2009)
18. Powell, M.J.D.: The bobyqa algorithm for bound constrained optimization without derivatives. Tech. Rep. DAMTP 2009/NA06, Centre for Mathematical Sciences, University of Cambridge, UK (2009)
19. Preuss, M., Rudolph, G., Wessing, S.: Tuning optimization algorithms for real-world problems by means of surrogate modeling. In: Proc. 12th Annual Conference on Genetic and Evolutionary Computation (GECCO 2010), pp. 401–408. ACM, New York (2010)
20. Rasmussen, C.E., Williams, C.K.I.: Gaussian processes for machine learning. Springer (2006)
21. Sacks, J., Welch, W.J., Mitchell, T.J., Wynn, H.P.: Design and analysis of computer experiments. Statistical Science 4(4), 409–423 (1989)
22. Tenne, Y., Izui, K., Nishiwaki, S.: Dimensionality-reduction frameworks for computationally expensive problems. In: Fogel, G., Ishibuchi, H. (eds.) Proc. 2010 IEEE Congress on Evolutionary Computation (CEC 2010), pp. 1–8. IEEE Press, Piscataway (2010)
23. Wang, G.G., Shan, S.: Review of metamodeling techniques in support of engineering design optimization. Journal of Mechanical Design 129(4), 370–380 (2007)

Pilot, Rollout and Monte Carlo Tree Search Methods for Job Shop Scheduling

Thomas Philip Runarsson[1], Marc Schoenauer[2,4], and Michèle Sebag[3,4]

[1] School of Engineering and Natural Sciences, University of Iceland
[2] TAO, INRIA Saclay Île-de-France, Orsay, France
[3] TAO, LRI, UMR CNRS 8623, Orsay, France
[4] Microsoft Research-INRIA Joint Centre, Orsay, France

Abstract. Greedy heuristics may be attuned by looking ahead for each possible choice, in an approach called the rollout or Pilot method. These methods may be seen as meta-heuristics that can enhance (any) heuristic solution, by repetitively modifying a master solution: similarly to what is done in game tree search, better choices are identified using lookahead, based on solutions obtained by repeatedly using a greedy heuristic. This paper first illustrates how the Pilot method improves upon some simple well known dispatch heuristics for the job-shop scheduling problem. The Pilot method is then shown to be a special case of the more recent Monte Carlo Tree Search (MCTS) methods: Unlike the Pilot method, MCTS methods use random completion of partial solutions to identify promising branches of the tree. The Pilot method and a simple version of MCTS, using the ε-greedy exploration paradigms, are then compared within the same framework, consisting of 300 scheduling problems of varying sizes with fixed-budget of rollouts. Results demonstrate that MCTS reaches better or same results as the Pilot methods in this context.

1 Introduction

In quite a few domains related to combinatorial optimization, such as constraint solving [1], planning or scheduling [2], software environments have been designed to achieve good performances in expectation over a given distribution of problem instances. Such environments usually rely on a portfolio of heuristics, leaving the designer with the issue of finding the best heuristics, or the best heuristics sequence, for his particular distribution of problem instances.

The simplest solution naturally is to use the default heuristics, assumedly the best one on average on all problem instances. Another approach, referred to as *Pilot or rollout* method, iteratively optimizes the option selected at each choice point [3,4], while sticking to the default heuristics for other choice points. Yet another approach, referred to as *Monte-Carlo Tree Search* (MCTS) [5] and at the origin of the best current computer-Go players [6], has been proposed to explore the search space while addressing the exploration *versus* exploitation dilemma in a principled way; as shown by [7], MCTS provides an approximation to optimal Bayes decision. The MCTS approach, rooted in the multi-Armed bandit (MAB)

Y. Hamadi and M. Schoenauer (Eds.): LION 6, LNCS 7219, pp. 160–174, 2012.
© Springer-Verlag Berlin Heidelberg 2012

setting [8], iteratively grows a search tree through tree walks. For each tree walk, in each node (choice point) the selection of the child node (heuristics) is handled as a MAB problem; the search tree thus asymmetrically grows to explore the most promising tree paths (the most promising sequences of heuristics).

Whereas increasingly used today in sequential decision making algorithms including games [9,10], to our best knowledge MCTS methods have rarely been used within the framework of combinatorial optimization, with the recent exception of [11]. This paper investigates the application of MCTS to job-shop scheduling, an NP-hard combinatorial optimization problem. As each job-shop scheduling problem instance defines a deterministic optimization problem, the standard *upper confidence bound applied to tree* (UCT) framework used in [11] does not apply. More precisely, the sought solution is the one with best payoff (as opposed to the one with best payoff on average); the MAB problem nested in the UCT thus is a max k-armed bandit problem [12,9]. Along this line, the randomized aspects in UCT must be addressed specifically to fit deterministic problems. Specifically, a critical difficulty lies in the randomized default handling of the choice points which are outside the current search tree (in contrast, these choice points are dealt with using the default heuristics in the pilot methods). Another difficulty, shared with most MCTS applications, is to preserve the exploration/ exploitation trade-off when the problem size increases. A domain-aware randomized default handling is proposed in this paper, supporting a MCTS-based scheduling approach called *Monte-Carlo Tree Scheduling* (MCS). MCS is empirically validated, using well established greedy heuristics and the pilot methods based on these heuristics as baselines. The empirical evidence shows that Pilot methods significantly outperform the best-known default heuristics; MCS significantly outperforms on the Pilot methods for small problem sizes. For larger problem sizes, however, MCS is dominated by the best Pilot methods, which is partly explained from the experimental setting as the computational cost of the Pilot methods is about 4 times higher than that of the MCS one. That is, heuristic scheduling is more costly than random scheduling.

The paper is organized as follows. Job-shop scheduling is introduced in Section 2, together with some basic greedy algorithms based on domain-specific heuristics called *dispatching rules*. The generic pilot method is recalled in section 3. The general MCTS ideas are introduced in section 4; its adaptation to the job-shop scheduling problem is described and an overview of MCS is given. and together with its application to combinatorial problems, and to the job-shop scheduling problem. Section 5 is devoted to the empirical validation of the proposed approach. After describing the experimental setting, the section reports on the MCS results on different problem sizes, with the dispatching rules and the Pilot methods as baselines. The paper concludes with a discussion of these results and some perspectives for further research.

2 Job Shop Scheduling and Priority Dispatching Rules

Scheduling is the sequencing of the order in which a set of *jobs* $j \in J := \{1, .., n\}$ are processed through a set of *machines* $a \in M := \{1, .., m_j\}$. In a *job shop*, the order in which a job is processed through the machines is predetermined. In a *flow shop* this order is the same for all jobs, and in a *open shop* the order is arbitrary. We will consider here only the *job shop*, where the jobs are strictly-ordered sequences of operations. A job can only be performed by one type of machine and each machine processes one job at a time. Once a job is started it must be completed. The performance metric for scheduling problems is generally based on flow or dues date. Here we will consider the completion time for the last job or the so called makespan.

Each job has a specified processing time $p(j, a)$ and the order through the machines is given by the permutation vector σ ($\sigma(j, i)$ is the ith machine for job j). Let $x(j, a)$ be the start time for job j on machine a, then

$$x(j, \sigma(j, i)) \geq x(j, \sigma(j, i-1)) + p(j, \sigma(j, i-1)) \quad j \in \{1, .., n\}, \ i \in \{2, .., m_j\} \quad (1)$$

The disjunctive condition that each machine can handle at most one job at a time is the following:

$$x(j, a) \geq x(k, a) + p(k, a) \quad \text{or} \quad x(k, a) + p(k, a) \geq x(j, a) \quad (2)$$

for all $j, k \in J, j \neq k$ and $a \in M$. The makespan can then be formally defined as

$$z = \max\{x(j, \sigma(j, m_j)) + p(j, m_j) \mid j \in J\}. \quad (3)$$

Smaller problems can be solved using a specialized branch and bound procedure [13] and an algorithmic implementation may be found as part of LiSA [14]. Jobs up to 14 jobs and 14 machines can still be solved efficiently, but at higher dimensions, the problems rapidly become intractable. Several heuristics have been proposed to solve job shop problems when their size becomes too large for exact methods. One such set of heuristics are based on *dispatch rules*, i.e. rules to decide which job to schedule next based on the current state of all machines and jobs. A survey of over 100 such rules may be found in [15]. Commonly used priority dispatch rules have been compared on a number of benchmark problems in [16]. When considering the makespan as a performance metric, the rule that selects a job which has the *Most WorK Remaining* (MWKR, the job with the longest total remaining processing time) performed overall best. It was followed by the rule that selects a job with the *Shortest Processing Time* (SPT), and by the rule that selects a job which the *Least Operation Number* (LOPN). These rules are among the simplest ones, and are by no means optimal. However, only these 3 rules will be considered in the remaining of this paper. In particular, experimental results of the corresponding 3 greedy algorithms can be found in Section 5.

The simplest way to use any of these rules is to embed them in a greedy algorithm: the jobs are processed in the order given by the repeated application of the chosen rule. Algorithm 1 gives the pseudo-code of such an algorithm. The variable t_j represents which machine is next in line for job j (more precisely machine $\sigma(j, t_i)$). When starting with an empty schedule, one would set $t_j \leftarrow 1$ for $j \in J$ and $\mathcal{S} = \emptyset$. At each step of the algorithm, one job is chosen according to the dispatching rule **R** (line 2), and the job is scheduled on the next machine in its own list, i.e., the pair (job, machine) is added to the partial schedule $\mathcal{S}$ (line 3) ($\oplus$ denotes the concatenation of two lists).

Algorithm 1. Greedy (Pilot) heuristic

---

    input : Partial sequence $\mathcal{S}_0$, $\mathbf{t} = (t_1, \ldots, t_n)$, and heuristic **R**
    output: An objective to maximize, for example negative makespan

1 **while** $\exists j \in J$; $t_j < m_j$ **do**
2      $b = \mathbf{R}(\mathcal{S}, t_j ; t_j < m_j)$;   // *Apply* **R** *to current partial schedule, get next job*
3      $\mathcal{S} \leftarrow \mathcal{S} \oplus \{(b, \sigma(b, t_b))\}$;   // *Schedule job on its next machine*
4      $t_b \leftarrow t_b + 1$;   // *Point to next machine for job b*
5 **end**

---

3 Pilot Method

The *pilot method* [3,17] or equivalently the *Rollout algorithm* [18] can enhance any heuristic by a simple look-ahead procedure. The idea is to add one-step look-ahead and hence apply greedy heuristics from different starting points. The procedure is applied repeatedly, effectively building a tree. This procedure is not unlike strategies used in game playing programs, that search a game trees for good moves. In all cases the basic idea is to examine all possible choices with respect to their future advantage. An alternative view is that of a sequential decision problem or dynamic programming problem where a solution is built in stages, whereby the components (in our cases the jobs) are selected one-at-a-time. The first k components form a so called k-solution [18]. In the same way as a schedule was built in stages in Algorithm 1, where the k-solution is the partial schedule $\mathcal{S}$. However, for the Pilot method the decisions made at each stage will depend on a look-ahead procedure. The Pilot method is then described in Algorithm 2. The algorithm may seem a little more complicated than necessary, however, as will be seen in the next section this algorithm is a special case of Monte Carlo tree search. The heuristic rollout is performed B times and each time adding a node to the tree. Clearly if all nodes can be connected to a terminal node, the repetition may be halted before the budget B is reached. This is not shown here for clarity. Furthermore, a new leaf on the tree is chosen such that those closer to the root have priority else branches are chosen arbitrarily with equal probability. In some version of the Pilot method, the tree is not expanded breadth first manner but with some probability allows for depth first search. This would be equivalent to executing line 8 with some probability. This is also commonly used in MCTS and is called progressive widening.

The greedy algorithm 1 is then used as the Rollout algorithm on line 23. As will be seen in the following section, the key difference between the MCTS and Pilot method is in the way a node is found to expand in the tree and the manner in which a rollout is performed. Other details of the Algorithm 2 will also become clearer.

Algorithm 2. Pilot or rollout algorithm

---

input : Budget B, partial sequence $\mathcal{S}_0$, $\mathbf{t} = (t_1, \ldots, t_n)$, and heuristic $\mathbf{R}$
output: Decision, job to dispatch next b

1   $root \leftarrow node$; // *initialize the root node*
2   $node.n \leftarrow 0, node.\mathbf{t} \leftarrow \mathbf{t}, node.child \leftarrow \emptyset$;
3   **for** $n \leftarrow 1$ **to** B **do**
4       $\mathcal{S} \leftarrow \mathcal{S}_0$;   // *set state to root node state and climb down the tree*
5   **while** $node.child \neq \emptyset$ **do**
6   **for** $j \in J; node.t_j < m_j$ **do**
7   **if** $node.n = 0$ **then**
8                   $Q(j) = \infty$
9   **else**
10                  $Q(j) = U(0, 1)$;   // *random value between 0 and 1*
11   **end**
12   **end**
13          $j' = \arg\max_{j \in J; t_j < m_j} Q(j)$;   // *largest Q value, break ties randomly*
14          $\mathcal{S} \leftarrow \mathcal{S} \oplus \{(j', \sigma(j', t_{j'}))\}$;   // *dispatch job j'*
15   **end**
        ; // *expand node if possible, i.e. $\mathcal{S}$ is not the complete schedule*
16   **for** $j \in J; node.t_j < m_j$ **do**
17          $node.child[j].parent \leftarrow node$;   // *keep pointer to parent node*
18          $node.child[j].child \leftarrow \emptyset$;   // *this node has not been expanded*
19          $node.child[j].n \leftarrow 0$;   // *and has not been rolled out*
20          $node.child[j].\mathbf{t} \leftarrow node.\mathbf{t}$;   // *copy machine counter from parent node*
21          $node.child[j].t_j \leftarrow node.t_j + 1$;   // *increment machine counter for job*
22   **end**
23      $R = \texttt{Rollout} (\mathcal{S}, node[\mathcal{S}].\mathbf{t}, \mathbf{R})$; // *Complete the solution via Pilot heuristic*
24   **repeat** *propagate result of rollout up the tree*
25          $node.n \leftarrow node.n + 1$;   // *number of visits incremented by one*
26          $node.Q \leftarrow \max(node.Q, R)$;   // *best found solution*
27          $node \leftarrow node.parent$;   // *climb up the tree to parent node*
28   **until** $node \neq root$;
29 **end**
30 $\arg\max_{j \in J; t_j < m_j} root.child(j).Q$

---

4   MCTS for Combinatorial Optimization

4.1   Monte Carlo Tree Search

Monte-Carlo Tree Search inherits from the so-called Multi-Armed Bandit (MAB) framework [8]. MAB considers a set of independent k arms, each with a different payoff distribution. Here each arm corresponds to selecting a job to be dispatched and the payoff the results returned by a rollout or greedy heuristic. Several goals have been considered in the MAB setting; one is to maximize the cumulative payoff gathered along time (k-arm bandit) [19]; another one is to identify the arm with maximum payoff (max-k arm) [20,12,9]. At one extreme is the exploitation-only strategy (selecting the arm with best empirical reward);

at the other extreme is the exploration-only strategy (selecting an arm with uniform probability).

When it comes to find a sequence of options, the search space is structured as a tree[1]. In order to find the best sequence, a search tree is iteratively used and extended, growing in an asymmetric manner to focus the exploration toward the best regions of the search space. In each iteration, a tree path a.k.a simulation is constructed through three building blocks: the first one is concerned with navigating in the tree; the second one is concerned with extending the tree and assessing the current tree path (reward); the third one updates the tree nodes to account for the reward of the current tree path.

Descending in the Tree. The search tree is initialized to the root node (current partial schedule). In each given node until arriving at a leaf, the point is to select among the child nodes of the current node (Fig. 1, left). For deterministic optimization problems, the goal is to maximize the maximum (rather than the expected) payoff. For this aim, a sound strategy has been introduced in [12] and used in [9]. This approach, referred to as Chernoff rule, estimates the upper bound on the maximum payoff of the arm, depending on its number of visits and the maximum value gathered.

However, the main goal of this work is to bridge the gap between the Pilot method and MCTS algorithms. Indeed, the Pilot method, as presented in algorithm 2, can be viewed as an MCTS algorithm in which the strategy used to chose next child to explore is to choose the best child after one deterministic rollout using the dispatch rule at hand – a rather greedy exploitation-oriented strategy. Such strategy is very close to a simple rule to balance exploration and exploitation known in the MCTS world as ϵ-greedy: with probability $1 - \epsilon$, one selects the empirically best child node[2] (i.e. the one with maximum empirical value); otherwise, another uniformly selected child node is retained. Furthermore, similar to the Pilot method described in the previous section, unexplored nodes (line: 8 in Algorithm2) will have priority. However, line: 10 should be replaced by

$$Q(j) \leftarrow node.child[j].Q.$$

Extending the Tree and Evaluating the Reward. Upon arriving in a leaf, a new option is selected and added as child node of the current one; the tree is thus augmented of one new node in each simulation (Fig. 1, right). The simulation is resumed until arriving in a final state (e.g., when all jobs have been processed). As already mentioned, the choices made in the further choice points in the Pilot method rely on the default heuristics (and the rollout is hence deterministic). In the MCTS method however, these choices must rely on randomized heuristics

---

[1] Actually, the search space may be structured as a graph if different paths can lead to a same state node. In the context of job-shop scheduling however, only a tree-structured search space needs be considered.

[2] Typically $\varepsilon = 0.1$.

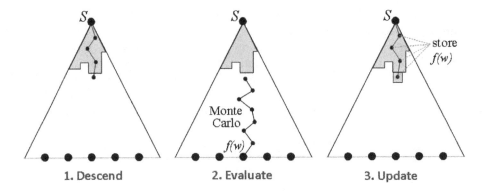

Fig. 1. Monte-Carlo Tree Search: the tree is asymmetrically grown toward the most promising region (in grey, the known part of the tree). Each simulation is made of a MAB phase (1); then the simulation is completed until arriving at a final state (2); finally, the first node of the Monte-Carlo path is added to the known tree, and the reward is computed and back-propagated to all nodes of the path that are in the known tree (3).

out of consistency with the MAB setting. The question thus becomes which heuristics to use in the so-called random phase (see section 4.2). Upon arriving in a final state, the reward associated to the simulation is computed (e.g., the makespan of the current schedule).

Updating the Tree. The number of visits of the nodes of the current tree that were in the path is incremented by 1; likewise, the cumulative reward associated to these nodes is incremented by the reward associated to the current path. Note that other statistics can be maintained in each node, such as the RAVE values [6], and must be updated there too.

4.2 Monte-Carlo Tree Schedule (MCS)

As already pointed out, solving a combinatorial problem can be viewed as a sequential decision making process, incrementally building a solution by choosing an element of a partial solution at a time (e.g., next town for TSP problems, of next machine to schedule for the job shop scheduling problem). In order to solve this sequential decision problem through a MCTS algorithm, several specific issues must be considered.

A first issue concerning the reward design has a significant impact on the exploration *versus* exploitation dilemma; it might require some instance-dependent parameter to reach a proper balance (see e.g., [21]). Indeed, in the case of job-shop scheduling for instance, different instances will have very different makespans.

A second issue concerns the heuristics to be used in the random phase (section 4.1). The original MCTS method [5] advocates pure random choices. Domain knowledge could however be used to propose a smarter procedure, e.g. using random selected dispatching rules for job-shop scheduling. Still, a lesson consistently learned from MCTS applications [6,9] is that doing many simulations completed with a brute random phase is more effective than doing less simulations completed with a smart final phase. For instance in the domain of computer-Go, the overall results were degraded by using gnuGo in the random phase, compared to a uniform move selection. Likewise, the use of the three dispatching rules (described in Section 2) in the random phase was outperformed by a pure random strategy, uniformly selecting the next job to be considered. Here we have chosen to follow this path and our rollout phase consists of the purely random dispatching of jobs. Replacing this with line 23 in Algorithm 2, along with the ε–greedy policy, and the Pilot method is transformed into MCTS.

Another important detail must be taken into account: When performing random rollouts, we might forget the best choices found previously during the building of a schedule. For this reason, a global best found sequence is kept throughout the scheduling procedure. If a suboptimal choice is found in a later partial schedule (k-solution), the globally better choice previously found during a random rollout will be forced. In some sense the idea here is similar to that of the fortified rollouts used by the Pilot method [17].

Finally, in MCTS, the stopping criterion is defined by the total number of simulations (i.e., rollouts here), and one single decision is taken after a complete tree exploration, chosen as the child of the root node with the maximum expected reward [6]. The situation is rather different here, and, in this first approach to MCTS for combinatorial optimization, similarly to the Pilot method, a single tree exploration is done, with a limited budget in terms of number of rollouts. A complete schedule is then built by descending the tree and always choose the child with maximum expected reward (makespan).

5   Experimental Results

This section reports on the experimental validation of the proposed approaches. After detailing the experimental setting, the results of the MCS method is reported and compared to the baseline methods, the greedy application of the three dispatching rules MWKR, SPT and LOPN (section 2), and the pilot methods built on these three dispatching rules (section 3).

5.1   Experimental Settings

All experiments have been conducted using the set of test instances proposed in [22]. The machine orders for the jobs are randomly generated and the processing times are discrete values uniformly distributed between 1 and 200. Three different $n \times m$ problem sizes were generated using this setup, 6×6, 10×10 and 14×14. The optimal makespans for one hundred instances generated of each size was then found using Brucker's branch and bound algorithm [13]. A further four instance of size 20×20 are also tested [23] and compared with their best known solution [24].

Table 1. Greedy algorithms: Performance statistics of the MWKR, SPT and LOPN rules on three problem sizes

Instance	Heuristic	min	mean	median	stdev	max	#opt
6 × 6	MWKR	1.000	1.155	1.151	0.084	1.384	2
	SPT	1.137	1.399	1.390	0.150	1.816	0
	LOPN	1.017	1.176	1.178	0.083	1.369	0
10 × 10	MWKR	1.096	1.228	1.222	0.069	1.430	0
	SPT	1.303	1.654	1.644	0.166	2.161	0
	LOPN	1.103	1.216	1.208	0.061	1.369	0
14 × 14	MWKR	1.159	1.264	1.261	0.052	1.399	0
	SPT	1.584	2.012	2.015	0.244	2.721	0
	LOPN	1.150	1.253	1.250	0.048	1.376	0

For all methods except the greedy ones, the time budget is varied to assess the convergence behavior of the pilot and MCS optimization methods, considering a total of 100, 1,000 and 5,000 rollouts a.k.a. simulation. Each rollout corresponds to designing and evaluating a complete solution (computing its total makespan). It is worth noting that not all rollouts are equally expensive; the rollout based on a dispatching rule (as used in the pilot methods) is more computationally demanding than the random rollout used in MCS, all the more so as the size of the problem instance increases. Nevertheless, the fixed rollout budget is meant to allow CPU-independent comparisons and assess the empirical behavior of the methods under restricted computational resources (e.g. in real-world situations).

For each method, each problem size and each time budget, the result is given as the average over 100 problem instances of the normalized makespan (1. being the optimal value), together with the minimum, maximum and median values, and the standard deviation; the number of times where the optimal value was found is additionally reported.

While the greedy and pilot algorithms actually are deterministic[3], MCS is not. The usual way to measure the performance of a stochastic algorithm on a given problem domain is through averaging the result out of a few dozen or hundred independent runs. For the sake of computational convenience however, MCS was run only once on each problem instance and the reported result is the average over the 100 independent instances.

5.2   The Greedy Algorithms

The results of the greedy algorithms are depicted in Table 1 for the three dispatching rules MWKR, SPT and LOPN (section 2), showing that MWKR and LOPN behave similarly and significantly outperform SPT. Further, the performances of SPT significantly decrease with the problem size, whereas MWKR and LOPN demonstrate an excellent scalability in the considered size range.

---

[3] Up to ties between jobs.

Table 2. Pilot algorithms. Performance statistics for Pilot algorithm using 3 different Pilot heuristics: MWKR, SPR, and LOPN, on the three different problem sizes.

Instance	Heuristic	min	mean	median	stdev	max	#opt
6 × 6	Pilot(MWKR,100)	1.000	1.049	1.050	0.038	1.167	14
	Pilot(MWKR,1000)	1.000	1.035	1.030	0.034	1.180	23
	Pilot(MWKR,5000)	1.000	1.025	1.014	0.029	1.104	33
	Pilot(SPT,100)	1.000	1.100	1.093	0.066	1.293	3
	Pilot(SPT,1000)	1.000	1.065	1.060	0.050	1.287	7
	Pilot(SPT,5000)	1.000	1.052	1.049	0.045	1.265	14
	Pilot(LOPN,100)	1.000	1.058	1.057	0.044	1.172	12
	Pilot(LOPN,1000)	1.000	1.046	1.036	0.040	1.134	17
	Pilot(LOPN,5000)	1.000	1.034	1.024	0.032	1.127	22
10 × 10	Pilot(MWKR,100)	1.032	1.109	1.109	0.039	1.217	0
	Pilot(MWKR,1000)	1.006	1.097	1.096	0.039	1.215	0
	Pilot(MWKR,5000)	1.004	1.082	1.083	0.035	1.158	0
	Pilot(SPT,100)	1.102	1.222	1.221	0.063	1.427	0
	Pilot(SPT,1000)	1.066	1.188	1.188	0.055	1.332	0
	Pilot(SPT,5000)	1.063	1.172	1.168	0.048	1.296	0
	Pilot(LOPN,100)	1.044	1.117	1.114	0.041	1.219	0
	Pilot(LOPN,1000)	1.028	1.106	1.105	0.042	1.212	0
	Pilot(LOPN,5000)	1.022	1.096	1.092	0.032	1.171	0
14 × 14	Pilot(MWKR,100)	1.081	1.156	1.155	0.036	1.256	0
	Pilot(MWKR,1000)	1.046	1.142	1.138	0.036	1.247	0
	Pilot(MWKR,5000)	1.046	1.129	1.129	0.034	1.230	0
	Pilot(SPT,100)	1.239	1.389	1.380	0.080	1.595	0
	Pilot(SPT,1000)	1.136	1.316	1.319	0.065	1.508	0
	Pilot(SPT,5000)	1.153	1.286	1.283	0.060	1.517	0
	Pilot(LOPN,100)	1.076	1.160	1.161	0.036	1.285	0
	Pilot(LOPN,1000)	1.080	1.149	1.152	0.037	1.264	0
	Pilot(LOPN,5000)	1.078	1.145	1.142	0.033	1.248	0

5.3   The Pilot Method

The results of the pilot method (section 3) related to the three above dispatching rules and three time budgets (100, 1000 and 5000) are displayed in table 2 for all three problem sizes 6 × 6, 10 × 10, and 14 × 14. For the sake of easy comparison with the greedy algorithm, the median performances respectively obtained on the same problem sizes displayed on Table 3.

As was expected, and demonstrated on some TSP instances by [3], the pilot method does improve on the greedy algorithm. With a time budget 100, a significant improvement is observed for all three methods, and confirmed by the number of times the optimal solution is discovered on problem size 6 x 6.

It is worth noting that the performance only very slightly improves when the time budget increases from 100 to 1000, and from 1000 to 5000, despite the significant increase in the computational effort. In particular, the optimal solutions are never discovered for higher problem sizes.

Table 3. Comparison of median performances of Greedy and Pilot with different budgets, for the 3 heuristics (from Tables 1 and 2)

Instance	Heuristic	Algorithm (budget)		
		Pilot (100)	Pilot (1000)	Pilot (5000)
6 × 6	MWKR	1.151	1.050	1.030	1.014
	SPT	1.390	1.093	1.060	1.049
	LOPN	1.178	1.057	1.036	1.024
10 × 10	MWKR	1.222	1.109	1.096	1.083
	SPT	1.644	1.221	1.188	1.168
	LOPN	1.208	1.114	1.105	1.092
14 × 14	MWKR	1.261	1.155	1.138	1.129
	SPT	2.015	1.380	1.319	1.283
	LOPN	1.250	1.161	1.152	1.142

Lastly, the performance order of the three rules is not modified when using the pilot method: MWKR and LOPN significantly outperform SPT. As a consequence the Pilot method can be quite sensitive to the Pilot heuristic chosen.

5.4   Monte-Carlo Tree Schedule

Table 4 reports on the results of the MCS approach described in (section 4). On problem size 6x6, MCS significantly improves on the best Pilot method (MKWR with 5000 rollout budget), as also witnessed by the number of times the optimal solution is found. On problem size 10x10, the average and median performances are comparable; still, the optimal solution is found twice by MCS with a 5000 rollout budget, whereas it is never found by the Pilot method. On problem size 14x14, Pilot (MWKR,5000) is significantly better than MCS. A first explanation for this fact relies on the computational effort: As already mentioned, the computational time required for a 5,000 rollout Pilot is circa 4 times higher than for a 5,000 rollout MCS. A second explanation is the fact that, as the tree depth increases with the problem size, it becomes necessary to adjust the parameters controlling the branching factor of the MCS tree. This can be achieved by introducing progressive widening (on-going work).

5.5   Larger Instances

The best known solutions (BKS) for the 20 × 20 benchmark problem [23] are taken from [24]. Here we demonstrate the performance of the MCS on problems that cannot be solved using exact methods. Only four instances are considered here, and the MCS is run 30 times on each instance. The method is unable to find any best known solution. Nevertheless, the performance does not degrade significantly when compared to the results obtained on the 10 × 10 problems.

Table 4. Monte-Carlo Tree Schedule. Performance statistics using ε-greedy and random scheduling.

Instance	Heuristic	min	mean	median	stdev	max	#opt
6 × 6							
	MCS(ε-greedy,100)	1.000	1.026	1.020	0.027	1.097	28
	MCS(ε-greedy,1000)	1.000	1.014	1.001	0.025	1.150	50
	MCS(ε-greedy,5000)	1.000	1.007	1.000	0.017	1.082	72
10 × 10							
	MCS(ε-greedy,100)	1.029	1.141	1.140	0.057	1.378	0
	MCS(ε-greedy,1000)	1.014	1.095	1.098	0.036	1.199	0
	MCS(ε-greedy,5000)	1.000	1.070	1.070	0.032	1.135	2
14 × 14							
	MCS(ε-greedy,100)	1.173	1.346	1.331	0.083	1.634	0
	MCS(ε-greedy,1000)	1.127	1.284	1.280	0.074	1.552	0
	MCS(ε-greedy,5000)	1.064	1.232	1.223	0.065	1.471	0

Table 5. Results for the MCS on four 20 × 20 instances. Performance statistics is given for 30 independent runs on each instance, yn01,..., yn04.

Instance	Heuristic	min	mean	median	stdev	max	#BKS
yn01	MCS(ε-greedy,100)	1.166	1.211	1.210	0.020	1.245	0
	MCS(ε-greedy,1000)	1.144	1.186	1.183	0.021	1.235	0
	MCS(ε-greedy,5000)	1.128	1.167	1.166	0.023	1.209	0
yn02	MCS(ε-greedy,100)	1.135	1.185	1.181	0.024	1.228	0
	MCS(ε-greedy,1000)	1.123	1.156	1.153	0.023	1.218	0
	MCS(ε-greedy,5000)	1.090	1.130	1.130	0.023	1.174	0
yn03	MCS(ε-greedy,100)	1.156	1.205	1.204	0.022	1.267	0
	MCS(ε-greedy,1000)	1.131	1.178	1.180	0.020	1.222	0
	MCS(ε-greedy,5000)	1.110	1.151	1.148	0.023	1.199	0
yn04	MCS(ε-greedy,100)	1.063	1.108	1.107	0.025	1.160	0
	MCS(ε-greedy,1000)	1.039	1.084	1.090	0.025	1.133	0
	MCS(ε-greedy,5000)	1.023	1.067	1.064	0.019	1.102	0

6   Discussion and Perspectives

The main contribution of this paper is to demonstrate the feasibility of using MCTS to address job-shop scheduling problems. This result has been obtained by using the simplest exploration/exploitation strategy in MCTS, the ε-greedy strategy, defining the *Monte-Carlo Tree Scheduling* approach (MCS). The empirical evidence gathered from the preliminary experiments presented here shows that MCS significantly outperforms its competitors on small and medium size problems. For larger problem sizes however, the Pilot method with the best dispatching rule outperforms this first verions of MCS. This fact is blamed on our adversary experimental setting, as we compared methods based on the number of rollouts, whereas the computational cost of a rollout is larger by almost an order of magnitude in the Pilot framework, as compared to that of the MCS.

The MCS scalability can also be improved through reconsidering the exploration vs exploitation trade-off, ever more critical in larger-sized problem instances. First of all, the MAX-k-arm strategy should be tried in lieu of the simple ε-greedy rule. Furthermore, this tradeoff can be also adjusted by avoiding the systematic first trial of all possible children, as this becomes harmful for large number or arms (jobs here). It is possible to control when a new child node should be added in the tree, and which one. Regarding the former aspect, a heuristics referred to as *Progressive Widening* has been designed to limit the branching factor of the tree, e.g. [10] . Regarding the second aspect, the use of a Rapid Action Value Estimate (RAVE), first developed in the computer-Go context [6] can be very efficient to aggregate the various rewards computed for the same option (*Queen Elisabeth*), and guide the introduction of the most efficient rules/jobs in average.

7   Conclusion and Outlook

This work has shown how the Pilot method may be considered a special case of MCTS, with an exploratory-only strategy to traversing the tree and using a deterministic rollout driven by the Pilot heuristic. It has demonstrated that the Pilot method can be sensitive to the chosen Pilot heuristic. As the chosen Pilot heuristic becomes more effective, so too may its computational costs. An extension of the Pilot method in the realm of MCTS algorithms, the MCS, has been proposed, using a simple ϵ–greedy strategy to traverse down the tree. However, more sophisticated strategies, such as the one based on the max-k bandit problem [12,20], need now be investigated. For larger problems, progressive widening should be an avenue for further research, as similar strategies have already been investigated in the Pilot framework. Finally, Rapid Action Value Estimates may not only be used to bias how the tree is traversed, possibly replacing the exploration term in the bandit formulas, but can also help to improve over the random rollouts.

References

1. Xu, L., Hutter, F., Hoos, H.H., Leyton-Brown, K.: Satzilla: Portfolio-based algorithm selection for sat. Journal of Artificial Intelligence Research 32, 565–606 (2008)
2. Burke, E., Hyde, M., Kendall, G., Ochoa, G., Ozcan, E., Qu, R.: Hyper-heuristics: A Survey of the State of the Art. Technical report, NOTTCS-TR-SUB-0906241418-2747, University of Nottingham (2010),
http://www.cs.nott.ac.uk/~gxo/papers/hhsurvey.pdf
3. Duin, C., Voß, S.: The Pilot method: A strategy for heuristic repetition with application to the Steiner problem in graphs. Networks 34(3), 181–191 (1999)

4. Lagoudakis, M.G., Parr, R.: Reinforcement learning as classification: Leveraging modern classifiers. In: Fawcett, T., Mishra, N. (eds.) Proc. 20th Int. Conf. on Machine Learning (ICML 2003), pp. 424–431. AAAI Press (2003)

5. Kocsis, L., Szepesvári, C.: Bandit Based Monte-Carlo Planning. In: Fürnkranz, J., Scheffer, T., Spiliopoulou, M. (eds.) ECML 2006. LNCS (LNAI), vol. 4212, pp. 282–293. Springer, Heidelberg (2006)

6. Gelly, S., Silver, D.: Combining online and offline knowledge in uct. In: Proc. of the 24th Int. Conf. on Machine learning (ICML 2007), pp. 273–280. ACM Press (2007)

7. Asmuth, J., Littman, M.: Learning is planning: near Bayes-optimal reinforcement learning via Monte-Carlo tree search. In: Proceedings of The 27th Conference on Uncertainty in Artificial Intelligence, UAI 2011 (2011)

8. Lai, T., Robbins, H.: Asymptotically efficient adaptive allocation rules. Advances in Applied Mathematics 6, 4–22 (1985)

9. De Mesmay, F., Rimmel, A., Voronenko, Y., Püschel, M.: Bandit-Based Optimization on Graphs with Application to Library Performance Tuning. In: Danyluk, A., Bottou, L., Littman, M. (eds.) Proc. Int. Conf. on Machine Learning (ICML 2009). ACM International Conference Proceeding Series, vol. 382, pp. 729–736. ACM (2009)

10. Rolet, P., Sebag, M., Teytaud, O.: Boosting Active Learning to Optimality: A Tractable Monte-Carlo, Billiard-Based Algorithm. In: Buntine, W., Grobelnik, M., Mladenić, D., Shawe-Taylor, J. (eds.) ECML PKDD 2009, Part II. LNCS, vol. 5782, pp. 302–317. Springer, Heidelberg (2009)

11. Matsumoto, S., Hirosue, N., Itonaga, K., Ueno, N., Ishii, H.: Monte-Carlo Tree Search for a reentrant scheduling problem. In: 40th Intl. Conf. on Computers and Industrial Engineering (CIE 2010), pp. 1–6. IEEE (2010)

12. Streeter, M.J., Smith, S.F.: A Simple Distribution-Free Approach to the Max k-Armed Bandit Problem. In: Benhamou, F. (ed.) CP 2006. LNCS, vol. 4204, pp. 560–574. Springer, Heidelberg (2006)

13. Brucker, P.: Scheduling algorithms. Springer (2007)

14. Bräsel, H., Dornheim, L., Kutz, S., Mörig, M., Rössling, I.: LiSA – Library of Scheduling Algorithms. Otto-von-Guericke Universität Magdeburg (2011), http://lisa.math.uni-magdeburg.de

15. Panwalkar, S., Iskander, W.: A Survey of Scheduling Rules. Operations Research 25(1), 45–61 (1977)

16. Kawai, T., Fujimoto, Y.: An efficient combination of dispatch rules for job-shop scheduling problem. In: 3rd IEEE Intl Conf. on Industrial Informatics (INDIN 2005), pp. 484–488 (2005)

17. Voß, S., Fink, A., Duin, C.: Looking ahead with the pilot method. Annals of Operations Research 136(1), 285–302 (2005)

18. Bertsekas, D., Tsitsiklis, J., Wu, C.: Rollout algorithms for combinatorial optimization. Journal of Heuristics 3(3), 245–262 (1997)

19. Auer, P., Cesa-Bianchi, N., Fischer, P.: Finite-time analysis of the multiarmed bandit problem. Machine Learning 47(2-3), 235–256 (2002)

20. Cicirello, V., Smith, S.: The max k-armed bandit: A new model of exploration applied to search heuristic selection. In: Veloso, M.M., Kambhampati, S. (eds.) Proc. Nat. Conf. on Artificial Intelligence, pp. 1355–1361. AAAI Press / The MIT Press (2005)

21. Fialho, Á., Da Costa, L., Schoenauer, M., Sebag, M.: Dynamic Multi-Armed Bandits and Extreme Value-Based Rewards for Adaptive Operator Selection in Evolutionary Algorithms. In: Stützle, T. (ed.) LION 3. LNCS, vol. 5851, pp. 176–190. Springer, Heidelberg (2009)
22. Taillard, E.: Benchmarks for basic scheduling problems. European Journal of Operational Research 64(2), 278–285 (1993)
23. Yamada, T., Nakano, R.: A genetic algorithm applicable to large-scale job shop Problems. In: Männer, R., Manderick, B. (eds.) Parallel Problem Solving from Nature (PPSN II), pp. 283–292. Elsevier (1992)
24. Banharnsakun, A., Sirinaovakul, B., Achalakul, T.: Job Shop Scheduling with the Best-so-far ABC. In: Engineering Applications of Artificial Intelligence (2006), pp. 1–11 (2011)

Minimizing Time When Applying Bootstrap to Contingency Tables Analysis of Genome-Wide Data

Francesco Sambo and Barbara Di Camillo

Department of Information Engineering, University of Padova, Italy
francesco.sambo@dei.unipd.it

Abstract. Bootstrap resampling is starting to be frequently applied to contingency tables analysis of Genome-Wide SNP data, to cope with the bias in genetic effect estimates, the large number of false positive associations and the instability of the lists of SNPs associated with a disease. The bootstrap procedure, however, increases the computational complexity by a factor B, where B is the number of bootstrap samples.

In this paper, we study the problem of minimizing time when applying bootstrap to contingency tables analysis and propose two levels of optimization of the procedure. The first level of optimization is based on an alternative representation of bootstrap replicates, *bootstrap histograms*, which is exploited to avoid unnecessary computations for repeated subjects in each bootstrap replicate. The second level of optimization is based on an ad-hoc data structure, the *bootstrap tree*, exploited for reusing computations on sets of subjects which are in common across more than one bootstrap replicate. The problem of finding the best bootstrap tree given a set of bootstrap replicates is tackled with best improvement local search. Different constructive procedures and local search operators are proposed to solve it.

The two proposed levels of optimization are tested on a real Genome-Wide SNP dataset and both are proven to significantly decrease computation time.

Keywords: Bootstrap, Contingency tables analysis, Genome-Wide SNP Data, Local Search.

Introduction

In the past few years, the genetic basis of disease susceptibility has started to be explored through the novel paradigm of Genome Wide Association Studies (GWASs). A GWAS searches for patterns of genetic variation between a population of affected individuals (cases) and a healthy control population, for complex diseases arising from the interaction of a genetic predisposition with environmental risk factors [11].

The most common form of genetic variation among individuals is Single Nucleotide Polymorphism (SNP), a point variation at a single DNA locus across

Y. Hamadi and M. Schoenauer (Eds.): LION 6, LNCS 7219, pp. 175–189, 2012.
© Springer-Verlag Berlin Heidelberg 2012

members of the same species. Diploid individuals, such as human, have two homologous copies of each chromosome and the genetic variation can occur at the same locus in either of the two chromosomes: human SNPs, thus, are ternary variables, encoding the three possible configurations of nucleotide pairs, or *genotypes*, at a certain locus (AA, BB and AB).

Current technologies allow the simultaneous measurement of $O(10^6)$ SNPs for each individual and the usual number of individuals involved in a GWAS is $O(10^3)$: the size of the resulting dataset has thus induced the vast majority of studies to search only for univariate association between each single SNP and the disease [15, 16] or to rely on univariate SNP association as a ranking and/or pre-filtering step [6, 12].

All the methodologies proposed in the literature to test for the association between a SNP and a disease condition require the computation, for each SNP, of a 2×3 contingency table, containing the number of case and control individuals for each of the three genotypes of the SNP. Test statistics can then be exploited to select SNPs significantly associated with the disease or to rank SNPs in decreasing order of genetic effect on the disease [1].

The large number of tests involved in a GWAS, together with the low sample size relative to the number of variables tested for association, can give rise to bias in genetic effect estimates, to a large number of false positive associations and to instability in the ranked list of SNPs [4]. To cope with these limitations, one of the strategies adopted in the literature consists in coupling bootstrap with contingency tables creation [4, 12–14].

Bootstrap [2] is a data-based simulation method for statistical inference: given a dataset X, consisting of n observations of p variables, and a statistic $s(X)$, the bootstrap method consists in (i) generating B *bootstrap replicates* of the original dataset $(X^1, \ldots, X^B)$, each one obtained by sampling with replacement n observations from X, (ii) computing the test statistic for each bootstrap replicate and (iii) exploiting the B results for estimating some properties of the statistic, such as standard error or confidence intervals.

In the context of GWAS data analysis, SNPs are the variables and subjects are the observations; bootstrap is thus used to obtain B replicates of the dataset, each with the same set of SNPs and with subjects sampled with replacement from the original set. The statistic of association is then computed for each SNP in each sample replicate. In [4], bootstrap is applied to contingency tables analysis for computing point estimates and confidence intervals of the genetic effect of each SNP, exploiting the relative rank of each SNP in each bootstrap replicate. The same approach is further exploited in [14] for estimating the minimum sample size needed in replication studies. With the aim of estimating the total number of susceptibility SNPs of a complex disease from GWAS data, bootstrap is applied in [13] to contingency table analysis for computing confidence intervals of the estimate. Finally, in [12] an ensemble of Naïve Bayes classifiers is trained on as many bootstrap replicates of a GWAS dataset, and SNP ranking through test statistics is exploited for learning classification probabilities and for selecting the attributes of each Naïve Bayes classifier.

The bootstrap method has the appealing feature that it can be used "on top" of the statistic to be computed, exploiting the statistic computation as a black box and simply iterating it through the various bootstrap replicates. Suggested values for B are 50-100 when estimating bias or standard error and 1000 when estimating confidence intervals [2], thus the main drawback is the $O(B)$ increase in computational complexity due to the replication of the statistic computation.

Even though the time needed for acquiring and pre-processing a GWAS dataset is much longer than the time needed for computing a statistic of association between all SNPs and the disease, an acquired dataset is seldom processed just once, both because multiple statistics are usually computed on the same dataset and because separate, smaller datasets are often joined together and re-processed in larger meta-analyses. Any attempt in reducing the computation time of data processing is thus definitely worth the effort.

In this paper, we explore the problem of minimizing computation time when applying bootstrap to contingency table analysis of Genome-Wide SNP data. The main contributions of the paper are two levels of optimization of the computational procedure: the first level of optimization derives from an alternative representation of bootstrap replicates as *bootstrap histograms*, which is exploited to avoid repeating computations for repeated subjects in each bootstrap replicate. The second level of optimization is based on an ad-hoc data structure, the *bootstrap tree*, exploited for reusing computations on sets of subjects which are in common across more than one bootstrap replicate. The problem of finding the best bootstrap tree, given a set of bootstrap replicates, is tackled with a best improvement local search approach [7] and different constructive procedures and local search operators are proposed to solve it.

We tested our optimized computational procedure on the WTCCC case-control study on Type 1 Diabetes [15] and indeed observed a significant decrease in computation time for both levels of optimization, with respect to a standard bootstrap approach.

The remainder of the paper is organized as follows: Section 1 describes in details the problem of applying bootstrap to contingency table analysis of GWAS data and presents the two levels of optimization, Section 2 describes the experimental dataset and reports performance results and Section 3 draws conclusions and presents some possible future directions.

1   Methods

Given a GWAS dataset X, consisting of p SNPs measured for $n = n_{cases} + n_{controls}$ subjects, and a binary vector of class labels Y of size n, computing a contingency table like the one in Table 1 for each SNP involves scanning all subjects and counting the occurencies of the three possible variants of the SNP. Iterating the process on the whole SNP set has thus computational complexity $O(pn)$.

The frequency counts in each contingency table are exploited to test for an association between the corresponding SNP and the disease condition, with a

Table 1. Example of a 2×3 contingency table for a particular SNP

	AA	AB	BB
cases	a	b	c
controls	d	e	f

certain statistic s. SNPs can then be ranked according to the computed statistics and the topmost SNPs, or the SNPs whose statistic pass a certain threshold, are identified as associated with the disease.

The reliability of the statistic and the robustness of the list of associated SNPs can be assessed with bootstrap.

If we define $I = \{i_1, \ldots, i_n\}$ the original patient set (Figure 1(a)), the bootstrap procedure generates B bootstrap replicates $\{I^1, \ldots, I^B\}$, each of size n and sampled with replacement from I (Figure 1(b)). For each SNP, B contingency tables and B corresponding statistics are then computed. The $B \times p$ resulting statistics can be exploited either for calculating bias, squared error or confidence intervals for each SNP [4, 14], or for computing B ranked lists of SNPs, which can be merged to obtain a single, more robust list [10].

The pseudocode of the classic bootstrap procedure is given in what follows. As it is clear from the pseudocode, the computational complexity of the algorithm is $O(Bpn)$.

CLASSICBOOTSTRAP(X, Y, B)

```
1   I = {i₁,...,iₙ}, original patient set
2   Generate the sets {I¹,...,Iᴮ}, each sampled with replacement from I
    // results of the statistics for each SNP and each replicate
3   S = p × B matrix of zeros
4   for b in {1,...,B}
5       for k in {1,...,p}
            // contingency table
6           CT = 2x3 matrix of zeros
7           for j in {1,...,n}
8               iⱼ = Iᵇ[j]
9               CT[ Y[iⱼ], X[k,iⱼ] ]+=1
            // statistic of the association between SNP k and the disease
            // for the bootstrap replicate b
10          S[k,b] = s(CT)
11  Use S to estimate properties of the statistic or to obtain a robust SNP ranking
```

Bootstrap replicates I^b belong to the class of *multisets*, *i.e.* sets that allow the repetition of elements. In the next section, we introduce a convenient representation for multisets, which will lead to a first level of optimization of the bootstrap procedure.

$$I: i_1 \ i_2 \ i_3 \ i_4 \ i_5 \ i_6 \ i_7 \ i_8 \ i_9 \ i_{10}$$
(a)

$$I^b: i_1 \ i_1 \ i_3 \ i_5 \ i_5 \ i_5 \ i_7 \ i_8 \ i_9 \ i_9$$
(b)

$$\overline{I^b}: 2\ 0\ 1\ 0\ 3\ 0\ 1\ 1\ 2\ 0$$
(c)

Fig. 1. (a) example of a patient set I with 10 patients. (b) example of a bootstrap replicate of I. (c) boostrap histogram of the replicate I^b.

1.1   Level I Optimization: Bootstrap Histograms

Given a multiset I^b with m elements, drawn from an underlying set $I = \{i_1, \ldots, i_n\}$ of n distinct elements, a *histogram* representation of the multiset is a vector $\overline{I^b}$ of length n containing, for each element $i_j \in I$, the number of times it appears in the multiset I^b (Figure 1(c)). We define *bootstrap histogram* the histogram representation of a bootstrap replicate.

Bootstrap histograms can be conveniently exploited to avoid unnecessary operations: when computing contingency tables for each bootstrap replicate $\overline{I^b}$, in fact, one needs only to process the elements j such that $\overline{I^b}[j] > 0$. Furthermore, given that each nonzero element of the boostrap histogram counts multiple copies of the same subject, one needs only to evaluate once the SNPs of the j-th subject and then add $\overline{I^b}[j]$ to the corresponding elements of the contingency tables. The pseudocode of the histogram-based bootstrap procedure is given in what follows.

HISTOGRAMBOOTSTRAP(X, Y, B)

```
1   I = {i₁,...,iₙ}, original patient set
2   Generate {Ī¹,...,Ī^B}, sampled with replacement from I
3   S = p × B matrix of zeros
4   for b in {1,...,B}
        // tmp storage of indices, values and labels of nonzero elements of Ī^b
5       tmpInd = ∅, tmpVal = ∅, tmpY = ∅
6       for j in {1,...,n}
7           if Ī^b[j] > 0
8               tmpInd = tmpInd ∪ j
9               tmpVal = tmpVal ∪ Ī^b[j]
10              tmpY = tmpY ∪ Y[Ī^b[j]]
11      for k in {1,...,p}
12          CT = 2x3 matrix of zeros
13          for j in {1,...,length(tmpInd)}
14              CT[ tmpY[j], X[k, tmpInd[j]] ]+=tmpVal[j]
15          S[k, b] = s(CT)
16  Use S to estimate properties of the statistic or to obtain a robust SNP ranking
```

The preprocessing routine at lines 5–10 extracts, for each bootstrap histogram $\overline{I^b}$, the indices, values and corresponding disease condition of the nonzero elements of $\overline{I^b}$; its computational complexity, $O(Bn)$, can be considered negligible if $n \ll p$, as in our case.

Thanks to the preprocessing routine, the summation at line 14 is now executed only $B \cdot p \cdot \text{length}(tmpInd)$ times, resulting in an expected relative gain of computation time equal to the average proportion of zero elements in a bootstrap histogram: for n sufficiently large, the relative gain approaches $e^{-1} \simeq 0.368$ [2].

A further level of optimization, obtained by exploiting the presence of common elements across multiple bootstrap histograms, is presented in the next section.

1.2 Level II Optimization: Bootstrap Tree

For our second level of optimization, the aim is to group together bootstrap histograms sharing common elements, so to be able to reuse the results of contingency table computations for multiple bootstrap histograms. To this purpose, we define *intersection* of two bootstrap histograms $\overline{I^x}$ and $\overline{I^y}$ the boostrap histogram $\overline{I^z}$ such that:

$$\overline{I^z}[j] = \overline{I^x}[j] \cap \overline{I^y}[j] = \begin{cases} \overline{I^x}[j] & \text{if } \overline{I^x}[j] = \overline{I^y}[j], \\ 0 & \text{otherwise} \end{cases} \qquad \text{for } j = 1 \ldots n.$$

Furthermore, we define *size* of a bootstrap histogram the number of its nonzero elements and *similarity* between two bootstrap histograms the size of their intersection.

Given a set of bootstrap histograms, we define *bootstrap tree* a data structure with the following features:

1. the bootstrap tree is a balanced binary tree,
2. each leaf of the tree, *i.e.* each node at level 0, contains one of the original bootstrap histograms,
3. each internal node of the tree, at level $l > 0$, contains the intersection of its two *children*, *i.e.* the pair of nodes connected to it at level $l - 1$.

An example of bootstrap tree is given in Figure 2.

For each node, we define *unique elements* the nonzero elements of its bootstrap histogram which are zero in the histogram of its parent node. Unique elements are marked in bold in the example tree of Figure 2.

A bootstrap tree can be effectively exploited to obtain a further decrease in computation time. Each bootstrap replicate can be processed by visiting the tree in a depth-first, left-first traversal, backtracking once leaves are reached. The intuition is that computations can be carried out while descending the tree, exploiting the bootstrap histogram of each internal node for computing partial results, which can then be reused for all the nodes in the corresponding subtree. The pseudocode of such an algorithm is given in what follows.

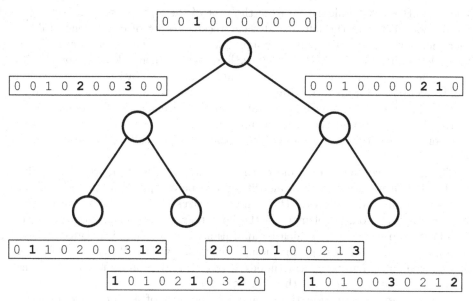

Fig. 2. Example of a bootstrap tree, for $B = 4$ bootstrap replicates. The leaves contain the bootstrap histograms of the 4 bootstrap replicates and each internal node contains the intersection of its children. Unique elements of each node, *i.e.* nonzero elements of its bootstrap histogram which are zero in the histogram of its parent node, are marked in bold.

TREEBOOTSTRAP($X, Y, tree$)

1   CTs = set of p contingency tables, all elements initialized to zero
2   $S = p \times B$ matrix of zeros
3   RECTREEBOOTSTRAP($X, Y, tree.root, tree.height, CTs$)
4   Use S to estimate properties of the statistic or to obtain a robust SNP ranking

RECTREEBOOTSTRAP($X, Y, node, level, CTs$)

1   Increment CTs according to the unique elements of *node*
2   **if** *level* > 0
        // Proceed to the children
3       $newCTs$ = copy of CTs
4   RECTREEBOOTSTRAP($X, Y, node.leftChild, level - 1, newCTs$)
5   RECTREEBOOTSTRAP($X, Y, node.rightChild, level - 1, CTs$)
6   **else**
        // A leaf has been reached, compute the statistic
7       $b = node.b$ // Index of the bootstrap replicate
8   **for** k **in** $\{1, \ldots, p\}$
9           $S[k, b] = s(CTs[k])$

The TREEBOOSTRAP algorithm creates a set CTs of p zero-valued contingency tables, allocates space for the results of the statistic and launches the recursive

RECTREEBOOSTRAP algorithm from the root of the tree, passing the set CTs. Each node visited in the left-first descent receives a set of contingency tables, increments them according to its unique elements and passes a copy of them to its left child (RECTREEBOOTSTRAP, lines 1-4). When a leaf is reached, its corresponding contingency tables are complete and the statistic of association can be computed for all SNPs (lines 7-9). When backtracking from a left child to a right child, the set of contingency tables at the parent node is directly passed to the right child rather then copied (line 5), since it does not need to be stored anymore. The algorithm terminates when the last right child, *i.e.* the rightmost leaf, is visited.

The gain in computation time of the TREEBOOTSTRAP algorithm, relative to the HISTOGRAMBOOTSTRAP algorithm, can be computed as the sum of the sizes of the internal nodes over the sum of the sizes of the leaves. We prove this intuitively: each nonzero element in the bootstrap histograms of nodes at level 1 indicates an element for which computations can be spared, because it is in common between two histograms; nonzero elements at level 2 stands for further spared elements when computing nodes at level 1, and so on up to the root. The gain of the tree in Figure 2 is thus $7/24 \simeq 0.29$[1].

The gain in computation time comes at the cost of an increased memory occupation: the TREEBOOTSTRAP algorithm needs to keep in memory a number of contingency tables, for each SNP, equal to the height of the bootstrap tree plus one. Memory occupation is often critical when dealing with GWAS data. We thus impose the constraint on the bootstrap tree of being a balanced binary tree: among all possible binary trees of B nodes, balanced binary trees have the lowest height, $log(B)$.

Having defined the bootstrap tree, the TREEBOOTSTRAP algorithm and the concept of gain of a tree, we can now formulate the optimization problem of searching for the bootstrap tree with the maximum gain, given a GWAS dataset and a set of B bootstrap histograms. We chose to tackle the problem with a best improvement local search approach [7]: starting from an initial bootstrap tree, we generate a neighbourhood of trees by applying a local search operator, choose the tree with the highest gain among the neighbourhood and iterate the process until a local maximum is reached.

We explore the use of different constructive procedures for building the initial tree and of different local search operators for generating the neighbourhood. Constructive procedures and local search operators are described in the next sections.

1.3 Constructive Procedures for Bootstrap Trees

We begin this section with a remark: since our final objective is to minimize computation time, which includes both the time for searching for the optimal

---

[1] To be precise, one should also consider the time spent for copying contingency tables. Copying p contingency tables, one for each SNP, has complexity O(p) and the whole set of contingency tables is copied $B - 1$ times: the computational complexity does not depend on n and can thus be considered negligible.

bootstrap tree and the time for the TREEBOOTSTRAP algorithm, simple but fast constructive heuristics can stand a chance against more complex but slower heuristics and should thus be considered.

The first constructive procedure we propose is a greedy agglomerative constructive heuristic, inspired by the literature on hierarchical clustering [5]: the GREEDYAGGLOMERATIVE heuristic builds the bootstrap tree starting from the leaves, *i.e.* the original boostrap histograms, by computing the mutual similarity between all pairs of histograms. Similarity between two bootstrap histograms, as defined in Section 1.2, is the size of the intersection between the two histograms. The two histograms with the highest similarity are joined as children of the first node at level 1, whose histogram is computed as the intersection of the two histograms. The heuristic keeps joining the pair of remaining leaves with the highest similarity, until no more leaves remain. The procedure is then iterated up to the root, by computing the mutual similarity between all pairs of nodes at level l and by iteratively joining the nodes with the highest similarity as children of the nodes at level $l + 1$. The computational complexity of the GREEDYAGGLOMERATIVE heuristic is $O(B^2 n)$.

The other constructive procedure we consider is RANDOMBUILD, which builds the bootstrap tree by joining pairs of nodes at random up to the root. Despite the lower expected gain with respect to the GREEDYAGGLOMERATIVE heuristic, we choose to try also the RANDOMBUILD procedure because of its lower computational complexity, $O(Bn)$.

1.4   Local Search Operators for Bootstrap Trees

The first local search operator we define, TREEOPT, can be applied to all nodes at level $l \geq 2$ and operates by testing the two possible swaps between grandchildren of a node G (which stands for Grandparent), *i.e.* between the four nodes whose parents are the two children of G (Figure 3). The total number of nodes at level $l \geq 2$ is $B/2 - 1$, thus the size of the neighbourhood of the TREEOPT operator is $B - 2$. The cost of a swap is the cost of updating the boostrap histograms for G's children, G itself and all the nodes in the path from G up to the root. The total cost of evaluating a TREEOPT neighbourhood is thus $O(nB \log B)$.

The second operator we define, 2OPT, is inspired by the homonymous operator for the TSP problem [7]: the 2OPT operator tests all possible swaps between two leaves of the tree, excluding the swaps between the two children of the same node. The size of the neighbourhood of the 2OPT operator is thus $B^2/2 - B$. The cost of a swap is the cost of updating the boostrap histograms of the nodes on the paths from the two swapped leaves up to the root: the total cost of evaluating a 2OPT neighbourhood is thus $O(nB^2 \log B)$.

2   Experimental Results

In this section, we present experimental results on the computational performance of the algorithms CLASSICBOOTSTRAP, HISTOGRAMBOOTSTRAP and

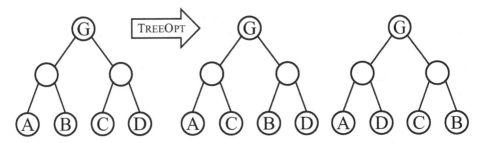

Fig. 3. Example of use of the TREEOPT operator: applied to the node G of the leftmost tree, the operator generates the two possible swaps of the nodes A, B, C and D, whose parents are the two children of G

TREEBOOTSTRAP. Different combinations of the constructive procedures and of the local search operators are tested for the bootstrap tree of the TREE-BOOTSTRAP algorithm.

As benchmark to assess the performance of the different algorithms, we choose the WTCCC case-control study on Type 1 Diabetes [15]: the dataset consists of 458376 SNPs, measured for 1963 T1D cases and 2938 healthy controls (after the application of all Quality Control filters reported in [15]). The numbers of SNPs and subjects involved are in line with the ones usually encountered in a GWAS [11], thus making the dataset a meaningful benchmark for our algorithms.

We measure the computation time spent by the following procedure: generate B bootstrap replicates, compute a contingency table and a univariate statistic of association (described in [12]) for each SNP in each replicate and rank SNPs according to the computed statistic. To remove a possible source of noise, we exclude from the measurements the time needed for loading the dataset in RAM.

The number of bootstrap replicates, B, is varied among all powers of 2 in the range $\{2^0 \ldots 2^{10}\}$: for each B, we generate 20 sets of B bootstrap replicates and repeat the whole procedure on each set.

All algorithms are written in C++ and all computations are carried out on a single 3.00 GHz Intel Xeon Processor E5450.

We first assess the effectiveness of the two levels of optimization by comparing the computation time of CLASSICBOOTSTRAP, HISTOGRAMBOOTSTRAP and TREEBOOTSTRAP, the latter tested with either the GREEDYAGGLOMERATIVE or the TREEOPT constructive procedure and without local search.

Results are shown in Figure 4, top panel: the figure reports, for each value of B, the median over the 20 runs of the average time needed to process one bootstrap replicate, computed as the total time over B. For each point, whiskers extend from the first to the third quartile. We preferred to plot the median rather than the mean because of the presence of a small number of random outliers, which were however included in all the tests for significance. As it is clear from the figure, both levels of optimization result in a significant decrease in computation time (p-values of CLASSICBOOTSTRAP *vs* HISTOGRAMBOOTSTRAP

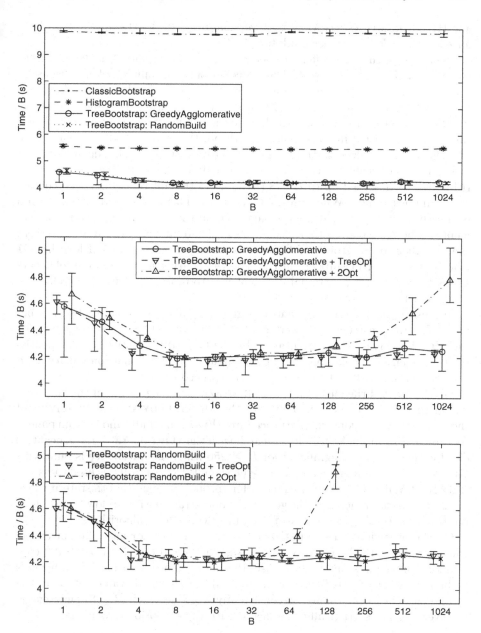

Fig. 4. Medians across 20 runs of the time for processing one bootstrap replicate (total time / B) versus the number of replicates B. Whiskers extend from first to third quartile. Top panel: NAIVEBOOTSRAP, HISTOGRAMBOOTSTRAP and TREEBOOTSTRAP with the two constructive procedures and without local search. Middle panel: TREE-BOOTSTRAP with the GREEDYAGGLOMERATIVE constructive procedure, without local search and with the two local search operators. Bottom panel: TREEBOOTSTRAP with the RANDOMBUILD constructive procedure, without local search and with the two local search operators.

and HISTOGRAMBOOTSTRAP *vs* both versions of TREEBOOTSTRAP $< 8.9 \times 10^{-5}$ for each B, Wilcoxon signed-rank test)[2].

Oberving a significant difference between HISTOGRAMBOOTSTRAP and TREE-BOOTSTRAP for $B = 1$ is somehow unexpected, since the number of operations executed by the two algorithms is practically the same. The only notable difference is that HISTOGRAMBOOTSTRAP allocates just one contingency table and reuses it, while TREEBOOTSTRAP has to allocate the whole set of p contingency tables: rather than resulting in an increased overhead, the second strategy seems to be more fit to be optimized by the compiler and thus results in a significant gain.

Further, significant increases in the time gain of the TREEBOOTSTRAP algorithm can be observed up to $B = 8$ (p-values of the differences between consecutive samples < 0.02, Wilcoxon rank-sum test). This means that the algorithm is effectively able to avoid unnecessary computations, by reusing common elements of the bootstrap histograms. For $B \geq 16$, no further gain can be observed: the identification of groups of 16 or more bootstrap histograms of length 4901 (the total number of subjects in the dataset) sharing many common elements seems to have a computational cost which is too high to result in an effective improvement of the overall procedure.

Interesting is also the fact that no significant difference can be observed between the two constructive procedures for TREEBOOTSTRAP (p-value > 0.21 for each B). The higher tree gain otained by the GREEDYAGGLOMERATIVE heuristic, thus, seems to be compensated by its higher computational complexity, with respect to the RANDOMBUILD constructive procedure.

We then tested the effect on the TREEBOOTSTRAP algorithm of the two local search operators, when combined with either the GREEDYAGGLOMERATIVE and the RANDOMBUILD constructive procedures (Figure 4, middle and bottom panel). As it is clear from the figures, local search with the 2OPT operator has a negative effect on performance, significant for $B \geq 256$ when the initial tree is built with GREEDYAGGLOMERATIVE and for $B \geq 64$ when RANDOMBUILD is used (p-values $< 6 \times 10^{-3}$, Wilcoxon signed-rank test). This probably means that the rate at which the 2OPT operator increases the gain of the bootstrap tree is too slow to be able to improve the overall performance of the TREEBOOTSTRAP algorithm.

On the other hand, the TREEOPT local search operator has different effects when coupled with the GREEDYAGGLOMERATIVE and the RANDOMBUILD constructive procedures: in the first case, the performance with local search tends to be consistently better than without local search for all values of B; in the second case, the performance is consistently worse for each $B \geq 8$. Differences, however, are not statistically significant, with the exception of one case.

3   Conclusions and Future Directions

In this paper, we studied the problem of minimizing computation when applying bootstrap to contingency table analysis of Genome-Wide SNP data. We proposed two levels of optimization of the procedure, which both result in a significant

---

[2] For all tests throughout the paper, we consider significant a p-value < 0.05.

improvement in computation time: altogether, the optimization strategies presented in this paper allow us to reduce computation time by a factor of approximately 2.5, a result which can be considered definitely valuable in the context of GWAS data analysis.

The first level of optimization, implemented in the HISTOGRAMBOOTSTRAP algorithm, is based on an alternative representation of bootstrap replicates as bootstrap histograms. The bootstrap histogram representation is not new in the literature: for example, in [3] the same representation is proven more effective than the standard representation for estimating the bias of order invariant statistics. As far as we know, however, the idea of exploiting the bootstrap histogram representation for reducing computation time has never been proposed in the literature.

The second level of optimization, implemented in the TREEBOOTSTRAP algorithm, is based on an ad-hoc data structure, the *bootstrap tree*, which is exploited for reusing partial results on sets of subjects shared by multiple replicates.

Once defined the bootstrap tree and the algorithm for exploiting it, we formulated the optimization problem of finding the tree leading to the highest gain in computation time and tackled the problem with a best improvement local search approach. Two constructive procedures and two local search operators were specifically designed for the problem and several combinations of them were tested. Experimental results show that simpler but faster approaches to tree construction and refinement are competitive with (and, in some cases, significantly more effective than) more powerful, yet slower approaches. As far as we know, the idea of exploiting common elements of the boostrap samples for reducing computation time of boostrap has never been proposed in the literature.

The algorithms designed for the second level of optimization require the number of bootstrap replicates, B, to be a power of two. We do not see this requirement as a big limitation: guidelines for the choice of the number of bootstrap replicates are present in the literature [2], but they usually give indications on the order of magnitude of B rather than on its exact value, the choice of which is left to the experimenter.

Concerning future directions, we intend to further explore the design of other constructive heuristics and local search operators, together with other stochastic local search techniques, to search for the optimal bootstrap tree.

The extent to which bootstrap performance can be improved with our two levels of optimization has, however, a theoretical limit. In each bootstrap replicate, all observations are sampled with equal probability $1/n$: the expected number of elements in common among B bootstrap histrograms is thus limited. Several authors, however, have proposed to exploit importance sampling for reducing the variance of certain bootstrap estimates [8, 9, 17]. The idea beyond importance sampling is to sample observations with nonuniform weights: such an approach dramatically increases the number of elements in common among subsets of bootstrap histograms and can thus further benefit from our optimization strategies. One of our future directions is thus to study how to adapt our two levels of optimization to importance sampling in bootstrap estimates.

Finally, the enhancement of bootstrap with the two levels of optimization can be extended to other application domains, as long as the function to be iterated on each bootstrap replicate has the two following features: i) the sets of observations to be processed can be split in m distinct subsets and the function can be independently applied to each subset, ii) the computational cost of processing each subset is considerably higher than the cost of assembling the m results. One of our future directions is thus to study the application of our optimization strategies to different problem domains, characterized by the two aforementioned features.

Acknowledgements. This study makes use of data generated by the Wellcome Trust Case-Control Consortium. A full list of the investigators who contributed to the generation of the data is available from www.wtccc.org.uk. Funding for the project was provided by the Wellcome Trust under award 076113 and 085475.

Francesco Sambo would like to thank Prof. Silvana Badaloni for her precious advices and support.

References

1. Balding, D.J.: A tutorial on statistical methods for population association studies. Nature Reviews Genetics 7(10), 781–791 (2006)
2. Efron, B., Tibshirani, R.J.: An Introduction to the Bootstrap. Chapman & Hall, New York (1993)
3. Efron, B.: More Efficient Bootstrap Computations. Journal of the American Statistical Association 85(409), 79–89 (1990)
4. Faye, L., Sun, L., Dimitromanolakis, A., Bull, S.: A flexible genome-wide bootstrap method that accounts for ranking- and threshold-selection bias in GWAS interpretation and replication study design. Statistics in Medicine 30(15), 1898–1912 (2011)
5. Hastie, T., Tibshirani, R., Friedman, J.: The Elements of Statistical Learning: Data Mining, Inference, and Prediction, 2nd edn. Springer Series in Statistics. Springer (February 2009)
6. He, Q., Lin, D.Y.: A variable selection method for genome-wide association studies. Bioinformatics 27, 1–8 (2011)
7. Hoos, H.H., Stützle, T.: Stochastic Local Search: Foundations & Applications (The Morgan Kaufmann Series in Artificial Intelligence). Morgan Kaufmann (September 2004)
8. Hu, J., Su, Z.: Short communication: Bootstrap quantile estimation via importance resampling. Computational Statistics and Data Analysis 52, 5136–5142 (2008)
9. Johns, M.: Importance sampling for bootstrap confidence intervals. Journal of the American Statistical Association 83, 709–714 (1988)
10. Jurman, G., Merler, S., Barla, A., Paoli, S., Galea, A., Furlanello, C.: Algebraic stability indicators for ranked lists in molecular profiling. Bioinformatics 24(2), 258–264 (2008)
11. Ku, C.S., Loy, E.Y., Pawitan, Y., Chia, K.S.: The pursuit of genome-wide association studies: where are we now? Journal of Human Genetics 55(4), 195–206 (2010)

12. Sambo, F., Trifoglio, E., Di Camillo, B., Toffolo, G., Cobelli, C.: Bag of Naïve Bayes: biomarker selection and classification from genome-wide SNP data. BMC Bioinformatics 13(S14), S2 (2012)
13. So, H.C., Yip, B.H.K., Sham, P.C.: Estimating the total number of susceptibility variants underlying complex diseases from genome-wide association studies. PLoS One 5(11), e13898 (2010)
14. Sun, L., Dimitromanolakis, A., Faye, L., Paterson, A., Waggott, D., Bull, S.: Br-squared: A practical solution to the winner's curse in genome-wide scans. Human Genetics 129(5), 545–552 (2011)
15. The Wellcome Trust Case Control Consortium: Genome-wide association study of 14,000 cases of seven common diseases and 3,000 shared controls. Nature 447(7145), 661–678 (2007)
16. Zeggini, E., et al.: Meta-analysis of genome-wide association data and large-scale replication identifies additional susceptibility loci for type 2 diabetes. Nature Genetics 40(5), 638–645 (2008)
17. Zhou, H., Lange, K.: A fast procedure for calculating importance weights in bootstrap sampling. Computational Statistics and Data Analysis 55, 26–33 (2011)

Quantifying Homogeneity of Instance Sets
for Algorithm Configuration

Marius Schneider[1] and Holger H. Hoos[2]

[1] University of Potsdam
manju@cs.uni-potsdam.de
[2] Unversity of British Columbia
hoos@cs.ubc.ca

Abstract. Automated configuration procedures play an increasingly prominent role in realising the performance potential inherent in highly parametric solvers for a wide range of computationally challenging problems. However, these configuration procedures have difficulties when dealing with inhomogenous instance sets, where the relative difficulty of problem instances varies between configurations of the given parametric algorithm. In the literature, instance set homogeneity has been assessed using a qualitative, visual criterion based on heat maps. Here, we introduce two quantitative measures of homogeneity and empirically demonstrate these to be consistent with the earlier qualitative criterion. We also show that according to our measures, homogeneity increases when partitioning instance sets by means of clustering based on observed runtimes, and that the performance of a prominent automatic algorithm configurator increases on the resulting, more homogenous subsets.

Keywords: Quantifying Homogeneity, Empirical Analysis, Parameter Optimization, Algorithm Configuration.

1 Introduction

The automated configuration of highly parametric solvers has recently lead to substantial improvements in the state of the art in solving a broad range of challenging computational problems, including propositional satisfiability (SAT) [1–3], mixed integer programming (MIP) [4] and AI planning [5]. Broader adoption of this approach is likely to lead to a fundamental change in the way effective algorithms for NP-hard problems are designed [6], and the design and application of automated algorithm configuration techniques is a very active area of research (see, e.g., [7–9]).

One fundamental challenge in automated algorithm configuration arises from the fact that the relative difficulty of problem instances from a given set or distribution may vary between different configurations of the algorithm to be configured. This poses the risk that an iterative configuration process is misguided by the problem instances considered at early stages. For this reason, the performance of ParamILS [10, 9], one of the strongest and most widely configuration procedures currently available, significantly depends on the ordering of the problem instances used for training [1, 9], and the same can be expected to hold for other algorithm configuration techniques. Therefore, the

Y. Hamadi and M. Schoenauer (Eds.): LION 6, LNCS 7219, pp. 190–204, 2012.
© Springer-Verlag Berlin Heidelberg 2012

question to which degree the relative difficulty of problem instances varies between configurations of a parametric algorithm is of considerably interest. Indeed, precisely this question has been addressed in recent work by Hutter et al. [11], who refer to instance sets for which the same instances are easy and hard for different configuration as *homogeneous* and ones for which this is markedly not the case as *inhomogeneous*. They state that inhomogeneous instance sets are "problematic to address with both manual and automated methods for offline algorithm configuration" [11] and list three approaches for addressing this issue: clustering of homogeneous instance sets [12, 13], portfolio-based algorithm selection [14, 15] and per-instance algorithm configuration [16, 17]. They furthermore use a heat map visualization to qualitatively assess homogeneity.

In the work presented in the following, we introduce two *quantitative* measures of instance set homogeneity and explore to which extent these may provide a basis for assessing the efficacy of state-of-the-art algorithm configuration approaches. Like the heat maps of Hutter et al., our homogeneity measures are solely based on the performance of a given set of configurations and do not make use of problem-specific instance features. We show that on well-known algorithm configuration scenarios from the literature, our new measures are consistent with previous qualitative assessments. We further demonstrate that clustering instance sets can produce more homogenous subsets, and we provide preliminary evidence that automated algorithm configuration applied to those subsets tends to produce better results, which indicates that our homogeneity measures behave as intended.

The remainder of this paper is structured as follows: After a brief survey of related work (Section 2), we present our homogeneity measures (Section 3). This is followed by a brief description of clustering methods, which we subsequently use to partition instance set into more homogenous subsets (Section 4). We then present an experimental evaluation of our new measures, in terms of their consistency with the earlier qualitative assessment by Hutter et al., and in terms of the extent to which they are affected by clustering based on run-time measurements and on problem-dependent features (Section 5). Finally, we provide some general insights as well as a brief outlook on future work (Section 6).

2   Related Work

We are not aware of any existing work focused on homogeneity of instance sets given a solver as a central theme. However, there is a close conceptual relationship with prior work on portfolio-based algorithm selection and feature-based clustering of instances.

2.1   Portfolio-Based Algorithm Selection

The key idea behind portfolio-based algorithm selection is, given a set S of solvers for a given problem, to map any given problem instance to a solver from S that can be expected to solve it most effectively (see [18]). Perhaps the best-known realization of this idea is the *SATzilla* approach [15], which has produced SAT solvers that won numerous medals in the 2007 and 2009 SAT competitions. *SATzilla* uses so-called empirical hardness models [19, 20] for predicting performance based on cheaply computable instance features, and selects the solver to be applied to a given instance based on these performance predictions. Similar techniques have been successfully applied to several other problems (see, e.g., [4, 21–23]).

In general, the problem of learning a mapping from solvers to instances can be seen as a multiclass classification problem and attacked using various machine learning techniques (see, e.g., [24, 25]). Regardless of how the mapping is learned and carried out, a portfolio-based algorithm selector partitions any given set of instances Ω into subsets Ω_j, such that each Ω_j consists of the instances for which one particular solver from the given portfolio, say, $s_j \in S$, is selected. Ideally, each instance gets mapped to the solver that performs best on it; in this case, s_j dominates all other solvers in subset Ω_j. Although, unlike portfolio-based algorithm selection methods such as *SATzilla*, our approach does not use any problem-specific instance features, one of the homogeneity measures we introduce in Section 3.1 is based on the intuition that a given instance set is perfectly homogeneous if, and only if, a single solver (or in our case: solver configuration) dominates all others on all instances from the set.

2.2 Feature-Based Instance Clustering

A rather different approach to algorithm selection underlies the more recent *ISAC* procedure [12]. Here, instances are clustered based on problem-dependent instance features. Next, an automated configuration procedure is used to find a good configuration for each of the instance subsets thus obtained. Unlike the clustering approaches we consider in our work, *ISAC* does not use runtimes of certain configurations of the given target algorithm for the clustering. The algorithm selector produced by *ISAC* maps each instance to the configuration associated with the nearest cluster, as determined based on instance features.

CluPaTra [13] also partitions instance sets based on instance features and subsequently optimizes parameters of the given target solver for each cluster. Lindawati et al. [13] have shown that clustering instance sets in this way leads to better performance than using random clustering of the same instances. This supports the idea that a parametric solver has a higher configuration potential on clustered instance sets if the clustering improves the homogeneity of these sets with respect to the parametric solver.

Our approach to clustering instance sets, explained in detail in Section 4, differs from *ISAC* and *CluPaTra* by not using any problem-specific instance features. While such features can be quite cheap to compute, expert knowledge is required to define and implement them. In addition, their use is based on the assumption that "[...] instances with alike features behave similarly under the same algorithm" [12]. We are not aware of any published work that provides stringent support for this hypothesis, and existing work (such as [12, 13]) does not directly investigate it.[1] The homogeneity measures we introduce here offer a way of assessing to which extent clustering based on

---

[1] The fact that problem instances that are indistinguishable with respect to simple syntactic features, such as critically constrained Random-3-SAT instances with a fixed number of variables, have been observed to vary substantially in difficulty for state-of-the-art solvers for the respective problems seems to contradict this hypothesis; however, one could conjecture that such instances might differ in more sophisticated, yet still cheaply computable features.

problem-dependent features produces sets of instances for which different configurations of the same target algorithm show consistent performance rankings.

3   Homogeneity Measures

In this section, we deal with theoretical considerations to develop homogeneity measures for analyzing instance sets as motivated in Section 1. Intuitively, we characterize homogeneity as follows: An instance set Ω is homogeneous for a given set Φ of configurations of a parametric solver if the relative performance of the configurations in Φ does not vary across the instances in Ω. In many cases, there will be deviations from perfect homogeneity, and because the degree to which these variations occur is of interest, we want to consider real-valued measures of instance set homogeneity.

Unfortunately, for most interesting parametric solvers, the configuration spaces are far too big to permit the evaluation of all configurations. For example, the discretized configuration space of the highly parametric SAT and ASP solver *Clasp* [26] is of size $\approx 10^{18}$. Therefore, following Hutter et al. [11], we consider sets of randomly sampled configurations as a proxy for the entire space. Somewhat surprisingly, even for relatively small samples, this rather simplistic approach turns out to be quite effective for optimizing the homogeneity of instance sets, as will become evident from the empirical results presented in Section 5.

To formally define homogeneity measures, we use Φ to denote the space of all configurations ($\phi \in \Phi$ for individual configurations), $\Phi_r \subset \Phi$ for a subset of n configurations sampled uniformly at random from Φ, and Ω for an instance set ($\omega \in \Omega$ for individual instances).

3.1   Ratio Measure - Similarity to the Oracle

Our first measure is motivated by our practical approach to determine whether a portfolio solver approach is useful for an instance set. The runtimes of all configurations (or solvers) in the portfolio are measured for each instance. Based on the sum of their runtimes over the given instance set, we compare the performance of the best configuration and that of the oracle (sometimes also called *virtual best solver*)[2] constructed from all the sampled configurations; if their performance is equal, there is one dominant configuration for the entire instance set, and portfolio-based selection offers no advantage over statically choosing this dominant configuration. We call such an instance set homogenous w.r.t. the given set of configurations. This approach corresponds to the interpretation of portfolio solvers given in Section 2.1.

Following this intuition, we define the *ratio measure*, Q_{Ratio}, to measure homogeneity based on the ratio of the runtimes between the best configuration and the oracle, as shown in Equations 1 to 3, where $t'_{Oracle(\Phi_r)}(\Omega)$ represents the performance of the oracle and $t'_{\phi*}(\Omega)$ that of the best configuration in Φ_r.

---

[2] The performance of the oracle solver on a given instance is the minimum runtime over all given configurations/solvers.

$$Q_{Ratio}(\Phi_r, \Omega) = 1 - \frac{t'_{Oracle(\Phi_r)}(\Omega)}{t'_{\phi^*}(\Omega)} \text{ with } Q_{Ratio} \in [0, 1[\tag{1}$$

s.t.

$$t'_{\phi}(\Omega) = \sum_{i}^{|\Omega|} t(\omega_i, \phi) \text{ and } t'_{Oracle(\Phi_r)}(\Omega) = \sum_{i}^{|\Omega|} \min_{\phi \in \Phi_r} t(\omega_i, \phi) \tag{2}$$

$$\phi^* \in \arg\min_{\phi \in \Phi_r} t'_{\phi}(\Omega) \tag{3}$$

Q_{Ratio} is defined such that a value of 0 corresponds to minimal inhomogeneity, and higher values characterize increasingly inhomogeneous instance sets.

3.2   Variance Measure - Performance Similarity

The intuition behind our second measure is closely related to the question whether different evaluators rate a set of products similarly. More precisely, we want to determine whether m products (configurations) are rated similarly by n evaluators (instances) based on a given evaluation measure (runtime). This setting is similar to that addressed by the Friedman hypothesis test; however, the Friedman test is not directly applicable in our context, in part because we are typically dealing with noisy and censored runtimes.

Our *variance measure* is based on the general idea of assessing instance set homogeneity by means of the variances in runtimes over instances for each given configuration. An instance set is perfectly homogenous, if (after compensating for differences in instance difficulty independent of the configurations considered, i.e., for situations in which certain instances are solved faster than others by all configurations) for every given configuration, all instances are equally difficult.

To account for differences in instance difficulty that are independent of the configurations considered, we perform a standardized z-score normalization of the performance of configurations on instances such that for any given instance, the distribution of performance (here: log-transformed runtime) over configurations has mean zero and variance one. As we will see in Section 5.2, these distributions are often close to lognormal, which justifies standardized z-score transformation on log-transformed runtime measurements.

Formally, if $Var(t^*_{\phi}(\Omega))$ is the variance of the log-transformed, standardized z-score normalized runtimes of configuration $\phi \in \Phi_r$ over the instances in the given set Ω, we define the *variance measure* Q_{Var} as follows:

$$Q_{Var}(\Phi_r, \Omega) = \frac{1}{|\Phi_r|} \sum_{\phi \in \Phi_r} Var(t^*_{\phi}(\Omega)) \tag{4}$$

As in the case of Q_{Ratio}, $Q_{Var} \geq 0$, where $Q_{Var} = 0$ characterizes perfectly homogenous instance sets, while higher values correspond to increasingly inhomogeneous sets.

4 Clustering of Homogeneous Subsets

Based on the homogeneity measures introduced in Section 3, instances within the given set Ω can be clustered with the goal of optimizing homogeneity of the resulting subsets of Ω.

We note that in order to calculate the values of Q_{Ratio} and Q_{Var}, runtimes for each configuration on each instance have to be measured; each of these runtimes can be interpreted as an observation on the behavior of the given parametric algorithm on the instance. Under this interpretation of the data, classical clustering approaches can be applied, in particular, K-Means [27], Gaussian Mixtures [28] and Hierarchical Agglomerative Clustering [29]. While many more clustering approaches can be found in the literature, these methods are amongst the most prominent and widely used classical clustering approaches based on observations.

Agglomerative Hierarchical Clustering[29] iteratively merges the clusters with the lowest distance, where distance between clusters can be defined in various ways. As an alternative to clustering based on distances between the observation vectors, we also explored a variant that always merges the two clusters resulting in the best homogeneity measure (of the merged cluster). Clusters will be merged until a termination criterion is satisfied (e.g., a given number of desired clusters is reached). Unfortunately, the property $\Omega_1 \subset \Omega_2 \Rightarrow Q(\Omega_1) < Q(\Omega_2)$ cannot be guaranteed for arbitrary instance sets Ω_1 and Ω_2, where Q is either of our homogeneity measures. This means that merging two clusters of instances does not necessarily result in a strict improvement in homogeneity. Therefore, we analyzed how our homogeneity measures vary as the number of clusters increases (see Section 5.4).

5 Experiments

In this section, we evaluate empirically how our approach can be used to analyze the homogeneity of instance sets on different kinds of solvers. First, we explain our experimental setting. Next, we characterize the distributions of runtimes on a given instance over solver configurations, which matter in terms of the standardized z-score normalization underlying our variance-based homogeneity measure, Q_{Var}. Then, we investigate to which degree our homogeneity measures agree with the earlier, qualitative analysis of homogeneity by Hutter et al. [11]. Finally, we investigate the question whether algorithm configurators, here ParamILS, perform better on more homogeneous instance sets, as obtained by clustering instances from large and diverse sets.

5.1 Data and Solvers

We used the runtime measurements produced by Hutter et al. [11][3]; their data includes runtimes of the mixed integer programming (MIP) solver *CPLEX* on the instance sets *Regions100* (CPLEX-Regions100: 2000 instances, 5 sec cutoff) and *Orlib* (CPLEX-Orlib: 140 instances, 300 sec cutoff); the local search SAT solver *Spear*

---

[3] See http://www.cs.ubc.ca/labs/beta/Projects/
AAC/empirical_analysis/index.html

on the instance sets *IBM* (SPEAR-IBM: 100 instances, 300 sec cutoff) and *SWV* (SPEAR-SWV: 100 instances, 300 sec cutoff); and the SAT solver *Satenstein* on the instance sets *QCP* (SATenstein-QCP: 2000 instances, 5 sec cutoff) and *SWGCP* (SATenstein-SWGCP: 2000 instances, 5 sec cutoff). In each case, runtime measurements were provided for 1000 solver configurations chosen uniformly at random. Instances that could not be solved by any configuration were excluded from further analysis of our homogeneity measures, as were configurations that could not solve any of our instances. (However, for the clustering performed in later experiments, these instances and configurations were *not* eliminated.)

We augments this extensive data set with addition runtime data for the successful ASP [30] and SAT solver *Clasp* [26] (in version 2.0.2). *Clasp* won the system competitions in the ASP competitions 2009 and 2011 and several gold medals in the 2009 and 2011 SAT competitions. As an open source project[4], *Clasp* is freely available. It also is a highly parametric solver with over fifty parameters, 38 of which we considered in this work (these all influence the solving process for SAT instances).

We applied *Clasp* to two subsets of the crafted and industrial/application benchmarks used in the SAT competitions between 2003 and 2009, dubbed CLASP-Crafted and CLASP-Industrial. Furthermore, we used a set of SAT-encoded bounded model checking problems [31] dubbed CLASP-IBM. We removed all instances for which the running time of every configurations of *Clasp* from a manually chosen set required less than 3 seconds or more than 600 seconds; this was done in order to avoid problems with inaccurate runtime measurements and excessive occurrence of timeouts as well as to ensure that all experiments could be completed within reasonable time. After this filtering step, we were left with 505 instances in the CLASP-Crafted set, 552 instances in CLASP-Industrial, and 148 instances in CLASP-IBM.

We measured runtimes for 32 configurations of *Clasp* chosen uniformly at random (from a total of $\approx 10^{18}$) for each instance, using a cutoff of 600 seconds per run.[5] These runtime measurements required a total of about 350 CPU hours. In the same way as done with the data of Hutter et al., instances that could not be solved by any of our 32 *Clasp* configuration were excluded from further analysis of our homogeneity measures, as were configurations that could not solve any of our instances.

All runs of *Clasp* were carried out on a Dell PowerEdge R610 with an Intel Xeon E5520 (2.26GHz), 48GB RAM running 64-bit Scientific Linux, while the runtime data of Hutter et al. [11] was measured on a 3.2GHz Intel Xeon dual core CPUs with 2GB RAM running Open SuseLinux 10.1.

5.2 Normalization and Distributions

Clearly, the distribution of runtimes over target algorithm configurations on a given problem instance depends on the semantics on the given target algorithm's parameters. As motivated in Section 3.2, our variance-based homogeneity measure requires normalization. The approach we have chosen for this normalization is based on our finding that

---

[4] http://potassco.sourceforge.net/

[5] This runtime data is available at
http://www.cs.uni-potsdam.de/wv/clusteredHomogeneity

Table 1. Quality of fit for distributions of runtime on given problem instances over randomly sampled sets of algorithm configurations, assessed using the Kolmogorov-Smirnov goodness-of-fit test: average rejection rate of test over instances (low values are good) and average p-values (for details see text)

Instance Sets	normal	log-normal	exponential	Weibull
CPLEX-Orlib	0.988(0.002)	0.494(0.160)	0.906(0.016)	0.859(0.036)
SPEAR-IBM	0.911(0.048)	0.533(0.155)	0.911(0.025)	0.867(0.055)
SPEAR-SWV	0.473(0.419)	0.243(0.496)	0.689(0.134)	0.351(0.421)
SATenstein-QCP	0.992(0.003)	0.063(0.474)	0.840(0.033)	0.055(0.413)

the distributions of log-transformed running times tends to be normal, described in the following.

We used the Kolmogorov-Smirnov goodness-of-fit test (KS test) with a significance level of 0.05 to evaluate for each instance, whether the empirical distribution of runtimes over configurations was consistent with a log-normal, normal, exponential or Weibull distribution. We excluded the *Clasp* data from this analysis, since each distribution was only based on 32 data points, resulting in very low power of the KS test. Since the occurrence of a significant numbers of timeouts for a given instance renders the characterization of the underlying distributional family via a KS test impossible, we also eliminated all instances from our test on which more than half the given configuration timed out; since this would have left very few instances in the sets CPLEX-Regions100 and SATenstein-SWGCP, we did not consider these sets in our distributional analysis.

Table 1 shows the averaged test results over all remaining instances, where a result of 1 was recorded, if the respective KS test rejected the null hypothesis of distribution of the given type, and 0 otherwise; we also reported average p-values for each set. As can be seen from these results, in most cases, the distributions tend to be log-normal, whereas the three other types of distributions have much weaker support.

5.3   Evaluation of Homogeneity

Runtimes of instance sets on a set of configurations can be visualized with heat maps, as illustrated in Figure 1; following Hutter et al. [11], we have sorted configurations and instances according to their average PAR-10 scores and represented log-transformed runtimes using different shades of gray (where darker grays correspond to shorter runtimes). As noted in their work, cases where the relative difficulty of the instances varies between different configurations give rise to checkerboard patterns in these plots, and using this qualitative criterion, the CPLEX-Orlib and CLASP-IBM configuration scenarios appear to be rather inhomogeneous, in contrast to the homogeneous instance sets CPLEX-Regions100 and SATenstein-QCP.

As can be seen from the column labeled *unclustered* in Table 2, our variance-based measure is consistent with these earlier qualitative observations. (The remaining columns are discussed later.) In particular, the values for the homogenous sets CPLEX-Regions100 and SATenstein-QCP are low compared to the remaining

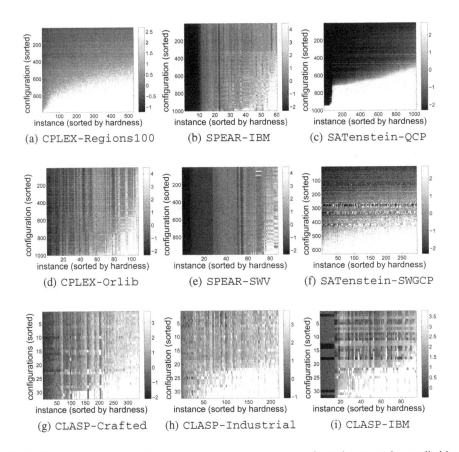

Fig. 1. Heat maps of log transformed runtime data from the configuration scenarios studied by Hutter et al. [11] and three new scenarios using the *Clasp* ASP solver. (The diagrams were generated with the Matlab code provided by Hutter et al.)

instance sets, which show all clear signs of qualitative inhomogeneity. On the other hand, coarser checkerboard patterns do not always correspond to instance sets with higher variance measures for three reasons: (1) the data for *Clasp* is based on far fewer configuration, leading necessarily to coarser patterns in the plots; (2) deviations from uniformity appear more prominent towards the middle of our gray scale than towards the dark and light ends of the spectrum; and (3) large local deviations can have a large influence on the variance measure, but are not necessarily visually as prominent as smaller global variations.

Our *ratio measure* also shows the lowest values for the qualitatively homogenous sets CPLEX-Regions100 and SATenstein-QCP, but ranks the remaining sets differently from the variance measure. In fact, considering the definition of *ratio measure*, it becomes clear that in extreme cases, it may substantially disagree with the qualitative visual measure of Hutter et al.: For example, if one configuration dominates all others,

Table 2. Homogeneity measures Q_{Ratio} and Q_{Var} on the entire instance set (*unclustered*), after configuration-based Gaussian Mixture clustering in four sets (*configuration-based*), and after feature-based K-Means clustering also in four sets (*feature-based*). All measures are averages based on 4-fold cross validation (for details see text).

Instance Sets	unclustered		configuration-based		feature-based	
	Q_{Ratio}	Q_{Var}	Q_{Ratio}	Q_{Var}	Q_{Ratio}	Q_{Var}
CPLEX-Regions100	0.41	0.23	0.11	0.23	–	–
CPLEX-Orlib	0.50	0.71	0.35	0.40	–	–
SPEAR-IBM	0.68	0.75	0.40	0.64	0.58	0.64
SPEAR-SWV	0.74	0.89	0.20	0.72	0.43	0.77
SATenstein-QCP	0.30	0.20	0.25	0.11	0.37	0.17
SATenstein-SWGCP	0.62	0.35	0.62	0.35	0.68	0.41
CLASP-Crafted	0.86	0.39	0.82	0.41	0.79	0.35
CLASP-Industrial	0.81	0.58	0.71	0.50	0.71	0.54
CLASP-IBM	0.57	0.41	0.37	0.30	0.40	0.36

but the remaining configurations are highly inconsistent with each other in terms of their relative performance over the given instance set, the *ratio measure* would be very low, yet, the corresponding heat map would display a prominent checker-board pattern. This illustrates that reasonable and interesting measures of homogeneity, such as the *ratio measure* provide information that is not easily apparent from the earlier qualitative criterion. It also indicates that a single quantitative measure of homogeneity, such as our variance measure, may not capture all aspects of instance set homogeneity of interest in a given context.

5.4   Comparison of Different Clustering Algorithms

We now turn our attention to the question whether partitioning a given instance set into subsets by means of clustering techniques leads to more homogenous subsets according to our ratio and variance measures, as one would intuitively expect. To investigate this question, we used the clustering approaches from Section 4, based on the observed runtimes in conjunction with Gaussian Mixtures and Agglomerative Hierarchical Clustering as well as for the direct optimization of the homogeneity measures using Agglomerative Hierarchical Clustering.

Inspired by *ISAC* [12], we also clustered our instance sets based on cheaply computable instance features [15], using ten runs of the K-Means algorithm for each set. (Preliminary experiments suggested that Gaussian Mixtures clustering on instances features does not yield results better than those produced by K-Means.) In addition, the instances were clustered uniformly at random to obtain a baseline against which the other clustering results could be compared. The SAT instance features were generated with the instance feature generator of *SATzilla* 2011 [15], which provides features based on graph representations of the instance, LP relaxation, DPLL probing, local search probing, clause learning, and survey propagation. Since we did not have feature computation code for MIP instances, we did not perform feature-based clustering on CPLEX-Orlib and CPLEX-Regions100.

To assess the impact of configuration-based clustering on instance set homogeneity, we used a 4-fold cross validation approach, where 3/4 of the configurations were used as a basis for clustering the instance set, and the remaining 1/4 was used for measuring instance homogeneity. (More then 4 folds could not be used, since that would have left too few configurations for measuring homogeneity.) The results in Table 2 and Figure 2 are averaged over the 4 folds, where within each fold, we combined the homogeneity measures for each cluster in the form of an average weighted by cluster size.

Figure 2 shows how our homogeneity measures vary with the number of clusters for instance sets CLASP-Crafted, CLASP-Industrial, and CLASP-IBM (the results for the other instance sets are qualitatively similar and have been omitted due to limited space). In most cases, Gaussian Mixture($\triangleright$) and the feature-based clustering ($\triangledown$) lead to considerable improvements in the *ratio measure* (Figure 2(a)) compared to random clustering. The same holds w.r.t. the *variance measure* (Figure 2(b)), which also tends to be improved by agglomerative clustering ($\times$). The reasons for the oscillations seen for Gaussian Mixture clustering on CLASP-Crafted are presently unclear. Overall, with the exception of CLASP-Crafted, configuration-based Gaussian Mixture clustering tends to produce the biggest improvements in instance set homogeneity.

Interestingly, agglomerative clustering in which we directly optimized the *variance measure* or *ratio measure* tended to give good results on our training sets, but those results did not generalize well to our testing scenarios (in which a disjoint set of configurations was used for measuring homogeneity).

In Table 2, we present numerical results for Gaussian Mixture clustering of our instance sets into four subsets. (We chose four subsets, because the efficiency, measured as the number of clusters in proportion to the optimization of our homogeneity measures, peaked around this number of clusters.) As can be seen from these results, configuration- and feature-based clustering resulted in improvements in homogeneity for almost all instance sets, and configuration-based clustering, although computationally considerably more expensive, tends to produce more homogenous subsets than feature-based clustering. (Preliminary observations from further experiments currently underway suggest that even better results can be obtained from configuration-based clustering using K-Means with multiple restarts.) The fact that these results were obtained using 4-fold cross-validation on our configuration sets indicates that improved homogeneity w.r.t. the configurations considered in the clustering process generalizes to previously unseen configurations.

5.5   Evaluation of Configuration Improvement

The goal of our final experiment was to investigate the hypothesis that automatic algorithm configuration yields better results on more homogenous instance sets. Therefore, we compared the results from applying the same standard configuration protocol to some of our original instance sets and to their more homogenous subsets obtained by clustering. This should not be misunderstood as an attempt to design a practically useful configuration strategy based on homogeneity-improving clustering, which, in order to be practical, would have to use cheaply computable features rather than the ones based on runtimes of a set of configurations used here (see, e.g., [12, 17]).

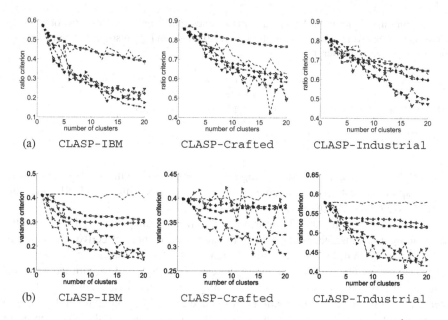

Fig. 2. Q_{Ratio} (a) and Q_{Var} (b) for a varying number of clusters; − random, × Agglomerative Clustering, ▷ Gaussian Mixture, □ Agglomerative Clustering with Q_{Var} optimization, ◇ Agglomerative Clustering with Q_{Ratio} optimization, ▽ feature-based K-Means clustering

For this experiment, we configured *Clasp* on the instance sets of CLASP-Crafted, CLASP-Industrial, and CLASP-IBM with the FocusedILS variant of the *ParamILS* framework (version 2.3.5) [9]. For each of CLASP-Crafted and CLASP-IBM, we conducted four independent runs of FocusedILS with a total time budget of 48 CPU hours per run, and for CLASP-Industrial, we performed 10 runs of FocusedILS of 72 CPU hours each, since this scenario was considerably more challenging. For each set, we then compared the default configuration of *Clasp* for SAT solving (*Default*) against a configuration optimized on the entire instance set (*Entire*), configurations optimized for each of the four subsets obtained by clustering (4 *Clusters*), the oracle performance (also called virtual best solver) over those four configurations (*Oracle*), and the performance obtained when running on each instance a version of *Clasp* optimized specifically for that instance by means of a single, 3 CPU hour run of FocusedILS (*Single Configuration*). With the exception of the last of these scenarios, the performance measurements reported in Table 3 were based on a set of test instances disjoint from the instances used for configuration, and those sets were obtained by random stratified splitting of each original clustered set into equal training and test sets.

The clustering method used in the context of these experiments was Gaussian Mixture clustering, based on the assumption that clusters should be normally distributed [12, 32] and the of the clustering methods we considered, Gaussian Mixtures performed best on average in five out of six cases in Figure 2(b). The target number of clusters was chosen to be four for the reasons explained in Section 5.4. Each instance subset thus obtained was split into a training and test set as previously explained, and *Clasp* was

Table 3. Runtimes in terms of PAR10 in CPU seconds (and number of timeouts) obtained by various configurations of *Clasp* along with (idealized) oracle and per-instance configuration performance (for details, see text)

Instance Set	#	Default	Entire Set	4 Clusters	Oracle	Single Configuration
CLASP-Crafted	254	883(34)	422(15)	361(12)	51(9)	35(0)
CLASP-Industrial	276	1607(70)	1310(56)	1164(50)	721(30)	83(0)
CLASP-IBM	75	2125(26)	1220(15)	1216(15)	1210(13)	135(0)

then configured for each of these training sets. After evaluating these configurations on the corresponding test sets, we aggregated the performance using weighted averaging, where the weights were given by the cluster sizes.

The results shown in Table 3 confirm that automated configuration of *Clasp* on the more homogenous instance sets obtained by clustering is more effective than configuration on the original, less homogenous instance sets. Not too surprisingly, algorithm selection between the resulting configurations can in principle yield additional improvements (as seen from the oracle results), and configuration on individual instances has the potential to achieve further, dramatic performance gains. We note that single-instance sets are, intuitively and by definition, completely homogenous and therefore represent an idealistic best-case scenario for automated algorithm configuration. Furthermore, the oracle performance provides a good estimate of the performance of a parallel portfolio of the respective set of configurations, whose performance is evaluated solely based on wallclock time.

6   Conclusions and Future Work

In this work, we introduced two quantitative measures of instance set homogeneity in the context of automated algorithm configuration. Our measures provide an alternative to an earlier qualitative visual criterion based on heat maps [11]; one of them, the *variance measure*, gives results that are highly consistent with the visual criterion, and both of them capture aspects of instance set homogeneity not easily seen from heat maps. Furthermore, we provided evidence that our measures are consistent with the previously informal intuition that more homogenous instance sets are more amenable to automated algorithm configuration (see, e.g., [13]).

The proposed homogeneity measures can be used directly to assess whether automated configuration of a given parametric algorithm using a particular instance set might be difficult due to instance set inhomogeneity. In addition, the *ratio measure* helps to assess the specific potential of portfolio-based approaches in a given configuration scenario, including instance-based algorithm configuration [15, 12], portfolio multithreading [33] and sequential portfolio solving [34, 35].

Unfortunately, like the previous qualitative approach, our quantitative homogeneity measures are computationally expensive. In future work, we plan to investigate how this computational burden can be reduced, for example, by using promising configurations encountered during algorithm configuration instead of randomly sampled ones.

Acknowledgements. We thank F. Hutter for providing the data and Matlab scripts used in [11], L. Xu for providing the newest version of the feature generator of *SATzilla*, and A. König and M. Möller for valuable comments on an early version of this work.

This work was partially funded by the DFG under grant SCHA 550/8-2 and by NSERC through a discovery grant to H. Hoos.

References

1. Hutter, F., Babić, D., Hoos, H., Hu, A.: Boosting verification by automatic tuning of decision procedures. In: Procs. FMCAD 2007, pp. 27–34. IEEE Computer Society Press (2007)
2. KhudaBukhsh, A., Xu, L., Hoos, H., Leyton-Brown, K.: SATenstein: Automatically building local search sat solvers from components. In: Boutilier, C. (ed.) Procs. IJCAI 2009, pp. 517–524. AAAI Press/The MIT Press (2009)
3. Tompkins, D.A.D., Balint, A., Hoos, H.H.: Captain Jack: New Variable Selection Heuristics in Local Search for SAT. In: Sakallah, K.A., Simon, L. (eds.) SAT 2011. LNCS, vol. 6695, pp. 302–316. Springer, Heidelberg (2011)
4. Hutter, F., Hoos, H.H., Leyton-Brown, K.: Automated Configuration of Mixed Integer Programming Solvers. In: Lodi, A., Milano, M., Toth, P. (eds.) CPAIOR 2010. LNCS, vol. 6140, pp. 186–202. Springer, Heidelberg (2010)
5. Vallati, M., Fawcett, C., Gerevini, A., Hoos, H.H., Saetti, A.: Generating fast domain-specic planners by automatically conguring a generic parameterised planner. In: Procs. PAL 2011, pp. 21–27 (2011)
6. Hoos, H.H.: Programming by optimisation. Communications of the ACM (to appear, 2012)
7. Ansótegui, C., Sellmann, M., Tierney, K.: A Gender-Based Genetic Algorithm for the Automatic Configuration of Algorithms. In: Gent, I.P. (ed.) CP 2009. LNCS, vol. 5732, pp. 142–157. Springer, Heidelberg (2009)
8. Birattari, M., Stützle, T., Paquete, L., Varrentrapp, K.: A racing algorithm for configuring metaheuristics. In: Kaufmann, M. (ed.) Procs. GECCO 2009, pp. 11–18. ACM Press (2002)
9. Hutter, F., Hoos, H.H., Leyton-Brown, K., Stützle, T.: ParamILS: An automatic algorithm configuration framework. Journal of Artificial Intelligence Research 36, 267–306 (2009)
10. Hutter, F., Hoos, H.H., Stützle, T.: Automatic algorithm configuration based on local search. In: Procs. AAAI 2007, pp. 1152–1157. AAAI Press (2007)
11. Hutter, F., Hoos, H.H., Leyton-Brown, K.: Tradeoffs in the Empirical Evaluation of Competing Algorithm Designs. Annals of Mathematics and Artificial Intelligenc (AMAI), Special Issue on Learning and Intelligent Optimization 60(1), 65–89 (2011)
12. Kadioglu, S., Malitsky, Y., Sellmann, M., Tierney, K.: ISAC - Instance-Specific Algorithm Configuration. In: Coelho, H., Studer, R., Wooldridge, M. (eds.) Procs. ECAI 2008, pp. 751–756. IOS Press (2010)
13. Lindawati, Lau, H.C., Lo, D.: Instance-Based Parameter Tuning via Search Trajectory Similarity Clustering. In: Coello, C.A.C. (ed.) LION 5. LNCS, vol. 6683, pp. 131–145. Springer, Heidelberg (2011)
14. Gomes, C., Selman, B.: Algorithm portfolios. Journal of Artificial Intelligence 126(1-2), 43–62 (2001)
15. Xu, L., Hutter, F., Hoos, H., Leyton-Brown, K.: SATzilla: Portfolio-based algorithm selection for SAT. Journal of Artificial Intelligence Research 32, 565–606 (2008)
16. Hutter, F., Hamadi, Y., Hoos, H.H., Leyton-Brown, K.: Performance Prediction and Automated Tuning of Randomized and Parametric Algorithms. In: Benhamou, F. (ed.) CP 2006. LNCS, vol. 4204, pp. 213–228. Springer, Heidelberg (2006)

17. Xu, L., Hoos, H.H., Leyton-Brown, K.: Hydra: Automatically configuring algorithms for portfolio-based selection. In: Fox, M., Poole, D. (eds.) Procs. AAAI 2010, pp. 210–216. AAAI Press (2010)
18. Rice, J.: The algorithm selection problem. Advances in Computers 15, 65–118 (1976)
19. Xu, L., Hoos, H.H., Leyton-Brown, K.: Hierarchical Hardness Models for SAT. In: Bessière, C. (ed.) CP 2007. LNCS, vol. 4741, pp. 696–711. Springer, Heidelberg (2007)
20. Leyton-Brown, K., Nudelman, E., Shoham, Y.: Empirical hardness models: Methodology and a case study on combinatorial auctions. Journal of the ACM 56(4), 1–52 (2009)
21. Gebser, M., Kaminski, R., Kaufmann, B., Schaub, T., Schneider, M.T., Ziller, S.: A Portfolio Solver for Answer Set Programming: Preliminary Report. In: Delgrande, J.P., Faber, W. (eds.) LPNMR 2011. LNCS, vol. 6645, pp. 352–357. Springer, Heidelberg (2011)
22. Hutter, F., Hoos, H.H., Leyton-Brown, K.: Sequential Model-Based Optimization for General Algorithm Configuration. In: Coello, C.A.C. (ed.) LION 5. LNCS, vol. 6683, pp. 507–523. Springer, Heidelberg (2011)
23. Smith-Miles, K.: Cross-disciplinary perspectives on meta-learning for algorithm selection. ACM Computing Surveys 41(1), 6:1–6:25 (2008)
24. Guerri, A., Milano, M.: Learning Techniques for Automatic Algorithm Portfolio Selection. In: de Mántaras, R.L., Saitta, L. (eds.) Procs. ECAI 2004, pp. 475–479. IOS Press (2004)
25. Kotthoff, L., Gent, I., Miguel, I.: A Preliminary Evaluation of Machine Learning in Algorithm Selection for Search Problems. In: Borrajo, D., Likhachev, M., López, C. (eds.) Procs. SoCS 2011, pp. 84–91. AAAI Press (2011)
26. Gebser, M., Kaufmann, B., Neumann, A., Schaub, T.: Conflict-driven answer set solving. In: Veloso, M. (ed.) Procs. IJCAI 2007, pp. 386–392. AAAI Press/The MIT Press (2007)
27. MacQueen, J.B.: Some methods for classification and analysis of multivariate observations. In: Cam, L., Neyman, J. (eds.) Procs. Mathematical Statistics and Probability, vol. 1, pp. 281–297. University of California Press (1967)
28. Bishop, C.: Pattern Recognition and Machine Learning (Information Science and Statistics), 1st edn. 2006. corr. 2nd printing edn. Springer (2007)
29. Ward, J.: Hierarchical Grouping to Optimize an Objective Function. Journal of the American Statistical Association 58(301), 236–244 (1963)
30. Baral, C.: Knowledge Representation, Reasoning and Declarative Problem Solving. Cambridge University Press (2003)
31. Zarpas, E.: Benchmarking SAT Solvers for Bounded Model Checking. In: Bacchus, F., Walsh, T. (eds.) SAT 2005. LNCS, vol. 3569, pp. 340–354. Springer, Heidelberg (2005)
32. Hamerly, G., Elkan, C.: Learning the k in k-means. In: Procs. NIPS 2003, MIT Press, Cambridge (2003)
33. Hamadi, Y., Jabbour, S., Sais, L.: ManySAT: a parallel SAT solver. Journal on Satisfiability, Boolean Modeling and Computation 6, 245–262 (2009)
34. Streeter, M., Golovin, D., Smith, S.: Combining Multiple Heuristics Online. In: Procs. AAAI 2007, pp. 1197–1203. AAAI Press (2007)
35. Kadioglu, S., Malitsky, Y., Sabharwal, A., Samulowitz, H., Sellmann, M.: Algorithm Selection and Scheduling. In: Lee, J. (ed.) CP 2011. LNCS, vol. 6876, pp. 454–469. Springer, Heidelberg (2011)

Automatically Configuring Algorithms for Scaling Performance

James Styles[1], Holger H. Hoos[1], and Martin Müller[2]

[1] University of British Columbia, 2366 Main Mall, Vancouver, BC, V6T 1Z4, Canada
{jastyles,hoos}@cs.ubc.ca
[2] Computing Science, University of Alberta, Edmonton, AB, T6G 2E8, Canada
mmueller@ualberta.ca

Abstract. Automated algorithm configurators have been shown to be very effective for finding good configurations of high performance algorithms for a broad range of computationally hard problems. As we show in this work, the standard protocol for using these configurators is not always effective. We propose a simple and computationally inexpensive modification to this protocol and apply it to state-of-the-art solvers for two prominent problems, TSP and computer Go playing, where the standard protocol is unable or unlikely to yield performance improvements, and one problem, mixed integer programming, where the standard protocol is known to be effective. We show that our new protocol is able to find configurations between 4% and 180% better than the standard protocol within the same time budget.

1 Introduction

Many high performance algorithms for computationally hard problems have numerous parameters, some exposed to end users and others hidden as hard-coded design choices and magic constants, that control their behaviour and performance. Recent work on automated configurators has proven to be very effective at finding good values for these parameters [4, 6, 7, 8, 12, 13, 14, 16, 17]. The standard protocol for using automated configurators, such as ParamILS [14], to optimize the performance of a parametric algorithm for a given problem is as follows:

1. Identify the intended use case of the algorithm (*e.g.*, structure and size of expected problem instances, resource limitations) and define a metric to be optimized (*e.g.*, runtime).
2. Construct a training set/scenario which is representative of the intended use case. The performance of the configurator depends on being able to evaluate a large number, ideally thousands, of configurations. Training instances/scenarios must be chosen to permit this.
3. Perform multiple independent runs of the configurator, typically 10-25 [14].
4. Validate the final configurations found by each run on the training set.
5. Select from the final configurations found by the independent runs the one with the best performance on the training set.

Y. Hamadi and M. Schoenauer (Eds.): LION 6, LNCS 7219, pp. 205–219, 2012.
© Springer-Verlag Berlin Heidelberg 2012

While this protocol has been successfully applied to many problems and solvers, we have observed that it is not always feasible. In particular, for Step 2, choosing a training set becomes problematic if the time taken to evaluate a configuration on the training settings is too large. This is the case for Fuego [9], a state-of-the-art computer Go player based on Monte Carlo tree search (MCTS). The time taken to evaluate a single configuration of Fuego for competition level play can take hours or even days (see Section 5.3). For ParamILS to have any hope of finding a good configuration, each run would have to be allowed to take several years. In situations of this nature, a training set significantly easier than the intended use case must be used. Unfortunately, the use of easier training sets may lead to configurations whose performance may not scale up to the intended use case.

In this paper, we explore a simple modification to the standard protocol for using automated configurators that attempts to resolve this problem in a generic manner. We apply our new protocol to three well-known problems and configuration scenarios. The first of these is the traveling salesperson problem (TSP), a widely studied combinatorial optimization problem with numerous industrial applications, for which we configure Keld Helsgaun's implementation of the Lin-Kerninghan algorithm (LKH) [11], the best incomplete solver for TSP currently known (and far superior to any complete solver in terms of finding optimal or near-optimal solutions fast). In an early stage of our work, described in Section 2, we found that the standard protocol is ineffective for this configuration scenario. Our second scenario concerns computer Go playing, a grand challenge in artificial intelligence, using the state-of-the-art Monte Carlo tree search (MCTS) based player Fuego [9]. Evaluating configurations for this scenario requires playing hundreds of games (see Section 4.3) which becomes prohibitively expensive for the intended use case. This makes using the standard protocol infeasible. The third scenario we consider involves solving mixed integer programming (MIP) problems, which are widely used for representing constrained optimization problems in academia and industry, using the state-of-the-art commercial solver CPLEX [2]. Unlike for the other two scenarios, the standard protocol has been proven to be very effective for this configuration scenario [12]; our primary motivation for studying it here is to verify that our new protocol does not lead to compromised configuration performance in cases where the standard protocol is already effective.

The remainder of this paper is structured as follows. Section 2 illustrates the problems we have encountered using the standard protocol for configuring LKH. Section 3 presents our new protocol. Section 4 describes the three configuration scenarios we consider in this work in more detail. Section 5 explains the experimental setup we used for evaluating our new configuration protocol, and Section 6 presents the empirical results we obtained. Section 7 provides conclusions and an overview of ongoing and future work.

2   A First Attempt at Configuring LKH

The starting point for this work was an attempt to configure LKH [11] using ParamILS. In particular, we were interested in reducing the time taken to find optimal (or near optimal) solutions for structured instances like those found in the well-known TSPLIB benchmark collection [15], with a focus on instances

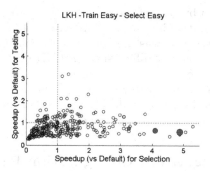

Fig. 1. Comparison between the selection and testing (PAR10) speedup, relative to the default configurations, for 300 configurations of LKH found by ParamILS using the easy instances, see Section 4.1, for both training and selection. Configurations shown in the CPU time column of Table 1 are filled in and coloured red. The size of the points for these configurations corresponds to the time required for finding them.

containing several thousand cities, on which LKH's default configuration can take several CPU hours on our reference machines (see Section 5) to find near-optimal solutions. As ParamILS requires thousands of evaluations [14] of an algorithm to reliably achieve good results, using such instances directly for training would result in individual configuration experiments with a duration of up to one year. Since this is infeasible, we decided to perform training on similarly structured, but significantly smaller (200–1500 node) instances which takes less than a CPU minute for the default configuration of LKH to solve.

Since LKH is an incomplete solver, there is no guarantee on the quality of solution found by a given run. We are therefore interested in the runtime of a configuration on an particular instance as well as the quality of solution found. To avoid constructing a Pareto front, we combine these two raw performance metrics using penalized average runtime (PAR10). This metric uses the total running time of a given run and then penalizes runs which are unable to achieve a target solution quality within some time cutoff. As we do not know the optimal solution quality for every instance used, we determine the target solution quality by using the final solution found by a single long run of the default configuration. On instances with a known optimal solution, the target quality chosen is often equivalent to the optimal for small instances and within 1% of the optimal for large instances.

Following the standard protocol, we performed multiple independent 24-hour runs of ParamILS using this easier training set and optimizing for penalized average runtime (PAR10). The configurations found by these experiments performed very well on the training set, but often turned out to be worse than the default configuration when evaluated on the testing set, consisting of the larger instances we were ultimately interested in solving.

To further explore the reasons for this apparent failure of the standard protocol, we expanded our experiment to include 300 independent 24-hour runs of ParamILS using the same metric and the same training set and evaluated the

final configuration found by each of these runs on the entire testing set. As seen
in Figure 1, we found that while the standard protocol for ParamILS was able
to find good configurations, it was unlikely to select them. This is due to the
fact that the performance of a configuration of LKH on the training set is not
a good predictor of that configuration's performance on the testing set. Despite
being ineffective as a predictor of testing set performance, the training set was
able to guide runs of ParamILS to configurations with a speedup factor of up to
3.19, which suggests it still has value in the configuration process.

3   Automated Algorithm Configuration for Scalable Performance

To address the problem encountered in Section 2, we devised a simple modification to the standard protocol for using configurators such as ParamILS. Instead
of selecting between the final configurations found by independent configurator
runs based on their performance on the training set, we select based on their
performance on a set of intermediate instances that are harder than the training set, but easier than the testing set. For this work, we define intermediate
difficulty based on percentiles of the distribution of running time for the default
configuration of a given solver over the testing set. This protocol has three advantages over alternative approaches: (1) it does not require any modifications
to the underlying configurator; (2) it can reuse the results of existing configuration experiments; and (3) it can be set up to require only a moderate amount
of additional processing time (in our experiments, the overhead is always below 50% of the total time budget). To assess this protocol, which we dubbed
Train-Easy, Select-Intermediate (TE-SI), we compare it to the original protocol, *Train-Easy, Select-Easy (TE-SE)*, and to an alternative approach, in which
training is directly performed on the harder instances used for selection, *Train-
Intermediate, Select-Intermediate (TI-SI)*, always correctly accounting for the
overhead required for evaluating configurations at the selection stage.

4   Configuration Scenarios

4.1   Solving TSP Using LKH

LKH [11] is a two-phase, incomplete solver for the TSP. It first performs deterministic preprocessing using subgradient optimization, which modifies the cost
function of the given TSP instance while preserving the total ordering of solutions by tour length. The main goal of this first phase, which can sometimes
already reach the desired solution quality, is to make an instance easier for the
subsequent phase to solve. The second phase consists of a stochastic local search
procedure based on chaining together so-called k-opt moves.

For the following experiments, we used a version of LKH 2.02, which we have
extended to allow several parameters to scale with instance size and to make
use of a simple dynamic restart mechanism to prevent stagnation behaviour we
had observed in preliminary experiments. The original configuration space is

preserved by these modifications (*i.e.*, it is possible to replicate the behaviour of any configuration for the original LKH 2.02 using our extended version).

Training and testing were done using instances from the well-known TSPLIB benchmark collection [15]. TSPLIB is a heterogeneous set consisting mostly of industrial and geographic instances. The original TSPLIB set contains only 111 instances; since we consider this too small to allow for effective automated configuration and evaluation, we generated new TSP instances based on existing TSPLIB instances by randomly selecting 10%, 20%, or 30% of the existing instance's nodes to be removed. These TSPLIB-like instances retain most of the original structure and are comparable in difficulty to the original instance, ranging from requiring a factor of 30 less time to a factor of 900 more time for the default configuration of LKH to solve.

The modified version of LKH 2.02 and the TSPLIB-like instances will be made available on our website upon publication.

4.2   Solving MIP Using CPLEX

CPLEX is one of the best-performing and most widely used solvers for mixed integer programming problems. It is based on a highly parameterized branch-and-cut procedure that generates and solves a large number of linear programming (LP) subproblems. While most details of this procedure are proprietary, at least 76 parameters which control CPLEX's performance while solving MIP problems are exposed to end users.

Our work on this scenario aims to mirror recent work by Hutter *et al.* [12] for configuring CPLEX 12.1 on the CORLAT instance set, for which the standard protocol for using ParamILS was able to achieve a 52-fold speedup over the CPLEX default settings. The CORLAT instance set consists of 2000 instances based on real data modeling wildlife corridors for grizzly bears in the Northern Rockies [10]. Our goal in considering this scenario is to show that our new configuration protocol is effective even in scenarios where the default protocol is known to work well.

Hutter *et al.* [12] used CPLEX 12.1, the most recent version available at the time of their study. CPLEX 12.3, the current version at the time of this writing, performs significantly better on the CORLAT instances, achieving a speedup factor of up to 90 on the hardest instances in the set. To compensate for this significant improvement of the default configuration, we performed a 1/50th time scale replica of the configuration experiments conducted by Hutter *et al.* [12]. Reducing the runtime of ParamILS and the per-instance cutoffs preserves both the percentage of training instances that the default configuration is capable of solving within the time cutoff as well as the number of evaluations ParamILS is able to perform within the total configuration time budget. The results of our experiments depend only on the ratio of per-instance runtime to total configuration time and are invariant with respect to the overall time scale. While we used significantly reduced configuration times, we believe that our results should generalize to longer configuration runs using harder instances.

The metric being optimized for this scenario is penalized average runtime (PAR10). We measure the total running time of given run and penalize if it is unable to find the optimal solution within a cutoff of 1 hour for testing, 6 seconds for training on easy and 24 seconds for training on intermediate. CPLEX is a complete solver, so every run of a configuration that does not exhibit errant behaviour is guaranteed to find the optimal solution to every instance given enough computational resources. For this scenario, cutoffs and penalties are only used to limit the total computational effort of performing these experiments.

We also applied the TE-SI protocol to CPLEX 12.1. We found configurations that offered significant speedups (> 52) compared to the default configuration of 12.1, improving upon the results found in [12]; however, the overall result remained qualitatively similar to our work with CPLEX 12.3. We do not present our results for CPLEX 12.1 due to space limitations.

4.3   Playing Go Using Fuego

Developing programs for the game of Go has been a topic of intense study over the last five decades. Only recently, with the advent of Monte Carlo Tree Search (MCTS), has the strength of Go programs caught up with top human players, at least on small board sizes (up to 9×9). MCTS combines position evaluation by randomized *playouts* of the remainder of a game with a new, selective search approach that balances exploration and exploitation: the algorithm combines exploration of parts of a game tree that are still underdeveloped with exploitation by deep search of the most promising lines of play. The open-source project Fuego [9] contains both a game-independent framework for MCTS and a state-of-the-art Go program. The program was the first to beat a top human professional player in an even game on the 9×9 board and has won numerous computer competitions [1].

Like the other configuration scenarios we study in this work, Fuego has a large number of configurable parameters. The performance metric to be optimized is the win rate of a configuration when played against the default configuration. Note that the baseline is not necessarily 50%: For certain board sizes and playout limits, the default configuration is stronger playing black than white, while for other board sizes and playout limits, the opposite holds.

Noisy Evaluations. Since Fuego uses a randomized playout strategy in its core MCTS procedure, the win rate of any set of test games played with Fuego varies. This introduces a significant source of noise when evaluating configurations of Fuego. This noise must be compensated for by playing additional games; otherwise, the observed win rates are meaningless (*e.g.*, with 10 games played, there is a more than 40% chance that the observed win rate of a configurations differs from its true win rate by at least 10%). The exact number of games needed depends on the true win rates of the configurations being compared, but often hundreds, if not thousands, of games are required to reduce the chance of incorrectly ranking two configurations to less than 1%. A key point is that the closer two configurations are in true win rate, the more games are needed to correctly rank them. We note that while in principle, similar concerns arise for many other configuration scenarios involving randomised algorithms, the amount of evaluation noise in the case of Fuego (and other randomised game players) is particularly large, due to the fact that individual games have binary outcome.

This is particularly problematic for automatic configurators like ParamILS, which often rely on a sequence of small incremental improvements to a configuration. If the improvement is too small, then it will be dwarfed by the noise in the evaluations and it is impractical to play a sufficient number of games, potentially thousands, to adequately compensate. We compromise by playing as many games as are necessary to evaluate a configuration, up to a limit of 200, during training. We note that when comparing two configurations with true win rates (as determined in playing against some reference configuration) within 1% of each other, there is a 20.7% chance of incorrectly ranking them based on a set of 200 games.

5   Experimental Setup

For each configuration scenario, we defined three instance sets of distinct difficulties: an easy instance set, designed to allow ParamILS to perform at least several hundred evaluations of candidate configurations; a hard instance set, designed to represent the difficulty of instances/situations that we are interested in optimizing the target algorithms performance for; and a set of intermediate instances with a difficulty between the easy and hard instances. The exact definition of easy, intermediate and hard is specific to each the configuration scenario.

Using these sets, we performed three sets of configuration experiments using independent runs of ParamILS. (We chose ParamILS, because it is the only readily available algorithm configuration procedure that has been demonstrated to work well on configuration scenarios of the difficulty considered here.) In the first set of experiments, we used the easy instances during training and then selected a configuration, from the set of the final configurations produced across a number of independent runs of ParamILS, according to its performance on the same (easy) set (this is the standard protocol, TE-SE). The second set used the easy instances set for training, but intermediate instances for selection (this is our new protocol, TE-SI). The third set used the intermediate instances for both training and selection (TI-SI). All testing was performed on the hard instances. (Recall that we are interested in the case where the hard instances are too difficult to be used in training.)

LKH and CPLEX Experiments were performed on the 384 node DDR partition of the Westgrid Orcinus cluster; Orcinus runs 64-bit Red Hat Enterprise Linux Server 5.3, and each node has two quad-core Intel Xeon E5450 64-bit processors running at 3.0 GHz with 16GB of RAM.

Fuego Experiments were performed on the 512 node Westgrid Lattice cluster. Lattice runs 64-bit Linux CentOS 5.5, and each node has two quad-core Intel Xeon L5520 64-bit processors running at 2.27 GHz with 12 GB of RAM.

5.1   Solving TSP Using LKH

The Hard Instance Set consists of 3192 instances containing up to 6000 cities, drawn from both the original TSPLIB and TSPLIB-like instances. The default configuration of LKH takes approximately 214 CPU hours on our reference machines to run on the entire set. The 99th percentile difficulty is 2900 CPU seconds.

The Intermediate Instance Set consists of instances which take the default configuration between 350 and 580 CPU seconds to solve; this range corresponds to between 12.5 and 20 percentile difficulty found in the hard instance set. The default configuration takes approximately 20 CPU hours to run on the entire intermediate instance set. All instances in the intermediate set are drawn from a set of TSPLIB-like instances disjoint from the hard instance set. When used for training, a per-instance cutoff of 780 CPU seconds is used.

The Easy Instance Set consists of instances which take the default configuration between 1 and 52 CPU seconds to solve. The default configuration takes 19 minutes to run on the entire easy instance set. All instances in the easy set are drawn from a set of TSPLIB-like instances disjoint from those used in the hard and intermediate sets. When used for training, a per-instance cutoff of 120 seconds is used.

Using these sets, we performed two sets of configuration experiments. The first set consists of 300 independent 24-hour runs of ParamILS using the easy set for training. The second set consists of 100 independent 24-hour runs of ParamILS using the intermediate set for training.

The TE-SE protocol requires 1459 CPU minutes per run of ParamILS. The TE-SI and TI-SI protocols require 2640 minutes per run of ParamILS.

5.2   Solving MIP Using CPLEX

The Hard Instance Set consists of 1650 instances drawn from the set of CORLAT instances used in [12]. The default configuration takes approximately 11.5 CPU hours to evaluate the entire instance set. The 99th percentile difficulty is 448 CPU seconds.

The Intermediate Instance Set consists of instances which take the default configuration between 54 and 90 seconds to evaluate; this range corresponds to between 12.5 and 20 percentile difficulty found in the hard instance set. The default configuration takes approximately 1.1 CPU hours to evaluate the entire intermediate instance set. The instances in the intermediate instance are disjoint from the hard instance set. When used for training, a per-instance cutoff of 24 CPU seconds is enforced.

The Easy Instance Set consists of instances which take the default configuration between 1 and 10 seconds to evaluate. The default configuration takes approximately 18 CPU minutes to evaluate the entire easy instance set. The easy instance set is disjoint from both the hard and intermediate instance sets. When used for training, a per-instance cutoff of 6 CPU seconds is enforced.

Using these sets, we performed two sets of configuration experiments. The first set consists of 100 independent 24-hour runs of ParamILS using the easy set for training. The second set consists of 100 independent 24-hour runs of ParamILS using the intermediate set for training.

The TE-SE protocol requires 4531 CPU seconds per run of ParamILS. The TE-SI and TI-SI protocols require 7456 seconds per run of ParamILS.

5.3 Playing Go Using Fuego

The Hard Setting consists of playing 5000 games on a 7×7 board with 300 000 playouts. For 5000 games there is a [100%, 99.8%, 91%] chance of correctly determining the true win rate of a configuration to within [3%, 2%, 1%].

The Intermediate Setting consists of playing 1000 games on a 7×7 board with 100 000 playouts for selection and 5000 such games for testing. For 1000 games there is a [97%, 88%, 60%] chance of correctly determining the true win rate of a configuration to within [3%, 2%, 1%].

The Easy Setting consists of playing 1000 games on a 7×7 board with 10 000 playouts for selection and 5000 such games for testing,

Using these sets, we performed two sets of configuration experiments. The first set consists of 80 independent 24-hour runs of ParamILS using the easy set for training. The second set consists of 80 independent 24-hour runs of ParamILS using the intermediate set for training. Each set of configuration experiments is split evenly across configuring for playing black or playing white.

The TE-SE protocol requires 5904 CPU hours per run of ParamILS. The TE-SI and TI-SI protocols require 7200 hours per run of ParamILS.

6 Results

We are interested in how effective the TE-SI protocol is in a typical setting where 10–25 independent runs of the configuration procedure are performed (tyically in parallel). To assess the variation in the results of such configuration experiments, we have performed a significantly higher number of configurator runs for each of our configuration scenarios and then performed a bootstrap analysis based on the data thus obtained.

For a specific protocol and a target number n of ParamILS runs, we generated 100 000 bootstrap samples by selecting, with replacement, the configurations obtained from the n runs. For each such sample R, we chose a configuration according to the selection criteria of the protocol under investigation and used the performance of that configuration on the testing set as the result of R.

We present the results from these analyses in two ways. In Table 1, we show the median performance of the bootstrap samples for each protocol when using different numbers of independent ParamILS runs and overall CPU time budget. In Figure 2, we show the median performance and the ranged spanned by the 5th and 95th percentile performance of bootstrap samples versus total CPU time budget. For reference, we also show the quality of the default and of the best known configuration for each scenario. The data in Table 1 thus represents several time slices from Figure 2.

6.1 Results for Configuring LKH

The TE-SI protocol was able to reliably improve upon the default configuration (see Fig. 2). The other two protocols tend to either produce configurations with quality similar to the default (TI-SI) or notably worse than the default (TE-SE).

Table 1. Overview of the speedup versus the default, for LKH and CPLEX, and the win rate versus the default, for Fuego, during testing for the configurations found by the three protocols. The performance of each configuration on the instances/settings used for selection is shown in parentheses. Values presented are the medians over 100 000 bootstrapped samples. The best known performance on easy, intermediate and testing instance sets / settings are provided for reference. Configurations shown in the CPU Time column are highlighted in the scatter plots in Figures 1, 3 and 4.

	Speedup Factor (PAR10) vs Default Configuration					
LKH	best easy: 5.29, best intermediate: 2.58, best testing: 3.19					
	Runs of ParamILS			CPU Time		
	10	25	50	20 Days	50 Days	100 Days
train easy select easy	0.82 (2.78)	0.73 (3.47)	0.67 (4.09)	0.76 (3.26)	0.67 (4.09)	0.61 (4.88)
train easy select inter.	**1.29** (1.29)	**1.52** (1.59)	**1.71** (1.84)	**1.29** (1.31)	**1.52** (1.61)	**1.71** (1.98)
train inter. select inter.	0.92 (1.25)	0.97 (2.06)	0.97 (2.08)	0.92 (1.25)	0.97 (2.06)	0.97 (2.08)
CPLEX	best easy: 1.53, best intermediate: 2.77, best testing: 3.03					
	Runs of ParamILS			CPU Time		
	10	25	50	1 Day	2.5 Days	5 Days
train easy select easy	1.63 (1.20)	1.63 (1.26)	1.61 (1.38)	1.63 (1.23)	1.61 (1.26)	2.36 (1.53)
train easy select inter.	**1.94** (1.54)	**2.24** (1.83)	**2.64** (1.92)	**2.00** (1.54)	**2.36** (1.83)	**2.64** (1.92)
train inter. select inter.	1.65 (1.63)	1.96 (1.88)	1.98 (1.99)	1.87 (1.71)	1.98 (1.88)	1.98 (1.99)

	Relative Win Rate (Configuration Win Rate / Default Win Rate)					
Fuego - Black	best easy: 1.04, best intermediate: 1.21, best testing: 1.22					
	Runs of ParamILS			CPU Time		
	10	15	30	50 Days	75 Days	150 Days
train easy select easy	1.08 (1.02)	1.08 (1.02)	0.94 (1.04)	1.08 (1.02)	1.08 (1.02)	0.94 (1.04)
train easy select inter.	**1.12** (1.17)	**1.13** (1.20)	**1.17** (1.21)	**1.12** (1.17)	**1.13** (1.20)	**1.17** (1.21)
train inter. select inter.	1.10 (1.10)	1.06 (1.21)	1.06 (1.21)	1.10 (1.10)	1.06 (1.21)	1.06 (1.21)
Fuego - White	best easy: 1.07, best intermediate: 1.13, best testing: 1.45					
	Runs of ParamILS			CPU Time		
	10	15	30	50 Day	75 Days	150 Days
train easy select easy	1.13 (1.05)	1.13 (1.05)	1.25 (1.07)	1.13 (1.05)	1.13 (1.05)	1.25 (1.07)
train easy select inter.	1.27 (1.08)	**1.41** (1.13)	**1.41** (1.13)	1.27 (1.08)	**1.41** (1.13)	**1.41** (1.13)
train inter. select inter.	**1.32** (1.12)	1.32 (1.12)	1.34 (1.12)	**1.32** (1.12)	1.32 (1.12)	1.34 (1.12)

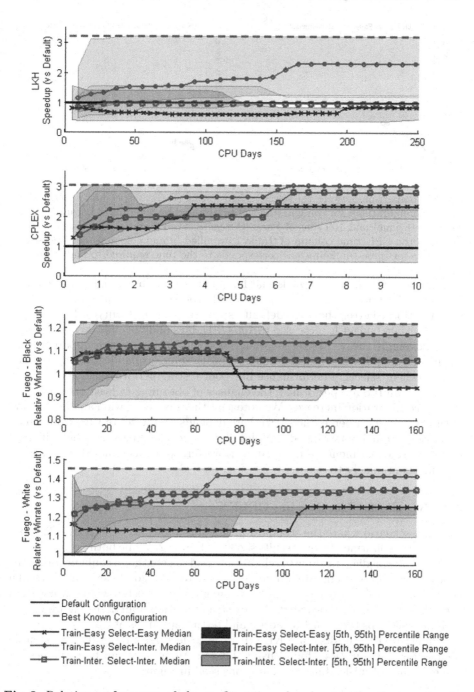

Fig. 2. Relative performance of the configurations found using the three protocols versus overall CPU time budget spent, including the median and [5th,95th] percentiles over 100 000 bootstrapped samples for every protocol as well as the quality of the default and best known configurations for reference. For all three configuration scenarios, the TE-SI protocol yields the best overall results.

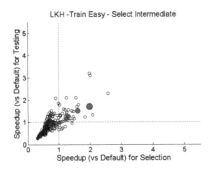
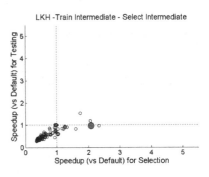

Fig. 3. Comparison between the selection and testing (PAR10) speedup, relative to the default configurations, for configurations of LKH found by 300 runs of ParamILS using TE-SI (left pane) and 100 runs of ParamILS using TI-SI (right pane). Configurations shown in the CPU time column of Table 1 are filled in and coloured red. The size of the points for these configurations corresponds to the time required to find them.

Our bootstrap analysis reveals that for small overall time budgets, there is a $\geq$ 5% chance for both the TE-SE and the TI-SI protocols to produce configurations which perform better than the default (see 95th percentile curve). However, as the CPU time budget is increased to 100 CPU days and beyond, this probability decreases significantly. The reason underlying this phenomenon is apparent from Figures 1 and 3: Both protocols encounter, with some probability, configurations with excellent selection performance but poor testing performance, and as more runs of ParamILS are performed, the chances of obtaining at least one such misleading configuration increases. We note that the precise location and magnitude of the drop in 95th percentile shown here depends on the set of runs from which we obtained our bootstrapped samples and would likely be somewhat different if the entire experiment were repeated. However, we expect that drops of some magnitude are likely to occur.

6.2 Results for Configuring CPLEX

This is a scenario where the standard protocol is known to be effective [12], and this result is confirmed by our results shown in Figure 2. While both protocols that select on intermediate are able to reliably find and selected good configurations, the protocol we propose generally provides the best results. For TE-SE there is still a significant ($\geq$ 5%) chance that the final configuration selected will be worse than the default; this is can be attributed to two configurations found, see Figure 4, with training speedups between 1.3 and 1.4 and testing speedups of 0.5.

Similar to the results for LKH, there is a decrease in the 95th percentile quality for configurations found using TE-SE. Again, this can be explained by the existence of misleading configurations seen in Figure 4.

6.3 Results for Configuring Fuego

Like the previous scenarios, using the TE-SI protocol provides the best overall performance when configuring Fuego for either playing black or white (see

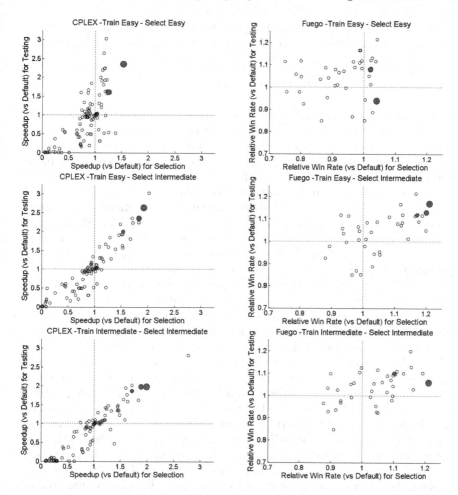

Fig. 4. Comparison between the selection and testing performance of CPLEX (left side), measuring PAR10 speedup, and Fuego (right side) trained for playing black, measuring relative winrate, found by multiple independent runs (100 for CPLEX and 40 for Fuego) of ParamILS using the TE-SE (top), TE-SI (middle) and TI-SI (bottom) protocols. Configurations shown in the CPU Time column of Table 1 are filled in and coloured red. The size of the points for these configurations corresponds to the time required to find them.

Figure 2). Interestingly, our results indicate that it is much easier to improve upon the default configuration of Fuego for playing white, and the majority of configurations found by all three protocols for playing white were indeed better than the default, see Figure 2.

Similar to the other scenarios, the TE-SE protocol suffered degrading performance when given additional computational resources, but surprisingly, the TI-SI protocol suffered from this as well when configuring for playing black. Looking at Figure 4, we can see that this is due to the presence of one outlier of particularly good quality (w.r.t. testing quality).

We are only presenting the scatter plots for configurations of Fuego trained to play black due to limited space. The results were qualitatively similar.

7   Conclusions and Future Work

In this paper we have shown that the TE-SI protocol provides benefit over the alternatives whenever it is infeasible to train directly for the intended use case of an algorithm given the available computational resources. Our simple modification to the standard protocol for using automated configurators does not require any modification to the underlying configurator and allows existing experiments to be reused. We have then demonstrated, through a large empirical study, the effectiveness of the TE-SI protocol across three different configuration scenarios. For solving MIP with CPLEX, a scenario where the standard TE-SE protocol is known to be effective, the TE-SI protocol was able to improve upon the results of the standard protocol for short configurator runs; we believe it will continue to provide benefit for longer runs using harder instances and are currently investigating this hypothesis. In the other two scenarios, where the standard protocol is either unable (for solving TSP using LKH) or unlikely (for playing Go using Fuego) to yield good configurations, the TE-SI protocol is reliably able to produce better configurations than both the TE-SE and TI-SI protocols and facilitates substantial improvements over the default configurations.

We see three main avenues for future work. First, we are currently extending the analysis of our new protocol by testing it on additional configuration scenarios, including CPLEX 12.3 applied to a harder set of MIP instances, based on real-world data modeling the spread of red-cockaded woodpeckers [3], as well as Concorde [5], the state-of-the-art complete TSP solver, on TSPLIB instances. We also plan to evaluate how well our new protocol works in conjunction with other algorithm configuration procedures, in particular, the latest version of SMAC [13]. Second, we have begun to investigate the use of predictive models in improving the effectiveness of selecting configurations. Finally, we plan to apply the methods presented in this paper as well as any that result from future work to configuring new versions of Fuego for upcoming Go competitions.

References

[1] Fuego, http://fuego.sourceforge.net/ (version visited last in October 2011)

[2] IBM ILOG CPLEX optimizer, http://www-01.ibm.com/software/integration/optimization/cplex-optimizer/ (version visited last in October 2011)

[3] Ahmadizadeh, K., Dilkina, B., Gomes, C.P., Sabharwal, A.: An Empirical Study of Optimization for Maximizing Diffusion in Networks. In: Cohen, D. (ed.) CP 2010. LNCS, vol. 6308, pp. 514–521. Springer, Heidelberg (2010)

[4] Ansótegui, C., Sellmann, M., Tierney, K.: A Gender-Based Genetic Algorithm for the Automatic Configuration of Algorithms. In: Gent, I.P. (ed.) CP 2009. LNCS, vol. 5732, pp. 142–157. Springer, Heidelberg (2009)

[5] Applegate, D., Bixby, R.E., Chvátal, V., Cook, W.J.: Concorde TSP solver, http://www.tsp.gatech.edu/concorde.html (version visited last in October 2011)

[6] Birattari, M., Stützle, T., Paquete, L., Varrentrapp, K.: A racing algorithm for configuring metaheuristics. In: GECCO 2002, pp. 11–18 (2002)

[7] Birattari, M., Yuan, Z., Balaprakash, P., Stützle, T.: F-Race and Iterated F-Race: An Overview. In: Experimental Methods for the Analysis of Optimization Algorithms, pp. 311–336. Springer (2010)

[8] Chiarandini, M., Fawcett, C., Hoos, H.H.: A modular multiphase heuristic solver for post enrolment course timetabling. In: Proceedings of the 7th International Conference on the Practice and Theory of Automated Timetabling, Montréal, pp. 1–6 (2008)

[9] Enzenberger, M., Müller, M., Arneson, B., Segal, R.: Fuego - an open-source framework for board games and Go engine based on Monte Carlo tree search. IEEE Transactions on Computational Intelligence and AI in Games 2, 259–270 (2010), Special issue on Monte Carlo Techniques and Computer Go

[10] Gomes, C.P., van Hoeve, W.-J., Sabharwal, A.: Connections in Networks: A Hybrid Approach. In: Trick, M.A. (ed.) CPAIOR 2008. LNCS, vol. 5015, pp. 303–307. Springer, Heidelberg (2008)

[11] Helsgaun, K.: An effective implementation of the Lin-Kernighan traveling salesman heuristic. European Journal of Operational Research 126, 106–130 (2000)

[12] Hutter, F., Hoos, H.H., Leyton-Brown, K.: Automated Configuration of Mixed Integer Programming Solvers. In: Lodi, A., Milano, M., Toth, P. (eds.) CPAIOR 2010. LNCS, vol. 6140, pp. 186–202. Springer, Heidelberg (2010)

[13] Hutter, F., Hoos, H.H., Leyton-Brown, K.: Sequential Model-Based Optimization for General Algorithm Configuration. In: Coello, C.A.C. (ed.) LION 2011. LNCS, vol. 6683, pp. 507–523. Springer, Heidelberg (2011)

[14] Hutter, F., Hoos, H.H., Leyton-Brown, K., Stützle, T.: ParamILS: An Automatic Algorithm Configuration Framework. Journal of Artificial Intelligence Research 36, 267–306 (2009)

[15] Reinelt, G.: TSPLIB, http://www.iwr.uni-heidelberg.de/groups/comopt/software/TSPLIB95 (version visited last in October 2011)

[16] Tompkins, D.A.D., Hoos, H.H.: Dynamic Scoring Functions with Variable Expressions: New SLS Methods for Solving SAT. In: Strichman, O., Szeider, S. (eds.) SAT 2010. LNCS, vol. 6175, pp. 278–292. Springer, Heidelberg (2010)

[17] Xu, L., Hutter, F., Hoos, H.H., Leyton-Brown, K.: Hydra-MIP: Automated algorithm configuration and selection for mixed integer programming. In: RCRA Workshop on Experimental Evaluation of Algorithms for Solving Problems with Combinatorial Explosion at the International Joint Conference on Artificial Intelligence, IJCAI (2011)

Upper Confidence Tree-Based Consistent Reactive Planning Application to MineSweeper

Michèle Sebag[1] and Olivier Teytaud[2]

[1] TAO-INRIA, LRI, CNRS UMR 8623,
Université Paris-Sud, Orsay, France
[2] OASE Lab, National University of Tainan, Taiwan

Abstract. Many reactive planning tasks are tackled through myopic optimization-based approaches. Specifically, the problem is simplified by only considering the observations available at the current time step and an estimate of the future system behavior; the optimal decision on the basis of this information is computed and the simplified problem description is updated on the basis of the new observations available in each time step. While this approach does not yield optimal strategies *stricto sensu*, it indeed gives good results at a reasonable computational cost for highly intractable problems, whenever fast off-the-shelf solvers are available for the simplified problem.

The increase of available computational power − even though the search for optimal strategies remains intractable with brute-force approaches − makes it however possible to go beyond the intrinsic limitations of myopic reactive planning approaches.

A consistent reactive planning approach is proposed in this paper, embedding a solver with an Upper Confidence Tree algorithm. While the solver is used to yield a consistent estimate of the belief state, the UCT exploits this estimate (both in the tree nodes and through the Monte-Carlo simulator) to achieve an asymptotically optimal policy. The paper shows the consistency of the proposed *Upper Confidence Tree-based Consistent Reactive Planning* algorithm and presents a proof of principle of its performance on a classical success of the myopic approach, the MineSweeper game.

1 Introduction

This paper focuses on reactive planning, of which power plants maintenance [20] or the MineSweeper game [13,5,18] are typical problem instances. The difficulty of reactive planning is due to the great many uncertainties about the problem environment, hindering the search for optimal strategies. For this reason, most reactive planners are content with selecting the current move based on their only current knowledge.

However, cheap and ever cheaper computational power makes it possible nowadays to aim at the best of both worlds, combining an approximate model of

Y. Hamadi and M. Schoenauer (Eds.): LION 6, LNCS 7219, pp. 220–234, 2012.
© Springer-Verlag Berlin Heidelberg 2012

the problem under examination with knowledge-based solvers. Taking inspiration of earlier work devoted to the combination of domain knowledge-based heuristics with Monte Carlo Tree Search and Upper Confidence Trees [21,22], this paper shows how fast solvers combined with Upper Confidence Tree approaches can be boosted to optimal planning. A proof of principle of the approach is given on the MineSweeper game, an NP-complete partially observable game. The choice of this game, despite its moderate complexity (indeed there exist many games with EXP or 2EXP computational complexity) is motivated as there exists efficient MineSweeper player algorithms, while many partially observable games still lack efficient algorithms.

The paper is organized as follows. Section 2 introduces the formal background of partially observable Markov decision processes (POMDP). The MineSweeper game is described in section 3. POMDP algorithms are discussed in Section 4 and illustrated on the MineSweeper problem. Section 5 describes our contribution and gives an overview of the proposed UC-CRP (*Upper Confidence Belief Estimation-based Solver*) algorithm, combining the Upper Confidence Tree algorithm with existing MineSweeper algorithms. Section 6 discusses the consistency of existing MineSweeper algorithms and establishes UC-CRP consistency. Section 7 reports on the comparative experimental validation of UC-CRP and the paper concludes with some perspectives for further work.

2   Formal Background

This section introduces the main notations used through the paper. After the standard terminology [24,6], a Markov Decision Process (MDP) is described from its space of states $\mathcal{S}$, its state of actions $\mathcal{A}$, the probabilistic transition model $p(s, a, s')$ describing the probability of arriving in state s' upon triggering action a in state s (with $\sum_{s'} p(s, a, s') = 1$), and a reward function r ($r : \mathcal{S} \times \mathcal{A} \mapsto \mathbb{R}$). A policy π maps a (finite sequence of) state(s) onto an action, possibly in a stochastic manner ($\pi : \mathcal{S}^i \times \mathcal{A} \mapsto \mathbb{R}$, with $\sum_a \pi(s, a) = 1$). Given initial state s_0, a probabilistic policy π thus defines a probability distribution on the cumulative rewards.

A partially observable MDP (POMDP) only differs from a MDP as each state s is partially visible through an observation o. Letting $\mathcal{O}$ denote the space of observations, a strategy π maps a (finite sequence of) observation(s) onto an action, possibly in a stochastic manner ($\pi : \mathcal{O}^i \times \mathcal{A} \mapsto \mathbb{R}$, with $\sum_a \pi(o, a) = 1$). In the following, we shall refer to s as *complete state* (including the partially hidden information), whereas observation o is referred to as *state* for the sake of consistency with the terminology used in the MineSweeper literature.

The feasible set $\mathcal{F}(o)$ denotes the set of complete states compatible with (a sequence of) observation(s) o. A consistent belief state estimation $p(s|o)$ is an algorithm sampling the uniform probability distribution on the feasible set $\mathcal{F}(o)$, given observation o. An asymptotically consistent belief state estimation is an algorithm yielding a probability distribution on $\mathcal{S}$, and converging toward the uniform probability distribution on $\mathcal{F}(o)$ as the computational effort goes to infinity.

3   The MineSweeper Game

MineSweeper is a widely spread game, which has motivated a number of studies as a model for real problems [12], or a challenge for machine learning [7] or genetic programming [15], or for pedagogical reasons [4] (including a nice version aimed at teaching quantum mechanics [11]).

3.1   The Rule

MineSweeper is played on a $h \times w$ board. The player starts by choosing a location on the board; thereafter, M mines are randomly placed on the board, anywhere except on the chosen location in order to avoid the immediate death out of bad luck. The location of the mines defines the (unknown) complete state of the game. The level of difficulty of the game is defined by (h, w, M). At each turn, the player selects an uncovered location ℓ (Fig. 1). Then,

- Either there is a mine in ℓ (immediate death) and the player has lost the game;
- Or there exists no mine and the player is informed of the number of mines in the 8-neighborhood of ℓ. Usually, when there is no mine in the 8-neighborhood of ℓ the player is assumed to automatically play all neighbors of ℓ, which are risk-less moves, in order to save up time[1].

The player wins iff she plays all non-mine $h \times w - M$ locations and avoids the immediate death termination. In each time step the current observation is made of the locations selected so far and the number of mines in their 8-neighborhood; by construction, the observation is consistent with the complete state s (the actual locations of the M mines).

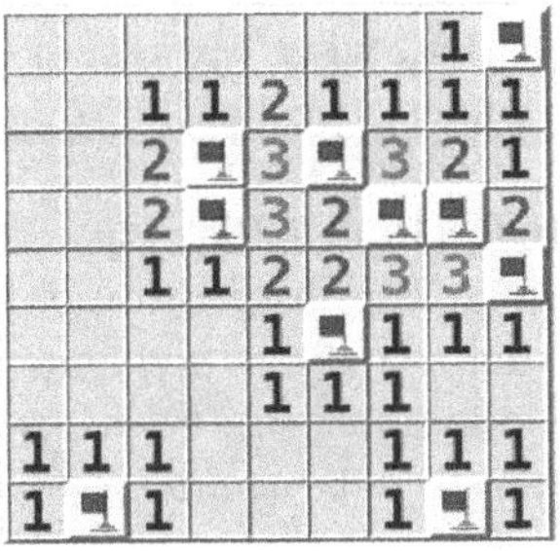

Fig. 1. The MineSweeper game, an observation state

---

[1] In some variants, e.g. the Linux Gnomine, the initial location is such that there is no mine in its 8-neighborhood. While this rule efficiently reduces the probability of loss out of bad luck, it is not widely used in the literature. For this reason, and for the sake of fair comparisons with the state of the art, this rule will not be considered in the remainder of the paper unless stated otherwise.

MineSweeper algorithms fall in two categories. In the former category are artificial intelligences (AIs) tackling the underlying Markov Decision Process, which is quite hard to analyze and solve [18]. In the latter category are AIs based on heuristics, and relying on Constraint Solving algorithms (CSP) [23]. In particular, a widely used heuristic strategy is based on evaluating the feasible set made of all complete states consistent with the current observation, counting for each location the percentage of states including a mine in this location (referred to as probability of immediate death), and playing the location with lowest probability of immediate death.

Two facts need be emphasized. On the one hand, CSP is a consistent belief state estimator [18]. On the other hand, the strategy of selecting the location with minimal probability of immediate death is suboptimal [9], irrespective of the computational effort involved in computing the feasible set (more in section 6).

3.2   State of the Art

The MineSweeper game is solved until 4x4 board [18]. Although it is only NP-complete in the general case [5] as already mentioned, it is more complex than expected at first sight [13,5,18].

The best CSP-based published methods have a probability of success of 80% at the beginner level (9x9, 10 mines), 45% at the intermediate level (16x16, 40 mines) and 34% in expert mode (16x30, 99 mines). Another, limited search method proposed by Pedersen [19] achieves a breakthrough with 92.5% success (respectively 67.7% success) at the beginner (resp. intermediate) level.

The importance of the initial move must be emphasized, although it is not specifically considered in CSP approaches. Our claim is that the good performances of some MineSweeper players can be attributed to good initialization heuristics or opening books, combined with standard CSP. However, as shown in [9] the initial move is not the only reason for CSP approaches being suboptimal; many small patterns lead to suboptimal moves with non-zero probability.

4   POMDP Algorithms

Many POMDP algorithms feature two components. The first one is in charge of estimating the belief state, i.e. the probability distribution on the complete state space conditioned by the current observation. The second component is in charge of move selection, based on the estimated belief state. Both components are tightly or loosely coupled depending on the problem and the approach.

CSP-based MineSweeper approaches feature a consistent belief state estimation and a trivial move selection algorithm (select the move with lowest immediate death probability). Let us briefly illustrate the main approaches proposed for belief state estimation in the MineSweeper context.

4.1   Rejection Method

The simplest approach is the rejection method; its pseudo-code is given in Alg. 1. It proceeds by uniformly drawing a complete state and rejecting it if not consistent with the observation. By construction, the rejection method thus is consistent (i.e. it proposes an exact belief state estimation) and slow (moderate to large computational efforts are required to ensure a correct estimate).

Algorithm 1. The rejection method for consistent belief state estimation.

Input: observations.
Output: a (uniformly sampled among consistent states) complete state.
$s \leftarrow$ random complete state
                                                                //*uniform selection in [0,1]*
while $U[0,1] \geq C \times Likelihood(s|o)$ **do**
    $s \leftarrow$ random complete state
end while
Return s

Parameter C is such that $C \times Likelihood(s|o) \leq 1$. In the case where the likelihood of an observation o given complete state s, is binary (0 or c), then $C = 1/c$ is the optimal choice, where state s is accepted iff it is consistent with observation o. In the MineSweeper case, $c = 1$: testing whether a complete state is consistent with the current observation is trivial.

Despite its simplicity, the rejection method is reported to outperform all other methods in [9], which is attributed to the fact that it is the only consistent method. In particular, we shall see that the Markov-Chain Monte-Carlo method (below) is only *asymptotically* consistent.

4.2   Constraint Solving (CSP)

Constraint Solving approaches usually rely on enumerating or estimating the whole feasible set, including all complete states which are consistent with the observations; their pseudo-code is given in Alg. 2. Indeed many implementations thereof have been proposed. Note that as far as the feasible set is exhaustively determined, CSPs are optimal belief state estimation algorithms in the MineSweeper context. The feasible set enables to compute for each location its probability of being a mine. However CSP-based approaches involve an intrinsically myopic move selection, optimizing some 1-step ahead reward (the probability of immediate death). As already mentioned, this greedy selection policy is not optimal.

Note that CSP-based approaches also enable to compute the exact probability of transition to a given state (which we shall use in Section 5), which has never been exploited within a tree search to our best knowledge.

Algorithm 2. The CSP algorithm for playing MineSweeper.
***
Input: observation o; total number M of mines
for each location ℓ on the $h \times w$ board **do**
   $Nb(\ell) \leftarrow 0$
end for
for each feasible state s (positioning of the M mines consistent with o) **do**
   for each location ℓ **do**
      If s includes a mine in ℓ, $Nb(\ell) \leftarrow Nb(\ell) + 1$
   end for
end for
Select move ℓ uniformly chosen among the set of (uncovered) locations with minimum $Nb(\ell)$.
***

4.3   Markov-Chain Monte-Carlo Approaches

Markov-Chain Monte-Carlo (MCMC) algorithms, of Metropolis-Hastings (MH) is a widely studied example, achieve the asymptotically consistent estimation of the belief state. In the MineSweeper context, and more generally when observations are deterministic with binary likelihood depending on the belief state, Metropolis-Hastings boils down to Alg. 3, under the assumption of a symmetric transition kernel (that is, the probability of mutating from state s to state s' is equal to the probability of mutating from state s' to state s).

Note that the asymptotic properties of MH are not impacted by the initialization to a consistent complete state (line 3 of Alg. 3), which only aims at saving up time. According to [9] and for the mutation used in the MineSweeper context however, the MH distribution is significantly biased by the choice of the initial state, making the overall MH algorithm weaker than the simple rejection algorithm (Alg. 1).

Algorithm 3. The Metropolis-Hastings algorithm with symmetric transition kernel in the case of deterministic binary observations.
***
Input: observation o, number T of iterations.
Output: a (nearly uniformly sampled) complete state consistent with o.
Init: Find state s consistent with o by constraint solving
Select a number T of iterations by heuristic methods
          // in [9], T depends on the number of UCT-simulations.
for $t = 1 \ldots T$ **do**
   Let s' be a mutation of s
   if s' is consistent with o **then**
      $s \leftarrow s'$
   end if
end for
***

Several MH variants have been investigated by [9], with disappointing results compared to the rejection method. This partial failure is blamed by the authors on the asymptotic MH consistency (as opposed to the "real" consistency of the rejection method). Indeed one cannot exclude that some MH variant, e.g. with some optimal tuning of its many parameters, may yield much better results. Still, it is suggested that a robust problem-independent MH approach is yet to be found.

In summary, our goal is to propose a fully consistent belief state estimation algorithm, such that it is faster than the rejection method and easier to adjust than MCMC.

5   Upper Confidence Tree-Based Consistent Reactive Planning

The contribution of the present paper is the UC-CRP algorithm, combining the consistent CSP-based belief state estimation and the consistent MDP solver UCT to achieve a non-trivial move selection. The Upper Confidence Tree algorithm is first briefly reminded for the sake of self-containedness, before detailing UC-CRP.

5.1   Upper Confidence Tree

Referring the reader to [10,14,17] for a comprehensive introduction, let us summarize the main steps of the Upper Confidence Tree algorithm (Fig. 2). UCT gradually constructs a search tree initialized to a single root node, by iterating tree-walks also referred to as *episodes*. Each episode involves two phases. In the so-called bandit part, the current action is selected after the Multi-Armed Bandit setting, using the Upper Confidence Bound formula [3] on the basis of the average reward and number of trials associated to each action, and the current state is updated accordingly until i) reaching a new state s (not already stored in the tree); or ii) reaching a final state. In the first case, UCT switches to the so-called Monte-Carlo or random phase, where an action is randomly selected until reaching a final state. Upon reaching a final state, the reward of the episode is computed, and the counter and average reward indicators attached to every stored node of the episode are updated.

UCT is well known for its ability to work on problems with little or no expert information. In particular, it does not require any heuristic evaluation function, although it enables to use prior knowledge encoded in heuristic strategies. These heuristic strategies can be used in the bandit or random parts of the algorithm, or in the episode evaluation. In particular, long-term effects of the decisions taken in an episode can be accounted for through simulating complete runs (as in $TD(1)$ methods, see e.g. [6]).

Indeed, a number of UCT variants have been considered in the literature, involving how to choose the actually selected action (last line in Fig. 2), or controlling the number of actions considered in each step of the bandit part, e.g. using progressive widening.

– Let the tree root be initialized to the current state s_0
– Repeat   // *simulate an episode*
  • Bandit part   // *bandit part of episode*
    * $s = s_0$ // *start at the root*
    * Iterate :
      Select $a(s)$ using the UCB formula and draw $s' \sim p(s, a, s')$.
      If s' is not yet stored in the tree, goto Monte-Carlo part.
      If s' is a final state, goto Scoring part $s \leftarrow s'$
  • Monte-Carlo part   // *Monte-Carlo part of episode*
    * Add s' as son node of s, a. $s \leftarrow s'$
    * Repeat until s is a final state
      Select action a randomly; $s' \sim p(s, a, s'); s \leftarrow s'$
  • Scoring part   // *scoring part of episode.*
    Let r be the overall reward associated to the episode. For each node (s, a) of
    the episode which is stored in the tree,
    * Update average reward $r(s, a)$ by r $(r(s, a) \leftarrow \frac{n(s,a)*r(s,a)+r}{n(s,a)+1})$ and likewise
      update the reward variance $v(s, a)$.
    * Increment counter $n(s, a)$ by 1;
  until reaching the allowed number of episodes/the computational budget.

– Play the action played most often from the root node s_0.

Fig. 2. Sketch of the UCT algorithm

5.2 UC-CRP = Belief State Estimation + Upper Confidence Trees

UC-CRP is an UCT variant embedding several CSP-based MineSweeper modules.

The first module is the single point strategy (SPS) found in PGMS[2]. SPS achieves a limited constrained propagation. Assuming that k out of the n neighbors of some location ℓ are mines, with m non-mine neighbor locations and $n - m - p$ yet uncovered locations, then two particular cases are considered. In the first case, $k = p$: it is easily seen that none of the uncovered neighbors is a mine, therefore all these neighbors can be played automatically to save up time (as already mentioned, most standard MineSweeper AIs include this constraint propagation heuristics). In the second case, $k = n - m - p$ and one likewise sees that all uncovered neighbors are mines. This case can also be automatically handled through the constraint propagation (which only involves risk-less moves).

The second module is a constraint solver, determining the feasible set of all complete states consistent with the current observation. Note that least constrained variables are used first; whereas most constrained variables should be used first when looking for one single solution, it is better to use least constrained variables first when looking for all solutions. CSP provides useful information at a moderate computational cost, including: i) the probability for each location to cover a mine, and thus the set of risk-less moves if any; ii) the exact probability distribution of the next state (i.e. how will be the board after the next

---

[2] http://www.ccs.neu.edu/home/ramsdell/pgms/

Table 1. The *Upper Confidence Tree-based Consistent Reactive Planning* combines UCT with problem domain-based solvers CSP and SPS (section 5.2). The CSP components are used to both select risk-less or optimal moves, and provide a belief state estimate (*). The notion of risk-less moves and moves with minimal risk depend on the considered problem. In the MineSweeper context, a risk-less move is one with 0 probability of immediate death.

Feature	In UC-CRP-Mine Sweeper
Node creation	One per simulation
Progressive widening	Double
Leaf evaluation	Monte-Carlo simulation
Bandit phase	Upper Confidence Bound formula with variance estimates [1]
Random phase Single point strategy (SPS)	if possible, returns a risk-less move/action
Constraint Satisfaction Solver (CSP)	If possible select a risk-less move (CSP can find all such moves if any)
Otherwise	with probability 0.7 Select move with minimum risk
Otherwise (*)	Randomly draw a complete state s from observation o and select a risk-less move after s
Forced moves [25]	Select a risk-less move (estimated by CSP)

move), i.e. the probabilistic transition model $p(o, a, o')$. Notably, with $p(o, a, o')$ the POMDP setting boils down to an MDP.

Finally, UC-CRP involves several UCT variants together with the above modules, depicted in Table 1. The variance-based UCB formula [1] is given as, where $n(s)$ is the number of visits to state s, $r(s, a)$ and $v(s, a)$ respectively are the empirical mean and variance of the reward gathered over episodes visiting (s, a), and $\alpha \geq 2$ is an integer value:

$$\text{Select argmax} \left\{ r(s, a) + \sqrt{\frac{2v(s, a) \log (4n(s)^\alpha)}{n(s)}} + \frac{16 \log (4n(s)^\alpha)}{3n(s)} \right\}$$

The double progressive widening detailed in [8] is used to control the branching factor of the tree, i.e. the number of considered moves in each node.

Note that UC-CRP requires more than assessing the probability of mine in every location conditioned by the current observation; it requires the full probability distribution over the complete states conditioned by the current observation.

6 Consistency Analysis

This section, devoted to the analytical study of the considered algorithms, establishes the consistency of UC-CRP in the general case and the inconsistency of CSP-based MineSweeper approaches.

6.1 Consistency of UC-CRP

Let us first examine the consistency of UCT approaches in the MDP setting.

UCT (Upper Confidence Trees) is known to be a consistent MDP solver [14] in the finite case. While the presented approach uses heuristics in the evaluation of leaves by Monte-Carlo simulations, this does not affect the consistency proof in [14]. While heuristics are also used in the tree part of UCT, the upper-confidence-bound proof from [16,2], used in [14], is independent of heuristics too. A last additional component of UC-CRP compared to UCT [14] is progressive widening [8]. Progressive widening however has no asymptotic impact since the number of actions and the number of random outcomes is finite. The UCT-variant involved in UC-CRP thus is a consistent solver in the MDP case.

Indeed MineSweeper is not an MDP: as far as the complete state is unknown it is a POMDP. However, as shown by [18], the hidden information can be exactly, estimated by CSP. Indeed the CSP-based move selection (select the move with lowest probability of immediate death) is not optimal as shown by [9]; nevertheless the estimation of the hidden information soundly enables to cast the POMDP MineSweeper problem into an MDP one.

As a consequence, UC-CRP is a consistent MDP solver, and it inherits from its CSP component the property of being an asymptotically optimal reactive planner.

6.2 Inconsistency of CSP-Based Approaches

Let us consider the 3×3 MineSweeper problem with 7 mines. As stated above, UC-CRP is asymptotically consistent (since UCT is an asymptotically consistent MDP solver and the CSP or rejection method provides an exact transition model). Therefore, UC-CRP finds the optimal strategy in the $(3, 3, 7)$ MineSweeper setting. Quite the contrary, the CSP cannot be optimal due to the uniform selection of the initial move.

The success rate can be analytically computed on this toy MineSweeper problem. If the initial move is the center location, this location necessarily has 7 neighbor mines. The probability of winning thus is $\frac{1}{8}$ as there are 7 mines in the 8 neighbors (Fig. 3).

If the initial move is located on a side of the board, then the number of mines in its neighborhood is: 5 with probability $\frac{3}{8}$ (with probability of winning $\frac{1}{3}$) and 4 with probability $\frac{5}{8}$ (with probability of winning $\frac{1}{5}$). Overall, the probability of winning when playing a side location as initial move thus is $\frac{1}{4} = \frac{3}{8} \times \frac{1}{3} + \frac{5}{8} \times \frac{1}{5}$.

Finally, if the initial move is located in a corner of the board, then the number of mines in its neighborhood is 2 with probability $\frac{3}{8}$ (with probability of winning $\frac{1}{3}$) and 3 with probability $\frac{5}{8}$ (with probability of winning $\frac{1}{5}$). The overall

Fig. 3. A bad initial move in 3x3 with 7 mines

probability of winning when playing a corner location as initial move thus is $\frac{1}{4} = \frac{3}{8} \times \frac{1}{3} + \frac{5}{8} \times \frac{1}{5}$.

An optimal strategy thus selects a side or corner location as first move, yielding a probability $\frac{1}{4}$ of winning. Quite the contrary, CSP-based MineSweeper algorithms (rightly) considers that all moves have the same probability of immediate death and thus the first move is uniformly selected. The probability of winning thus is $\frac{1}{9} \times \frac{1}{8} + \frac{8}{9} \times \frac{1}{4} = \frac{17}{72}$, which is less than the probability $\frac{1}{4}$ of winning of an optimal strategy (and asymptotically reached by UC-CRP).

7 Experimental Validation

The goal of the experiments is to study both the asymptotic optimality and the comparative performances of UC-CRP. The optimality of UC-CRP is studied using the GnoMine Custom mode (the first move has no mine in its 8-neighborhood), as this setting facilitates the analytical study. The standard MineSweeper setting (where the first move is only assumed to be not a mine) is also considered for the fair comparison of UC-CRP with the state of the art CSP-PGMS algorithm.

All experiments are done on a 16-cores Xeon 2.93GHz Ubuntu 2.6-32-34. The computational effort is measured using a single core per experiment.

7.1   A Gnomine Custom Mode: 15 Mines on a 5x5 Board

Let us consider the 5x5 board with 15 mines (Fig. 4). Under the Gnomine custom mode, playing the center location implies that all mines are located on the sides, thus yielding a sure win. Interestingly, UC-CRP finds this optimal strategy (that humans easily find as well), while CSP-based approaches still uniformly select their first move. In such case, the probability of playing location $(2,2)$ (up to rotational symmetry) is $\frac{4}{25}$, yielding a loss probability of $\frac{1}{2}$. Likewise, the probability of playing $(2,3)$ (up to rotational symmetry) is $\frac{4}{25}$, yielding a loss probability of $\frac{1}{4}$. The overall probability of loosing the game is at least $\frac{1}{2} \times \frac{4}{25} + \frac{1}{4} \times \frac{4}{25} = \frac{3}{25}$ (indeed, the actual loss probability is bigger since the side

Table 2. Comparative performances of UC-CRP and CSP-PGMS

Format	CSP-PGMS	UC-CRP
4 mines on 4x4	64.7 %	**70.0% ± 0.6%**
1 mine on 1x3	100 %	100% (2000 games)
3 mines on 2x5	22.6%	**25.4 % ± 1.0%**
10 mines on 5x5	8.20%	9% (p-value: 0.14)
5 mines on 1x10	12.93%	**18.9% ± 0.2%**
10 mines on 3x7	4.50%	**5.96% ± 0.16%**
15 mines on 5x5	0.63%	**0.9% ± 0.1%**

locations are even worse initial moves), showing that CSP approaches do not find the optimal strategy as opposed to UC-CRP.

Experimentally, UC-CRP (with no expert knowledge besides the gnomine rule) finds the best move in all out of 500 independent runs. The computational cost is 5 seconds on a 16-cores Xeon 2.93GHz Ubuntu 2.6-32-34 (using 1 core only).

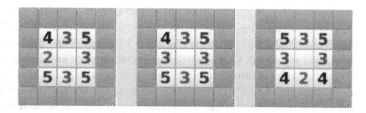

Fig. 4. The Gnomine version of the 5x5 board with 15 mines is a sure win: in each case the position of all mines can be deduced as there exists only one non-mine location. The three reported cases cover all possible cases by rotational symmetry.

7.2 Standard MineSweeper Setting

UC-CRP is compared to a CSP-based MineSweeper player[3], selecting corner locations as initial moves.

UC-CRP is allotted a computational budget of 10s per move, except for 10 mines in 5x5 (300 seconds per move) and 10 mines in 3x7 (30s per move). The average winning rate is reported together with the standard deviation on Table 2 for several board sizes. The winning rate of CSP-PGMS is estimated on 100000 games. Note that, while CSP-PGMCS is significantly faster than UC-CRP, its performances do not increase with additional computational time. Overall, UC-CRP outperforms CSP-PGMCS in all cases with a p-value .05, except for the 5×5 with 10 mines MineSweeper, where the p-value is .14.

---

[3] The CSP-PGMS implementation from
http://www.ccs.neu.edu/home/ramsdell/pgms/ is used.

8   Conclusion and Perspectives

This paper has investigated the tight coupling of Upper Confidence Trees and Constraint Solving to tackle partially observable Markov decision making problems, defining the UC-CRP algorithm. The role of the CSP component in UC-CRP is twofold. On the one hand, it consistently estimates the belief state, turning a POMDP setting into an MDP one; on the other hand, the belief state estimate is exploited to sort out risk-less or heuristically good moves.

The first contribution of the paper is a generic methodology for improving a myopic solver, through its combination with UCT. Our claim is that such approaches, combining consistent asymptotic behaviors and myopic efficient heuristics, will provide artificial intelligences with the best of both worlds: relevant heuristics are used to enforce computational efficiency and yield tractability guarantees; UCT provides asymptotic optimality guarantees, enabling to boost the available heuristics to optimality through computational efforts. Improving a decent heuristic policy to reach optimality with UCT seems to be a simple though general and potentially very efficient strategy.

The second contribution of the paper is to formally establish the consistency of the proposed UC-CRP approach. The third contribution is the empirical validation of the approach, using the difficult MineSweeper problem as proof of concept. It has been emphasized that MineSweeper is a particularly challenging problem. Indeed the CSP approach, which ignores long term effects, is very effective as uncertainties are so big that long-term effects are very unreliable. On this difficult problem, significant improvements on small boards have been obtained compared to [9]. Further, UC-CRP results improve as the computational budget increases while CSP does not benefit from additional computational resources due to its intrinsic myopic limitations.

A key and very promising feature of UC-CRP is that no (manually acquired or programmed) opening book needed be considered; quite the contrary, the good performances of advanced CSP approaches requires specific heuristics to handle the first move selection. Along the same line, UC-CRP flexibly accommodates MineSweeper variants through modifying the only rule module (transition model).

A research perspective aimed at the game community is concerned with a faster implementation of UC-CRP; at the moment UC-CRP relies on a much slower CSP module than e.g. CSP-PGMS. Further assessment of UC-CRP, e.g. comparatively to [19] will be facilitated by using a CSP implementation as fast as CSP-PGMS, enabling to consider expert MineSweeper modes.

Acknowledgements. The authors acknowledge the support of NSC (funding NSC100-2811-E-024-001) and ANR (project EXPLO-RA ANR-08-COSI-004). We thank Laurent Simon, LRI Université Paris-Sud, for fruitful discussions, and we thank the authors of PGMS and PGMS-CSP for making their implementations freely available.

References

1. Audibert, J.-Y., Munos, R., Szepesvari, C.: Use of variance estimation in the multi-armed bandit problem. In: NIPS 2006 Workshop on On-line Trading of Exploration and Exploitation (2006)
2. Auer, P.: Using confidence bounds for exploitation-exploration trade-offs. The Journal of Machine Learning Research 3, 397–422 (2003)
3. Auer, P., Cesa-Bianchi, N., Fischer, P.: Finite-time analysis of the multiarmed bandit problem. Machine Learning 47(2/3), 235–256 (2002)
4. Becker, K.: Teaching with games: the minesweeper and asteroids experience. J. Comput. Small Coll. 17, 23–33 (2001)
5. Ben-Ari, M.M.: Minesweeper as an NP-complete problem. SIGCSE Bull. 37, 39–40 (2005)
6. Bertsekas, D., Tsitsiklis, J.: Neuro-dynamic Programming. Athena Scientific (1996)
7. Castillo, L.P.: Learning minesweeper with multirelational learning. In: Proc. of the 18th Int. Joint Conf. on Artificial Intelligence, pp. 533–538 (2003)
8. Couëtoux, A., Hoock, J.-B., Sokolovska, N., Teytaud, O., Bonnard, N.: Continuous Upper Confidence Trees. In: Coello, C.A.C. (ed.) LION 5. LNCS, vol. 6683, pp. 433–445. Springer, Heidelberg (2011)
9. Couetoux, A., Milone, M., Teytaud, O.: Consistent belief state estimation, with application to mines. In: Proc. of the TAAI 2011 Conference (2011)
10. Coulom, R.: Efficient Selectivity and Backup Operators in Monte-Carlo Tree Search. In: Ciancarini, P., van den Herik, H.J. (eds.) Proc. of the 5th Int. Conf. on Computers and Games, pp. 72–83 (2006)
11. Gordon, M., Gordon, G.: Quantum computer games: quantum Minesweeper. Physics Education 45(4), 372 (2010)
12. Hein, K.B., Weiss, R.: Minesweeper for sensor networks–making event detection in sensor networks dependable. In: Proc. of the 2009 Int. Conf. on Computational Science and Engineering, CSE 2009, vol. 01, pp. 388–393. IEEE Computer Society (2009)
13. Kaye, R.: Minesweeper is NP-complete. Mathematical Intelligencer 22, 9–15 (2000)
14. Kocsis, L., Szepesvári, C.: Bandit Based Monte-Carlo Planning. In: Fürnkranz, J., Scheffer, T., Spiliopoulou, M. (eds.) ECML 2006. LNCS (LNAI), vol. 4212, pp. 282–293. Springer, Heidelberg (2006)
15. Koza, J.R.: Genetic Programming II: Automatic Discovery of Reusable Programs. MIT Press (1994)
16. Lai, T., Robbins, H.: Asymptotically efficient adaptive allocation rules. Advances in Applied Mathematics 6, 4–22 (1985)
17. Lee, C.-S., Wang, M.-H., Chaslot, G., Hoock, J.-B., Rimmel, A., Teytaud, O., Tsai, S.-R., Hsu, S.-C., Hong, T.-P.: The Computational Intelligence of MoGo Revealed in Taiwan's Computer Go Tournaments. IEEE Transactions on Computational Intelligence and AI in Games (2009)
18. Nakov, P., Wei, Z.: Minesweeper, #minesweeper (2003)
19. Pedersen, K.: The complexity of Minesweeper and strategies for game playing. Project report, univ. Warwick (2004)
20. ROADEF-Challenge. A large-scale energy management problem with varied constraints (2010), http://challenge.roadef.org/2010/

21. Rolet, P., Sebag, M., Teytaud, O.: Boosting Active Learning to Optimality: A Tractable Monte-Carlo, Billiard-Based Algorithm. In: Buntine, W., Grobelnik, M., Mladenić, D., Shawe-Taylor, J. (eds.) ECML PKDD 2009, Part II. LNCS, vol. 5782, pp. 302–317. Springer, Heidelberg (2009)
22. Rolet, P., Sebag, M., Teytaud, O.: Optimal robust expensive optimization is tractable. In: GECCO 2009, Montréal Canada, 8 p. ACM Press (2009)
23. Studholme, C.: Minesweeper as a constraint satisfaction problem. Unpublished project report (2000)
24. Sutton, R., Barto, A.G.: Reinforcement learning. MIT Press (1998)
25. Teytaud, F., Teytaud, O.: On the Huge Benefit of Decisive Moves in Monte-Carlo Tree Search Algorithms. In: IEEE Conf. on Computational Intelligence and Games, Copenhagen, Denmark (2010)

Bounding the Effectiveness of Hypervolume-Based (μ + λ)-Archiving Algorithms

Tamara Ulrich and Lothar Thiele

Computer Engineering and Networks Laboratory, ETH Zurich,
8092 Zurich, Switzerland
firstname.lastname@tik.ee.ethz.ch

Abstract. In this paper, we study bounds for the α-approximate effectiveness of non-decreasing $(\mu+\lambda)$-archiving algorithms that optimize the hypervolume. A $(\mu + \lambda)$-archiving algorithm defines how μ individuals are to be selected from a population of μ parents and λ offspring. It is non-decreasing if the μ new individuals never have a lower hypervolume than the μ original parents. An algorithm is α-approximate if for any optimization problem and for any initial population, there exists a sequence of offspring populations for which the algorithm achieves a hypervolume of at least $1/\alpha$ times the maximum hypervolume.

Bringmann and Friedrich (GECCO 2011, pp. 745–752) have proven that all non-decreasing, locally optimal $(\mu + 1)$-archiving algorithms are $(2+\epsilon)$-approximate for any $\epsilon > 0$. We extend this work and substantially improve the approximation factor by generalizing and tightening it for any choice of λ to $\alpha = 2 - (\lambda - p)/\mu$ with $\mu = q \cdot \lambda - p$ and $0 \leq p \leq \lambda - 1$. In addition, we show that $1 + \frac{1}{2\lambda} - \delta$, for $\lambda < \mu$ and for any $\delta > 0$, is a lower bound on α, i.e. there are optimization problems where one can not get closer than a factor of $1/\alpha$ to the optimal hypervolume.

Keywords: Multiobjective Evolutionary Algorithms, Hypervolume, Submodular Functions.

1 Introduction

When optimizing multiple conflicting objectives, there usually is no single best solution. Instead, there are incomparable tradeoff solutions, where no solution is strictly better than any other solution. *Better* in this case refers to Pareto-dominance, i.e. one solution is said to be better than another, or dominate it, if it is equal or better in all objectives, and strictly better in at least one objective. The set of non-dominated solutions is called the Pareto-optimal set. Usually, this Pareto-optimal set can contain a large number of solutions, and it is infeasible to calculate all of them. Instead, one is interested in finding a relatively small, but still *good* subset of this Pareto-optimal set.

It is not a priori clear how a good subset should look like, i.e. how the goodness of a subset can be measured. One of the most popular measures for subset quality

Y. Hamadi and M. Schoenauer (Eds.): LION 6, LNCS 7219, pp. 235–249, 2012.
© Springer-Verlag Berlin Heidelberg 2012

is the hypervolume indicator, which measures the volume of the dominated space. Therefore, one possibility to pose a multiobjective optimization problem is to look for a solution set $\mathcal{P}^*$ of fixed size, which maximizes the hypervolume.

Algorithms that optimize the hypervolume face several problems. First, the number of possible solutions can become very large, so it is not possible to select from all solutions. Second, even if all solutions are known and the non-dominated solutions can be identified, the number of subsets explodes and not all of them can be enumerated for comparison.

In this paper, we consider $(\mu + \lambda)$-evolutionary algorithms, or $(\mu + \lambda)$-EAs. They iteratively improve a set of solutions, where the set is named 'population' and the iteration is denoted as 'generation'. In particular, they maintain a population of size μ, generate λ offspring from the μ parents and then select μ solutions from the μ parents and the λ offspring that are to survive into the next generation. Note that we here only consider non-decreasing algorithms, i.e. algorithms whose hypervolume cannot decrease from one generation to the next.

Several questions arise in this setting. First, what are upper and lower bounds on the hypervolume that a population of a fixed size will achieve? Is it possible to prove that a set of size μ with the maximal hypervolume can be found, without explicitly testing all possible sets? To answer these questions, the term *effectiveness* has been defined. An algorithm is effective if for any optimization problem[1] and for any initial population[2], there is a sequence of offspring[3] which leads to the population with maximum hypervolume. Obviously, $(\mu + \mu)$-EAs are always effective: We just choose the first set of offspring to be exactly the population with the maximal hypervolume and then we select this set as the new population. It has also been shown by Zitzler et al.[4] that $(\mu + 1)$-EAs, on the other hand, are ineffective. Recently, it has been shown by Bringmann and Friedrich [1] that all $(\mu + \lambda)$-EAs with $\lambda < \mu$ are ineffective.

Bringmann and Friedrich then raised the follow-up question: If it is not possible to reach the optimal hypervolume for all optimization problems and all initial populations, is it at least possible to give a lower bound on the achieved hypervolume? To this end, they introduced the term α-*approximate effectiveness*. An algorithm is α-approximate if for any optimization problem and for any initial population there is a sequence of offspring with which the algorithm achieves at least $1/\alpha \cdot H^{\max}$, where $H^{\max}$ is the maximum achievable hypervolume of a population of size μ. They proved in their paper that a $(\mu + 1)$-EA is 2-approximate and conjectured that for larger λ, a $(\mu + \lambda)$-EA is $O(1/\lambda)$-approximate.

---

[1] We only consider finite search spaces here, such that mutation operators exist which produce offspring with a probability larger than zero. Note that any search space coded on a computer is finite.

[2] Note that the term *for any initial population* implies that at any point during the algorithm, there exists a sequence of offspring with which an effective algorithm can achieve the optimal hypervolume.

[3] Note that the term *there is a sequence of offspring* assumes that we are given variation operators that produce any sequence of offspring with probability greater than zero.

On the other hand, we might also be interested in upper bounds on the achievable hypervolume. Bringmann and Friedrich [1] have found an optimization problem where no algorithm can achieve more than $1/(1 + 0.1338(1/\lambda - 1/\mu) - \epsilon)$ of the optimal hypervolume, i.e. there is no $(1 + 0.1338(1/\lambda - 1/\mu) - \epsilon)$-approximate archiving algorithm for any $\epsilon > 0$.

Why is knowledge of the bounds of the α-approximate effectiveness useful? Assume that we are using an exhaustive mutation operator, which produces any offspring with a probability larger than zero. Therefore, the probability of generating an arbitrary sequence of offspring is also larger than zero. The $\frac{1}{2}$-approximate effectiveness of $(\mu + 1)$-EAs now tells us that if we execute the evolutionary algorithm for a sufficiently large number of generations, we will end up with a population that has at least half of the maximal hypervolume. In case of a $(\mu + \mu)$-EA, on the other hand, we know that we will finally achieve a population with maximum hypervolume, i.e. $\alpha = 1$. We are therefore interested in deriving bounds on the effectiveness of evolutionary algorithms.

This paper extends the work of Bringmann and Friedrich by (a) computing the α-approximate effectiveness of $(\mu + \lambda)$-EAs for general choices of λ, (b) tightening the previously known upper bound on α, and (c) tightening the previously known lower bound on α. The results for (a) and (b) are based on the theory of submodular functions, see [2]. For (c) we show that for $\lambda < \mu$, there exist optimization problems where any $(\mu + \lambda)$-EA does not get closer than a factor of $1/\alpha$ to the optimal hypervolume with $\alpha = 1 + \frac{1}{2\lambda} - \delta$, for any $\delta > 0$.

The paper is organized as follows: The next section presents the formal setting, including the definition of the hypervolume, the algorithmic setting, definitions for the effectiveness and approximate effectiveness and an introduction into submodular functions. In Section 3 we determine an upper bound on α for general choices of μ and λ, thereby giving a quality guarantee in terms of a lower bound of the achievable hypervolume. Finally in Section 4, we will determine a lower bound on α for general choices of μ and λ.

2   Preliminaries

Consider a multiobjective minimization problem with a decision space $\mathcal{X}$ and an objective space $\mathcal{Y} \subseteq \mathbb{R}^m = \{f(x) | x \in \mathcal{X}\}$, where $f : \mathcal{X} \to \mathcal{Y}$ denotes a mapping from the decision space to the objective space with m objective functions $f = \{f_1, ..., f_m\}$ which are to be minimized.

The underlying preference relation is weak Pareto-dominance, where a solution $a \in \mathcal{X}$ weakly dominates another solution $b \in \mathcal{X}$, denoted as $a \preceq b$, if and only if solution a is better or equal than b in all objectives, i.e. iff $f(a) \leqslant f(b)$, or equivalently, iff $f_i(a) \leq f_i(b), \forall i \in \{1, ..., m\}$. In other words, a point $p \in \mathcal{X}$ weakly dominates the region $\{y \in \mathbb{R}^m : f(p) \leqslant y\} \subset \mathbb{R}^m$.

2.1   Hypervolume Indicator

The hypervolume indicator of a given set $\mathcal{P} \subseteq \mathcal{X}$ is the volume of all points in $\mathbb{R}^m$ which are dominated by at least one point in $\mathcal{P}$ and which dominate at least one point of a reference set $\mathcal{R} \subset \mathbb{R}^m$.[4] Roughly speaking, the hypervolume measures the size of the dominated space of a given set. Sets with a larger hypervolume are considered better. More formally, the hypervolume indicator can be written as

$$H(\mathcal{P}) := \int_{y \in \mathbb{R}^m} A_{\mathcal{P}}(y)\, dy$$

where $A_{\mathcal{P}}(y)$ is called the *attainment function* of set $\mathcal{P}$ with respect to a given reference set $\mathcal{R}$, and is defined as follows:

$$A_{\mathcal{P}}(y) = \begin{cases} 1 & \text{if } \exists p \in \mathcal{P},\, r \in \mathcal{R} : f(p) \leqslant y \leqslant r \\ 0 & \text{else} \end{cases}$$

The goal of a $(\mu + \lambda)$-EA is to find a population $\mathcal{P}^* \subseteq \mathcal{X}$ of size μ with the maximum hypervolume:

$$H(\mathcal{P}^*) = \max_{\mathcal{P} \subseteq \mathcal{X}, |\mathcal{P}| = \mu} H(\mathcal{P}) = H_{\mu}^{\max}(\mathcal{X})$$

2.2 Algorithmic Setting

The general framework we are considering here is based on a $(\mu + \lambda)$ evolutionary algorithm (EA) as shown in Algorithm 1. The selection step of Line 5 is done by a $(\mu + \lambda)$-archiving algorithm[5]. We here assume that the archiving algorithm is non-decreasing, i.e. $H(\mathcal{P}^t) \geq H(\mathcal{P}^{t-1})$, $1 \leq t \leq g$. We use the following formal definition (as given in [1]) to describe an archiving algorithm:

Algorithm 1. General $(\mu + \lambda)$-EA framework: μ denotes the population size; λ the offspring size; the algorithm runs for g generations.

```
1: function EA(μ, λ, g)
2:      P⁰ ← initialize with μ random solutions
3:      for t = 1 to g do
4:          Oᵗ ← generate λ offspring
5:          Pᵗ ← select μ solutions from Pᵗ⁻¹ ∪ Oᵗ
6:      end for
7:      return Pᵍ
8: end function
```

Definition 1. *A $(\mu + \lambda)$-archiving algorithm A is a partial mapping $A : 2^{\mathcal{X}} \times 2^{\mathcal{X}} \to 2^{\mathcal{X}}$ such that for a μ-population $\mathcal{P}$ and a λ-population $\mathcal{O}$, $A(\mathcal{P}, \mathcal{O})$ is a μ-population and $A(\mathcal{P}, \mathcal{O}) \subseteq \mathcal{P} \cup \mathcal{O}$.*

---

[4] No assumptions on the reference set have to be made, as our results have to hold for any objective space, including the one only containing solutions that dominate at least one reference point. If that set is empty, all algorithms are effective, as the hypervolume is always zero.

[5] We use the term *archiving algorihm* here to be compliant with [1]. It does not mean that we keep a separate archive in addition to the population $\mathcal{P}^t$.

Using this definition, the for-loop in Algorithm 1 can be described as follows, see also [1]:

Definition 2. *Let* $\mathcal{P}^0$ *be a* μ*-population and* $\mathcal{O}^1, ..., \mathcal{O}^N$ *a sequence of* λ*-populations. Then*

$$\mathcal{P}^t := A(\mathcal{P}^{t-1}, \mathcal{O}^t) \quad \text{for all } t = 1, ..., N$$

We also define

$$\begin{aligned} A(\mathcal{P}^0, \mathcal{O}^1, ..., \mathcal{O}^t) &:= A(A(\mathcal{P}^0, \mathcal{O}^1, ..., \mathcal{O}^{t-1}), \mathcal{O}^t) \\ &= A(...A(A(\mathcal{P}^0, \mathcal{O}^1), \mathcal{O}^2), ..., \mathcal{O}^t) \\ &= \mathcal{P}^t \quad \text{for all } t = 1, ..., N \end{aligned}$$

As mentioned above, we only consider non-decreasing archiving algorithms which are defined as follows, see also [1]:

Definition 3. *An archiving algorithm* A *is non-decreasing, if for all inputs* $\mathcal{P}$ *and* $\mathcal{O}$*, we have*

$$H(A(\mathcal{P}, \mathcal{O})) \geq H(\mathcal{P})$$

2.3 Effectiveness and Approximate Effectiveness

Following Bringmann and Friedrich [1], we here assume a worst-case view on the initial population and a best-case view on the choice of offspring. This means that we would like to know for any optimization problem, starting from any initial population, whether there exists a sequence of offspring populations such that the EA is able to find a population with the maximum possible hypervolume. If so, the archiving algorithm is called *effective*:

Definition 4. *A* $(\mu + \lambda)$*-archiving algorithm* A *is effective, if for all finite sets* $\mathcal{X}$*, all objective functions* f *and all* μ*-populations* $\mathcal{P}^0 \subseteq \mathcal{X}$*, there exists an* $N \in \mathbb{N}$ *and a sequence of* λ*-populations* $\mathcal{O}^1, ..., \mathcal{O}^N \subseteq \mathcal{X}$ *such that*

$$H(A(\mathcal{P}^0, \mathcal{O}^1, ..., \mathcal{O}^N)) = H_\mu^{max}(\mathcal{X})$$

Similarly, we use the following definition for the approximate effectiveness, which quantifies the distance to the optimal hypervolume that can be achieved:

Definition 5. *Let* $\alpha \geq 1$*. A* $(\mu + \lambda)$*-archiving algorithm* A *is* α*-approximate if for all finite sets* $\mathcal{X}$*, all objective functions* f *and all* μ*-populations* $\mathcal{P}^0 \subseteq \mathcal{X}$*, there exists an* $N \in \mathbb{N}$ *and a sequence of* λ*-populations* $\mathcal{O}^1, ..., \mathcal{O}^N$ *such that*

$$H(A(\mathcal{P}^0, \mathcal{O}^1, ..., \mathcal{O}^N)) \geq \frac{1}{\alpha} H_\mu^{max}(\mathcal{X})$$

Of course, an effective archiving algorithm is 1-approximate. Here, we are interested in deriving bounds on α for any choice of μ and λ.

2.4 Submodular Functions

The theory of submodular functions has been widely used to investigate problems where one is interested in selecting optimal subsets of a given size. But what exactly is a submodular function? At first, they map subsets of a given base set to real numbers, just like the hypervolume indicator defined above. In addition, submodular functions show a diminishing increase when adding points to sets that become larger. In other words, let us define the set function $z : 2^{\mathcal{X}} \to \mathbb{R}$, where $2^{\mathcal{X}}$ is the power set of the decision space. Then the contribution of a point $s \in \mathcal{X}$ with respect to set $\mathcal{A} \subset \mathcal{X}$ is $c(s, \mathcal{A}) = z(\mathcal{A} \cup \{s\}) - z(\mathcal{A})$. When z is a submodular function, the contribution $c(s, \mathcal{A})$ gets smaller when $\mathcal{A}$ becomes larger. More formally, a submodular function z is defined as follows:

$$\forall \mathcal{A} \subseteq \mathcal{B} \subseteq \mathcal{X}, \forall s \in \mathcal{X} \backslash \mathcal{B} : z(\mathcal{A} \cup \{s\}) - z(\mathcal{A}) \geq z(\mathcal{B} \cup \{s\}) - z(\mathcal{B}) \qquad (1)$$

i.e. if set $\mathcal{A}$ is contained in set $\mathcal{B}$, the contribution of adding a point s to $\mathcal{A}$ is larger or equal than the contribution of adding s to $\mathcal{B}$. A submodular function is non-decreasing if it is monotone in adding points:

$$\forall \mathcal{B} \subseteq \mathcal{X}, \forall s \in \mathcal{X} \backslash \mathcal{B} : z(\mathcal{B} \cup \{s\}) \geq z(\mathcal{B})$$

Now, we show that the hypervolume indicator as defined above is non-decreasing and submodular.

Theorem 1. *The hypervolume indicator $H(\mathcal{P})$ is non-decreasing submodular.*

Proof. At first, we define the contribution of a solution s to a set $\mathcal{B}$ as

$$H(\mathcal{B} \cup \{s\}) - H(\mathcal{B}) = \int_{y \in \mathbb{R}^m} C(\mathcal{B}, s, y) \, dy$$

with

$$C(\mathcal{B}, s, y) = A_{\mathcal{B} \cup \{s\}}(y) - A_{\mathcal{B}}(y)$$

Using the definition of the attainment function A we find

$$C(\mathcal{B}, s, y) = \begin{cases} 1 & \text{if } (\exists r \in \mathcal{R} : f(s) \leqslant y \leqslant r) \wedge (\nexists p \in \mathcal{B} : f(p) \leqslant y) \\ 0 & \text{else} \end{cases}$$

As $C(\mathcal{B}, s, y)$ is non-negative, the hypervolume indicator is non-decreasing.

Consider two arbitrary sets $\mathcal{A}, \mathcal{B} \subseteq \mathcal{X}$ with $\mathcal{A} \subseteq \mathcal{B}$, and an arbitrary solution $s \in \mathcal{X}$, $s \notin \mathcal{B}$. To prove that the hypervolume indicator is submodular, we have to show that

$$H(\mathcal{A} \cup s) - H(\mathcal{A}) \geq H(\mathcal{B} \cup s) - H(\mathcal{B}) \qquad (2)$$

or equivalently

$$\int_{y \in \mathbb{R}^m} C(\mathcal{A}, s, y) \, dy \geq \int_{y \in \mathbb{R}^m} C(\mathcal{B}, s, y) \, dy \qquad (3)$$

for $\mathcal{A} \subseteq \mathcal{B}$, $s \notin \mathcal{B}$.

To this end, we will show that for all $y \in \mathbb{R}^m$ the inequality $C(\mathcal{A}, s, y) \geq C(\mathcal{B}, s, y)$ holds. As $C(\cdot, \cdot, \cdot)$ can only assume the values 0 and 1, we have to show that for all $y \in \mathbb{R}^m$, $s \notin \mathcal{B}$ we have

$$C(\mathcal{A}, s, y) = 0 \quad \Rightarrow \quad C(\mathcal{B}, s, y) = 0$$

Following the definition of C, there are the following three cases where $C(\mathcal{A}, s, y) = 0$:

1. ($\nexists r \in \mathcal{R} : y \leqslant r$): In this case, we also have $C(\mathcal{B}, s, y) = 0$ as the condition is the same for $C(\mathcal{A}, s, y)$ and $C(\mathcal{B}, s, y)$.
2. ($f(s) \nleqslant r$): Again, we find $C(\mathcal{B}, s, y) = 0$ as the condition is the same for $C(\mathcal{A}, s, y)$ and $C(\mathcal{B}, s, y)$.
3. ($\exists p \in \mathcal{A} : f(p) \leqslant y$): In other words, there exists a solution $p \in \mathcal{A}$ in $\mathcal{A}$ which weakly dominates y. But as $\mathcal{A} \subseteq \mathcal{B}$, we also have $p \in \mathcal{B}$ and therefore, ($\exists p \in \mathcal{B} : f(p) \leqslant y$). Therefore, we find $C(\mathcal{B}, s, y) = 0$.

As a result, (3) holds and the hypervolume indicator is submodular. $\qquad\square$

3   Upper Bound on the Approximate Effectiveness

In this section, we will provide quality guarantees on the hypervolume achieved by an EA in terms of the α-approximate effectiveness, i.e. we will provide an upper bound on α for all population sizes μ and offspring set sizes λ.

In the previous section, we showed that the hypervolume is non-decreasing submodular. Nemhauser, Wolsey and Fisher [3] have investigated interchange heuristics for non-decreasing set functions and showed approximation properties in case of submodular set functions. We will first show that the interchange heuristic in [3] is execution-equivalent to the previously defined $(\mu+\lambda)$-EA framework. Then, the approximation properties for the R-interchange heuristics are used to determine upper bounds on α.

The heuristic described in [3] is shown in Algorithm 2 where we deliberately changed the variable names to make them fit to the notations introduced so far. It makes use of the difference between sets, which is defined as follows: Given two sets $\mathcal{A}$ and $\mathcal{B}$, the difference between $\mathcal{A}$ and $\mathcal{B}$ is $\mathcal{A} - \mathcal{B} = \{x : x \in \mathcal{A} \wedge x \notin \mathcal{B}\}$, i.e. the set of all solutions which are contained in $\mathcal{A}$ but not in $\mathcal{B}$.

The heuristic in Algorithm 2 is of a very general nature. No assumptions are made about the starting population $\mathcal{P}^0$, and the method of searching for $\mathcal{P}^t$. For example, we can set the function $z(\mathcal{P}) = H(\mathcal{P})$ and then choose the following strategy for Line 5:

1. Determine a set $\mathcal{O}^t$ of offspring of size λ.
2. Select μ solutions from $\mathcal{P}^{t-1} \cup \mathcal{O}^t$ using an archiving algorithm A, i.e. $\mathcal{S} = A(\mathcal{P}^{t-1}, \mathcal{O}^t)$.
3. Execute the above two steps until $H(\mathcal{S}) > H(\mathcal{P}^{t-1})$ and then set $\mathcal{P}^t = \mathcal{S}$, or until no such $\mathcal{S}$ can be found.

Algorithm 2. Interchange heuristic: μ is the size of the final set; λ the maximum number of elements which can be exchanged.

```
 1: function HEURISTIC(μ, λ)
 2:     𝒫⁰ ← initialize with an arbitrary set of size μ
 3:     t ← 1
 4:     while true do
 5:         determine a set 𝒫ᵗ of size μ with |𝒫ᵗ − 𝒫ᵗ⁻¹| ≤ λ such that z(𝒫ᵗ) > z(𝒫ᵗ⁻¹)
 6:         if no such a 𝒫ᵗ exists then
 7:             break
 8:         end if
 9:         t ← t + 1
10:     end while
11:     return 𝒫ᴳ ← 𝒫ᵗ⁻¹
12: end function
```

Following Algorithm 2, the above steps need to guarantee that a set $\mathcal{P}^t$ with $H(\mathcal{P}^t) > H(\mathcal{P}^{t-1})$ is found if it exists. For example, we can use an exhaustive offspring generation, i.e. every subset of size λ of the decision space $\mathcal{X}$ can be determined with a probability larger than zero. Moreover, the archiving algorithm A must be able to determine an improved subset of $\mathcal{P}^{t-1} \cup \mathcal{O}^t$ if it exists. In other words, we require from A that $H(A(\mathcal{P}, \mathcal{O})) > H(\mathcal{P})$ if there exists a subset of $\mathcal{P} \cup \mathcal{O}$ of size μ with a larger hypervolume than $H(\mathcal{P})$. For example, A may in turn remove all possible subsets of size λ from $\mathcal{P}^{t-1} \cup \mathcal{O}^t$ and return a set that has a better hypervolume than $\mathcal{P}^{t-1}$. Note that this instance of the interchange heuristic can be easily rephrased in the general $(\mu + \lambda)$-EA framework of Algorithm 1 with an unbounded number of generations.

Nemhauser et al. [3] have proven the following result for the interchange heuristic:

Theorem 2. *Suppose z is non-decreasing and submodular. Moreover, define the optimization problem $z^* = \max_{\mathcal{P} \subseteq \mathcal{X}, |\mathcal{P}| \leq \mu} z(\mathcal{P})$. If $\mu = q \cdot \lambda - p$ with q a positive integer, and p integer with $0 \leq p \leq \lambda - 1$, then*

$$\frac{z^* - z(\mathcal{P}^G)}{z^* - z(\emptyset)} \leq \frac{\mu - \lambda + p}{2\mu - \lambda + p}$$

where $z(\mathcal{P}^G)$ is the value of the set obtained by Algorithm 2 and $z(\emptyset)$ is the value of the empty set.

We have shown that the hypervolume indicator is non-decreasing submodular. Therefore, if we set the function $z(\mathcal{P}) = H(\mathcal{P})$ and note that $H(\emptyset) = 0$, we can easily obtain the following bound on the approximation quality of Algorithm 2:

Proposition 1. *If $\mu = q \cdot \lambda - p$ with an integer $0 \leq p \leq \lambda - 1$, then*

$$H(\mathcal{P}^G) \geq \frac{1}{2 - \frac{\lambda - p}{\mu}} \cdot H_\mu^{\max}(\mathcal{X}) \tag{4}$$

This bound can be compared to the definition of the approximate effectiveness, see Definition 5, i.e. it bounds the achievable optimization quality in terms of the hypervolume if a certain algorithm structure is used. But whereas Definition 5 and the corresponding value of $\alpha = 2 + \epsilon$ from [1] is related to Algorithm 1, the above bound with $\alpha = 2 - \frac{\lambda-p}{\mu}$ is related to Algorithm 2.

We will now show that the improved approximation bound of $\alpha = 2 - \frac{\lambda-p}{\mu}$ is valid also in the case of Algorithm 1, thereby improving the results in [1].

Theorem 3. *Suppose a non-decreasing $(\mu + \lambda)$-archiving algorithm which satisfies in addition*

$$\exists \mathcal{S} : (\mathcal{S} \subset \mathcal{P} \cup \mathcal{O}) \wedge (|\mathcal{S}| = \mu) \wedge (H(\mathcal{S}) > H(\mathcal{P})) \quad \Rightarrow \quad H(A(\mathcal{P}, \mathcal{O})) > H(\mathcal{P})$$

Then for all finite sets $\mathcal{X}$, all objective functions f and all μ-populations $\mathcal{P}^0 \subseteq \mathcal{X}$ the following holds: For any run of an instance of Algorithm 2, one can determine a sequence of λ-populations $\mathcal{O}^1, ..., \mathcal{O}^N$ such that

$$H(A(\mathcal{P}^0, \mathcal{O}^1, ..., \mathcal{O}^N)) = H(\mathcal{P}^G)$$

Proof. The proof uses the special instance of Algorithm 2 that has been introduced above. Line 5 is implemented as follows: (1) Determine a set $\mathcal{O}^t$ of offspring of size λ using an exhaustive generation, i.e. each subset of $\mathcal{X}$ is determined with non-zero probability. (2) Use the archiving algorithm A to determine a set $\mathcal{S} = A(\mathcal{P}^{t-1}, \mathcal{O}^t)$. (3) Repeat these two steps until $H(\mathcal{S}) > H(\mathcal{P}^{t-1})$ or no such $\mathcal{S}$ can be found. Due to the required property of A, no such $\mathcal{S}$ can be found if it does not exist.

Algorithm 2 yields as final population $\mathcal{P}^G = \mathcal{P}^{t-1}$ which can be rewritten as $\mathcal{P}^{t-1} = A(\mathcal{P}^0, \mathcal{O}^1, ..., \mathcal{O}^{t-1})$ The sets of offspring $\mathcal{O}^i$ are generated as described above. Using $N = t-1$ yields the required result $H(A(\mathcal{P}^0, \mathcal{O}^1, ..., \mathcal{O}^N)) = H(\mathcal{P}^G)$. $\qquad\square$

As a direct consequence of the execution equivalence between Algorithm 1 and Algorithm 2 according to the above theorem, the Definition 5 and (4), we can state the following result:

Proposition 2. *A non-decreasing $(\mu + \lambda)$-archiving algorithm $A(\mathcal{P}, \mathcal{O})$, which yields a subset of $\mathcal{P} \cup \mathcal{O}$ of size μ with a better hypervolume than that of $\mathcal{P}$ if there exists one, is $(2 - \frac{\lambda-p}{\mu})$-approximate where $\mu = q \cdot \lambda - p$ with an integer $0 \leq p \leq \lambda - 1$.*

It is interesting to note two special cases of the above proposition:

1. $\mu = \lambda$: In this case, we have a $(\mu + \mu)$-EA. It holds that $p = 0$ and therefore, the formula evaluates to $\alpha = 1$, which means that this algorithm actually is effective. This corresponds to the obvious result mentioned in the introduction.

2. $\lambda = 1$: In this case, we have a $(\mu + 1)$-EA. It holds that $p = 0$ and $q = \mu$ and therefore, the formula evaluates to $\alpha = 2 - \frac{1}{\mu}$, which is tighter than the bound of Bringmann and Friedrich [1].

Figure 1 shows the relation between λ and α for several settings of μ. As can be seen, it is a zigzag line which corresponds to the modulo-like definition of p and q. The local maxima of each line are located where μ is an integer multiple of λ.

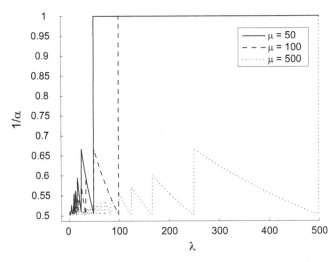

Fig. 1. Quality guarantees for the hypervolume achieved by a $(\mu + \lambda)$-EA. For a given μ and a given λ, there is a sequence of offspring such that at least $\frac{1}{\alpha} \cdot H_\mu^{max}(\mathcal{X})$ can be achieved, irrespective of the optimization problem and the chosen initial population.

4   Lower Bound on the Approximate Effectiveness

In the previous section we gave an upper bound on α. In this section, on the other hand, we will give a lower bound on α. This lower bound is tight for $\mu = 2$, i.e. is equal to the upper bound. To find this bound, we will show that there exist optimization problems and initial populations, such that any non-decreasing archiving algorithm will end up with a hypervolume that is at most $1/(1 + \frac{1}{2\lambda})$ of the optimal hypervolume. Whereas a first particular example has been shown in [4], a more general lower bound was shown in [1], where Bringmann and Friedrich found a problem where any non-decreasing archiving algorithm ends up with a hypervolume that is at most $1/(1 + 0.1338(1/\lambda - 1/\mu) - \epsilon)$ of the optimal hypervolume, for any $\epsilon > 0$. The new bound substantially tightens the result of [1], but relies on the general definition of the hypervolume indicator which uses a reference set $\mathcal{R}$ instead of a single reference point.

Theorem 4. *Let $\lambda < \mu$. There is no α-approximate non-decreasing $(\mu + \lambda)$-archiving algorithm for any $\alpha < 1 + \frac{1}{2\lambda}$.*

Proof. We proof this theorem by finding a population $\mathcal{P}^0 = \{s_0, ..., s_{\mu-1}\}$ whose hypervolume indicator $H(\mathcal{P}^0)$ can not be improved by any non-decreasing $(\mu + \lambda)$-archiving algorithm, i.e. it is locally optimal. At the same time, the optimal population $\mathcal{P}^* = \{o_0, ..., o_{\mu-1}\}$ has a hypervolume indicator value of $H(\mathcal{P}^*)$ which satisfies $H(\mathcal{P}^*) = (1 + \frac{1}{2\lambda} - \delta)H(\mathcal{P}^0)$ for any $\delta > 0$.

The setting we are considering for the proof is shown in Figure 2. There are $2 \cdot \mu$ points, where the initial population is set to $\mathcal{P}^0 = \{s_0, ..., s_{\mu-1}\}$ and the optimal population would be $\mathcal{P}^* = \{o_0, ..., o_{\mu-1}\}$. We consider a setting with multiple reference points $\{r_0, ..., r_{2\mu-2}\}$, such that the areas contributing to the hypervolume calculation are A_i (areas only dominated by the initial population), B_i (areas only dominated by the optimal population), and C_i and D_i (areas dominated by one solution of the initial population and one solution of the optimal population), see Figure 2. The objective space is the union of all points, i.e. $\mathcal{Y} = \mathcal{P}^0 \cup \mathcal{P}^*$.

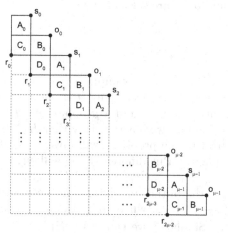

Fig. 2. Schematic drawing of the example setting in the proof of Theorem 4

In our example, we set these areas as follows, assuming $\lambda < \mu$:

$$A_i = \epsilon \text{ for } 0 \leq i < \mu \quad , \quad B_i = \begin{cases} \epsilon & \text{for } 0 \leq i < \lambda \\ 1 & \text{for } \lambda \leq i < \mu \end{cases}$$

$$C_i = \sum_{i-\lambda \leq j < i} B_{j \bmod \mu} \quad , \quad D_i = \sum_{i+1 \leq j < i+1+\lambda} B_{j \bmod \mu}$$

Note that for any choice of areas A_i, B_i, C_i, and D_i, corresponding coordinates can be found for all s_i and o_i and r_i by using the following recursions:

$$s_i^x = o_{i-1}^x + \frac{A_i}{s_i^y - o_i^y}, \quad o_i^x = s_i^x + \frac{B_i}{o_i^y - s_{i+1}^y}$$

$$s_i^y = o_{i-1}^y - \frac{C_{i-1}}{s_{i-1}^x - o_{i-2}^x}, \quad o_i^y = s_i^y - \frac{D_{i-1}}{o_{i-1}^x - s_{i-1}^x}$$

$$r_i^x = \begin{cases} o_{i/2-1}^x & i \text{ even} \\ s_{(i-1)/2-1}^x & i \text{ odd} \end{cases}, \quad r_i^y = \begin{cases} o_{i/2+1}^y & i \text{ even} \\ s_{(i-1)/2}^y & i \text{ odd} \end{cases}$$

where s_i^x, o_i^x, s_i^y, o_i^y and r_i^y, r_i^y are the x-axis and y-axis coordinates of s_i, o_i and r_i, respectively. While s_0^x, s_0^y, and r_0^x with $r_0^x < s_0^x$ can be chosen arbitrarily, the coordinates for o_0^y and s_1^y are set as follows:

$$o_0^y = s_0^y - \frac{A_0}{s_0^x - r_0^x}, \quad s_1^y = o_0^y - \frac{C_0}{s_0^x - r_0^x}$$

Furthermore, $r_{2\mu-2}^y$ and $o_{\mu-1}^x$ are set as follows:

$$r_{2\mu-2}^y = o_{\mu-1}^y - \frac{C_{\mu-1}}{s_{\mu-1}^x - o_{\mu-2}^x}, \quad o_{\mu-1}^x = s_{\mu-1}^x + \frac{B_{\mu-1}}{o_{\mu-1}^y - r_{2\mu-2}^y}$$

First, we want to show that for the example, $\mathcal{P}^0$ is a local optimum, i.e. $H(\mathcal{P}^0)$ can not be improved by any non-decreasing $(\mu + \lambda)$-archiving algorithm. To do so consider a λ-population $\mathcal{O} \subset \mathcal{Y}$ and a μ-population $\mathcal{P}^1 \subset \mathcal{P}^0 \cup \mathcal{O}$. In order for $\mathcal{P}^0$ to be a local optimum, we have to show that $H(\mathcal{P}^0) \geq H(\mathcal{P}^1)$.

Note that for the rest of the proof, we will always use the indices modulo μ without writing it explicitly. Put differently, we will write A_i, B_i, C_i, and D_i as a short form of $A_{i \bmod \mu}$, $B_{i \bmod \mu}$, $C_{i \bmod \mu}$, and $D_{i \bmod \mu}$.

The hypervolume of the initial population can be written as $H(\mathcal{P}^0) = H - \sum_{0 \leq i < \mu} B_i = H - (\mu - \lambda) - \lambda\epsilon$, where H is the hypervolume of all solutions, i.e. $H = H(\mathcal{P}^0 \cup \mathcal{P}^*)$. Similarly, we can write $H(\mathcal{P}^1) = H - \sum_{i:s_i,o_i \notin P^1} C_i - \sum_{i:s_{i+1},o_i \notin P^1} D_i - \sum_{i:o_i \notin P^1} B_i - \sum_{i:s_i \notin P^1} A_i$. Using these expressions, we get the following set of equivalent inequalities:

$$H(\mathcal{P}^0) \geq H(\mathcal{P}^1)$$
$$H - (\mu - \lambda) - \lambda\epsilon \geq H - \sum_{i:s_i,o_i \notin P^1} C_i - \sum_{i:s_{i+1},o_i \notin P^1} D_i$$
$$- \sum_{i:o_i \notin P^1} B_i - \sum_{i:s_i \notin P^1} A_i$$
$$(\mu - \lambda) + \lambda\epsilon \leq \sum_{i:s_i,o_i \notin P^1} C_i + \sum_{i:s_{i+1},o_i \notin P^1} D_i$$
$$+ ((\mu - \lambda) + \lambda\epsilon - \sum_{i:o_i \in P^1} B_i) + \sum_{i:s_i \notin P^1} A_i$$

$$\sum_{i:o_i \in P^1} B_i \leq \sum_{i:s_i,o_i \notin P^1} C_i + \sum_{i:s_{i+1},o_i \notin P^1} D_i + \sum_{i:s_i \notin P^1} A_i \qquad (5)$$

To prove this inequality (5), we need to consider all possible μ-populations $\mathcal{P}^1 \subset \mathcal{P}^0 \cup \mathcal{O}$, i.e. the results of all possible $(\mu + \lambda)$-archiving algorithms. To go from $\mathcal{P}^0$ to $\mathcal{P}^1$, λ solutions s_i of the initial set are discarded and the same number of solutions o_i from the optimal set are added. We call these discarded s_i and added o_i affected solutions.

In the following, we consider blocks of affected solutions. To this end, we first mark all solutions in $\mathcal{P}^0 \cup \mathcal{P}^*$ that are either removed from $\mathcal{P}^0$ or added to $\mathcal{P}^0$ when going from $\mathcal{P}^0$ to $\mathcal{P}^1$. This set of marked solutions is then partitioned into the minimal number of subsets, such that each subset contains all solutions in index range $[i, i+k]$. Depending on whether the first and last solutions in such a subset are from set $\mathcal{P}^0$ or $\mathcal{P}^*$ we call it an (s, s)-, (s, o)-, (o, s)- or (o, o)-block, respectively. For example, an (o, s)-block with index range $[2, 5]$ contains solutions $\{o_2, s_3, o_3, s_4, o_4, s_5\}$. The rationale is that non-neighboring solutions do not influence each other, as they do not dominate any common area. As for the blocks, there are two cases which will be considered separately.

Blocks of even length: There are two types of blocks of even length: Those starting with an added solution from the optimal set, i.e. (o, s)-blocks, and those starting with a discarded solution from the initial set, i.e. (s, o)-blocks. The first case can be formalized as follows: The (o, s)-block with index range $[i, i+k]$ exists iff $(o_l \in \mathcal{P}^1, i \leq l < i+k) \wedge (o_{i+k} \notin \mathcal{P}^1) \wedge (s_i \in \mathcal{P}^1) \wedge (s_l \notin \mathcal{P}^1, i+1 \leq l < i+k+1)$. For this block, (5) evaluates to:

$$\sum_{i:o_i \in \mathcal{P}^1} B_i \leq \sum_{i:s_i, o_i \notin \mathcal{P}^1} C_i + \sum_{i:s_{i+1}, o_i \notin \mathcal{P}^1} D_i + \sum_{i:s_i \notin \mathcal{P}^1} A_i$$
$$\sum_{i \leq l < i+k} B_l \leq C_{i+k} + 0 + \sum_{i+1 \leq l < i+k+1} A_l$$
$$\sum_{i \leq l < i+k} B_l \leq \sum_{i+k-\lambda \leq l < i+k} B_l + k\epsilon$$
$$0 \leq \sum_{i+k-\lambda \leq l < i} B_l + k\epsilon$$

The last step is true because we know that $k \leq \lambda$. As all B_l as well as ϵ are larger than zero, (5) holds.

The second case can be formalized as follows: The (s, o)-block with index range $[i, i+k]$ exists iff $(o_{i-1} \notin \mathcal{P}^1) \wedge (o_l \in \mathcal{P}^1, i \leq l < i+k) \wedge (s_l \notin \mathcal{P}^1, i \leq l < i+k) \wedge (s_{i+k} \in \mathcal{P}^1)$. For this block, (5) evaluates to:

$$\sum_{i:o_i \in \mathcal{P}^1} B_i \leq \sum_{i:s_i, o_i \notin \mathcal{P}^1} C_i + \sum_{i:s_{i+1}, o_i \notin \mathcal{P}^1} D_i + \sum_{i:s_i \notin \mathcal{P}^1} A_i$$
$$\sum_{i \leq l < i+k} B_l \leq 0 + D_{i-1} + \sum_{i \leq l < i+k} A_l$$
$$\sum_{i \leq l < i+k} B_l \leq \sum_{i \leq l < i+\lambda} B_l + k\epsilon$$
$$0 \leq \sum_{i+k \leq l < i+\lambda} B_l + k\epsilon$$

Again, we can see that the last inequality holds, and therefore, (5) holds.

Blocks of odd length: Such blocks consist of either a set of discarded solutions that enclose a set of added solutions or vice versa, i.e. (s, s)- or (o, o)-blocks. Due to $|\mathcal{P}^0| = |\mathcal{P}^1|$, the number of added solutions from the optimal set must be equal to the number of discarded solutions from the initial set. Directly following this, we know that for each block of discarded solutions enclosing added solutions, there must be another block of added solutions enclosing discarded solutions and vice versa. These two types of blocks can be formalized as follows: The (s, s)-block with index range $[i, i+k]$ exists iff $(o_l \in \mathcal{P}^1, i \leq l < i+k-1) \wedge (o_{i-1}, o_{i+k-1} \notin \mathcal{P}^1) \wedge (s_l \notin \mathcal{P}^1, i \leq l < i+k)$. The (o, o)-block with index range $[j, j+p]$ exists iff $(o_l \in \mathcal{P}^1, j \leq l < j+p) \wedge (s_l \notin \mathcal{P}^1, j+1 \leq l < j+p) \wedge (s_j, s_{j+p} \in \mathcal{P}^1)$. Also, we know that $1 \leq k, p \leq \lambda$ and $k + p \leq \lambda + 1$. Considering both of these blocks, (5) evaluates to:

$$\sum_{i:o_i \in \mathcal{P}^1} B_i \leq \sum_{i:s_i, o_i \notin \mathcal{P}^1} C_i + \sum_{i:s_{i+1}, o_i \notin \mathcal{P}^1} D_i$$
$$+ \sum_{i:s_i \notin \mathcal{P}^1} A_i$$
$$\sum_{i \leq l < i+k-1} B_l + \sum_{j \leq l < j+p} B_l \leq C_{i+k-1} + D_{i-1} + \sum_{i \leq l < i+k} A_l$$
$$+ \sum_{j+1 \leq l < j+p} A_l$$
$$\sum_{i \leq l < i+k-1} B_l + \sum_{j \leq l < j+p} B_l \leq \sum_{i+k-1-\lambda \leq l < i+k-1} B_l + \sum_{i \leq l < i+\lambda} B_l$$
$$+ (k+p-1)\epsilon$$
$$\sum_{j \leq l < j+p} B_l \leq p \leq \sum_{i+k-1-\lambda \leq l < i+\lambda} B_l + (k+p-1)\epsilon$$
$$p \leq \lambda\epsilon + \lambda - k + 1 + (k+p-1)\epsilon$$

The second last step can be done because we know that at most λ of the B_l's are set to ϵ and therefore, at least $\lambda - k + 1 \geq p$ of the B_l's remain which are set to 1. Also, because of $p \leq \lambda - k + 1$, the last inequality holds and with it (5) holds.

Combinations of blocks: As stated before, only neighboring solutions in $\mathcal{Y} = \mathcal{P}^0 \cup \mathcal{P}^*$ share a common dominated area. From the definition of the different types of blocks it can be seen that there are no adjacent blocks, because in this case, the two blocks would be combined into one. Therefore, each pair of blocks is separated by at least one solution from $\mathcal{Y}$ which is not affected by the transition from $\mathcal{P}^0$ to $\mathcal{P}^1$. As a result, the changes in hypervolume when going from $\mathcal{P}^0$ to $\mathcal{P}^1$ can be considered separately for each block. We have shown that for any block, (5) holds. From this we can conclude that $H(\mathcal{P}^0) \geq H(\mathcal{P}^1)$ and therefore, $\mathcal{P}^0$ is a local optimum.

Now that we've done the first part of the proof, i.e. showing that any nondecreasing $(\mu + \lambda)$-archiving algorithm will not be able to escape from $\mathcal{P}^0$, we would like to calculate how far the hypervolume of $\mathcal{P}^0$ is from the maximum achievable hypervolume. In other words, we would like to calculate $\frac{H(\mathcal{P}^*)}{H(\mathcal{P}^0)}$. The hypervolume of the initial population evaluates to:

$$H(\mathcal{P}^0) = \sum_{0 \leq l < \mu} C_l + \sum_{0 \leq l < \mu} D_l + \sum_{0 \leq l < \mu} A_l$$
$$= \sum_{0 \leq l < \mu} \left(\sum_{l-\lambda \leq j < l} B_j + \sum_{l+1 \leq j < l+1+\lambda} B_j \right) + \mu\epsilon$$
$$= \sum_{0 \leq l < \mu} \left(\sum_{l-\lambda \leq j < l+1+\lambda} B_j - B_l \right) + \mu\epsilon$$
$$= (2\lambda + 1) \sum_{0 \leq l < \mu} B_l - \sum_{0 \leq l < \mu} B_l + \mu\epsilon$$
$$= 2\lambda \sum_{0 \leq l < \mu} B_l + \mu\epsilon$$

The hypervolume of the optimal population, on the other hand, can be calculated as follows:

$$H(\mathcal{P}^*) = \sum_{0 \leq l < \mu} C_l + \sum_{0 \leq l < \mu} D_l + \sum_{0 \leq l < \mu} B_l$$
$$= \sum_{0 \leq l < \mu} \sum_{l-\lambda \leq j < l+1+\lambda} B_j$$
$$= (2\lambda + 1) \sum_{0 \leq l < \mu} B_l$$

Both sets of equations make use of $\sum_{0 \leq l < \mu} \sum_{l-\lambda \leq j < l+1+\lambda} B_j = (2\lambda + 1) \sum_{0 \leq l < \mu} B_l$. This is due to the fact that the inner sum of the left-hand term consists of $2\lambda + 1$ summands. Because all indices are taken modulo μ, we see that each B_j is summed up $2\lambda + 1$ times in the whole term.

Finally, this leads us to the following result, which holds for any $\delta > 0$ if $\epsilon \to 0$ and $\lambda < \mu$:

$$\frac{H(\mathcal{P}^*)}{H(\mathcal{P}^0)} = \frac{(2\lambda+1)\sum_{0 \leq l < \mu} B_l}{2\lambda \sum_{0 \leq l < \mu} B_l + \mu\epsilon}$$
$$= 1 + \frac{1}{2\lambda} - \delta$$

Note that in the case of $\lambda = \mu$, the equation evaluates to $\frac{H(\mathcal{P}^*)}{H(\mathcal{P}^0)} = 1$, which is very natural, since for $\mu = \lambda$, any non-decreasing $(\mu + \lambda)$-archiving algorithm is effective.   □

We may also interpret the above result in terms of the more practical interchange heuristic shown in Algorithm 2. One can conclude that for $z(\mathcal{P}) = H(\mathcal{P})$, i.e. we use the hypervolume indicator for archiving, we may end up with a solution that is not better than $1/\alpha$ times the optimal hypervolume with $\alpha > 1 + \frac{1}{2\lambda}$, even after an unlimited number of iterations.

5   Conclusion

In this paper, we investigated the α-approximate effectiveness of $(\mu + \lambda)$-EAs that optimize the hypervolume. The value of α gives a lower bound on the hypervolume which can always be achieved, independent of the objective space and the chosen initial population. While it is obvious that for $\mu = \lambda$, α is equal to 1, Bringmann and Friedrich have shown that for $\lambda = 1$, α is equal to 2. This paper strictly improves the currently known bound and finds that for arbitrary λ, the approximation factor α is equal to $2 - \frac{\lambda-p}{\mu}$, where $\mu = q \cdot \lambda - p$ and $0 \leq p \leq \lambda - 1$.

Furthermore, we improve the available lower bound on α for the general definition of the hypervolume indicator, i.e. $\alpha > 1 + \frac{1}{2\lambda}$. Upper and lower bounds only match for a population size of $\mu = 2$. It might be possible to further tighten the lower bound by extending the worst case construction in the proof of Theorem 4 to higher dimensions of the objective space.

References

1. Bringmann, K., Friedrich, T.: Convergence of hypervolume-based archiving algorithms i: Effectiveness. In: Genetic and Evolutionary Computation Conference (GECCO), pp. 745–752 (2011)
2. Edmonds, J.: Submodular functions, matroids and certain polyhedra. In: Guy, R. (ed.) Combinatorial Structures and their Applications, pp. 69–87. Gordon and Breach, New York (1971)
3. Nemhauser, G.L., Wolsey, L.A., Fisher, M.L.: An analysis of approximations for maximizing submodular set functions – i. Mathematical Programming 14, 265–294 (1978)
4. Zitzler, E., Thiele, L., Bader, J.: On set-based multiobjective optimization. IEEE Trans. on Evolutionary Computation 14, 58–79 (2010)

Optimization by ℓ_1-Constrained Markov Fitness Modelling

Gabriele Valentini[1], Luigi Malagò[2], and Matteo Matteucci[2]

[1] IRIDIA, CoDE, Université Libre de Bruxelles
gabriele.valentini@ulb.ac.be
[2] Department of Electronics and Information, Politecnico di Milano
{malago,matteucci}@elet.polimi.it

Abstract. When the function to be optimized is characterized by a limited and unknown number of interactions among variables, a context that applies to many real world scenario, it is possible to design optimization algorithms based on such information. Estimation of Distribution Algorithms learn a set of interactions from a sample of points and encode them in a probabilistic model. The latter is then used to sample new instances. In this paper, we propose a novel approach to estimate the Markov Fitness Model used in DEUM. We combine model selection and model fitting by solving an ℓ_1-constrained linear regression problem. Since candidate interactions grow exponentially in the size of the problem, we first reduce this set with a preliminary coarse selection criteria based on Mutual Information. Then, we employ ℓ_1-regularization to further enforce sparsity in the model, estimating its parameters at the same time. Our proposal is analyzed against the 3D Ising Spin Glass function, a problem known to be NP-hard, and it outperforms other popular black-box meta-heuristics.

Keywords: Estimation of Distribution Algorithms, Markov Fitness Model, DEUM, ℓ_1-constrained Linear Regression, Least Angle Regression.

1 Introduction

Black-box optimization consists of a set of meta-heuristics used to search for the optimum of a function when no information about its structure is available. Such approach to optimization can be used to define general purpose algorithms that do not depend on the function to be optimized, and it becomes the only possible approach when the mathematical formulation of the function is unknown. In particular model based meta-heuristics [25] introduce a statistical model to represent correlations among variables and to guide the search for the optimum.

Most of the model based meta-heuristics [25] use a probabilistic description of the problem to drive their search towards solutions with the best value. The most general model is the joint probability distribution p which characterizes the correlations among all the variables involved in the objective function f. In

Y. Hamadi and M. Schoenauer (Eds.): LION 6, LNCS 7219, pp. 250–264, 2012.
© Springer-Verlag Berlin Heidelberg 2012

the discrete setting, the probability simplex is able to capture any possible order of interactions among the variables; however, its dimension equals the cardinality of the search space, and the estimation of its parameters is unfeasible.

Many problems are characterized by a limited set of correlations among the variables, even NP-hard problems. It follows that model based search in a black-box context, to be really effective, must face the problem of selecting a lower dimensional model, which is computationally tractable, and would be able to capture all, or at least most, of relevant correlations. The family of the model and the way in which it is chosen define the particular class of meta-heuristics.

Once the model has been selected, model based algorithms implement different techniques to search for the optimal distribution in the model, for instance by applying estimation and sampling techniques, as in Estimation of Distribution Algorithm (EDA) [10], or by following the gradient of the expected value of f as in CMA-ES [8] and SNGD [11]. Within EDAs, a distinctive feature of the algorithms that belong to the Distribution Estimation Using Markov Networks (DEUM) [18] framework is the direct use of a probabilistic model of the objective function, which is sampled to search for a global minimum.

Selecting a model and estimating the parameters correspond, respectively, to a model selection and a model fitting problem, and in the general case are computationally expensive to address. On the other hand being able to recover the correct set of interactions, or at least a model that capture most of them, allows to work with tractable models with good properties, i.e., from any point the gradient of the expected value of the original function points in the direction of the global optimum, so that different optimization algorithms are less prone to end up with local minima, [12].

In DEUM, the joint probability distribution is represented using the formalism of Markov Networks (MNs) [22], also known as Markov Random Fields (MRFs), an example of undirected Probabilistic Graphical Models (PGMs). The structure of the MN, i.e., the set of conditional independences, can be either fixed a priori [18,19], in which case we refer to fixed structure DEUM algorithms, or learned from scratch using model selection criteria such as Mutual Information [17] or χ^2-independence test [4]. Once the structure is identified, the parameters of the model are estimated from a subset of points with least square method, and then the model is sampled to look for a global optimum.

A common hypothesis when learning a model in EDAs is to limit the search to pairwise interactions. This reduces the number of possible interactions to $\binom{n}{2}$. Other additional hypothesis [4,17] limit the maximum size of the neighborhood of each variable to force a sparse pattern of interactions. A different approach to model selection in DEUM has been proposed in [13] where ℓ_1-regularized logistic regression is employed to recover the neighborhood of each variable, cf. [16]. This choice allows to shrink the conditional probability distribution of a variable given its neighborhood through a regularization parameter. The approach showed promising results both in terms of model selection and optimization performance. However, the computational effort was still very expensive.

The aim of this paper is to provide a novel method to estimate the statistical model used in DEUM by introducing a sparse model selection approach when estimating the Markov Fitness Model, thus dealing with model selection and model fitting at the same time. To obtain this result, we formalize the estimation problem as an ℓ_1-constrained linear regression problem, also known as the Lasso [20]. In this formulation, the penalizing ℓ_1-constraint addresses model selection, while the least square error minimization allows to estimate the coefficients of the model. Since candidate interactions grows exponentially in the problem size in the general case and quadratically if we restrict to pairwise interactions, we firstly use a preliminary coarse selection criteria based on Mutual Information to reduce the size of this set, similarly to the approach in [17], but with no constraint on the size of the neighborhood.

The remaining of the paper is organized as follows. In Section 2 we describe the Markov Fitness Model underlying the DEUM framework. In Section 3 we introduce our approach based on ℓ_1-constrained linear regression to estimate the set of interactions and associated parameters of the model. In Section 4 we present the Sparsified DEUM algorithm (sDEUM), while in Section 5 we provide an empirical analysis of its performance using the well-known 3D Ising Spin Glass function [2] as a benchmark. The paper ends in Section 6 with conclusions and future directions of research.

2   Objective Function Modeling by Markov Networks

EDAs and more in general most model-based meta-heuristics make use of a statistical model, i.e., a set of probability distributions, to represent the interactions among the variables of an optimization problem. Usually the model is estimated from a subset of points, selected from a larger sample according to the value of f. The same applies for the algorithms in the DEUM framework, with the difference that the statistical model is employed to learn a model of f, rather than to estimate the correlations among its variables.

We consider the unconstrained optimization problem of minimizing a real-valued function f defined over a vector of n binary variables $X = (X_1, \ldots, X_n)$ with values in $\Omega = \{-1, +1\}^n$. Since the domain is finite, and $x^2 = 1$, any f can be written as a square-free polynomial

$$f(x) = \sum_{\alpha \in I} c_\alpha x^\alpha. \tag{1}$$

Here, we employ a notation based on the multi-index $\alpha = (\alpha_1, \ldots, \alpha_n) \in I \subset \{0,1\}^n$, with $x^\alpha = \prod_{i=1}^n x_i^{\alpha_i}$. The associated real coefficients $c_\alpha \in \mathbb{R} \setminus \{0\}$ are indexed by α. Each monomial represents an interaction among a set of variables in f. We say that the set of interactions in function f is sparse if $\#(I) \ll 2^n$, where $\#(I)$ represents the cardinality of I. Many well known functions belong to this class, and even if the number of interactions is limited the optimization of such functions can be an NP-hard problem. For instance, the energy function of an Ising Spin glass problem [2] defined over a 3D toroidal lattice has $\#(I) = 3n$

interactions, where $l = \sqrt[3]{n}$ is the size of the grid. In the maximum cut [23] problem the cardinality of I corresponds to the number of edges in the graph, and in general $\#(I) \leq \binom{n}{2}$.

In the DEUM framework, probabilities of points in the search space Ω are assigned under the hypothesis that the probability of x should be proportional to the value of f, i.e.,

$$p(x) \equiv \frac{f(x)}{Z}, \qquad \text{with } Z = \sum_{x \in \Omega} f(x). \tag{2}$$

In particular, DEUM uses the Gibbs distribution as a statistical model, which is an example in the exponential family of distributions that can be equivalently represented with the formalism of MNs. The Gibbs distribution is used to learn a model of the objective function, by means of the Markov Fitness Model [3].

2.1 Markov Networks and Gibbs Distribution

Most EDAs make use of PGMs to represent the statistical model they use. In particular, the algorithms in the DEUM framework employ undirected graphical models called MNs. One of the advantages of a PGM is that the graph represents the conditional independence structure of the random variables, and provides a way to factorize the joint probability distribution associated to the graph.

Given a vector $X = (X_1, \ldots, X_n)$ of random variables, a MN is defined by a pair $(\mathcal{G}, \Phi)$, where $\mathcal{G} = (\mathcal{V}, \mathcal{E})$ is an undirected graph and Φ is a set of local energy functions φ associated to the cliques. Each random variable X_i in X corresponds to a vertex $v_i \in \mathcal{V}$, while the edges $e_{ij} \in \mathcal{E}$ define the topology of the graph. We denote with $\mathcal{N}_i$ the *neighborhood* of a variable X_i, i.e., the set of vertices v_j such that $e_{ij} \in \mathcal{E}$. A set X_C of fully connected vertexes of $\mathcal{G}$ is called *clique*. A clique is *maximal* if it is not contained in the set of vertices of any other clique.

The topology of the MN determines a set of conditional independence statements according to the absence of edges in the graph. As stated in the Hammersley-Clifford theorem [7], a positive probability distribution satisfies all the Markov properties with respect to the graph $\mathcal{G}$ if and only if it factorizes according to the graph itself. This implies that the joint probability distribution of X can be expressed as the product of a set of non-negative functions φ_C, called *potential functions*, defined over the clique $C \in \mathcal{C}$, i.e.,

$$p(x) = \frac{1}{Z} \prod_{C \in \mathcal{C}} \varphi_C(x_C) \tag{3}$$

where Z is a normalization constant that ensures the probabilities sum to 1. Without loss of generality, by absorbing cliques in maximal cliques, in the rest of the paper we restrict the factorization to the product of potential functions defined over the maximal cliques of $\mathcal{G}$.

Moreover, the Hammersley-Clifford theorem implies the equivalence of the probability distribution p in (3) associated to $\mathcal{G}$ and the *Gibbs* (or *Boltzmann*) distribution of the form

$$p(x) = \frac{1}{Z} e^{-U(x)/T}, \qquad \text{with } Z = \sum_{x \in \Omega} e^{-U(x)/T}. \tag{4}$$

In statistical physics, the normalizing constant Z is called *partition function*, $T > 0$ is the *temperature* of the distribution, and $U(x)$ the *energy function*. The temperature parameter controls the sharpness of the distribution. Indeed, for $T \to \infty$, Equation (4) tends to the uniform distribution over Ω, while for $T \to 0$ the probability mass concentrates over the global minima of the energy function. The energy function of the Gibbs distribution is defined as the sum of local functions u_C associated to φ_C defined over the maximal cliques of $\mathcal{G}$, i.e.,

$$U(x) = \sum_{C \in \mathcal{C}} u_C(x_C). \tag{5}$$

In EDAs, the search space Ω is explored by sampling from a density in a statistical model. However, sampling from (4) is non trivial due to the presence of the partition function Z, whose computation requires a summation over the entire space Ω, and thus is unfeasible since it is exponential in n. Nevertheless, the Gibbs distribution can be sampled using a Gibbs sampler, and exploiting the local Markov property, so that the conditional probability of X_i only depends on its neighborhood $\mathcal{N}_i$. Moreover, due to the $\{\pm 1\}$ encoding, we have

$$p_i(x_i|\mathcal{N}_i) = \frac{p(x)}{\sum_{x_i \in \{\pm 1\}} p(x)} = \frac{e^{-U(x)/T}}{\sum_{x_i \in \{\pm 1\}} e^{-U(x)/T}} = \frac{1}{1 + e^{x_i \Delta_i U(x)/T}}, \tag{6}$$

where $\Delta_i U(x)$ is the difference between $U(x)$ and $U(\tilde{x}^i)$, where $\tilde{x}^i$ equals x except for the sign of x_i that has been changed. Since all terms in $U(x)$ and $U(\tilde{x}^i)$ agree except those containing x_i, and thus $\Delta_i U(x)$ only depends on $\mathcal{N}_i$, its computation can be further simplified.

2.2   The Markov Fitness Model

In the DEUM framework, probabilities are assigned to points in Ω proportionally to the value of f, and a model is chosen in the family of Gibbs distributions. By setting $T = 1$, in order to simplify the formulas, and combining Equations (2), (4) and (5), we have

$$p(x) \equiv \frac{f(x)}{\sum_\Omega f(x)} = \frac{e^{-\sum_{C \in \mathcal{C}} u_C(x_C)}}{\sum_\Omega e^{-\sum_{C \in \mathcal{C}} u_C(x_C)}},$$

that in particular is implied by setting

$$-\ln(f(x)) = \sum_{C \in \mathcal{C}} u_C(x_C), \tag{7}$$

i.e., when $U(x)$ is supposed to be a good model for f. This relationship between the energy function of the Gibbs distribution and f is called Markov Fitness Model (MFM) [3]. Notice that Equation (7) defines a probabilistic model of f.

Every u_C is defined over a subset of the variables in x according to the nodes in the maximal clique. Thus u_C admits a polynomial expansion as for f in Equation (1), and

$$- \ln(f(x)) = \sum_{C \in \mathcal{C}} \sum_{\alpha \in I_C} \theta_{\alpha,C} x^{\alpha}, \tag{8}$$

where the set of interactions identified by I_C depends on the variables in the maximal clique. Every $\theta_{\alpha,C} \in \mathbb{R}$ is a parameter associated to the expansion of u_C. By grouping similar terms and introducing a set M for all the monomials that appear in Equation (8), the expression can be simplified to

$$- \ln(f(x)) = \sum_{\alpha \in M} \theta_{\alpha} x^{\alpha}. \tag{9}$$

The statistical model used in the MFM in (9) can be written as an m-dimensional exponential family, with $m = \#(M)$,

$$p(x; \theta) = \exp \left\{ \sum_{\alpha \in M} \theta_{\alpha} x^{\alpha} - \psi(\theta) \right\}, \tag{10}$$

where $\psi(\theta) = \ln Z(\theta)$ is the normalizing factor and x^{α} are the sufficient statistics.

In order to reduce the number of parameters of the statistical model, further assumptions can be made in the choice of the monomials that appear in u_C in Equation (8). For instance, in the Ising DEUM algorithm [19], where the $\mathcal{G}$ is a toroidal 2D lattice and all maximal cliques have size 2, every $u_{ij}(x_i, x_j)$ takes the form of $\theta_{ij} x_i x_j$, so that all linear terms are not included among the sufficient statistics of the exponential model since they are not required to capture such class of Ising Spin Glass functions.

3 Sparse Learning of the Markov Fitness Model

To make the estimation of the MFM computationally feasible, we need to consider a reduced set of monomials as support statistics in (10) by imposing sparsity on the interactions pattern of the variables. This can be done a priori by making proper assumptions on the model, for instance limiting the neighborhood size of each variable or the total number of interactions in the graph. On the other hand sparsity can be obtained by employing machine learning techniques such as ℓ_1-regularization in the estimation of the model. For instance, Ravikumar et al. [16] address sparse model selection by solving a set of n ℓ_1-constrained logistic regression problems. Other approaches, such as [9], solve the problem of sparse structure learning by evaluating pseudo-likelihoods. In the literature of discrete EDAs, some related methods have been applied in L1BOA [24] and DEUM$_{\ell_1}$ [13].

3.1 Problem Statement and Theoretical Approach

Let consider the MFM in Equation (8), where the set of monomials identified by indices in M defines a set of interactions among the variables in f. In the

DEUM framework the coefficients θ are estimated by solving a linear system of equations. More in general the estimation of θ can be seen as a linear regression problem where, given a sample of observations, $-ln(f(x))$ corresponds to the response variable, and x^α to the covariates. By introducing a shrinkage regression technique in estimating the value of the parameters we can perform model selection by zeroing a subset of coefficients, and thus obtaining a sparse model. As a consequence, by applying a shrinkage technique in estimation, we can perform model selection at the same time of model fitting.

In particular we learn the MFM by solving an ℓ_1-constrained linear regression problem, also known as the Lasso [20]. The solution of the Lasso gives a sparse estimation of θ, hence, it selects a set of sufficient statistics for the statistical model of f in (10). The ℓ_1-constrained linear regression problem can be formalized as the minimization problem

$$\min_{\theta \in \mathbb{R}^m} \left\{ \frac{1}{2} || - \ln(f(x)) - \sum_{\alpha \in M} \theta_\alpha x^\alpha ||_2^2 + \lambda ||\theta||_1 \right\}, \tag{11}$$

where the first term represents the residual sum of squares, and the second term is an ℓ_1-constraint weighted by a control parameter λ, called *regularization parameter*. The value of the regularization parameter strongly affects the sparsity pattern of the vector of coefficients. Indeed, for $\lambda \to \infty$ all coefficients will vanish, while, for $\lambda \to 0$ the solution of the Lasso corresponds to the usual least square estimation of the MFM, which in general is not sparse.

To correctly dimension the value of the regularization parameter λ we refer to the asymptotic results presented in [5]. In particular dimensioning λ as

$$\lambda = K \sqrt{\frac{\log(m)}{N}}, \tag{12}$$

where K is a constant, m is the number of monomials in the exponential family, and N is the size of the sample used for the regression, guarantees that the correct correlations can be identified as $N \to \infty$. The same result has been applied in [16], where the authors show how N may depend on the topology of the graph. Such result is obtained under the hypothesis that the sample is i.i.d. from to an unknown probability distribution. Usually such hypothesis cannot be satisfied in black-box optimization, since f is unknown. In order to deal with this issue, we propose to perform ℓ_1-constrained linear regression over a subset of samples selected from a randomly generated initial sample according to the value of f. This procedure can only approximate an i.i.d. sample, but from our experiments it was sufficient to correctly reconstruct the topology of the MN.

A solution of the minimization problem defined in Equation (11) gives an estimation of the MFM that approximates a statistical model of f. However, the number of potential covariates in the regression problem grows exponentially with n, making the minimization problem computationally unfeasible. Indeed its complexity is bounded by $\mathcal{O}(m^3)$. Even under the hypothesis of limiting the maximum order of interactions to the second, we have $m = \binom{n}{2}$ and the problem does not scale very well. For this reason, we propose to apply a rough selection procedure to reduce the set of covariates in the MFM before solving the Lasso.

3.2   Taking Care of Dimensionality: Candidate Edges Reduction

In order to reduce the complexity of the ℓ_1-constrained linear regression problem we only consider pairwise interactions among variables, so that the MFM in Equation (8) can be represented as a complete pairwise graph $\mathcal{G}(\mathcal{V}, \mathcal{E})$, such that $(i, j) \in \mathcal{E}$ for every $j > i$. However, the number of terms to consider still grows quadratically with n. In order to further reduce the number of edges before solving the Lasso, we select first a subset with a computationally lighter but yet less accurate method based on a measure of correlation among the variables.

Similarly to [17], we evaluate Mutual Information (MI) for each pair of random variables in the original function. MI is a metric that measures the mutual dependence between random variables. Given a pair of discrete random variables X_i and X_j, their Mutual Information $\mathcal{I}$ is defined as

$$\mathcal{I}(X_i, X_j) = \sum_{x_i, x_j \in \{\pm 1\}} p_{ij}(x_i, x_j) \log \left(\frac{p_{ij}(x_i, x_j)}{p_i(x_i)p_j(x_j)} \right), \tag{13}$$

where p_i and p_j are the marginal probabilities, and p_{ij} is their joint probability. If the MI between X_i and X_j is higher than a given threshold, then we include the associated monomial during the solution of the Lasso; otherwise, we remove the edge (i, j) from the graph.

The overall procedure can be summarized as follow. Given a sample we compute the Mutual Information matrix A, which is symmetric and has dimension $n \times n$. Then, we proceed by removing from the initial complete graph every edge (i, j) whose Mutual Information a_{ij} is lower than the threshold $b \cdot \bar{a}$, where b is a weight coefficient and $\bar{a}$ is the average Mutual Information of variables in X.

Such procedure allows us to reduce the candidate set of interactions in the regression problem. The optimal value of b such that only real interactions are recovered, strongly depends from both the original function f and the sample. In principle, correctly dimensioning b represents a hard task to address. Higher values of b cut most of the edges, while less restrictive choices give a dense network, that in turn, is more like to contain all the relevant interactions of the problem together with many undesired ones. We choose non-restrictive values for b since the purpose of this preliminary selection is to reduce the number of edges rather then selecting a good model for X. Shakya et al. [17] reduce furthermore the density of the network by making hypothesis on the maximum neighborhood size of the nodes. We do not apply such step here, since we would like model selection to be as much independent as possible on prior knowledge about f, and leave to the Lasso the task of identifying the correct interactions.

Algorithm 1. sDEUM(P, b, s_{mi}, s_{ℓ_1})

---

1 Let $\mathcal{E}$ be the set of edges of a fully connected pairwise MN
2 Randomly generate initial sample $\mathcal{P}$ of size P
3 Evaluate f for each point in $\mathcal{P}$
4 Select a subset $\mathcal{P}_{mi}$ from $\mathcal{P}$ of size $s_{mi}P$
5 Compute the MI matrix A and the average MI $\bar{a}$ given $\mathcal{P}_{mi}$
6 Select a subset of edges $\mathcal{E}_{mi}$ from $\mathcal{E}$ according to A and the threshold $b\bar{a}$
7 Let $m = \#(\mathcal{E}_{mi})$
8 Select a subset $\mathcal{P}_{\ell_1}$ from $\mathcal{P}$ of size $N = s_{\ell_1}P$
9 Estimate a distribution p in the MFM, by solving the Lasso with covariates
   associated to $\mathcal{E}_{mi}$ and observations in $\mathcal{P}_{\ell_1}$, with $\lambda = K\sqrt{\frac{\log m}{N}}$, as in Eq. (11)
10 Sample p with the Gibbs sampler by evaluating conditional probabilities in Eq. (6)

---

4 Shrinkage DEUM Optimization Algorithm

In this section, in the light of the machine learning techniques described in the first part of the paper, we present the *Shrinkage Distribution Estimation Using Markov Networks* algorithm (sDEUM). The sDEUM algorithm consists of a black-box meta-heuristics able to learn from scratch a sparse probabilistic model of the function to be minimized. The use of ℓ_1-penalized linear regression allows sDEUM to shrinkage the size of the θ parameters of the MFM. Due to the ℓ_1 constraint and according to the size of the λ parameter, some coefficients are fixed to zero with high probability, so that an implicit model selection is performed while solving the regression problem. The model is then sampled to generate new points, possibly with optimal values for f.

Algorithm 1 summarizes the procedure implemented in sDEUM. The meta-heuristic is characterized by some parameters: the population size P, the MI coefficient b, the percentages of selection s_{mi} and s_{ℓ_1}, and the constant K. Two different subsets are selected from the same initial random sample: $\mathcal{P}_{mi}$ for the computation of the MI matrix and $\mathcal{P}_{\ell_1}$ for solving the Lasso, respectively. This choice allows a better sizing of observations for the two different estimation tasks. In both cases a truncation selection operator is employed, other policies are possible, but they are not investigated here.

Once the MFM is estimated, next step in the DEUM framework consists of sampling the distribution to search for the optimum of f. In sDEUM, as in [19,13], we use a Gibbs sampler, i.e., a Monte Carlo Markov Chain sampling method. The Gibbs sampler allows to generate instances with minimum values for the energy U of the Gibbs distribution by cooling the temperature during sampling. Refer to [19] for a presentation of the sampling schema employed in DEUM.

When the estimated model is good enough, repeatedly sampling the model with an adequate cooling schema yields with high probability the global optima of f. As a consequence, as in most of the DEUM framework algorithms, model learning in sDEUM is performed only once, and the learned model is repeatedly sampled using the Gibbs sampler (single generation approach).

5 Empirical Performance Analysis

In this section we present the results of an empirical analysis of the performance of sDEUM. We set up a series of experiments in order to evaluate the ability of the algorithm to reconstruct the correct set of interactions among the variables in the model and to find the global minimum of the function. In all the experiments, we evaluated the performance using the 3D Ising spin glass problem [2] as a benchmark, whose interaction structure is known and can be used to determine a set of model selection statistics, such as precision, recall and F1 score. First we analyse the behavior of sDEUM when its parameters are changed, in order to find the best configuration, then we compare its performance in solving the energy minimization problem with those of DEUMce [17], Simulated Annealing [1] and hBOA [15]. DEUMce and Simulated Annealing have been tuned to achieve best performance, while results of hBOA are taken from [14]. Since the difficulty of the 3D Ising spin glass problem may depend from the particular instance, we averaged the results over 10 different instances, and for each of them we run 30 independent executions of every algorithm. In order to simplify the experimental comparison and evaluation, sDEUM, DEUMce and Simulated Annealing were implemented within the Evoptool toolkit [21]. The source code of the algorithms and the Ising spin glass instances can be found on the Evoptool homepage[1].

5.1 Experimental Setting and 3D Ising Spin Glass Problem

In statistical physics, the Ising spin glass problem is an energy minimization problem in the space of binary configurations of a set of spins $\sigma = (\sigma_1, \ldots, \sigma_n)$, where each spin can be either up, $\sigma_i = +1$, or down, $\sigma_i = -1$. The optimal solutions, i.e., the ground states of the spin glass, are those configurations that minimize the energy function

$$E(\sigma) = -\sum_{i \in L} h_i \sigma_i - \sum_{i < j \in L} J_{ij} \sigma_i \sigma_j, \qquad (14)$$

where L is a toroidal lattice of n sites, while h_i and J_{ij} are coupling constants respectively of a single spin σ_i and a pair of spins (σ_i, σ_j). The difficulty of the problem is strongly related to the dimensionality of the lattice. Indeed, even if with particular choices of h and J the problem in 1D and 2D can be solved in polynomial time, it becomes NP-hard for all kind of coupling constants, as soon as it reaches the third dimension, and in particular when the edge degree of each vertex equals 6, see [2].

In our experiments we use spin glasses defined over a 3D grid with periodic boundaries [2]. The contribution to the energy given by singleton spins is not taken into account, therefore $h_i = 0$ for every spin. The instances of the problem are randomly generated with couplings J_{ij} that takes values in $\{\pm 1\}$ with equal probability. Instances of the problem and their optimal solutions are generated using the spin glass ground states server by the group of Prof. Michael Jünger[2].

---

[1] Available at http://airwiki.ws.dei.polimi.it/index.php/Evoptool
[2] Available at http://www.informatik.uni-koeln.de/ls_juenger/research/sgs/

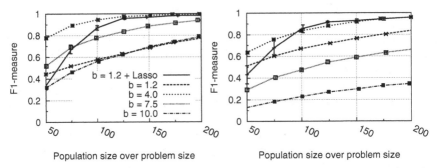

Fig. 1. F1 measure of model selection based on MI vs preliminary selection based on MI followed by ℓ_1-constrained linear regression for the Ising spin glass problem for $n = 64$, (left) 2D lattice; (right) 3D lattice

The sDEUM algorithm has been run for different values of its parameters: the sample size P, the threshold coefficient of MI b, and the percentages of selection s_{mi} and s_{l1}. After preliminary tests, the constant K in (12) has been fixed to the value of $K = 16$. In particular, to solve the ℓ_1-constrained linear regression problem, we employed the R package `lars` available on CRAN, implementing the Least Angle Regression (LARS) [6] algorithm.

The performance of sDEUM is compared with those of DEUMce, Simulated Annealing (SA) and the Hierarchical Bayesian Optimization Algorithm (hBOA). DEUMce is a DEUM algorithm with model learning capability based on the evaluation of the Mutual Information plus a structure refinement mechanism that bounds the maximum edge degree of each node. SA is a meta-heuristic characterized by the number P of starting points, by the initial temperature T and the cooling rate c of the Metropolis sampler. The hBOA algorithm is an optimization meta-heuristic belonging to the family of EDAs based on Bayesian Networks (BNs). At each generation, hBOA employs a niching mechanism to select individuals in the population. Once the BN has been estimated, the next population is sampled. Further details on the implementations of DEUMce, SA, and hBOA can be found in [17], [1], and [15], respectively.

The performance of the algorithm is evaluated according to a set of statistics that concern the F1 measure, the probability of success and the average number of evaluations of f required to find the first ground state at each execution. The F1 measure is defined as the harmonic mean of precision and recall, and the probability of success is computed as the rate of successful executions, i.e., the percentage of runs in which at least an optimal solution is found.

5.2   Impact of Learning Parameters

In order to consistently find the minimum of f, it is essential to recover a good statistical model for the variables in the problem, i.e., to learn most of the interactions present in f and to correctly estimate the value of their parameters.

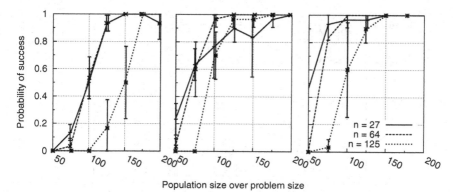

Fig. 2. Probability of success over normalized size of initial population (P/n). Benchmark: 3D Ising Spin Glass function, $n \in \{27, 64, 125\}$. sDEUM parameters: $s_{mi} = 0.3$; (left) $b = 1.5, s_{\ell_1} = 0.1$; (center) $b = 1.5, s_{\ell_1} = 0.3$; (right) $b = 1.1, s_{\ell_1} = 0.3$.

The threshold coefficient b of the preliminary selection based on MI, as well as the λ parameter of the Lasso, determine the sparsity level of the recovered structure. To correctly dimension b a preliminary tuning phase which depends on the problem is usually required, while, in contrast, the λ parameter can be chosen according to Equation (12) to ensure good theoretical performance.

In Fig. 1 we compared the model selection performances of our approach with those of the model selection based on MI, when solving the Spin Glass function with 2D and 3D structure and 64 variables. As we can see, in case of model selection based only on MI the results vary greatly according to the value of b. In contrast, when MI is followed by the ℓ_1-constrained regression, the choice of value for b is less problem dependent.

In Fig. 2 we can see the probability of success plotted against the size of the initial population P for problem size $n \in \{27, 64, 125\}$, and for different values of the threshold coefficient $b \in \{1.1, 1.5\}$, that determines how dense the MFM is after preliminary model selection based on MI. When $n = 27$ or $n = 64$, a less restrictive value of b provides better performance, see Fig. 2 (right); while when the size of the problem increases, $n = 125$, a higher value of the coefficient b results in earlier convergence, see Fig. 2 (center).

These results suggest that the value of b should increase with n. A possible explanation is given by the fact that the number of interactions grows linearly as $3n$ for a 3D lattice, while the number of total interactions is quadratic, for this reason a more restrictive choice of b helps to reduce the number of candidate interaction before the Lasso is solved.

In a black-box scenario an i.i.d. sample is not available to solve the lasso. Instead we choose a subset of the sample based on the value of the function f, and we compared the performance of the algorithm for different values of s_{ℓ_1}. In Fig. 2(left) and Fig. 2(center) we show the results for s_{ℓ_1} equal to 0.1 and 0.3, respectively. It is possible to notice that even if selection helps identify a good sample with respect to the random observations generated when the algorithms

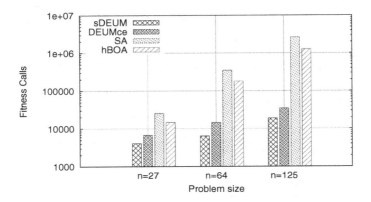

Fig. 3. Average number of evaluations of f (log scale) over problem size required to find first optimal solution with probability 1. Benchmark: 3D Ising Spin Glass function, $n \in \{27, 64, 125\}$. Algorithms: sDEUM, DEUMce, SA, hBOA.

starts, decreasing that percentage too much results in lower performances. This result suggests that if the sample after selection is not informative enough, then we have preliminary convergence and a larger population is necessary.

5.3   Analysis of Optimization Performance

In this section we compare the performance of sDEUM to find the ground states of the 3D Ising Spin Glass function with those of DEUMce, SA and hBOA. We analyze the results in terms of average number of function evaluations required to find the optimum with a probability of success equals to 1 for each algorithm on 10 instances of the problem. The parameters of sDEUM, DEUMce, and SA have been chosen after a preliminary tuning phase, while for the hBOA algorithm, the results are taken from [14][3].

The trend highlighted in Fig. 3 suggests that sDEUM algorithm requires a lower number of evaluations of f with respect to other meta-heuristics on this benchmark. Indeed, the overall number of evaluations for sDEUM appears to grow polynomially as $\mathcal{O}(n^{2.04})$, while the same metric grows as $\mathcal{O}(n^{2.16})$, $\mathcal{O}(n^{3.06})$ and $\mathcal{O}(n^{2.91})$, for DEUMce, SA and hBOA [14], respectively.

The lower requirements in terms of fitness evaluations of both sDEUM and DEUMce with respect to SA and hBOA are due to the single iteration approach characteristic of the DEUM algorithms. Indeed, most of the evaluations in DEUM concern the initial sample, before selection. Moreover, the shrinkage method used in sDEUM compared with the approach of DEUMce based on MI and structure refinement is able to recover a good model with a smaller sample and thus further reduce the number of function evaluations.

---

[3] The performance of hBOA in [14] are evaluated over a set of instances of the 3D Ising Spin Glass function different from our set but with the same setting: 3D toroidal lattice, $h_i = 0$, $J_{ij} \in \{\pm 1\}$.

6   Conclusions

In this paper we presented a novel approach to the estimation of the MFM based on ℓ_1-regularized linear regression. Our proposal allows to perform both model selection and model fitting at the cost of solving a single regularized linear regression problem. The advantage of this approach is due to theoretical results on the dimensioning of λ, that in contrast to the threshold parameter of Mutual Information, permits to be more robust and less problem dependent.

In the context of the DEUM framework, we developed a novel algorithm called sDEUM that estimates the MFM using an approach based on shrinkage regression. In order to make Lasso more efficient, sDEUM uses a preliminary σ coarse selection based on Mutual Information in order to find a candidate set of interactions for the MFM. This candidate set is then used to solve the regularized regression problem by means of Least Angle Regression (LARS). We showed that sDEUM is able to learn a probabilistic description of the objective function and to successfully use it to address optimization. We remark lower requirements in terms of number of evaluations of f to reach optimality with respect to other popular algorithms in the EDA framework. In particular, solving the Lasso defined on the MFM allows to reduce the sample size with respect to performing ℓ_1-regularized logistic regression on the conditional probability distribution of each variable, as previously done in DEUM_{ℓ_1} [13].

Acknowledgements. This work was supported by the META-X project, an Action de Recherche Concertée funded by the Scientific Research Directorate of the French Community of Belgium, and by the PRIN 2009 project ROAMFREE (Robust Odometry Applying Multisensor Fusion to Reduce Estimation Errors) funded by the Italian Ministry of University and Research.

References

1. Aarts, E., Korst, J.: Simulated annealing and Boltzmann machines: a stochastic approach to combinatorial optimization and neural computing. John Wiley & Sons, Inc., New York (1989)
2. Barahona, F.: On the computational complexity of Ising spin glass models. Journal of Physics A: Mathematical and General 15(10), 3241–3253 (1982)
3. Brown, D.F., Garmendia-Doval, A.B., McCall, J.A.W.: Markov Random Field Modelling of Royal Road Genetic Algorithms. In: Collet, P., Fonlupt, C., Hao, J.-K., Lutton, E., Schoenauer, M. (eds.) EA 2001. LNCS, vol. 2310, pp. 65–76. Springer, Heidelberg (2002)
4. Brownlee, A.E.I., McCall, J.A.W., Shakya, S.K., Zhang, Q.: Structure Learning and Optimisation in a Markov Network Based Estimation of Distribution Algorithm. In: Chen, Y.-P. (ed.) Exploitation of Linkage Learning. ALO, vol. 3, pp. 45–69. Springer, Heidelberg (2010)
5. Bunea, F., Tsybakov, A., Wegkamp, M.: Sparsity oracle inequalities for the lasso. Electronic Journal of Statistics 1, 169 (2007)
6. Efron, B., Hastie, T., Johnstone, I., Tibshirani, R.: Least angle regression. The Annals of Statistics 32(2), 407–499 (2004)

7. Hammersley, J., Clifford, P.: Markov fields on finite graphs and lattices (1971) (unpublished)
8. Hansen, N., Müller, S., Koumoutsakos, P.: Reducing the time complexity of the derandomized evolution strategy with covariance matrix adaptation (CMA-ES). Evolutionary Computation 11(1), 1–18 (2003)
9. Höfling, H., Tibshirani, R.: Estimation of sparse binary pairwise markov networks using pseudo-likelihoods. The Journal of Machine Learning Research 10, 883–906 (2009)
10. Larrañaga, P., Lozano, J.A. (eds.): Estimation of Distribution Algoritms. A New Tool for evolutionary Computation. Number 2 in Genetic Algorithms and Evolutionary Computation. Springer (2001)
11. Malagò, L., Matteucci, M., Pistone, G.: Optimization of pseudo-boolean functions by stochastic natural gradient descent. In: 9th Metaheuristics International Conference, MIC 2011 (2011)
12. Malagò, L., Matteucci, M., Pistone, G.: Towards the geometry of estimation of distribution algorithms based on the exponential family. In: Proceedings of the 11th Workshop on Foundations of Genetic Algorithms, FOGA 2011, pp. 230–242. ACM, New York (2011)
13. Malagò, L., Matteucci, M., Valentini, G.: Introducing ℓ_1-regularized logistic regression in Markov Networks based EDAs. In: Proceedings of the IEEE Congress on Evolutionary Computation, CEC 2011. IEEE Press (2011)
14. Pelikan, M., Goldberg, D., Ocenasek, J., Trebst, S.: Robust and scalable black-box optimization, hierarchy, and ising spin glasses. Technical report, Illinois Genetic Algorithms Laboratory, IlliGAL (2003)
15. Pelikan, M., Goldberg, D.E.: A hierarchy machine: Learning to optimize from nature and humans. Complexity 8(5), 36–45 (2003)
16. Ravikumar, P., Wainwright, M.J., Lafferty, J.D.: High-dimensional Ising model selection using ℓ_1-regularized logistic regression. The Annals of Statistics 38(3), 1287–1319 (2010)
17. Shakya, S., Brownlee, A., McCall, J., Fournier, F., Owusu, G.: A fully multivariate DEUM algorithm. In: IEEE Congress on Evolutionary Computation (2009)
18. Shakya, S., McCall, J.: Optimization by Estimation of Distribution with DEUM framework based on Markov random fields. International Journal of Automation and Computing 4(3), 262–272 (2007)
19. Shakya, S., McCall, J., Brown, D.: Solving the Ising spin glass problem using a bivariate EDA based on Markov random fields. In: IEEE Congress on Evolutionary Computation, pp. 908–915 (2006)
20. Tibshirani, R.: Regression shrinkage and selection via the lasso. Journal of the Royal Statistical Society. Series B (Methodological), 267–288 (1996)
21. Valentini, G., Malagò, L., Matteucci, M.: Evoptool: an extensible toolkit for evolutionary optimization algorithms comparison. In: Proceedings of IEEE World Congress on Computational Intelligence, pp. 2475–2482 (July 2010)
22. Winkler, G.: Image Analysis, Random Fields and Dynamic Monte Carlo Methods: A Mathematical Introduction, 2nd edn. Springer (2003)
23. Wolsey, L.A.: Integer Programming. Wiley Interscience (1998)
24. Yang, J., Xu, H., Cai, Y., Jia, P.: Effective structure learning for eda via l1-regularized bayesian networks. In: Proceedings of the 12th Annual Conference on Genetic and Evolutionary Computation, GECCO 2001, pp. 327–334. ACM (2010)
25. Zlochin, M., Birattari, M., Meuleau, N., Dorigo, M.: Model-based search for combinatorial optimization: A critical survey. Annals of Operations Research 131(1-4), 375–395 (2004)

Vehicle Routing and Adaptive Iterated Local Search within the *HyFlex* Hyper-heuristic Framework

James D. Walker[1], Gabriela Ochoa[1],
Michel Gendreau[2], and Edmund K. Burke[3]

[1] School of Computer Science, University of Nottingham, UK
[2] CIRRELT, University of Montreal, Canada
[3] Department of Computing Science and Mathematics, University of Stirling, UK

Abstract. HyFlex (Hyper-heuristic Flexible framework) [15] is a software framework enabling the development of domain independent search heuristics (hyper-heuristics), and testing across multiple problem domains. This framework was used as a base for the first *Cross-domain Heuristic Search Challenge*, a research competition that attracted significant international attention. In this paper, we present one of the problems that was used as a hidden domain in the competition, namely, the capacitated vehicle routing problem with time windows. The domain implements a data structure and objective function for the vehicle routing problem, as well as many state-of- the-art low-level heuristics (search operators) of several types. The domain is tested using two adaptive variants of a multiple-neighborhood iterated local search algorithm that operate in a domain independent fashion, and therefore can be considered as hyper-heuristics. Our results confirm that adding adaptation mechanisms improve the performance of hyper-heuristics. It is our hope that this new and challenging problem domain can be used to promote research within hyper-heuristics, adaptive operator selection, adaptive multi-meme algorithms and autonomous control for search algorithms.

1 Introduction

There is an increasing and renewed research interest in developing heuristic search methods that are more generally applicable [7,8]. The goal is to reduce the role of the human expert in the design of effective heuristic methods to solve hard computational search problems. Researchers in this field, however, are often constrained on the number of problem domains in which to test their adaptive, self-configuring algorithms. This can be explained by the inherent time and effort required to implement a new problem domain, including efficient data structures and search operators; initialisation routines; the objective function; and an varied a set of benchmark instances.

The HyFlex framework has been recently proposed to assist researchers in hyper-heuristics and autonomous search control. HyFlex features a common

Y. Hamadi and M. Schoenauer (Eds.): LION 6, LNCS 7219, pp. 265–276, 2012.
© Springer-Verlag Berlin Heidelberg 2012

software interface for dealing with different combinatorial optimisation problems, and provides the algorithm components that are problem specific.In this way, it simultaneously liberates algorithm designers from needing to know the details of the problem domains; and it prevents them from incorporating additional problem specific information in their algorithms. Efforts can instead be focused on designing high-level strategies to intelligently combine the provided problem-specific algorithmic components. The framework also served as the basis of an an international research competition: the first *Cross-domain Heuristic Search Challenge (CHeSC 2011)*[1] [5], which successfully attracted the interest and participation of over 40 researchers at universities and academic institutions across six continents. This competition differed from other challenges in heuristic search and optimisation in that the goal was to design a search algorithm that works well, not only across different instances of the same problem, but also across different problem domains. It can be considered as the first *Decathlon* challenge of search heuristics. For testing purposes, four domain modules were provided to the participants, each containing around 10 low-level heuristics of the types discussed below, and 10 instances of medium to hard difficulty. The domains provided were: permutation flowshop, one dimensional bin packing, Boolean satisfiability (MAX-SAT) and personnel scheduling.

For calculating the final competition scores, two additional (hidden) domains were implemented and used: the traveling salesman problem, and the capacitated vehicle routing problem with time windows. This paper describes the design of the vehicle routing domain. It also tests this domain using two adaptive variants of a multiple neighborhood iterated local search algorithm. The first variant was the best performing algorithm in [6], while the second is a modification that improves the local search stage of the algorithm by incorporating an adaptive mechanism.

The next section briefly overviews the HyFlex framework, whilst section 3 describes the design of the vehicle routing domain. Section 4 describes the adaptive iterated local search hyper-heuristics that have been developed. Section 5 summarises the experiments and results and section 6 concludes and discusses our contributions.

2   The HyFlex Framework

HyFlex (Hyper-heuristics Flexible framework) [15] is a Java object oriented framework for the implementation and comparison of different iterative general-purpose (domain independent) heuristic search algorithms (also called hyper-heuristics). The framework appeals to modularity and is inspired by the notion of a domain barrier between the low-level heuristics and the hyper-heuristic [9,7]. HyFlex provides a software interface between the hyper-heuristic and the problem domain layers, thus enabling a clearly defined separation and communication protocol between the domain specific and the domain independent algorithm components.

---

[1] http://www.asap.cs.nott.ac.uk/external/chesc2011/

HyFlex extends the conceptual framework discussed in [9,7] in that a population of solutions (instead of a single incumbent solution) is maintained in the problem layer. Also, a richer variety of low-level heuristics is provided. Another relevant antecedent to HyFlex is PISA [2], a text-based software interface for multi-objective evolutionary algorithms, which divides the implementation of an evolutionary algorithm into an application-specific part and an algorithm-specific part. HyFlex differs from PISA in that its interface is not text-based but is instead given by an abstract Java class. Moreover, HyFlex provides a rich variety of combinatorial optimisation problems including real-world instance data. Each HyFlex problem domain module consists of:

1. A user-configurable memory (a population) of solutions, which can be managed by the hyper-heuristic.
2. A routine to initialise randomised solutions in the population.
3. A set of heuristics to modify solutions classified into four groups:

   mutational : makes a (randomised) modification to the current solution.

   ruin-recreate : destroys part of the solution and rebuilds it using a constructive procedure.

   local search : searches in the *neighbourhood* of the current solution for an improved solution.

   crossover : takes two solutions, combines them and returns a new solution.
4. A varied set of instances that can be easily loaded.
5. A fitness function, which can be called to obtain the objective value of any member of the population. HyFlex problem domains are always implemented as minimisation problems, so a lower fitness is always superior. The fitness of the best solution found so far in the run is stored and can be easily obtained.
6. Two parameters: α and β, $(0 <= [\alpha, \beta] <= 1)$, which are the 'intensity' of mutation and 'depth of search', respectively, that control the behaviour of some search operators.

Currently, six problem domain modules are implemented (which can be downloaded from the CHeSC 2011 website [14]). These are the original four test domains: permutation flow shop, one-dimensional bin packing, maximum satisfiability (MAX-SAT) and personnel scheduling; and the two additional (hidden) domains used for the competition: the traveling salesman problem and the vehicle routing problem with time windows.

3   The Vehicle Routing HyFlex Domain

The vehicle routing problem was implemented using the HyFlex software framework interface[2]. Specifically, a java class derived from the HyFLex *ProblemDomain* class was implemented, following the descriptions below.

---

[2] The API documentation can be found at:
   http://www.asap.cs.nott.ac.uk/external/chesc2011/javadoc/help-doc.html

3.1   Problem Formulation

The vehicle routing problem can be described as the task of meeting the demand of all customers, using as few vehicles as possible, and satisfying all constraints, such as vehicle capacity. Furthermore, the variant of the vehicle routing problem which is modelled here is the vehicle routing problem with time windows. This variant includes extra time window constraints, whereby a customer must be served between two time points for a solution to be valid.

There is a base location, or depot, from where each vehicle must start and end its route. A route is a series of location visits for a single vehicle. The objective function for this domain balances the dual objectives of minimising the number of vehicles needed and minimising the total distance travelled. It was defined as follows:

$$objective function = c \times numVehicles + distance,$$

where c is a constant that we empirically set to 1000 to give higher importance to the number of vehicles in a solution.

The problem domain offers a set of operators to initialise and modify solutions which are commonly found in effective meta-heuristics and a set of benchmarks instances (due to [21]) that are readily available.

3.2   Solution Initialisation

The initialisation method is stochastic, generating solutions based upon the given seed. Customers are inserted into the solution one at a time, with the customer to be inserted being chosen by a metric measuring the proximity of a customer in terms of distance and time to the most recently inserted customer. The metric also includes a stochastic element to ensure different solutions are generated. If it is not possible to insert any customer into the current route, a new route is generated. This process is repeated until all customers have been scheduled.

3.3   Low Level Heuristics

The module includes 12 low level heuristics $h_1, \ldots, h_{12}$ across the four categories of heuristics, as specified within HyFlex. They are described below, sorted by category.

Mutational Heuristics

h_1: **Two-opt**[3]. Swaps two adjacent customers within a single route.

h_2: **Or-opt**[16]. Moves two adjacent customers to a different place, within a single route.

h_3: **Shift**[18]. Moves a single customer from one route to another.

h_4: **Interchange**[18]. Swaps two customers from different routes.

Ruin and Recreate Heuristics

h_5: **Time-based radial ruin**[19]. Chooses a number of customers to be removed from the solution, based upon the proximity of their time window to a given time. Each remaining customer is inserted into the best route possible, based on a metric of distance and time proximity. If it is not feasible to insert into any route, a new route is created.

h_6: **Location-based radial ruin**[19]. Chooses a number of customers to be removed from the solution, based upon the proximity of their location to a given location. Each remaining customer is inserted into the best route possible, based on a metric of distance and time proximity. If it is not feasible to insert into any route, a new route is created.

Local Search Heuristics. These heuristics implement 'first-improvement' local search operators. In each iteration, a neighbour is generated, and it is accepted immediately if it has superior or equal fitness. If the neighbor is worse, then the change is not accepted. The *depth-of-search parameter* (see section 2) controls the number of iterations to attempt to obtain an improved solution.

h_7: **Shift**[18]. Moves a customer from one Route, to another providing that the new position yields an improvement in objective function score.

h_8: **Interchange**[18]. Swaps two customers from different routes, providing that the new routes yield an improvement in objective function score.

h_9: **Two-opt***[17]. Takes the end sections of two routes, and swaps them to create two new routes.

h_{10}: **GENI**[11]. A customer is taken from one route, and placed into another route, between the two customers of that route which are closest to it. Re-optimisation is then performed on the route.

Crossover Heuristics

h_{11}: **Combine.** A random percentage of routes (between 25% and 75%) are kept from one of the solutions (chosen randomly.) Then all routes which don't contain any conflicts with the routes already chosen are taken from the other solution. Finally, all unrouted customers are inserted into the solution.

h_{12}: **Longest Combine.** All routes from both solutions are taken and ordered by length (here length is defined as the number of customers served in a route.) The routes are taken from longest to shortest, providing there are no customer conflicts. Then, all unrouted customers are inserted into the solution.

3.4   Problem Instances

The problem instances provided in this module are taken from two sources. The first is the Solomon data set of 100 customer problems. The second is the Gehring and Homberger data set of 1000 customer problems. For both data sets, there are three types of instances. These are:

R: **Random**. The customers' locations are determined in a uniformly random way.

C: **Clustered**. The customers' locations are grouped in a number of clusters.

CR: **Clustered Random**. The customers' locations are in a mix of random and clustered locations.

4   Adaptive Iterated Local Search Hyper-heuristics

Iterated local search is a relatively simple but successful algorithm. It operates by iteratively alternating between applying a move operator to the incumbent solution and restarting local search from the perturbed solution. This search principle has been rediscovered multiple times, within different research communities and with different names [1,13]. The term *iterated local search* (ILS) was proposed in [12]. The algorithms compared in this article can be considered as ILS with multiple perturbation heuristics and multiple local search heuristics. They can be considered to be hyper-heuristics as they both coordinate several low-level heuristics and operate in a domain-independent fashion. Three variants were considered as described below.

4.1   The Baseline ILS Hyper-heuristic

The ILS implementation proposed in [4] contains a perturbation stage during which a neighborhood move is selected uniformly at random (from the available pool of mutation and ruin-recreate heuristics) and applied to the incumbent solution. This perturbation phase is then followed by an improvement phase, which works as follows. Each of the local search heuristics is independently applied to the incumbent solution. Providing at least one of the applications has yielded an improvement, then the application resulting in the greatest improvement in objective function is kept. The process is then repeated until no improvement in objective function value is found. If the resulting new solution is better than the original solution then it replaces the original solution, otherwise the new solution is simply discarded. This last stage corresponds to a greedy (only improvements) acceptance criterion. The pseudo-code of this iterated local search algorithm is shown below (Algorithm 1), notice that this differs from traditional implementations of ILS in that multiple heuristics are used in both the improvement and perturbation stages. We refer to this algorithm as *Rnd-ILS*.

4.2   The Adaptive ILS Hyper-heuristics

The adaptive versions of the base-line ILS hyper-heuristic described above, incorporate adaptive mechanisms in the perturbation and/or the improvement stages. The most successful adaptive ILS hyper-heuristic suggested in [6] implements an online learning mechanisms for selecting the move operators in the perturbation stage, instead of selecting then uniformly at random at each iteration. Specifically, it implements an adaptive operator selection mechanisms. As discussed in

Algorithm 1. *Iterated Local Search Hyper-heuristic.*

s_0 = GenerateInitialSolution
s^* = ImprovementStage(s_0)
repeat
    s' = PerturbationStage (s^*)
    $s^{*'}$ = ImprovementStage(s')
    if $f(s^{*'}) < f(s*)$ **then**
        $s^* = s^{*'}$
    end if
until time limit is reached

[10], an adaptive operator selection scheme consists of two components: a *credit assignment* mechanisms and a *selection* mechanism. The algorithm proposed in [6] used *extreme value* credit assignment, which is based on the principle that infrequent, yet large, improvements in the objective score are likely to be more effective than frequent, small improvements [10]. It rewards operators which have had a recent large positive impact on the objective score, while consistent operators that only yield small improvements receive less credit, and ultimately have less chance of being chosen. Following the application of an operator to the problem, the change in objective score is added to a window of size W, which works on a FIFO mechanism. The credit for any operator is the maximum score within the window. Window size plays an important part in the mechanism. If it is too small then the range of information on offer is narrowed, meaning that useful operators are missed. If it is too large then information is considered from many iterations ago, when the position in the search space might have meant that the operator performed differently to how it would at the latest iteration. However, the window size is the only parameter that needs to be tuned, which is a desirable property when the goal is to achieve robust and general algorithms. After testing several values of (W), we decided upon a value of 25. The credit assignment mechanism is combined with a selection strategy that uses the accumulated credits to select the operator to apply in the current iteration. Operator selection strategies in the literature, generally assign a probability to each operator and use a roulette wheel-like process to select the operator according to them. We use here one of these rules, namely, *adaptive pursuit*, originally proposed for learning automata and adapted to the context of operator selection in [22]. With this method, at each time step, the operator with maximal reward is selected and its selection probability is increased (follows a winner-take-all strategy.), while the other operators have their selection probability decreased. We refer to this adaptive ILS hyper-heuristic as *Ad-ILS*.

The variant proposed in this article keeps the adaptive selection of operators in the perturbation stage described above, but modifies the improvement stage by incorporating a simple adaptive mechanisms that considers the past performance of the local search heuristics. The mechanism works as follows: each heuristic has a score attributed to it, which is updated after each application of that heuristic. The score corresponds to the mean improvement in objective function obtained from that heuristic's applications (from all applications across whole search). These scores are then used to order the local search heuristics, with the best performing heuristics being placed to the front of the list. The local search heuristics are then applied in sequence following this order. This new improvement stage is illustrated below (see Algorithm 2). We refer to the ILS hyper-heuristic with this modified component as *AdOr-ILS*.

Algorithm 2. *Ordered ImprovementStage.*

---

repeat
    $ls \leftarrow$ OrderLocalSearches(scores)
    for $i = 0 \rightarrow numLocalSearchers$, in the order ls **do**
      $s' =$ LocalSearch(s',i)
      $scores \leftarrow UpdateScores$
    end for
until no improvement found

---

5 Experiments and Results

For testing the three algorithm variants described above, *Rnd-ILS*, *Ad-ILS* and *AdOr-ILS*, 10 instances were chosen, representing a range of instance types from both the Solomon and Gehring-Homberger data sets (see Table 1). These 10 instances are those currently available in the version of the HyFlex software used for the competition (which can be downloaded from the CHeSC 2011 website [14]).

Twenty runs were performed for each instance and algorithm. Following the experimental set up used in the CHeSC competition, the running time was set to 10 CPU minutes. The machine running the tests has a 2.27GHz Intel(R) Core(TM) i3 CPU and 4GB RAM.

Table 1. Capacitated vehicle routing problem instances, taken from [20]

Instance	name	no. vehicles	vehicle capacity
0	Solomon/RC/RC207	25	1000
1	Solomon/R/R101	25	200
2	Solomon/RC/RC103	25	200
3	Solomon/R/R201	25	1000
4	Solomon/R/R106	25	200
5	Homberger/C/C1-10-1	250	200
6	Homberger/RC/RC2-10-1	250	1000
7	Homberger/R/R1-10-1	250	200
8	Homberger/C/C1-10-8	250	200
9	Homberger/RC/RC1-10-5	250	200

Table 2 shows the average and standard deviation of the best objective function value at the end of the run, from the ten runs per instance. The adaptive ILS hyper-heuristic that incorporates both adaptive operator selection and adaptive ordering of the local searchers (*AdOr-ILS*)outperforms the other two variants in 9 out otf the 10 instances. Only for one of of the smallest and less constrained instances (instance 1), it is the base-line ILS hyper-heuristic the one producing the best performance. It seems that the added complexity of the adaptive mechanisms does not help in this case. The experiments also suggest that for the smaller Solomon instances (instances 0 to 4), the difference in performance among the competing algorithms is less noticeable.

Table 2. Vehicle routing results for the 10 instances in Table 1. The entries account for the average and standard deviation of objective function values (out of 20 runs).

instance	*AdOr-ILS*	*Ad-ILS*	*Rnd-ILS*
0	**5281.71**$_{334.614}$	5406.48$_{404.159}$	5292.43$_{337.186}$
1	21291.89$_{482.56}$	21212.60$_{509.28}$	**21054.87**$_{500.73}$
2	**13605.03**$_{451.64}$	13932.67$_{616.29}$	13827.54$_{516.39}$
3	**6564.42**$_{554.77}$	7055.26$_{748.15}$	6760.62$_{597.41}$
4	**14280.79**$_{319.54}$	14549.22$_{449.1}$	14600.09$_{471.7}$
5	**155305.46**$_{6154.24}$	163041.76$_{11226.39}$	180301.07$_{2921.14}$
6	**77302.72**$_{3384.83}$	79175.63$_{3431.57}$	82316.66$_{2326.49}$
7	**163177.74**$_{2100.09}$	164341.16$_{1550.06}$	169729.31$_{1721.3}$
8	**158941.93**$_{2460.71}$	163332.72$_{4314.93}$	172007.42$_{2055.46}$
9	**149447.68**$_{1500.9}$	150276.89$_{1644.28}$	153648.66$_{1079.4}$

The boxplots shown in Figure 1 illustrate the magnitude and distribution of the best objective values for 4 of the harder Homberger instances (instances 5, 7, 8 and 9). Each plot summarises the result of 20 runs from each algorithm. It can be clearly observed that the best performing hyper-heuristic is *AdOr-ILS*, followed by *Ad-ILS*. The base-line non-adaptive ILS hyper-heuristic is the less competitive in these challenging instances.

To test statistical significance between the performances of Ad-ILS and the new variant, AdOr-ILS, the two sided Wilcoxon Signed Rank test has been used. The test is performed at the 95% confidence level, where a p value of less than 0.05 indicates a rejection of the null hypothesis - this being that there is no difference between the results. The following table shows the p values for each instance. From the table, we can see that in seven out of the ten instances, there is a statistical difference between the results.

Table 3. p-values resulting from comparisons of Ad-ILS and AdOr-ILS. Values of less than 0.05 (shown in bold) indicate statistical significance.

Instance	0	1	2	3	4	5	6	7	8	9
p-value	**0.017**	0.455	**0.023**	**0.021**	**0.005**	**0.04**	0.086	**0.048**	**0.005**	0.126

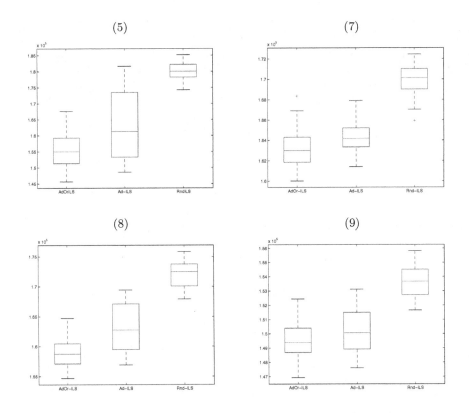

Fig. 1. Distribution of objective function values for the harder Homberger instances (instances 5, 7, 8 and 9 from Table 1)

6   Conclusions

This paper summarises the design of the capacitated vehicle routing domain for the HyFlex hyper-heuristic framework. The domain makes use of a large number of low-level heuristics, many of which are considered to be among the state-of the-art in current vehicle routing research. Considering this, the domain provides the opportunity to develop domain independent high-level algorithms or hyper-heuristics for the vehicle routing problem without the need to develop the low-level heuristics and underlying data structures. The domain was used as a hidden domain for the CHeSC competition and has been made freely available. It is our aim that it will be utilised within the hyper-heuristic, adaptive operator selection, adaptive multi-meme algorithms, and autonomous control for search algorithms research themes in intelligent optimisation.

The vehicle routing domain was used to test adaptive hyper-heuristics that can be considered as multiple-neighborhood iterated local search algorithms. This algorithmic scheme has proved to have good generalisation abilities. The inclusion of adaptive mechanisms both at the perturbation stage and at the improvement stage of the ILS framework led to an improved performance on the challenging vehicle routing instances.

Future work will extend the the vehicle routing domain by including new low-level heuristics, and additional problem instances. New hyper-heuristics can be implemented using HyFlex. For example, we are currently exploring the use of a population and the crossover heuristics, which were not employed by our ILS hyper-heuristics. We are also exploring mechanisms for adapting the heuristic parameters provided in the framework, namely, *intensity-of-mutation* and *depth-of-search*, which have an important impact in the hyper-heuristic performance.

Finally, HyFlex can be extended to include new domains, additional instances and operators in existing domains; and multi-objective and dynamic problems. The current software interface can also be extended to incorporate additional feedback information from the domains to guide the adaptive search controllers. It is our vision that the HyFlex framework will continue to facilitate and increase international interest in developing domain independent and adaptive heuristic search methodologies, that can find wider application in practice.

References

1. Baxter, J.: Local optima avoidance in depot location. Journal of the Operational Research Society 32, 815–819 (1981)
2. Bleuler, S., Laumanns, M., Thiele, L., Zitzler, E.: PISA – A Platform and Programming Language Independent Interface for Search Algorithms. In: Fonseca, C.M., Fleming, P.J., Zitzler, E., Deb, K., Thiele, L. (eds.) EMO 2003. LNCS, vol. 2632, pp. 494–508. Springer, Heidelberg (2003)
3. Braysy, O., Gendreau, M.: Vehicle routing problem with time windows, part i: Route construction and local search algorithms. Transportation Science (2005)
4. Burke, E.K., Curtois, T., Hyde, M., Kendall, G., Ochoa, G., Petrovic, S., Vazquez-Rodriguez, J.A., Gendreau, M.: Iterated local search vs. hyper-heuristics: Towards general-purpose search algorithms. In: IEEE Congress on Evolutionary Computation (CEC 2010), Barcelona, Spain, pp. 3073–3080 (July 2010)
5. Burke, E.K., Gendreau, M., Hyde, M., Kendall, G., McCollum, B., Ochoa, G., Parkes, A.J., Petrovic, S.: The Cross-Domain Heuristic Search Challenge – An International Research Competition. In: Coello, C.A.C. (ed.) LION 5. LNCS, vol. 6683, pp. 631–634. Springer, Heidelberg (2011)
6. Burke, E.K., Gendreau, M., Ochoa, G., Walker, J.D.: Adaptive iterated local search for cross-domain optimisation. In: Proceedings of the 13th Annual Conference on Genetic and Evolutionary Computation, GECCO 2011, pp. 1987–1994. ACM, New York (2011)
7. Burke, E.K., Hart, E., Kendall, G., Newall, J., Ross, P., Schulenburg, S.: Hyper-heuristics: An emerging direction in modern search technology. In: Glover, F., Kochenberger, G. (eds.) Handbook of Metaheuristics, pp. 457–474. Kluwer (2003)
8. Burke, E.K., Hyde, M., Kendall, G., Ochoa, G., Ozcan, E., Woodward, J.: A Classification of Hyper-heuristic Approaches. In: Handbook of Metaheuristics. International Series in Operations Research & Management Science, vol. 146, ch. 15, pp. 449–468. Springer (2010)
9. Cowling, P.I., Kendall, G., Soubeiga, E.: A Hyperheuristic Approach to Scheduling a Sales Summit. In: Burke, E., Erben, W. (eds.) PATAT 2000. LNCS, vol. 2079, pp. 176–190. Springer, Heidelberg (2001)

10. Fialho, Á., Da Costa, L., Schoenauer, M., Sebag, M.: Extreme Value Based Adaptive Operator Selection. In: Rudolph, G., Jansen, T., Lucas, S., Poloni, C., Beume, N. (eds.) PPSN X. LNCS, vol. 5199, pp. 175–184. Springer, Heidelberg (2008)
11. Gendreau, M., Hertz, A., Laporte, G.: A new insertion and postoptimization procedures for the traveling salesman problem. Operations Research (1992)
12. Lourenco, H.R., Martin, O., Stutzle, T.: Iterated Local Search, pp. 321–353. Kluwer Academic Publishers, Dordrecht (2002)
13. Martin, O., Otto, S.W., Felten, E.W.: Large-step Markov chains for the TSP incorporating local search heuristics. Operations Research Letters 11(4), 219–224 (1992)
14. Ochoa, G., Hyde, M.: The Cross-domain Heuristic Search Challenge, CHeSC 2011 (2011), http://www.asap.cs.nott.ac.uk/external/chesc2011/
15. Ochoa, G., Hyde, M., Curtois, T., Vazquez-Rodriguez, J.A., Walker, J., Gendreau, M., Kendall, G., McCollum, B., Parkes, A.J., Petrovic, S., Burke, E.K.: HyFlex: A Benchmark Framework for Cross-Domain Heuristic Search. In: Hao, J.-K., Middendorf, M. (eds.) EvoCOP 2012. LNCS, vol. 7245, pp. 136–147. Springer, Heidelberg (2012)
16. Or, I.: Traveling salesman-type combinatorial problems and their relation to the logistics of regional blood banking. PhD thesis, Northwestern University, Evanston, IL (1976)
17. Potvin, J.-Y., Rousseau, J.-M.: An exchange heuristic for routeing problems with time windows. The Journal of the Operational Research Society (1995)
18. Savelsbergh, M.W.P.: The vehicle routing problem with time windows: Minimizing route duration. Informs Journal on Computing 4(2), 146–154 (1992)
19. Schrimpf, G., Schneider, J., Stamm-Wilbrandt, H., Dueck, G.: Record breaking optimization results using the ruin and recreate principle. Journal of Computational Physics (2000)
20. SINTEF. VRPTW benchmark problems, on the SINTEF transport optimisation portal (2011), http://www.sintef.no/Projectweb/TOP/Problems/VRPTW/
21. Taillard, E.: Benchmarks for basic scheduling problems. European Journal of Operational Research 64(2), 278–285 (1993)
22. Thierens, D.: An adaptive pursuit strategy for allocating operator probabilities. In: Proceedings of the 2005 Conference on Genetic and Evolutionary Computation, GECCO 2005, pp. 1539–1546. ACM, New York (2005)

Quasi-elementary Landscapes
and Superpositions of Elementary Landscapes

Darrell Whitley[1] and Francisco Chicano[2]

[1] Dept. of Computer Science, Colorado State University, Fort Collins CO, USA*
whitley@cs.colostate.edu
[2] Dept. de Lenguajes y Ciencias de la Computación, University of Málaga, Spain**
chicano@lcc.uma.es

Abstract. There exist local search landscapes where the evaluation
function is an eigenfunction of the graph Laplacian that corresponds
to the neighborhood structure of the search space. Problems that display this structure are called "Elementary Landscapes" and they have a
number of special mathematical properties. The term "Quasi-elementary
landscapes" is introduced to describe landscapes that are "almost" elementary; in quasi-elementary landscapes there exists some efficiently
computed "correction" that captures those parts of the neighborhood
structure that deviate from the normal structure found in elementary
landscapes. The "shift" operator, as well as the "3-opt" operator for the
Traveling Salesman Problem landscapes induce quasi-elementary landscapes. A local search neighborhood for the Maximal Clique problem
is also quasi-elementary. Finally, we show that landscapes which are
a superposition of elementary landscapes can be quasi-elementary in
structure.

1 Introduction

Grover [4] originally observed that there exist neighborhoods for Traveling Salesman Problem (TSP), Graph Coloring, Min-Cut Graph Partitioning, Weight
Partitioning, as well as Not-All-Equal-SAT that can be modeled using a wave
equation borrowed from mathematical physics. Stadler [7] named this class of
problems "elementary landscapes" and showed that if a landscape is elementary,
the objective function is an eigenfunction of the Laplacian matrix that describes
the connectivity of the neighborhood graph representing the search space. Whitley and Sutton developed a "component" based model of elementary landscapes
that makes it easy to identify elementary landscapes [13]. In many cases, the

---

* This research was sponsored by the Air Force Office of Scientific Research, Air Force
Materiel Command, USAF, under grant number FA9550-11-1-0088. The U.S. Gov-
ernment is authorized to reproduce and distribute reprints for Governmental pur-
poses notwithstanding any copyright notation thereon.
** It was also partially funded by the Spanish Ministry of Science and Innovation and
FEDER under contract TIN2008-06491-C04-01 (the M* project) and the Andalusian
Government under contract P07-TIC-03044 (DIRICOM project).

Y. Hamadi and M. Schoenauer (Eds.): LION 6, LNCS 7219, pp. 277–291, 2012.
© Springer-Verlag Berlin Heidelberg 2012

components are weights in a cost matrix. In the case of pseudo-Boolean functions, the components can also be the weight coefficients of a polynomial form of the cost function. Let the set of components be denoted by C; a solution x will also denote the subset of components that contribute to the evaluation of x so that the sum of the components that contribute to x is the same as the evaluation of x, denoted by the evaluation function $f(x)$. Finally, let $C - x$ denote the subset of components that do not contribute to the evaluation of solution x. Note that the sum of the components in $C - x$ is computed by $(\sum_{w \in C} w) - f(x)$. Another precondition for a landscape to be elementary is:

$$\bar{f} = p_3 \sum_{w \in C} w,$$

where p_3 is the frequency of appearance of any component $w \in C$ in a random solution x.

We will denote a landscape as a triple (X, N, f) where f is the evaluation function $f : X \to \mathbb{R}$, the set of solutions X represents the discrete domain of f and $N(x)$ is the neighborhood operator that defines adjacency between elements $x \in X$ under some local search neighborhood. N can also be expressed in the form of an adjacency matrix $\mathbf{A}$. The elements of $\mathbf{A}$ are such that $\mathbf{A}_{x,y} = 1$ if $y \in N(x)$ and $\mathbf{A}_{x,y} = 0$ otherwise. When a neighborhood is regular, the *Laplacian operator* is $\Delta = \mathbf{A} - d\mathbf{I}$ and it acts as a type of difference operator on the fitness function f such that:

$$\Delta f(x) = \sum_{y \in N(x)} (f(y) - f(x)).$$

A landscape is said to be *elementary* if f is an eigenvector of $-\Delta$ up to a constant, formally: $-\Delta f = k(f - b)$ for a constant b and an eigenvalue k of $-\Delta$. As a direct consequence, we can compute the neighborhood average in an elementary landscape as follows:

$$\operatorname*{Avg}_{y \in N(x)}(f(y)) = f(x) + \frac{k}{d}(\bar{f} - f(x)), \tag{1}$$

where $d = |N(x)|$ is the neighborhood size, which we assume is the same for all the solutions. Often, this same result can be expressed another way.

$$\operatorname*{Avg}_{y \in N(x)}(f(y)) = f(x) - p_1 f(x) + p_2 \left(\sum_{w \in C} w - f(x) \right), \tag{2}$$

where $p_1 = \alpha/d$ is the (sampling) rate at which "components" that contribute to $f(x)$ are removed from solution x to create a neighboring solution $y \in N(x)$, and $p_2 = \beta/d$ is the rate at which components in the set $C - x$ are sampled to create a neighboring solution $y \in N(x)$. Said in another way, in order to build all the solutions in the neighborhood each component in x has been removed α times and each component in $C - x$ have been added β times. By simple algebra,

$$\text{Avg}_{y \in N(x)}(f(y)) = f(x) - p_1 f(x) + p_2 \left(\sum_{w \in C} w - f(x) \right) = f(x) + \frac{k}{d}(\bar{f} - f(x)), \quad (3)$$

where $k = \alpha + \beta$, $\bar{f}/p_3 = \sum_{w \in C} w$ and $p_3 = \beta/(\alpha + \beta)$ [12,13,14].

It should also be noted that some landscapes that are not elementary can nevertheless be expressed as a superposition of a small number of elementary landscapes. For example, a MAX-3SAT landscape under the Hamming-1 neighborhood is not elementary, but it can be re-expressed as the sum of three functions. Let $M(x)$ denote the MAX-3SAT evaluation function for a Boolean string x; there exists functions f_1, f_2 and f_3 such that the landscapes of f_1, f_2 and f_3 are elementary, and

$$M(x) = f_1(x) + f_2(x) + f_3(x).$$

This makes it possible to compute averages over the Hamming-1 neighborhood. Using Walsh analysis it is even possible to compute higher order statistical moments (variance, skew, kurtosis) in polynomial time over arbitrary Hamming balls in the landscape, even over regions that are exponentially large [9]. The same method can be applied to all k-bounded pseudo-Boolean functions, including NK-Landscapes [6].

Thus, there is a great deal that we can potentially compute about local search landscapes that is not being utilized by search algorithms. We continue to find new ways to model new problems using elementary landscape theory, and we continue to find new ways to compute statistical information even more efficiently. Quasi-elementary landscapes are another step in this direction. In the current paper we present new results for the 3-opt move operator for the Traveling Salesman Problem, as well as new more general results for a Max-Clique neighborhood. We also establish a connection between problems that are a superposition of elementary landscapes and the concept of quasi-elementary landscapes.

1.1   Quasi-Elementary Landscapes

Whitley [11] introduced the term *quasi-elementary landscape* to describe a landscape and neighborhood structure where a variant of Grover's wave equation can be used to compute the neighborhood average by adding a correction to the usual wave equation. Thus, a landscape is quasi-elementary if:

$$\text{Avg}_{y \in N(x)}(f(y)) = f(x) - p_1 f(x) + p_2 \left(\bar{f}/p_3 - f(x) \right) + g(x). \quad (4)$$

We will refer to g as an auxiliary function. To be quasi-elementary it is critical that the computational complexity of $g(x)$ be less than the computation complexity of enumerating and evaluating the neighbors of solution x. Like the cost function $f(x)$, the auxiliary function $g(x)$ can sample from the set of components (e.g., the cost matrix) and can compute a "correction" relative to solution x to account for the fact that the landscape is not elementary.

In principle, one might allow $g(x)$ to be computed as the sum of multiple subfunctions. We know, for example, that MAX-3SAT is a superposition of 3 elementary landscapes. In the current paper we will show the average of the neighborhood for MAX-3SAT can be computed using $g(x) = a_1 f_1(x) + a_2 f_2(x)$ where functions f_1 and f_2 are elementary landscapes, but $g(x)$ is not elementary. It is again critical that the complexity of $g(x)$ must be less than the complexity of enumerating and evaluating the neighbors in $N(x)$.

In some cases, we can provide additional information about the function $g(x)$. Assume that the set of components can be broken into 3 sets: $f(x), f'(x)$ and $\sum_{w \in C} w - f(x) - f'(x)$ where $f'(x)$ identifies (and sums) a set of components relative to x where those components that contribute to $f'(x)$ are sampled at a different rate relative to the other two subsets of components. Assume that

$$g(x) = p_4 f'(x) - p_2 f'(x),$$

where p_4 is the new sampling rate for the components that contribute to $f'(x)$. Assume the complexity of computing $f'(x)$ is no greater than the complexity of computing $f(x)$. We then obtain the following result establishing a quasi-elementary landscape.

$$\operatorname*{Avg}_{y \in N(x)}(f(y)) = f(x) - p_1 f(x) + p_4 f'(x) + p_2 \left(\frac{\bar{f}}{p_3} - f(x) - f'(x) \right). \quad (5)$$

We use this model to show that the "shift" operator and the "3-opt" operator for the Traveling Salesman Problem induce a quasi-elementary landscape. We can also use the model to show that a local search algorithm for the Maximal Clique problem also induces a quasi-elementary landscape; this same local search algorithm can also be used to find densely connected subgraphs in larger graphs.

Finally, we can also construct a quasi-elementary landscape that samples the functions $f(x)$ and $f'(x)$ and the constant $\bar{f}$. Let $s_1/d = (p_1 + p_2)$, $s_2/d = (p_2 - p_4)$ and $s_3/d = (p_2/p_3)$. We will say that a landscape is quasi-elementary if:

$$\operatorname*{Avg}_{y \in N(x)}(f(y)) = \left(1 - \frac{s_1}{d}\right) f(x) - \frac{s_2}{d} f'(x) + \frac{s_3}{d} \bar{f}. \quad (6)$$

In this form, we can show that landscapes which are a superposition of elementary landscapes are in fact quasi-elementary landscapes if $f'(x)$ can be efficiently computed. In the case of a superposition of two elementary landscapes we might reasonably expect the complexity of f' to be no greater than the complexity of f since f' is a subfunction that can be use to compute f. Quasi-elementary landscapes can also result from a superposition of more than two elementary landscapes. Assume we have a superposition of elementary landscapes where $f = f_1(x) + f_2(x) + f_3(x)$. Let $g(x) = f_1(x) + f_2(x)$. For some classes of problems one can still prove that g has the same complexity as evaluating $f(x)$.

2   Examples of Quasi-Elementary Landscape

We will first look at examples of quasi-elementary landscapes which can be described using the following equation.

$$\underset{y \in N(x)}{\text{Avg}(f(y))} = f(x) - p_1 f(x) + p_4 f'(x) + p_2 \left(\bar{f}/p_3 - f(x) - f'(x) \right). \qquad (7)$$

2.1   The Shift Operator for the TSP

We will assume a permutation representation is used. A shift operator works by deleting one vertex from the permutation, then that vertex is re-inserted at every other possible position in the permutation. When done at every possible position, this yields some duplicate neighbors. These duplicates can be eliminated by 1) shifting the vertex to be deleted to the beginning of the permutation, and 2) doing insertion to the next $n-3$ possible positions. Therefore, the neighborhood size is $n(n-3)$.

One might assume that the "shift" operator is not a commonly used TSP operator. However, the "shift" operator can also be modeled as a special 3-opt move where one of the segments is a single city. If the tour is then broken into a segment of 1 city, and then broken into two segments of 2 or more cities, reversing the two longer segments exactly yields a move under the "shift" operator.

Let $f'(x)$ denote an auxiliary function to $f(x)$. When a vertex is deleted in solution x the deletion removed 2 edges, and introduces 1 new edge: $f'(x)$ ignores the deleted edges, but counts the cost associated with the new edges. Since n vertices are deleted, $f'(x)$ is the sum of the n new edges. For example, consider the tour: 1 2 3 4 5. Then the cost function $f(x) = w_{1,2} + w_{2,3} + w_{3,4} + w_{4,5} + w_{5,1}$ while the auxiliary function $f'(x) = w_{1,3} + w_{2,4} + w_{3,5} + w_{4,1} + w_{5,2}$. In the case of the "shift" operator, the edges in $f'(x)$ appear with greater frequency in the neighborhood $N(x)$.

When a vertex is deleted and reinserted, this cuts 3 edges: it cuts the edge to the left and to the right of the deleted vertex, and it cuts the edge where the deleted vertex is reinserted. Because every vertex will be deleted and it will be inserted into all possible (non-duplicate) positions all the edges in x will be removed at the same frequency. We can group the neighbors into subsets of size $n-3$ where the same vertex is deleted from x, but the deleted vertex is reinserted into all feasible positions. In this subset of $n-3$ neighbors, the edge to the right of the deleted vertex is reinserted once (e.g., if AB represent the two consecutive vertices in the tour and A is deleted and re-inserted after B, the edge (A, B) is recovered). This happens 1 time for every edge in solution x across all neighbors.

We can calculate the rate at which edges are removed from x by assuming the symmetry case as a baseline, then correct for the 1 neighbor where a specific edge from x is reinserted. This means that the rate with which edges are removed from x is given by

$$-\frac{3}{n} f(x) + \frac{1}{n(n-3)} f(x) = \frac{-3n + 10}{n(n-3)} f(x).$$

The calculation of $\bar{f}$ is neighborhood independent; therefore we can use the calculation associated with the 2-opt neighborhood:

$$\sum_{w \in C} w - f(x) = \left(\frac{\beta + \alpha}{\beta}\bar{f}\right) - f(x) = \left(\frac{n-1}{2}\bar{f}\right) - f(x).$$

We next consider the edges in the set $C - x$. All of the edges are sampled in a symmetric fashion, but a subset of edges receive additional samples.

The symmetric case can be described as follows. New edges are created by inserting the deleted vertex in a new position. Since the insertion occurs in every position that does not produce a redundant neighbor, this case is symmetric. This can happen in only 4 ways: vertex P has been deleted and it is inserted before and after Q; or vertex Q is deleted and it is inserted before and after vertex P. Thus, the symmetric sampling rate over the entire neighborhood is $4/(n(n-3))$.

The non-symmetric case derives from the fact that when a vertex is deleted, the deletion also creates a new edge. Furthermore, the *same* edge is created $n-3$ times for each of the $n-3$ cases where the *same* vertex is deleted. However, one of these $n-3$ cases is also one of the symmetric cases previously counted. Removing this one case, there remains $n-4$ cases where the same edge is created when the same vertex is deleted.

The set of edges that are sampled an additional $n-4$ times can be found starting the with current solution x and then deleting each vertex in x to create a permutation (circuit) of $n-1$ vertices. As each vertex is deleted, one new edge is created: thus, there are n new edges that are created. Let $f'(x)$ be the sum of the n new edges that are created by deleting each vertex one at a time.

Note that $|C - x| = n(n-3)/2$. Each of the n edges that contributes to $f'(x)$ appears $n-4$ times across the entire neighborhood. Therefore, the symmetric and non-symmetric sampling from $C - x$ is given by

$$\frac{4}{n(n-3)}\left(\frac{n-1}{2}\bar{f} - f(x)\right) + \frac{n-4}{n(n-3)}f'(x).$$

Using the sample rate from x and $C - x$ yields the combined effect. Note that the operator is only well-defined when $n > 3$. Therefore:

$$\operatorname*{Avg}_{y \in N(x)}(f(y)) = f(x) - \frac{3n-10}{n(n-3)}f(x) + \frac{4}{n(n-3)}\left(\frac{n-1}{2}\bar{f} - f(x)\right) + \frac{n-4}{n(n-3)}f'(x).$$

This can be rearranged to yield:

$$\operatorname*{Avg}_{y \in N(x)}(f(y)) = f(x) - \frac{3n-10}{n(n-3)}f(x) + \frac{n}{n(n-3)}f'(x) + \frac{4}{n(n-3)}\left(\frac{n-1}{2}\bar{f} - f(x) - f'(x)\right).$$

While computing $f'(x)$ with no prior knowledge requires $O(n)$ time, if we currently know the evaluation of $f(x)$ and $f'(x)$ and we move to a point y such that $y \in N(x)$, then both $f(y)$ and $f'(y)$ can be computed as a partial update to $f(x)$

and $f'(x)$ respectively, and both partial updates can be computed in constant time. Thus, the resulting landscape is quasi-elementary.

In special cases (usually when n is small) the quasi-elementary landscape is actually elementary. For example, when $n = 4$ we find that

$$\mathrm{Avg}_{y \in N(x)}(f(y)) = f(x) + \frac{3}{2}(\bar{f} - f(x)).$$

2.2   The 3-opt Quasi-Elementary Landscape

The next example is a landscape for the classic 3-opt operator. This version of 3-opt does not include the "shift" operator and all segments must be of length 2 or greater. However, we will find that the results presented in this paper can be combined to characterize a more general 3-opt neighborhood that allows one of the segments to include the single city (shift operator) case.

Stattenberger et al. [8] give a general formula for counting the number of ways that a Hamiltonian Circuit can be cut into k segments corresponding to those used by a Lin-Kernighan k-opt operator. For $k = 3$ this quantity is $n(n-4)(n-5)/3!$. In principle, one could then reconfigure the tour by reversing one segment, two segments, or all three segments. However, note that reversing one segment results in a 2-opt move because the two segments that are not reversed can be concatenated into one segment. Thus, there are four patterns of reversal where either two segments are reversed or all three segments are reversed. Therefore:

$$d = 4n(n-4)(n-5)/3! = 2n(n-4)(n-5)/3.$$

In the following we will use $f(x)$ and $f'(x)$ with the same meaning as in the case of the "shift" neighborhood. However, the values of d, p_1, p_2, p_3 and p_4 could change. We have already presented the new value of d. Next we will present the new values of the p_i constants. We are searching for an expression of the following form:

$$\mathrm{Avg}_{y \in N(x)}(f(y)) = f(x) - p_1 f(x) + p_4 f'(x) + p_2 \left(\sum_{w \in C} w - f(x) - f'(x) \right).$$

In a 3-opt move, exactly 3 edges are removed from the current solution. Thus, $p_1 = 3/n$ as in the case of the "shift" neighborhood. Let us now consider the edges in $C - x$ that are included in a new neighbor.

Let P and Q be two cities that are distance 2 apart in the current solution x. In order to bring together these 2 cities (and include the edge $e_{P,Q} \in C - x$ in the neighbor) a segment of length 2 must be reversed. There are two ways this segment of length two can be chosen. It can be chosen to include the first city P, which is then moved adjacent to the second city when the segment is reversed, or it can be chosen to include the second city Q which is moved adjacent to the first city when the segment is reversed. The location of the third cut does not matter, and there are $n - 5$ possible locations for the third cut to occur. Two segments must be reversed to be a legal 3-opt move: one must be the segment of length 2, the other must be the segment that does not contain either P or Q. Three segments cannot be reversed. Thus, there are $2(n-5)$ ways to segment the tour to yield the desired result, and there is only one reversal pattern in each case that reverses two segments to yield the desired result (see Figure 1). The sum of the weights of the edges of cities that are distant 2 apart in x is exactly $f'(x)$. Thus, $f'(x)$ must be summed $2(n-5)$ times in the whole neighborhood, yielding $p_4 = 2(n-5)/d$. Note that the auxiliary function is *exactly* the same as the one used for the "shift" operator. If the tour is: 1 2 3 4 5 then the auxiliary function is $f'(x) = w_{1,3} + w_{2,4} + w_{3,5} + w_{4,1} + w_{5,2}$. This means the results for the two neighborhoods can be easily combined.

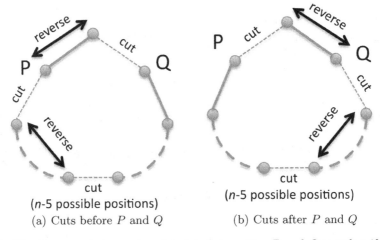

(a) Cuts before P and Q (b) Cuts after P and Q

Fig. 1. The two ways cuts can be placed to bring cities P and Q together if they are distance 2 apart

We next compute the value of p_2. Let P and Q be two cities in x which are distance $l > 2$ apart. We pick $l \le n/2$ so we consider the shortest path between P and Q in the tour. There are four ways cuts can be placed to bring P and Q together:

– Before P and before Q ($n-6$ other cuts are possible with only one reversal pattern). See Figure 2(a).
– After P and after Q ($n-6$ other cuts are possible with only one reversal pattern). See Figure 2(b).
– After P and before Q ($n-l-2$ cuts are possible with 2 reversal patterns). See Figure 2(c).
– Before P and after Q ($l-2$ cuts are possible with 2 reversal patterns). See Figure 2(d).

Therefore $p_2 = (n-6)/d + (n-6)/d + 2(n-l-2+l-2)/d = 4(n-5)/d$.
Finally, by substitution:

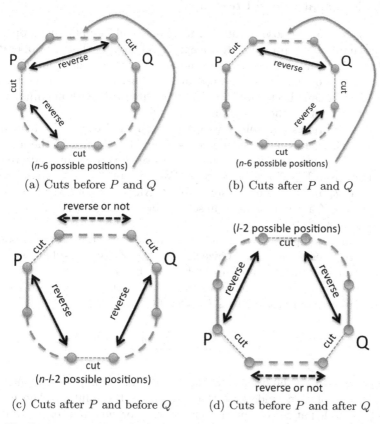

(a) Cuts before P and Q (b) Cuts after P and Q

(c) Cuts after P and before Q (d) Cuts before P and after Q

Fig. 2. The four ways cuts can be placed to bring cities P and Q together in a neighboring solution

$$\underset{y\in N(x)}{\mathrm{Avg}(f(y))} = f(x) - \frac{3}{n}f(x) + \frac{2(n-5)}{d}f'(x) + \frac{4(n-5)}{d}\left(\sum_{w\in C} w - f(x) - f'(x)\right)$$

$$= f(x) - \frac{3}{n}f(x) + \frac{2}{2n(n-4)/3}f'(x) + \frac{4}{2n(n-4)/3}\left(\sum_{w\in C} w - f(x) - f'(x)\right)$$

$$= f(x) - \frac{3}{n}f(x) + \frac{3}{n(n-4)}f'(x) + \frac{6}{n(n-4)}\left(\sum_{w\in C} w - f(x) - f'(x)\right). \qquad (8)$$

Recall that computing both $f(y)$ and $f'(y)$ can be done as a partial update of $f(x)$ and $f'(x)$, respectively.

While we could combine the results for the "shift" neighborhood and this 3-opt neighborhood result, it is more useful to leave then separated, since it provides information about which subset of the larger more general 3-opt neighborhood yields better moves on average. It is also notable that "shift" increases the relative sampling of $f'(x)$ while this form of 3-opt decreases the relative sampling of $f'(x)$.

2.3   The Maximal Clique Problem

Katayama *et al.* [5] propose a variable neighborhood "k-opt" move operator[1]
for the maximal clique problem. Katayama *et al.* report that this operator is
competitive with other heuristic search methods for generating solutions to the
maximal clique problem. While the operator has been applied to the maximal
clique problem, it can also be used to search for subgraphs with maximal density
independent of whether the subgraph is a clique or not.

The operator breaks a graph G with vertices V and edges E into two subgraphs, one with a set of vertices denoted by Z, and the other with the remaining
vertices, $V - Z$. Under the "k-opt" move operator, k vertices in Z are removed
and k vertices from $V - Z$ are added to Z. We will say that an edge $e_{i,j}$ *belongs*
to Z if vertices v_i and v_j are in the subgraph Z. All subsets of k vertices that are
not currently in Z are moved into subset Z; thus all edges that do not belong
to Z come into Z.

We must also compute $\bar{f}$ as a uniform sample over the cost components.
Assume we are attempting to maximize the number of edges in subgraph $Z \subset V$.
Let $f(Z)$ count the number of edges of E in Z. Let $r = |Z|$, $q = |V - Z|$ and
$n = |V|$. Using counting arguments, one can prove that the average of f over all
possible assignments of vertices to Z is given by:

$$\bar{f} = \frac{r(r-1)}{n(n-1)} \sum_{w \in C} w. \tag{9}$$

To have an elementary landscape, we need to divide "components" into those
that are in the solution, and those that are not in the solution. And each set
needs to be sampled at a uniform frequency corresponding to p_1 or p_2. But when
counting (and maximizing) the number of edges in Z and the number of edges
not in Z, we can classify edges into 3 different types that are uniformly sampled.

- Type 1: Edges that belong to Z and contribute to $f(Z)$.
- Type 2: Edges not in Z that connect two vertices not in Z.
- Type 3: Edges not in Z that connect a vertex in Z to a vertex in $(V - Z)$.

Therefore we can compute p_1 and p_3, but there are two sampling rates instead
of one for the edges normally accounted for p_2. As in the previous examples we
will denote the two different sampling rates by p_2 and p_4.

We should also note here that Katayama *et al.* limit their neighborhood moves
to those involving only the Type 1 and Type 2 edges. If a vertex in $(V - Z)$ did
not connect to a vertex in Z, then moving that vertex was not considered in the
Katayama implementation. However, we will analysis the full neighborhood. To
do this we generalize an early result by Whitley [11].

Theorem 1. *The "k-opt" operator for maximizing the density of edges in a
subgraph Z of graph G induces a quasi-elementary landscape. Let $r = |Z|$, $q =$*

---

[1] This is not the k-opt operator used for the Traveling Salesman Problem.

$|V - Z|$ and $n = |V| = r + q$. For $r > k$ and $q > k$ the neighborhood average for the "k-opt" operator is given by:

$$\operatorname*{Avg}_{Y \in N(Z)} (f(Y)) = f(Z) - \frac{k(r + q + 2q^2 r + k(q - q^2 - r + r^2) - (r + q)^2)}{q(q - 1)r(r - 1)} f(Z)$$

$$+ \frac{k(qr + k(1 - r - q))}{rq(q - 1)} f'(Z) + \frac{k(k - 1)n(n - 1)}{r(r - 1)q(q - 1)} \bar{f}, \qquad (10)$$

where the function $f'(Z)$ is the weighted sum of the edges having one vertex in Z and the other in $V - Z$.

Proof. We already know that $p_3 = \frac{r(r-1)}{n(n-1)}$ by Eq. (9).

We need to count 3 kinds of edges:

1. $f(Z)$ is the weighted sum of edges of E that are included in Z.
2. $f'(Z)$ is the weighted sum of edges with one vertex in Z and one in $V - Z$.
3. $\sum_{w \in C} w$ is the total weighted sum of edges.

Putting these together we obtain

$$\operatorname*{Avg}_{Y \in N(Z)} (f(Y)) = f(Z) - p_1 f(Z) + p_4 f'(Z) + p_2 \left(\sum_{w \in C} w - f(Z) - f'(Z) \right). \quad (11)$$

We first compute p_1, the probability that edge $e_{i,j}$ moves out in Z. If two vertices are randomly drawn from Z for exchange, there are $(r - k)/r$ ways the first vertex can stay in Z and $(r - k - 1)/(r - 1)$ that the second vertex can stay in Z. The probability of an edge that is currently in Z moving out of Z is

$$p_1 = 1 - \frac{(r - k)(r - k - 1)}{r(r - 1)}.$$

We define p_2 to be the probability that an edge "contained" within $V - Z$ moves to Z. There are k/q ways to select the first vertex, and $(k - 1)/(q - 1)$ ways that the second vertex can be selected. Thus,

$$p_2 = \frac{k(k - 1)}{q(q - 1)}.$$

Finally, we will compute p_4. Consider an edge $e_{i,j}$ such that $v_i \in Z$ and $v_j \in V - Z$. Then p_4 corresponds to the probability that v_i stays in Z and v_j moves to Z when any random neighbor is considered.

$$p_4 = \frac{(r - k)}{r} \cdot \frac{k}{q} = \frac{k(r - k)}{rq}.$$

Substituting into (11) we obtain:

$$\operatorname*{Avg}_{Y \in N(Z)}(f(Y)) = f(Z) - \left(1 - \frac{(r-k)(r-k-1)}{r(r-1)}\right) f(Z) + \frac{k(r-k)}{rq} f'(Z)$$

$$+ \frac{k(k-1)}{q(q-1)} \left(\frac{n(n-1)}{r(r-1)} \bar{f} - f(Z) - f'(Z)\right), \tag{12}$$

which reduces to (10).

Computing both $f(Z)$ and $f'(Z)$ can be done by enumerating the vertices in Z and checking the edges that are incident on vertices in Z. Therefore $f'(Z)$ can be computed as a side-effect of computing $f(Z)$. And for $Y \in N(Z)$, both $f(Y)$ and $f'(Y)$ can be computed as a partial update to $f(Z)$ and $f'(Z)$ respectively. Therefore computing $f'(Z)$ has complexity less than or equal to computing $f(Z)$ and the maximal clique problem is a quasi-elementary landscape.

These results could be expressed with respect to the neighborhood size. The total size of the neighborhood is given by

$$d = \binom{r}{k}\binom{q}{k}.$$

When $p_4 = p_2$ and $k \geq 2$ the landscape becomes elementary. This happens when $r = 3$, $q = 4$ and $k = 2$ for example. But this is a special case, and this does not happen in general.

3   Superpositions of Two Elementary Landscapes

Assume we have a landscape that is a superposition of two elementary landscapes, such that

$$f(x) = f_1(x) + f_2(x).$$

In this case, the evaluation function is not an elementary landscape, but f_1 and f_2 are elementary landscapes. A number of problems have been shown to be a superposition of two elementary landscapes. These include the asymmetric frequency assignment problem [2,12] and all pseudo-Boolean functions with 2-bounded complexity, like MAX-2SAT, Unconstrained Quadratic Optimization (UQO) [3], the Subset Sum [1] as well as all the NK-Landscapes when K=1 [9]. This means that

$$\operatorname*{Avg}_{y \in N(x)}(f(y)) = \operatorname*{Avg}_{y \in N(x)}\{f_1(y)\} + \operatorname*{Avg}_{y \in N(x)}\{f_2(y)\}$$

$$= f_1(x) + \frac{k_1}{d}\left(\bar{f}_1 - f_1(x)\right) + f_2(x) + \frac{k_2}{d}\left(\bar{f}_2 - f_2(x)\right). \tag{13}$$

Given that we know $f(x) = f_1(x) + f_2(x)$ and $f_2(x) = f(x) - f_1(x)$ we obtain:

$$\operatorname*{Avg}_{y \in N(x)}(f(y)) = f(x) + \frac{k_1}{d}\left(\bar{f}_1 - f_1(x)\right) + \frac{k_2}{d}\left(\bar{f}_2 - f_2(x)\right),$$

$$\mathrm{Avg}_{y\in N(x)}(f(y)) = f(x) - \frac{k_1}{d}f_1(x) - \frac{k_2}{d}(f(x) - f_1(x)) + \left\{\frac{k_1}{d}\bar{f}_1 + \frac{k_2}{d}\bar{f}_2\right\},$$

where $\frac{k_1}{d}\bar{f}_1 + \frac{k_2}{d}\bar{f}_2$ is a constant. Note that we can chose to eliminate either f_1 or f_2 and normally would select the simpler of the two functions to include in the computation. This means that every problem which is a superposition of two elementary landscapes is also a quasi-elementary landscape as long as the computational complexity of either f_1 or f_2 is less than the cost of enumerating the neighborhood. Generally, we would expect the computational cost of f_1 or f_2 to be less than the cost of computing the full evaluation function, $f = f_1 + f_2$.

We have previously shown in this paper that the "shift" operator and the "3-opt" operators for the TSP, as well as the Katayama's "k-opt" operator for Max-Clique can be captured by an equation of the following form:

$$\mathrm{Avg}_{y\in N(x)}(f(y)) = f(x) - p_1 f(x) + p_4 f'(x) + p_2 \left(\frac{\bar{f}}{p_3} - f(x) - f'(x)\right). \quad (14)$$

In all of these problems the auxiliary function $f'(x)$ is the linear sum of a subset of components drawn from the set $C - x$. This means that in these problems there are exactly 3 distinct sampling rates over component's in the set C. Having this knowledge also makes it easier to search for a superposition of elementary landscapes, because it limits the number of equivalence classes that must be constructed when attempting to construct a superposition of elementary landscapes.

4   MAX-3SAT: A Superposition of Three Elementary Landscapes

Now assume that we have a superposition of three elementary landscapes: $f(x) = f_1(x) + f_2(x) + f_3(x)$. Then the average can be computed as:

$$\mathrm{Avg}_{y\in N(x)}(f(y)) = \mathrm{Avg}_{y\in N(x)}\{f_1(y)\} + \mathrm{Avg}_{y\in N(x)}\{f_2(y)\} + \mathrm{Avg}_{y\in N(x)}\{f_3(y)\}$$

$$= f_1(x) + \frac{k_1}{d}\left(\bar{f}_1 - f_1(x)\right) + f_2(x) + \frac{k_2}{d}\left(\bar{f}_2 - f_2(x)\right)$$

$$+ f_3(x) + \frac{k_3}{d}\left(\bar{f}_3 - f_3(x)\right). \quad (15)$$

But for MAX-3SAT, we can include $\bar{f}$ in f_1 so that $\bar{f}_2 = \bar{f}_3 = 0$ and

$$\mathrm{Avg}_{y\in N(x)}(f(y)) = f(x) + \frac{k_1}{d}\left(\bar{f}_1 - f_1(x)\right) - \frac{k_2}{d}\left(f(x) - f_1(x) - f_3(x)\right) - \frac{k_3}{d}f_3(x).$$

One can express f, f_1, f_2 and f_3 as Walsh functions. Assume f_1 captures the linear interactions, f_2 pairwise, and f_3 the order-3 interactions [10]. There is only one 3-way interaction per clause. And there are only n linear terms, the number of variables of the instance. We remove f_2 because it is larger and less regular in structure than f_1 and f_3. If there are n bits and m clauses, then the number of components needed to compute the average (with f_2 out of the picture) is at most $n + m + 1$ since there are at most m f_3 coefficients and n linear f_1 coefficients and $\bar{f} = \bar{f_1}$.

However, evaluations can also be done by partial evaluation. Let y_b be a neighbor of solution x generated by flipping bit b. It is then easy to prove that the cost of the partial evaluation is constant on average for $f(y), f_1(y)$ and $f_3(y)$ given $f(x), f_1(x)$ and $f_3(x)$. If y_b is a neighbor of x, only one Walsh coefficient (w_p) changes in f_1. Thus, $f_1(y_p) = f_1(x) - 2(\psi_p(x)w_p)$. And on average, only a constant number of order-3 Walsh coefficients changes sign in f_3; the coefficients exactly map to those clauses that contain bit b. In expectation, a bit appears in a clause with probability $3/n$ and across all clauses a bit appears $3m/n = O(1)$ times. Thus, evaluating f_3 also has a partial update that is almost identical to the partial update for evaluating f.

Hence, $\text{Avg}(f(y))_{y \in N(x)}$ can also be computed in $O(1)$ time on average, and the MAX-3SAT landscape is quasi-elementary. This generalizes to all MAX-kSAT problems, as well as NK-Landscapes.

5   Conclusions

In this paper we present three examples of quasi-elementary landscapes: TSP with the "shift" and 3-opt neighborhoods and Maximal Clique with Katayama's k-opt neighborhood. We also show that functions which are a superposition of elementary landscapes can also be quasi-elementary landscapes. A direct application of the concept of quasi-elementary landscapes is the generalization of the Grover's wave equation to landscapes which are not elementary. In future work we plan to continue to explore the relationship between the elementary landscape decomposition of the problems and the quasi-elementary property.

References

1. Chicano, F., Whitley, D., Alba, E.: A methodology to find the elementary landscape decomposition of combinatorial optimization problems. Evolutionary Computation (2011)
2. Chicano, F., Whitley, D., Alba, E., Luna, F.: Elementary landscape decomposition of the frequency assignment problem. Theoretical Computer Science 412(43), 6002–6019 (2011)
3. Chicano, F., Alba, E.: Elementary landscape decomposition of the 0-1 unconstrained quadratic optimization. Journal of Heuristics xx, xx–xx (2011), doi:10.1007/s10732-011-9170-6
4. Grover, L.K.: Local search and the local structure of NP-complete problems. Operations Research Letters 12, 235–243 (1992)
5. Katayama, K., Hamamoto, A., Harihisa, H.: An effective local search for the maximum clique problem. Information Processing Letters 95(5), 503–511 (2005)

6. Kauffman, S., Levin, S.: Towards a general theory of adaptive walks on rugged landscapes. Journal of Theoretical Biology 128, 11–45 (1987)
7. Stadler, P.F.: Toward a theory of landscapes. In: Lopéz-Peña, R., Capovilla, R., García-Pelayo, R., Waelbroeck, H., Zertruche, F. (eds.) Complex Systems and Binary Networks, pp. 77–163. Springer (1995)
8. Stattenberger, G., Dankesreiter, M., Baumgartner, F., Schneider, J.J.: On the neighborhodd structure of the traveling salesman problem generated by local search moves. Journal of Statistical Physics 129, 623–648 (2007)
9. Sutton, A., Whitley, D., Howe, A.: Computing the moments of k-bounded pseudo-boolean functions over hamming spheres of arbitrary radius in polynomial time. Theoretical Computer Science (2011), doi:10.1016/j.tcs.2011.02.006
10. Sutton, A.M., Howe, A.E., Whitley, L.D.: A Theoretical Analysis of the k-Satisfiability Search Space. In: Stützle, T., Birattari, M., Hoos, H.H. (eds.) SLS 2009. LNCS, vol. 5752, pp. 46–60. Springer, Heidelberg (2009)
11. Whitley, D.: Quasi-elementary landscapes. In: Multi-Interdisciplinary Scheduling: Theory and Applications, MISTA (2011)
12. Whitley, D., Chicano, F., Alba, E., Luna, F.: Elementary landscapes of frequency assignment problems. In: GECCO 2010, pp. 1409–1416. ACM Press (2010)
13. Whitley, D., Sutton, A.M.: Partial neighborhoods of elementary landscapes. In: Proceedings of the Genetic and Evolutionary Computation Conference, Montreal, Canada, pp. 381–388 (July 2009)
14. Whitley, L.D., Sutton, A.M., Howe, A.E.: Understanding elementary landscapes. In: Proceedings of the Genetic and Evolutionary Computation Conference, Atlanta, GA, pp. 585–592 (July 2008)

Fast Permutation Learning

Tony Wauters[1], Katja Verbeeck[1],
Patrick De Causmaecker[2], and Greet Vanden Berghe[1]

[1] CodeS Group, KAHO Sint-Lieven, Gent, Belgium
{tony.wauters,katja.verbeeck,greetvb}@kahosl.be
[2] CodeS Group, K.U. Leuven Campus Kortrijk, Kortrijk, Belgium
patrick.decausmaecker@kuleuven-kortrijk.be

Abstract. Permutations occur in a great variety of optimization problems, such as routing, scheduling and assignment problems. The present paper introduces the use of learning automata for the online learning of good quality permutations. Several centralized and decentralized methods using individual and common rewards are presented. The performance, memory requirement and scalability of the presented methods is analyzed. Results on well known benchmark problems show interesting properties. It is also demonstrated how these techniques are successfully applied to multi-project scheduling problems.

Keywords: Permutations, online learning, learning automata, dispersion games.

1 Introduction

The process of creating a permutation, i.e. arrangement of objects or values into a particular order, is a recurring phenomenon in combinatorial optimization problems. The permutations can represent a full or a partial solution to such problems. Typical examples can be found in routing and scheduling. The traveling salesman problem (TSP) for instance, aims at finding a tour of minimum distance through a number of cities. A solution can be represented by a permutation, which defines the order in which cities are visited. Many solutions for scheduling problems also contain some permutation representation. A solution for the permutation flow shop scheduling problem (PFSP) is such an example. In the PFSP a number of jobs have to be sequenced in order to be processed on a predefined number of resources. All these problems have a search space exponential in the number of inputs n (cities, jobs, ...). Due to the very nature of permutations there are at least $n!$ different solutions. An objective function, which represents the quality of the solutions, has to be optimized. If a solution to a problem can be represented by a permutation, then the objective function states how good the permutation is.

In fact, we can imagine the following general problem (see Figure 1): given a permutation π, a function f can give a value for that permutation $f(\pi)$. Function f can be the optimization problem under study, and it is assumed that the

Y. Hamadi and M. Schoenauer (Eds.): LION 6, LNCS 7219, pp. 292–306, 2012.
© Springer-Verlag Berlin Heidelberg 2012

function is not known. It is a black box. Since all values can be normalized, we can assume that $f(\pi) \in [0,1]$, with a value $f(\pi) = 0$ meaning the worst permutation and $f(\pi) = 1$ the best or optimal permutation.

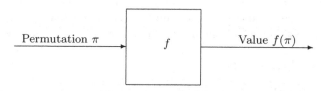

Fig. 1. General permutation problem seen as a black box function

In the present contribution, this general permutation problem is tackled using simple reinforcement learning devices, called Learning Automata (LA). It is shown how these methods are capable of learning permutations of good quality ($f(\pi)$ close to 1) without the use of problem specific information or domain knowledge. However, it is not the goal to outperform existing optimization methods for these problems, but only to show the strength of LA for online permutation learning.

The present paper is structured as follows. Section 2 shows related work on the learning of permutations and the use of learning automata for solving optimization problems. Section 3 gives a small overview on learning automata. In Section 4 different categories of permutation functions and their properties are discussed. Section 5 presents some centralized and decentralized methods based on learning automata for the online learning of permutations. In Section 6 the presented methods are analyzed on some well known benchmarks, and a successful application to project scheduling is demonstrated. A conclusion and final remarks are given in Section 7.

2   Related Work

Population based incremental learning (PBIL) [1] is a method for solving optimization problems, which is related to genetic algorithms. It however maintains a real-valued probability vector for generating solutions. PBIL is very similar to a cooperative system of finite learning automata where the learning automata choose their actions independently and update with a common reward. COMET [2] incorporates probabilistic modeling in conjunction with fast search algorithms for application to combinatorial optimization problems. The method tries to capture inter-parameter dependencies by creating a tree-shaped probabilistic network. PermELearn [7] is an online algorithm for learning permutations. The approach makes use of a doubly stochastic[1] weight matrix to represent estimations of the permutations, together with exponential weights and an iterative

---

[1] A matrix is doubly stochastic if all its elements are nonnegative, all the rows sum to 1, and all the columns sum to 1.

procedure to restore double stochasticity. [15] introduces a method using multiple learning automata to cooperatively find good quality schedules for the multi-mode resource-constrained project scheduling problem (MRCPSP). A common reward, based on the makespan of the scheduling solution is used. [14] present a method using learning automata combined with a dispersion game for solving the decentralized resource-constrained multi project scheduling problem (DR-CMPSP). To date the method belongs to the state-of-the-art for this problem[2].

3   Learning Automata

Learning Automata (LA) [11,13] are simple reinforcement learning components for adaptive decision making in unknown environments. An LA operates in a feedback loop with its environment and receives feedback (reward or punishment) for the actions taken. A single learning automaton maintains a probability vector p over its actions, which it updates according to a reinforcement scheme. Several reinforcement schemes with varying convergence properties are available in the literature. Examples of linear reinforcement schemes are linear reward-penalty, linear reward-inaction and linear reward-ϵ-penalty. The philosophy of these schemes is to increase the probability of selecting an action in the event of success and to decrease it when the response is a failure. The general update scheme is given by:

$$p_m(t + 1) = p_m(t) + \alpha_{reward}(1 - \beta(t))(1 - p_m(t))$$
$$- \alpha_{penalty}\beta(t)p_m(t) \tag{1}$$
$$\text{if } a_m \text{ is the action taken at time } t$$
$$p_j(t + 1) = p_j(t) - \alpha_{reward}(1 - \beta(t))p_j(t)$$
$$+ \alpha_{penalty}\beta(t)[(r - 1)^{-1} - p_j(t)] \tag{2}$$
$$\text{if } a_j \neq a_m$$

With $p_i(t)$ the probability of selecting action i at time step t. The constants α_{reward} and $\alpha_{penalty}$ are the reward and penalty parameters. When $\alpha_{reward} = \alpha_{penalty}$, the algorithm is referred to as linear reward-penalty (L_{R-P}), when $\alpha_{penalty} = 0$, it is referred to as linear reward-inaction (L_{R-I}) and when $\alpha_{penalty}$ is small compared to α_{reward}, it is called linear reward-ϵ-penalty ($L_{R-\epsilon P}$). $\beta(t)$ is the reward received by the reinforcement signal for an action taken at time step t. r is the number of actions. In the present paper, learning automata with finite actions and linear reinforcement schemes will be used. More specifically we will only use the linear reward-inaction update scheme, because it has nice theoretical convergence results.

4   Permutation Functions

A permutation function, i.e. a function mapping permutations to values, can take several forms. Most of them are highly non-linear. The most straightforward

---

[2] Multi project scheduling problem library: http://www.mpsplib.com ; accessed on September: 23, 2011

function is one that gives a value to each individual position. Take for example an assignment problem where a matrix defines the cost for assigning an agent to a task. A permutation, where task i at position j is performed by agent j, is a possible solution representation for this problem.

Table 1. Cost matrix for an example assignment problem of size $n = 4$

	T0	T1	T2	T3
A0	3	4	1	3
A1	3	2	3	1
A2	3	4	2	2
A3	2	3	4	4

The assignment problem with the cost matrix from Table 1 results in a permutation function as shown in Figure 2. The total cost for each permutation is plotted. The search space has an optimal solution $[2, 1, 3, 0]$ with a total cost of 7. This solution denotes that task 2 is performed by agent 0, task 1 is performed by agent 1, task 3 is performed by agent 2 and task 0 is performed by agent 3.

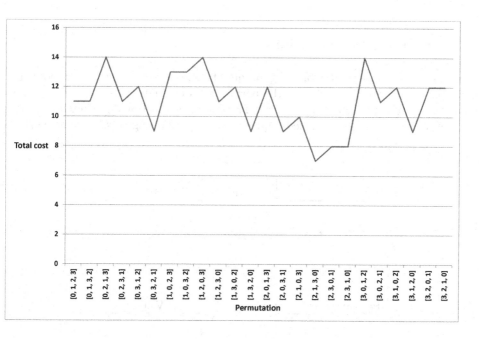

Fig. 2. Permutation function for an example assignment problem

For many problems using a permutation representation the total cost or value is determined by the adjacent elements. For example, if element A is directly followed by element B in the permutation, there is a cost of c_{AB}. These costs can be symmetric or asymmetric. In this category of permutation functions we can make a distinction between cyclic and non-cyclic functions. A typical example where the permutation function is based on adjacency costs and is also cyclic is a TSP.

To summarize, we distinguish the following default permutation function categories:

- **individual position**
- **adjacent positions (cyclic and non-cyclic)**

Many permutation functions use or combine elements from these default categories.

Some optimization problems have additional constraints on the permutations. For example, one can have precedence constraints, imposing that one element must occur before or after another element. Examples include the sequential ordering problem (SOP) which is a TSP with precedence constraints) and project scheduling problems. Yet another additional constraint can be that several elements must be adjacent to each other and form groups. One can incorporate these additional constraints by adding a high penalty to the cost value of the permutation.

5   Learning Permutations

In order to learn permutations of good quality one or more learning components are put in a feedback loop with the permutation evaluation function f (i.e. the environment), as is shown in Figure 3. The rest of this section describes a number of centralized and decentralized approaches for performing this learning task.

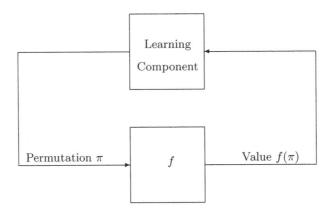

Fig. 3. Learning component in a feedback loop with the permutation learning problem

5.1   Naive Approach

A naive and centralized approach (Figure 4) to learning permutations using learning automata would be to assign one action per permutation. This results in a total of $n!$ actions, which is impractical for larger n both with respect to calculation time and memory usage.

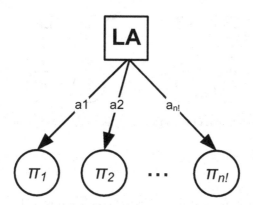

Fig. 4. A naive approach, one LA with one action per permutation

5.2   Hierarchical Approach

A better approach would be to divide the learning task among different learning automata, more specifically a tree of learning automata, called a hierarchical learning automaton [12]. An example of such a hierarchical learning automaton for $n = 3$ is shown in Figure 5. An LA at depth $d \in \{1, 2, \ldots, n\}$ in the tree is responsible for choosing the element at position d in the permutation. Each LA at depth d has $n + 1 - d$ actions, excluding all the actions chosen in the LA in the path from this LA to the root of the tree. The advantage of this hierarchical approach is that each individual LA has a smaller action space (maximum n). There is also a drawback. In case of a large exploration, the whole search tree is visited in the worst case, which results in $\sum_{d=1}^{n} \frac{n!}{d!}$ LAs. Since all action selection probabilities for each LA have to be stored, this can be very memory intensive.

5.3   Probability Matrix Approach

To deal with the large memory requirements of the hierarchical approach we developed an approach with a compact representation. Similar to the method in [7], we use a doubly stochastic matrix P with n rows and n columns. The column and row sums are always 1. P_{ij} is the probability for element i to be on position j in the permutation. The approach works as follows:

1. generate a uniform doubly stochastic matrix P with:
   $\forall_{i=1..n} \forall_{j=1..n} P_{ij} = \frac{1}{n}$

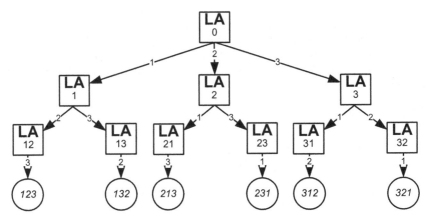

Fig. 5. An example of a hierarchical learning automaton for $n = 3$

2. select a permutation π using P
3. retrieve a reward $r = f(\pi)$ for the selected permutation
4. update P using reward r
5. repeat from step 2 until some stopping condition is met.

These steps are now described in more detail.

Selecting a Permutation from P: Several methods can be used to select a permutation from a doubly stochastic matrix. A first method is to uniformly select a row i and then use a probabilistic selection (e.g. roulette wheel selection) on this row for determining on which position j we have to put element i in the permutation. After that we reduce the matrix by removing row i and column j. Then we normalize the remaining rows, and repeat the process until all rows (and also all columns) have been selected once. We then have a complete permutation. Another permutation selection method is based on entropy. The method is similar to the explanation above, but the next row is now selected based on minimum entropy.

$$\underset{i}{argmin} \ H = - \sum_{j=1..n} P_{ij} log \left(P_{ij} \right)$$

Updating P with Reward r: The probability matrix P is updated with an LA update scheme for each row i, and the selected action is determined by j. After updating all rows, the matrix P remains doubly stochastic, which is not the case in the PermELearn algorithm [7].

5.4   Decentralized Approach

The following method is similar to the 'probability matrix method', but uses a more decentralized approach. Each position in the permutation is determined by an individual agent. An agent employs a learning automaton for choosing its individual position from the full set of positions. Thus, there are n agents (LA) with n actions each, resulting in the same memory footprint as the 'probability matrix approach'. The agents play a dispersion game [6] for constructing a permutation. In a dispersion game, the number of agents is equal to the number of actions. In order to form a permutation, all agents need to select a distinct action so that the assignment of actions to agents is maximally dispersed. For example, if three agents select the following distinct actions: agent 1 selects position 2, agent 2 selects position 3 and agent 3 selects position 1, then the permutation becomes $[3, 1, 2]$.

A Basic Simple Strategy (BSS) was introduced in [6], allowing agents to select maximally dispersed actions in a logarithmic (in function of the number of agents) number of rounds, where a naive approach would be exponential. BSS does not incorporate the agents' preferences, it uses uniform selection. To take the agents' preferences into account, we introduce a probabilistic version of BSS, which we call Probabilistic Basic Simple Strategy (PBBS). The PBBS works as follows. Given an outcome $o \in O$ (selected actions for all agents), and the set of all actions A, an agent using the PBBS will:

- select action a with probability 1 in the next round, if the number of agents selecting action a in outcome o is 1 ($n_a^o = 1$).
- select an action from the **probabilistic** distribution over actions $a' \in A$ for which $n_{a'}^o \neq 1$, otherwise.

The probabilistic distribution over actions is obtained from the agents' LA. Once a permutation is constructed by playing the game, a common reward (permutation function) or individual reward can be obtained. These common or individual rewards are then given to the agents' LA, which consequently update their probabilistic distribution. Experiments have shown that this decentralized approach has very similar performance characteristics as the 'probability matrix approach'.

6   Experiments

As an illustration of the methods' behaviour, some experiments were performed on a fictitious permutation function and simple benchmark problems, the TSP and an assignment problem. Subsequently, application to a more extensive multi-project scheduling problem is given, which shows the real advantage of the described methods. The following experiments report on the decentralized approach unless mentioned otherwise. Similar properties were observed for the hierarchical and probability matrix approach. All results are averaged over 100 runs and the experiments were performed on an Intel Core i7 2600 3.4Ghz processor, using the Java version 6 programming language.

6.1   Peaked Permutation Function

Consider the following permutation function definition. If the decimal number of the permutation π is $dec(\pi)$ according to the factorial number system (Lehmer code) [9]. Then the value of the permutation is defined as:

$$f(\pi) = \left(\frac{2dec(\pi)}{n!}\right)^{10} \text{ if } dec(\pi) \le \frac{n!}{2} \tag{3}$$

$$= \left(\frac{2(n! - dec(\pi))}{n!}\right)^{10} \text{ if } dec(\pi) > \frac{n!}{2} \tag{4}$$

This function has a peak value in the middle of the permutation range. Figure 6 shows this permutation function for $n = 9$.

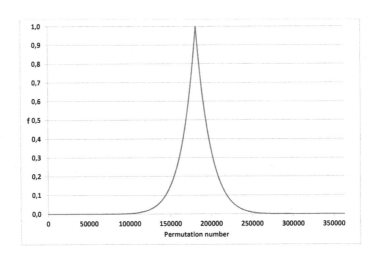

Fig. 6. Peaked permutation function landscape $n = 9$

Figure 7 compares the average calculation time (in milliseconds) of the presented approaches. For each size $n = 2..20$, 5000 iterations are performed on the peaked permutation function. A learning rate of 0.005 is used for each approach. All but the naive approach show good calculation time properties.

Figure 8 shows the maximum function value over a number of iterations (50,100,500) for different learning rates when applying the decentralized approach with common reward (i.e. the permutation function value). The results show that for a particular range of learning rates better permutations are learned, compared to random sampling (i.e. learning rate equal to 0). When more iterations are given for learning, the difference between random sampling and learning becomes smaller. Bad performance can be observed when the learning rates are too high, and thus premature convergence occurs.

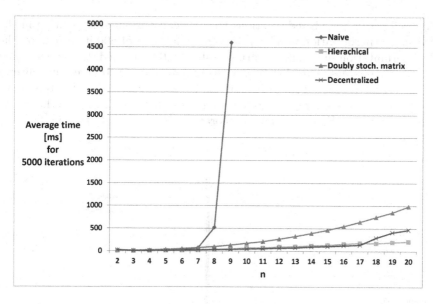

Fig. 7. Average calculation time in milliseconds for performing 5000 iterations of the presented approaches for different permutation sizes on the peaked permutation function

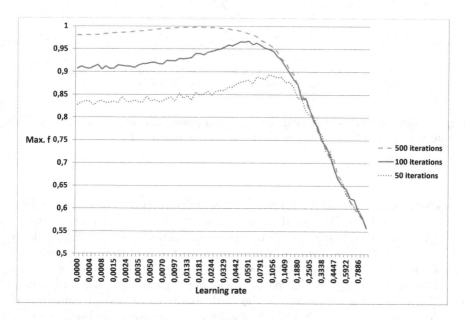

Fig. 8. Max. objective function value for different number of iterations on a peaked permutation function of size $n = 15$

Figure 9 shows the visited solutions with their corresponding position in the search space (permutation number) during a single run of 500 iterations on the peaked permutation function. A learning rate of 0.03 was used. In the beginning of the search a extensive exploration of the search space is observed, but after a while the search is focused towards high quality solutions (peak of the function). The duration of the exploration phase depends on the learning rate. In the experiments presented in Figure 9, most exploration disappears after approximately 150 iterations.

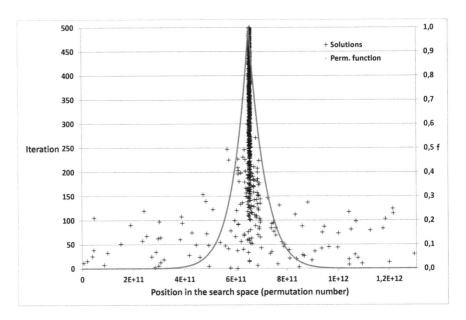

Fig. 9. Visited solutions with their corresponding position in the search space. Run of 500 iterations on the peaked permutation function of size $n = 15$.

6.2 TSP

We tested the decentralized approach with common reward on a number of TSP instances from TSPLIB[3]. The distance was scaled to a value between 0 and 1, such that a value of 1 corresponds to an optimal distance and 0 corresponds to an upper bound on the distance. Figure 10 shows the maximum function value over a number of iterations (1000, 5000, 10000, 50000) for different learning rates on a size $n = 17$ instance with name 'gr17'. For a particular range of learning rates better solution values can be observed, compared to random sampling. If more iterations are given, then the best solutions occur for lower learning rates. Again, too high learning rates lead to worse solutions.

---

[3] TSPLIB website: http://comopt.ifi.uni-heidelberg.de/software/TSPLIB95/ ; last check of address September: 23, 2011.

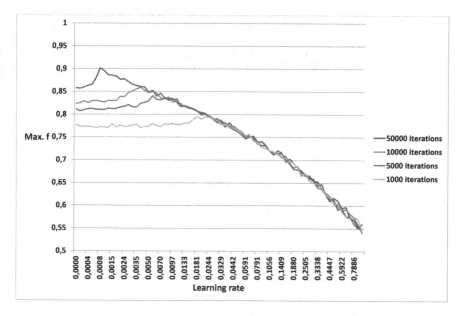

Fig. 10. Max. objective function value for different number of iterations on a TSP instance (gr17) of size $n = 17$

6.3   Assignment Problem

Assignment problems belong to the category of permutation functions where the total cost of the permutation is equal to the sum of the individual costs. Therefore both individual and global rewards can be given to the agents. Figure 11 shows a comparison of the maximum objective function value for individual and common rewards on a random assignment problem of size $n = 9$ with the cost matrix of Table 2. For each learning rate a run of 1000 iterations is performed where the maximal objective function value is measured. The objective function is scaled between 0 and 1 such that 1 corresponds to the optimal solution. The results show that individual rewards produce better solutions than common rewards. For a particular range of learning rates the method performs better than random sampling, and the optimal solution can be found by using individual rewards .

6.4   Multi-project Scheduling

The decentralized method introduced in the present paper, was used for solving the decentralized resource-constrained multi project scheduling problem (DR-CMPSP) [14]. The DRCMPSP was introduced in [4,5] and extended in [8]. It is a generalization of the Resource Constrained Project Scheduling Problem (RCPSP) [3,10] and can be stated as follows. A set of n projects have to be scheduled simultaneously using autonomous and self-interested decision makers. Each individual project contains a set of jobs or activities, precedence relations

Table 2. Cost matrix for a random assignment problem of size $n = 9$

	T0	T1	T2	T3	T4	T5	T6	T7	T8
A0	7	8	5	3	9	3	9	4	7
A1	3	6	9	3	2	9	6	5	7
A2	6	3	5	1	3	6	9	2	7
A3	8	1	9	3	3	6	3	6	3
A4	7	3	5	7	3	8	9	3	2
A5	4	2	8	2	7	5	4	6	4
A6	7	8	8	9	4	8	9	8	8
A7	7	4	7	8	9	8	1	3	5
A9	9	3	9	7	6	1	5	2	8

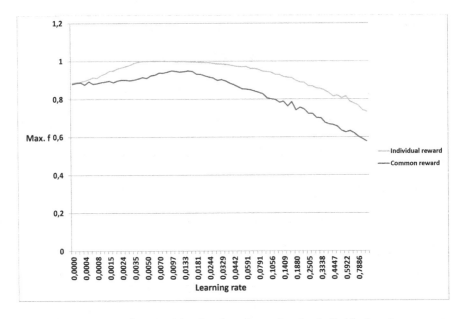

Fig. 11. Comparison of max. objective function value for individual and common reward on a assignment problem of size $n = 9$

between the jobs, and resource requirements for executing each job. These resource requirements constitute local renewable resources, and global renewable resources shared among all projects. A global objective value must be minimized, examples include but are not limited to: the average project delay(APD), the total makespan (TMS), and the deviation of the project delay (DPD). The project delay of a project is its makespan minus the critical path duration. The remainder of the current section will concentrate on the APD objective. A constructive schedule generation scheme was applied to solve the DRCMPSP, requiring the project order and the job order for each project as input. The project order is a permutation of projects, while the job order is a permutation with precedence constraints. The decentralized method with the dispersion game was used to find good quality project orders, leading to schedules with a low APD objective value. Each project was represented by one project order decision maker. Instances from the Multi project scheduling problem library[4] have been experimented on. To date, the method belongs to the state-of-the-art for this problem, showing 104 best solutions out of 140, with respect to the average project delay objective.

7 Conclusion

We have presented several centralized and decentralized methods using learning automata for the online learning of good quality permutations. Different permutation functions have been discussed. The capabilities of the methods have been analyzed and demonstrated on well known benchmark problems, and we demonstrated how to successfully apply these methods to a multi-project scheduling problem. The methods are very general because they do not use any problem specific information or domain knowledge, which makes them well suited for application within general optimization methods, like hyper-heuristics. In the future, the very same methods can be applied to other optimization problems and make them core elements of new intelligent optimization approaches.

Acknowledgment. Thanks to Erik Van Achter for his help on improving the quality of this text. Funded by IWT (IWT SBO DiCoMAS project).

References

1. Baluja, S.: Population-based incremental learning: A method for integrating genetic search based function optimization and competitive learning. School of Computer Science, Carnegie Mellon University (1994)
2. Baluja, S., Davies, S.: Fast probabilistic modeling for combinatorial optimization. In: Proceedings of the Fifteenth National/Tenth Conference on Artificial Intelligence/Innovative Applications of Artificial Intelligence, AAAI 1998/IAAI 1998, pp. 469–476. American Association for Artificial Intelligence, Menlo Park (1998), http://dl.acm.org/citation.cfm?id=295240.295718
3. Blazewicz, J., Lenstra, J., Rinnooy Kan, A.: Scheduling subject to resource constraints: classification and complexity. Discrete Applied Mathematics 5, 11–24 (1983)
4. Confessore, G., Giordani, S., Rismondo, S.: An auction based approach in decentralized project scheduling. In: Proc. of PMS 2002 International Workshop on Project Management and Scheduling, Valencia, pp. 110–113 (2002)
5. Confessore, G., Giordani, S., Rismondo, S.: A market-based multi-agent system model for decentralized multi-project scheduling. Annals of Operational Research 150, 115–135 (2007)

---

[4] Multi project scheduling problem library: http://www.mpsplib.com ; retrieved September: 23, 2011

6. Grenager, T., Powers, R., Shoham, Y.: Dispersion games: general definitions and some specific learning results (2002),
http://citeseer.ist.psu.edu/grenager02dispersion.html
7. Helmbold, D.P., Warmuth, M.K.: Learning permutations with exponential weights. Journal of Machine Learning Research 10, 1705–1736 (2009)
8. Homberger, J.: A multi-agent system for the decentralized resource-constrained multi-project scheduling problem. International Transactions in Operational Research 14, 565–589 (2007)
9. Knuth, D.E.: The Art of Computer Programming, Volume 3: Sorting and Searching. Addison-Wesley (1973)
10. Kolisch, R.: Serial and parallel resource-constrained project scheduling methods revisited: theory and computation. European Journal of Operational Research 90, 320–333 (1996)
11. Narendra, K., Thathachar, M.: Learning Automata: An Introduction. Prentice-Hall International, Inc. (1989)
12. Thathachar, M., Ramakrishnan, K.: A hierarchical system of learning automata. IEEE Transactions On Systems, Man, and Cybernetics 11(3), 236–241 (1981)
13. Thathachar, M., Sastry, P.: Networks of Learning Automata: Techniques for Online Stochastic Optimization. Kluwer Academic Publishers (2004)
14. Wauters, T., Verbeeck, K., De Causmaecker, P., Vanden Berghe, G.: A game theoretic approach to decentralized multi-project scheduling (extended abstract). In: Proc. of 9th Int. Conf. on Autonomous Agents and Multiagent Systems (AAMAS 2010), No. R24 (2010)
15. Wauters, T., Verbeeck, K., Van den Berghe, G., De Causmaecker, P.: Learning agents for the multi-mode project scheduling problem. Journal of the Operational Research Society 62(2), 281–290 (2011)

Parameter-Optimized Simulated Annealing for Application Mapping on Networks-on-Chip

Bo Yang[1,3], Liang Guang[1], Tero Säntti[1,2], and Juha Plosila[1]

[1] Department of Information Technology, University of Turku, Finland
[2] Academy of Finland, Research Council for Natural Sciences and Engineering
[3] Turku Center for Computer Science, Turku, Finland
{boyan,liagua,teansa,juplos}@utu.fi

Abstract. Application mapping is an important issue in designing systems based on many-core networks-on-chip (NoCs). Simulated Annealing (SA) has been often used for searching for the optimized solution of application mapping problem. The parameters applied in the SA algorithm jointly control the annealing schedule and have great impact on the runtime and the quality of the final solution of the SA algorithm. The optimized parameters should be selected in a systematic way for each particular mapping problem, instead of using an identical set of empirical parameters for all problems. In this work, we apply an optimization method, Nelder-Mead simplex method, to obtain optimized parameters of SA. The experiment shows that with optimized parameters, we can get an average 237 times speedup of the SA algorithm, compared to the work where the empirical values are used for setting parameters. For the set of benchmarks, the proposed parameter-optimized SA algorithm achieves comparable communication energy consumption using less than 1% of iterations of that used in the reference work.

1 Introduction

In the past decade, the multi- and many-core processors have been rapidly developing and widely used for processing a massive amount of data in increasingly complex systems [2]. As the number of cores and their processing capabilities are continuously increasing, the communicational aspect, instead of the traditional computational aspect, is becoming the major concern in designing systems-on-chip (SoCs). The underlying communication architecture in many-core systems plays a great role in improving the performance and decreasing the energy consumption of the system. To deal with the emerging communication challenge in many-core system, the NoC has been proposed as a promising alternative to the conventional bus-based and point-to-point communication architectures [1].

Figure 1 shows an example of an 8-core SoC on a 2×4 2D mesh NoC. The NoC is a communication infrastructure composed of a set of routers connected by inter-router communication channels. The processing elements (PEs) such as CPU or DSP modules, FPGAs, embedded memory blocks, are connected to a router via the network interface (NI). Typically, the term node or tile on a NoC refers to a PE and the corresponding router. The data generated by the source PE is first transformed into packets coupled with appropriate control information, and then transmitted to the destination PE by

Y. Hamadi and M. Schoenauer (Eds.): LION 6, LNCS 7219, pp. 307–322, 2012.
© Springer-Verlag Berlin Heidelberg 2012

traveling multiple routers and channels over the NoC. The routing decision is made on each router based on a specific routing protocol. By decentralizing the role of the arbitration into each router, the NoC architecture is suitable to deal with the great number of concurrent communications in modern many-core systems. The bandwidth on the NoC is also enhanced by sharing network channels among concurrent communications [4].

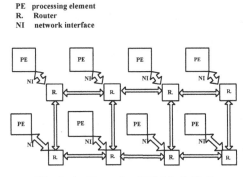

Fig. 1. An Example of 2D Mesh NoC

Based on the NoC communication architecture, there are set of problems related to many-core systems design. One of them is the *application mapping*. Given an application implemented by a set of tasks, and a many-core NoC, the problem of the application mapping is to decide how to map each task onto a node so that the predefined objectives and constraints can be met. The application mapping has a great impact on the system performance and energy consumption. The experiment in [6] shows that one optimized mapping algorithm achieves 51.7% communication energy savings compared to an ad hoc implementation. While exhaustive search is not possible for the *NP-hard* mapping problem, stochastic and heuristic searches are generally used for finding the near-optimal mapping solutions. As surveyed in [8], these stochastic and heuristic methods include, for example, simulated annealing (SA), tabu search (TS) and greedy incremental (GI) heuristic.

In these search heuristics, the SA has been often used since it is able to escape from the local minimum and find the global minimum of the dedicated cost function. The detail of the SA algorithm will be described in Section 4. To apply the SA algorithm, a set of parameters and functions needs to be specified, such as *initial temperature, final temperature, cooling ratio, temperature function, accept function*, etc. These parameters and functions determine how closely and how quickly SA can converge to the global minimum of the objective function. However, the fact is, there is no a straightforward way for specifying these parameters. In previous works, these parameters are set either by using empirical values [6] [13], or decided by the specific characteristics of a particular problem [11]. Apparently, we cannot make sure that these empirical or problem-specific values could be generally applicable to all other problems. Without a systematic way, the parameters of SA are usually randomly selected, and as a result, the quality of the set of parameters is not guaranteed.

A systematic method is necessary to generate the parameters of the SA algorithm for solving the application mapping problem, even though it has not so far been mentioned in literature. In this work, we propose to use the *Nelder-Mead simplex method*, which is originally introduced in [7], to automatically generate the optimized parameters of SA. The generated parameters are applied to the SA algorithm to find the optimized mapping solution which achieves minimized communication energy consumption on the NoC. In this work, by using the set of optimized parameters, we target to utilize less evaluations in SA while achieving comparably good quality of the final solution with the reference work.

The rest of the paper is organized as follows. Section 2 reviews previous works which use the SA algorithm for application mapping and outlines our considerations and contributions in this work. In Section 3, the problem of application mapping is modeled and the objective function is defined. Section 4 describes the general SA algorithm and Section 5 presents the Nelder-Mead simplex method. The proposed parameter-optimized SA algorithm is presented in Section 6. The experimental results are demonstrated in Section 7. Section 8 summarizes this paper.

2 Related Work

In [6], the authors developed an energy-aware mapping algorithm, namely Branch and Bound, for 2D mesh NoCs. The SA algorithm is implemented as a reference to evaluate the Branch and Bound algorithm. The comparison shows that SA can find the better mappings which achieve lower communication energy consumption than the ones found by the proposed Branch and Bound algorithm. However, the major drawback of SA is speed. The result in [6] shows that for a video/audio application, SA is 82 times slower than the Branch and Bound algorithm. The initial temperature was set 100 and the cooling ratio , 0.9. The final temperature was not set because a different termination criteria was used.

In [13], the authors tried to speed up SA by optimizing the number of iterations per temperature level and the core swapping process as well. A both NoC- and application-aware iteration number is used. In addition, to generate a neighboring solution, two cores are selected and swapped based on the possibility distribution function, instead of on an uniformly random possibility. The optimized SA is claimed to be 98% faster at the price of 13% memory consumption on average. In this work, the initial temperature, final temperature and cooling ratio are 1, 0.001 and 0.9 respectively.

In [11], the authors used SA to map an application on multiprocessor system-on-chip (MPSoC). The cooling ratio was set 0.95. Two functions, which are based on the execution time of each task defined in the task graph, were introduced to derive the initial and final temperature. It shows that by using parameter obtained by these two functions, the proposed method saves over half the optimization time and loses only 0.3% in performance.

We can see that for the application mapping problem, the parameters of SA were selected randomly by authors in these works. In this way, whenever we encounter a particular application mapping problem, we have to decide which set of parameters we should use. Since we are not able to determine which one is best suitable for our specific problem, the decision is difficult to make.

For the parameters used in the SA algorithm, we believe that:

1. The parameters of SA are problem-specific. One set of parameters would not be appropriate to other problems. The selection of the set of parameters has to be done with respect to the particular problem.
2. The parameters of SA have a joint impact on the performance of SA. This means that these parameters should be selected systematically, instead of being set independently.

Based on these two considerations, in this work, we apply the Nelder-Mead simplex method to systematically select the parameters of SA for the application mapping problem. The original Nelder-Mead simplex method is presented in [7] with purpose of finding the minimal value of a function of n variables. By using the Nelder-Mead simplex method, both the parameters to be selected and the cost function used in the selecting process can be defined with respect to the particular application mapping problem. This provides flexibility for us to use the SA algorithm for solving application mapping problems with single or multiple objectives. Moreover, since the set of parameters is selected by the systematic procedure for the particular problem, instead of being set by empirical values, the selected set of parameters is problem-specific and their qualities are guaranteed. To our best knowledge, this is the first work on investigating the systematic way of selecting optimized parameters of SA in the application mapping problem domain.

3   Application Mapping Problem

3.1   Application and NoC Model

The inputs of the mapping problem consist of two parts: one is the application and another is the many-core NoC. In this work, the application is modeled by a communication weighted graph (CWG), and the many-core NoC by a computation and communication resource graph (CCRG). For the sake of simplicity, in this paper, we use a 2D mesh NoC with homogeneous cores as the target computation and communication platform. We note that the method presented in this work is also applicable to other 2D NoCs with regular or irregular topologies. The X-Y deterministic routing is adopted by which a flit is first routed to the X direction and then the Y direction over the NoC.

Definition 1. A CWG is a directed graph $< V, E >$, where $V = \{v_1, v_2, \ldots, v_M\}$ represents the set of tasks of an application, corresponding to the set of CWG vertices, and $E = \{(v_i, v_j)|v_i, v_j \in V\}$ denotes the set of communications between tasks, corresponding to the set of CWG edges. Each edge (v_i, v_j), denoted by e_{ij} has weight vol_{ij} representing the total communication amount, in bits, transmitted from task v_i to v_j. M denotes the total number of tasks.

Definition 2. A CCRG is a directed graph $< TL, CH >$, where $TL = \{tl_1, tl_2, \ldots, tl_N\}$ denotes the set of tiles on the NoC, corresponding to the set of CCRG vertices. N denotes the total number of tiles. $CH = \{(tl_i, tl_j)|tl_i, tl_j \in TL\}$ designates the set of communication channels between tiles. The length of communication channel (tl_i, tl_j), denoted by $|ch_{ij}|$ is represented by the number of hops from tile tl_i to tl_j. On a 2D mesh

NoC with X-Y routing, the length of the communication channel between tiles tl_i and tl_j is calculated as follows:

$$|ch_{ij}| = |x_i - x_j| + |y_i - y_j| \tag{1}$$

where (x_i, y_i) and (x_j, y_j) are the coordinates of the tile tl_i and tl_j on a 2D mesh NoC respectively.

3.2 Objective Function

Using the preceding application and NoC model, the mapping of CWG to CCRG is defined by the one-to-one task-tile mapping function $map : V \to TL$:

$$map(v_i) = tl_i, \forall v_i \in V, \exists tl_i \in TL \tag{2}$$

When two tasks v_i and v_j of an edge e_{ij} in CWG are mapped on two tiles on the CCRG, an amount of vol_{ij} data will be transferred from the tile tl_i $(map(v_i))$ to the tile tl_j $(map(v_j))$. Based on the energy model proposed in [6], the communication energy consumption of e_{ij} is

$$E_{ij} = vol_{ij} \times (|ch_{ij}| \times E_{S_{bit}} + (|ch_{ij}| - 1) \times E_{L_{bit}}) \tag{3}$$

where $E_{S_{bit}}$ and $E_{L_{bit}}$ refer to the energy consumed by the switch and the link for transmitting one bit of data. And the total energy consumed by the application defined in CWG is

$$E_{app} = \sum_{\forall e_{ij} \in E} E_{ij} \tag{4}$$

From Equation 3 and 4, we can see that, given the constants $E_{S_{bit}}$ and $E_{L_{bit}}$, the communication energy consumption of an application is linearly proportional to the product of the data volume vol_{ij} and the length of communication channel $|ch_{ij}|$. In this work, since the objective of application mapping is to minimize the E_{app}, we need to minimize the sum of the product of vol_{ij} and $|ch_{ij}|$ for all communications in an application. Herein, we define the weighted communication of an application (WCA) as the objective function to evaluate the quality of each candidate mapping.

Definition 3. The Weighted Communication of an Application (WCA) is the sum of products of the data volume vol_{ij} and the length of communication channel $|ch_{ij}|$ for all communications in E.

$$WCA = \sum_{\forall e_{ij} \in E} vol_{ij} \times |ch_{ij}| \tag{5}$$

A mapping solution which can produce smaller WCA is considered to be a better solution because it will in turn yield a lower E_{app}.

4 Simulated Annealing

Simulated annealing is a stochastic search method for optimization problem. The pseudo-code of the general SA algorithm, derived from [10], is shown in Algorithm 1. The symbols and corresponding definitions used in Algorithm 1 are listed in Table 1. SA simulates the metallurgical process of heating up a solid and then cooling down slowly until it crystallizes. It starts from an initial, higher temperature and stops at a final, lower temperature. An initial solution and its cost are given at the initial temperature. Thereafter, at each temperature, SA tries L times of attempt mappings. In each attempt, a new mapping solution is generated from current one using the move function $Move(S, T)$. The cost of the new solution is compared with current cost. The algorithm always accepts a move with lower cost. Contrary to the greedy algorithm, SA accepts a worse move with higher cost by a changing possibility. This helps to avoid local minimum and find the global minimum. The accept possibility is decided by acceptance function $Accept(\Delta C, T)$ and decreases along with the temperature. As temperature cools down, SA gradually becomes greedy and converges to the global minimum.

Algorithm 1. General Simulated Annealing Algorithm

1 $S \leftarrow S_0$
2 $C \leftarrow Cost(S_0)$
3 $S_{best} \leftarrow S$
4 $C_{best} \leftarrow C$
5 $R \leftarrow 0$
6 **for** $i \leftarrow 0\ to\ \infty$ **do**
7    $T \leftarrow Temp(i)$
8    $S_{new} \leftarrow Move(S, T)$
9    $C_{new} \leftarrow Cost(S_{new})$
10   $\Delta C \leftarrow C_{new} - C$
11   **if** $\Delta C < 0\ or\ Accept(\Delta C, T)$ **then**
12   **if** $C_{new} < C_{best}$ **then**
13       $S_{best} \leftarrow S_{new}$
14       $C_{best} \leftarrow C_{new}$
15   **end if**
16     $S \leftarrow S_{new}$
17     $C \leftarrow C_{new}$
18     $R \leftarrow 0$
19   **else**
20     $R \leftarrow R + 1$
21   **if** Terminate(i, R) = True **then**
22       $break$
23   **end if**
24   **end if**
25 **end for**
26 return S_{best} and C_{best}

---

5   Nelder-Mead Simplex Method

The Nelder-Mead simplex method is proposed in [7] for the minimization of a function $f(p)$ with n variables $x_1, x_2, \ldots, x_n$. In this method, a number of $n + 1$ points (solutions) $p_0, p_1, \ldots, p_n$ are originally selected and form the so-called simplex. The set of points are then used to generate a new and better point which will replace the worst point in current simplex and forms a new simplex. Each point of the simplex is a n-tuple with n variables, i.e., $p_k = (x_1^k, x_1^k, \ldots, x_n^k)$. The Nelder-Mead simplex method compares the $n + 1$ function values $f(p_i)$ $(0 \leq i \leq n)$ and replaces the point with largest cost by the newly generated point. In each iteration, the replacement is realized by three operations: *reflection, expansion* and *contraction*. If it fails to do the replacement through these three operations, all points forming the simplex are updated with new values to generate a new simplex. The general Nelder-Mead simplex method is described in Algorithm 2.

Algorithm 2. Nelder-Mead Simplex Method for Minimizing $f(p)$

---

1   Select the initial $n + 1$ points p_i $(0 \leq i \leq n)$.
2   **while** ($!stop()$) **do**
3   Sort $f(p_i)$ $(0 \leq i \leq n)$ such that $f(p_0) \leq f(p_1) \leq \cdots \leq f(p_{n-1}) \leq f(p_n)$.
4   Let $\bar{p} = \sum\limits_{i=0}^{n-1} p_i/n$.
5   Generate *reflection point* $p_r = \bar{p} + \alpha * (\bar{p} - p_n)$.
6   **if** $f(p_r) \leq f(p_{n-1})$) **then**
7   Replace p_n by p_r.
8   Generate *expansion point* $p_e = \bar{p} + \beta * (p_r - \bar{p})$.
9   **if** $(f(p_r) < f(p_0)) \wedge (f(p_e) < f(p_r))$ **then**
10   Replace p_n by p_e.
11   **end if**
12   **else**
13   Let $f(p^*) = min(f(p_r), f(p_n))$.
14   Generate *contraction point* $p_c = \bar{p} + \gamma * (p^* - \bar{p})$.
15   **if** $f(p_c) \leq f(p^*)$ **then**
16   Replace p_n by p_c.
17   **else**
18   Update p_j with $(p_j + p_0)/2$ for $j = 0, 1, \ldots, n$.
19   **end if**
20   **end if**
21 **end while**
22 Return the point p_0.

As shown in Algorithm 2, the principle of the Nelder-Mead simplex method is, if $f(p_r) \leq f(p_{n-1})$, then the point p_n is replaced by its reflection point p_r. Thereafter, if $f(p_r) < f(p_0)$, the reflection point is expanded to the expansion point p_e and the point p_n is replaced by p_e. The procedure restarts when the expansion is done. In the case that $f(p_r) > f(p_{n-1})$, the contraction point p_c is generated. If $f(p_c) < min(f(p_r), f(p_n))$, the point p_n is replaced by contraction point p_c. Otherwise, all points in current simplex are updated by $p_j = (p_j + p_0)/2(j = 0, 1, \ldots, n)$ and a new simplex is generated. Thereafter the process restarts.

By continuously replacing the point p_n with a point which achieves smaller $f(p)$, the value of the function $f(p)$ converges to the minimum. The process terminates when the function $stop()$ becomes true. The state of function $stop()$ can be determined by whether the value of function $f(p)$ has converged to a final value [7], or whether the points forming the simplex have already converged to a final point [12]. In this work, because we try to find the optimized parameters for the SA algorithm, we adopt the latter way to define the function $stop()$. More precisely, in Algorithm 2, $stop()$ becomes true when $|x_i^k - x_j^k| \leq \varepsilon^k (i \neq j)$, for all i, j and k, where x_i^k and x_j^k are the kth element of point p_i and p_j respectively. Each element of vector ε, called *convergence degree* of variable x, is a predefined small positive value which determines the magnitude of the convergence.

In Algorithm 2, reflection coefficient α, expansion coefficient β and contraction coefficient γ give the factors by which the new simplex is generated by reflection, expansion and contraction respectively. These coefficients decide the speed of the convergence and the quality of the final point. In [7] and [12], different values of α, β and γ were used. In this work, we evaluated both sets of values by applying them in the Nelder-Mead simplex method for the same set of benchmarks. The result shows that both sets of parameters achieve comparable performance of the SA algorithm, but the Nelder-Mead simplex method using the coefficients in [12] can converge to the final point with 100 times less CPU time than that in [7]. Therefore, we use 1/3, 2.0 and 1.5 in [12] for α, β and γ respectively.

6   Parameter-Optimized Simulated Annealing

6.1   Parameters and Functions in SA

As shown in Algorithm 1, to apply SA to the application mapping problem, a number of parameters and functions have to be specified. In this section, we specify the parameters and functions used for implementing the SA algorithm in this work.

Cost Function. The objective function of application mapping, i.e., the WCA defined in Equation (5), is used as the cost function $Cost(S)$ in SA.

Annealing Schedule: $Temp(i)$ Function. The annealing schedule determines how the temperature is cooling down. At each step of annealing, a new temperature is generated by temperature function $Temp(i)$. We choose the geometric annealing schedule presented in [10] where the $Temp(i)$ is defined as:

$$Temp(i) = T_0 \times q^{\lfloor \frac{i}{L} \rfloor} \tag{6}$$

The new temperature is decided by the initial temperature T_0, the cooling ratio q, the accumulated number of iterations i and the number of iterations at each temperature L.

Number of Iterations L. The number of iterations at each temperature L is identically set as $M(N-1)$, where M and N are the number of tasks in CWG and that of tiles in CCRG respectively.

Acceptance Function: $Accept(\Delta C, T)$. While an improving move ($\Delta C < 0$) is always accepted, the function $Accept(\Delta C, T)$ determines whether a worse move

($\Delta C > 0$) should be accepted or not at the temperature T. The normalized inverse exponential form is chosen to implement the acceptance function in this work.

$$Accept(\Delta C, T) = True \Leftrightarrow random() < p$$

$$p = \frac{1}{1 + \exp\left(\frac{\Delta C}{KC_0 T}\right)} \tag{7}$$

With this acceptance function, the possibility of accepting a worse move, p, is less than 50%. On the basis of the original normalized inverse exponential form presented in [10], we add the normalizing ratio K in the acceptance function which works together with the initial cost C_0 to normalize the cost difference ΔC. This comes from the observation that using the original normalized inverse exponential form , in cases that the C_0 is huge, an accepting possibility close to 50% will be created even for a very small ΔC at a very low temperature. This makes SA inefficient at the last set of lower temperatures.

Initial and Final Temperature. The acceptance function in (7) defines the relation between the accepting possibility p, cost difference ΔC and temperature T. Equation (7) can be solved with respect to T as follows:

$$T = \frac{\Delta C}{\ln(\frac{1}{p} - 1)} \tag{8}$$

If we define P_s the possibility of accepting the maximal ΔC at initial temperature T_0, and P_f the possibility of accepting the minimal ΔC at final temperature T_f, then the initial and final temperature can be calculated as follows:

$$T_0 = \frac{\Delta C_{max}}{\ln(\frac{1}{P_0} - 1)}, T_f = \frac{\Delta C_{min}}{\ln(\frac{1}{P_f} - 1)} \tag{9}$$

In the way that the T_0 and T_f are set manually using empirical values (the cases in [6] [13]), only a numerical range is given by T_0 and T_f, there are no realistic meanings behind T_0 and T_f. On the contrary, in this work, the usage of P_s and P_f is more meaningful and understandable for designers to choose the T_0 and T_f by Equation (9).

Move Function:$Move(S, T)$. We use the random swapping as the move function. A task in current mapping is randomly selected and then it is swapped with another randomly selected task.

Termination Function:$Terminate(i, R)$. We add one criteria $N_{\Delta C=0}$ into the termination function of coupled temperature and rejection threshold which is presented in [10], to determine the stopping condition in this work.

$$Terminate(i, R) = True \Leftrightarrow (Temp(i) < T_f \wedge R \geq R_{max})$$
$$\vee (N_{\Delta C=0} = Z) \tag{10}$$

$N_{\Delta C=0}$ stands for the number of consecutive temperatures at which the lowest cost C_{best} has not been changed. Z is the maximal number of $N_{\Delta C=0}$ allowed in the SA algorithm. R is the number of consecutive rejections since last acceptance and R_{max} is the maximal number of rejections allowed in the SA algorithm. With this termination function, the annealing is stopped either when the temperature reaches to or below the final temperature and the moves in last R_{max} iterations are rejected, or in the last Z temperatures, no better solutions have been found. In this work, we set $R_{max} = L$ and $Z = 0.1N_T$, when N_T stands for the total number of temperatures from T_0 to T_f.

Initial Mapping. A random mapping in which each task is randomly mapped on a tile, is generated as the initial mapping.

Summary of Parameters. Table 1 summarizes the parameters and functions used in the SA algorithm in this work. We can see from Table 1, to apply the SA algorithm in

Table 1. Functions and Parameters for SA

Symbol	Definition	Value
S	Mapping solution (S_0: initial solution)	
$Cost(S)$	Cost function	WCA (Equation (5))
$Temp(i)$	Temperature function i	$T_0 \times q^{\lfloor \frac{i}{L} \rfloor}$
i	Accumulated number of iterations	
q	Geometric annealing schedule cooling ratio	Nelder-Mead Simplex Method
L	Number of iterations at each temperature	$M(N-1)$
N	Number of tiles in CCRG	
M	Number of tasks in CWG	
$Accept(\Delta C, T)$	Return accept (True) or reject (False) for a worse move	$random() < 1/(1 + \exp \frac{\Delta C}{KC_0 T})$
K	Normalizing ratio	Nelder-Mead Simplex Method
C_0	Initial cost	$Cost(S_0)$
T_0	Initial temperature	$\Delta C_{max}/\ln(\frac{1}{P_s} - 1)$
T_f	Final temperature	$\Delta C_{min}/\ln(\frac{1}{P_f} - 1)$
ΔC_{max}	The maximal ΔC at initial temperature T_0	Experiment
ΔC_{min}	The maximal ΔC at final temperature T_f	Experiment
P_0	The possibility of accepting the maximal ΔC at initial temperature T_0	Nelder-Mead Simple Method
P_f	The possibility of accepting the minimal ΔC at final temperature T_f	Nelder-Mead Simplex Method
$Move(S, T)$	Return a neighboring mapping of S	
$Terminate(i, R)$	Return terminate (True) or continue (False)	$Temp(i) < T_f \wedge R \geq R_{max} \vee N_{\Delta C=0} = Z$
R	Number of rejections	
R_{max}	Allowed maximal number of rejections	L
N_T	The total number of temperatures from T_0 to T_f	$\ln(\frac{T_0}{T_f})/\ln(q)$
Z	The allowed maximal number of temperatures with $\Delta C = 0$	$0.1N_T$

this work, the values of 6 parameters need to be specified: $q, K, P_s, P_f, \Delta C_{max}$ and ΔC_{min}. As long as these parameters are specified, other parameters such as T_0 and T_f can be decided and all functions can work properly. In these 6 parameters, the values of ΔC_{max} and ΔC_{min} can be obtained from a set of mapping trials generated from the original mapping using the move function (see details in Section 7). The other 4 parameters, labeled "Nelder-Mead Simplex Method" in column "Value" in Table 1, are the most important parameters of SA in this work and they are going to be optimized by the Nelder-Mead simplex method presented in Section 5.

6.2 Parameter Optimization

To apply the Nelder-Mead simplex method to get the optimized parameters q, K, P_s and P_f, we need to define the function $f(p)$ and specify the boundaries and convergence degree of each parameter from which the parameter is chosen.

Function $f(p)$. Since using various sets of q, K, P_s and P_f, the SA algorithm will produce different minimized values of WCA, we can define the output of the SA, i.e., the minimized WCA, as the function $f(p)$ of variables q, K, P_s and P_f. With this definition, it is possible to use the Nelder-Mead simplex method for finding the final point of variables q, K, P_s and P_f which produces the minimum WCA by SA.

Parameter Boundaries and Convergence Degrees. Contrary to the work in [7] where the variables x is unbounded, the parameters q, K, P_s and P_f in our specific application mapping problem are bounded. Among them, the parameters P_s and P_f are theoretically in the range $(0.0, 0.50]$ according to Equation (7). For the SA algorithm, it is reasonable to set a higher acceptance possibility at the initial temperature and a relatively lower acceptance possibility at the final temperature. In this work, we set the range of P_s and P_f by $[0.20, 0.49]$ and $(0.0, 0.10]$ respectively. And the convergence degrees ε_{P_s} and ε_{P_f} are set 0.01 and 0.005 respectively. The cooling ratio q is supposed to be in range $(0.0, 1.0)$. In this work, we set the range $[0.80, 0.99]$ for q. The convergence degree ε_q is set 0.005. The value of K is allowed in the range of $(0.0, 1.0]$ and the convergence degree ε_K is set 0.05.

At the beginning of the Nelder-Mead simplex method, 5 initial points are generated by choosing 4 elements, i.e, q, K, P_s and P_f, from their allowable range. During the process of the Nelder-Mead simplex method, whenever an element of a point exceeds its boundary, the bound value is used for the element. The function $stop()$ becomes true and the process is terminated when these 5 points converge to one point.

6.3 Parameter-Optimized Simulated Annealing Algorithm

Applying the Nelder-Mead simplex method, we develop the parameter-optimized simulated annealing algorithm for application mapping problems on many-core NoCs. The proposed algorithm is described in Algorithm 3 where the function $sa()$ and $simplex()$ apply the Algorithm 1 and 2 respectively. After defining the boundaries and convergence degree for four target parameters, i.e., q, K, P_s and P_f, the optimized set of parameters are obtained by the Nelder-Mead simplex method. The optimal mapping

---

Algorithm 3. Parameter-Optimized Simulate Annealing

---
1 Define the boundaries and convergence degree ε for parameters q, K, P_s and P_f.
2 Obtain the final point of the Nelder-Mead simplex method, $p_{opt} = simplex()$.
3 Set $q_{opt} = p_{opt}.q, K_{opt} = p_{opt}.K$.
4 Set $P_{s_{opt}} = p_{opt}.P_s, P_{f_{opt}} = p_{opt}.P_f$.
5 Find the best solution by applying the final point to SA,
   $S_{best} = sa(q_{opt}, K_{opt}, P_{s_{opt}}, P_{f_{opt}})$.
6 Return S_{best}.

---

solution with minimized WCA is then found by running the SA algorithm with the optimized set of parameters.

Note that, with various implementations of the SA algorithm, the parameters needed to be selected are different. Since the variables and objective functions applied in the Nelder-Mead method can be arbitrary, the Algorithm 3 is applicable for obtaining different sets of optimized parameters corresponding to different implementations. This makes the method proposed in this work viable for selecting optimized parameters of the SA algorithm which deals with diverse problems.

7 Experiment

To evaluate the efficiency of the proposed parameter-optimized simulated annealing (POSA) algorithm, we experiment POSA with a set of benchmarks and compare with the implementation of the SA algorithm in [6].

7.1 Setup

The implementation of the SA algorithm in [6] is available in the NoCmap project [3]. In the NoCmap, the geometric annealing schedule is used and the q is set 0.9. T_0 is fixed to 100 and the final temperature T_f is unbounded. The objective of the NoCmap is to minimized the total communication energy consumption and the energy model presented in [6] is adopted in the simulator. In this work, we also use the NoCmap simulator to obtain the communication energy consumption of the mappings generated by the POSA algorithm.

Four benchmark applications are selected for the comparison, including a video object plane decoder (VOPD) and a MPEG4 from SUNMAP [9], a multimedia systems application (MMS) [5] and a H.264 decoder (H264) [14]. The CWGs of these applications are derived from original descriptions in these works. The benchmarks and corresponding NoCs used in this work are summarized in Table 2.

The optimized mapping of each benchmark is found both by the NoCmap and the POSA algorithm. The communication energy of both mappings are produced by the NoCmap simulator. For POSA, the average ΔC_{max} and ΔC_{min} are obtained from $5 * L$ move trials starting from the original random mapping, which are used to calculate the T_0 and T_f with given parameters P_0, P_f, C_0 and K. Both algorithms were executed on a Desktop PC having a 3.0 GHz Intel Core2 Duo CPU and 8.0 GB of memory.

7.2 Results

Optimized Parameters. Applying the POSA algorithm, the optimized mapping solution with minimized WCA of each application is achieved. At the same time, the optimized parameters of the SA algorithm are obtained. Table 2 shows the optimized parameters of SA for mapping the four benchmarks. As mentioned in Section 2, the parameters of SA are problem-specific. Table 2 illustrates that, instead of using an identical set of parameters, to find an optimized mapping, different parameters should be used in SA for mapping different applications.

Table 2. Optimized Parameters of SA for Benchmarks

Benchmark	Cores	NoC	q	P_0	P_f	K
VOPD	16	4x4	0.91	0.44	0.05	0.72
MPEG4	12	3x4	0.95	0.34	0.05	0.36
MMS	25	5x5	0.94	0.36	0.05	0.62
H264	16	4x4	0.89	0.42	0.05	0.49

Iterations and Runtime. Table 3 shows the iterations (I_s) of NoCmap and POSA algorithm for finding the final mapping solution of each application. The column T_0 and T_f are the initial and final temperature respectively. T_t refers to the temperature at which SA terminates. I_s is the number of iterations that the SA algorithm has run until it terminates. pct is the percentage of the iterations of POSA to that of NoCmap. We can see that, since optimized parameters are applied, a much lower initial temperature is set in POSA. As a result, POSA uses significantly smaller number of iterations which is on average less than 1% of that used in NoCmap, to get the final mapping.

Table 3. Iterations of SA for Benchmarks

Benchmark	T_0		T_f		T_t		I_s		
	NoCmap	POSA	NoCmap	POSA	NoCmap	POSA	NoCmap	POSA	pct
VOPD	100	2.69	-	1.35e-4	2.28e-6	9.88e-5	4.30e6	2.74e4	0.64%
MPEG4	100	1.90	-	1.26e-4	5.80e-7	8.38e-5	2.61e6	2.77e4	1.06%
MMS	100	1.36	-	1.26e-5	5.80e-7	1.25e-5	1.14e7	1.18e5	1.04%
H.264	100	3.11	-	1.94e-4	0.15	1.43e-4	1.61e6	1.94e4	1.02%

Table 4 shows the runtimes of SA in NoCmap and POSA (in seconds) and the speedup achieved by POSA. POSA is, on average, 1.41 times faster than that in NoCmap. Note that, the runtime of POSA includes the time consumed by the Nelder-Mead simplex method in which the SA is run more than hundred times. In terms of the runtime of a single run of SA, a significant speedup is achieved by POSA due to less evaluating iterations. In Table 4, $POSA'$ and $Speedup2$ represent the runtime of a single run of SA applying the set of optimized parameters, and the speedup over that in the NoCmap respectively. We can see that in POSA, the SA with optimized parameters is on average 237 times faster than that in NoCmap. This indicates how important the selection of parameters is regarding to the runtime of the SA algorithm.

Table 4. Runtimes and Speedup for Benchmarks

Benchmark	NoCmap	POSA	Speedup1	POSA´	Speedup2
VOPD	31.69	15.50	2.04	0.087	364
MPEG4	15.74	9.67	1.63	0.059	267
MMS	171.74	181.75	0.94	1.17	147
H.264	12.34	11.90	1.04	0.072	171
Average	-	-	1.41	-	237

WCA and Energy Consumption. In this work, minimizing communication energy consumption on NoC is the objective of applying SA to solve the application mapping problem. Figure 2 shows the WCA achieved by NoCmap and POSA for each application respectively. The results of both algorithms vary slightly. The maximum of WCA variance is less than 4% in the case of application H.264.

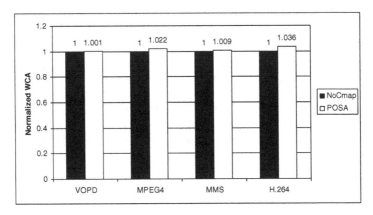

Fig. 2. Comparison of WCA

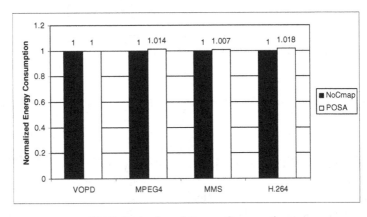

Fig. 3. Evaluation of Energy Consumption

As anticipated from the results of the minimized WCA, the communication energy consumptions achieved by NoCmap and POSA are almost same. Figure 3 shows that the maximal difference exists again in the case of application H.264, which is less than 2%. The comparable energy consumption verify the efficiency of the proposed POSA algorithm. Although a significantly smaller number of iterations is processed, POSA still can find the optimized mapping solution which is similar to the one found in NoCmap where a huge searching space is explored.

8 Conclusions

The set of parameters applied in the SA algorithm has great impact on the runtime and the quality of the final solution. A method to systematically select the parameters of the SA algorithm for the application mapping problem is proposed in this work. The Nelder-Mead simplex method, which is used to get the minimization of a function of n variables, is applied to find the optimized parameters of the SA algorithm. With the set of optimized parameters, less evaluations are performed and the SA algorithm is accelerated. In addition, this work also points out that the parameters of SA are problem-specific. Instead of using an identical set of empirical parameters, the proposed POSA algorithm provides a way to flexibly select various number of parameters with respect to different cost functions for different kinds of mapping problems.

The experiment shows that the proposed POSA algorithm is time- and energy-efficient. The POSA algorithm only uses on average less than 1% iterations of that used in NoCmap algorithm to converge to the final optimized solution. An average speedup of 1.4 times is achieved by POSA over NoCmap. With the optimized parameters, the SA instance in the POSA is 237 times faster than that in the NoCmap, while the optimal mapping is still found.

Acknowledgement. The authors would like to thank the Academy of Finland for the financial support for this work.

References

1. Benini, L., De Micheli, G.: Networks on chips: a new soc paradigm. Computer 35(1), 70–78 (2002)
2. Borkar, S.: Thousand core chips: a technology perspective. In: Proceedings of the 44th Annual Design Automation Conference, DAC 2007, pp. 746–749. ACM, New York (2007)
3. SLD: System Level Design Group @ CMU. Nocmap: an energy- and performance-aware mapping tool for networks-on-chip,
   http://www.ece.cmu.edu/sld/wiki/doku.php?id=shared:nocmap
4. Dally, W.J., Towles, B.: Route packets, not wires: on-chip inteconnection networks. In: Proceedings of the 38th Annual Design Automation Conference, DAC 2001, pp. 684–689. ACM, New York (2001)
5. Hu, J., Marculescu, R.: Energy-aware mapping for tile-based noc architectures under performance constraints. In: Proceedings of the Asia and South Pacific, Design Automation Conference ASP, DAC 2003. pp. 233–239 (January 2003)
6. Hu, J., Marculescu, R.: Energy- and performance-aware mapping for regular noc architectures. IEEE Transactions on Computer-Aided Design of Integrated Circuits and Systems 24(4), 551–562 (2005)

7. Nelder, J.A., Mead, R.: A simplex method for function minimization. Computer Journal 7, 308–313 (1965)
8. Marcon, C.A.M., Moreno, E.I., Calazans, N.L.V., Moraes, F.G.: Comparison of network-on-chip mapping algorithms targeting low energy consumption. IET Computers & Digital Techniques 2(6), 471–482 (2008)
9. Murali, S., De Micheli, G.: Sunmap: a tool for automatic topology selection and generation for nocs. In: Proceedings of the 41st Design Automation Conference, pp. 914–919 (July 2004)
10. Orsila, H., Salminen, E., Hämäläinen, T.D.: Best practices for simulated annealing in multiprocessor task distribution problems. In: Simulated Annealing, pp. 321–342. I-Tech Education and Publishing KG (2008)
11. Orsila, H., Salminen, E., Hämäläinen, T.D.: Parameterizing simulated annealing for distributing kahn process networks on multiprocessor socs. In: Proceedings of the 11th International Conference on System-on-Chip, SOC 2009, pp. 19–26. IEEE Press, Piscataway (2009)
12. Park, M.-W., Kim, Y.-D.: A systematic procedure for setting parameters in simulated annealing algorithms. Comput. Oper. Res. 25, 207–217 (1998)
13. Radu, C., Vinţan, L.: Optimized simulated annealing for network-on-chip application mapping. In: Proceedings of the 18th International Conference on Control Systems and Computer Science (CSCS-18), Bucharest, Romania, May 24-27, vol. 1, pp. 452–459. Politehnica Press (2011)
14. van der Tol, E.B., Jaspers, E.G.T., Gelderblom, R.H.: Mapping of h.264 decoding on a multiprocessor architecture. In: Image and Video Communications and Processing, pp. 707–718 (2003)

Learning Algorithm Portfolios for Parallel Execution

Xi Yun[1] and Susan L. Epstein[1,2]

[1] Department of Computer Science, The Graduate School of The City University of New York,
New York, NY 10016, USA
[2] Department of Computer Science, Hunter College of The City University of New York,
New York, NY 10065, USA
xyun@gc.cuny.edu, susan.epstein@hunter.cuny.edu

Abstract. Portfolio-based solvers are both effective and robust, but their prom-
ise for parallel execution with constraint satisfaction solvers has received relatively little attention. This paper proposes an approach that constructs algorithm
portfolios intended for parallel execution based on a combination of case-based
reasoning, a greedy algorithm, and three heuristics. Empirical results show that
this method is efficient, and can significantly improve performance with only a
few additional processors. On problems from solver competitions, the resultant
algorithm portfolios perform nearly as well as an oracle.

Keywords: constraint satisfaction, algorithm portfolio, parallel processing, machine learning.

1   Introduction

Given a set of solvers and a set of constraint satisfaction problems (*CSPs*), no one
solver may consistently outperform all the others on every problem (e.g., [1-5]). In-
formally, an algorithm portfolio is a set of algorithms that run according to some
schedule on a set of problems. The thesis of this work is that learning and parallelism
can improve the efficiency and effectiveness of algorithm portfolios, so that they outperform each of their constituents. This paper explores offline learning to construct
such portfolios for CSPs. Given the performance of several algorithms on a training
set, we seek an algorithm portfolio that executes on multiple processors to solve the
most problems within some time limit. The principal result reported here is that, given
several additional processors, our method can construct algorithm portfolios whose
performance is competitive with that of an *oracle*, a solver that always chooses the
best available algorithm for each problem.

For parallel execution, a portfolio could simply schedule the same CSP on many
processors, each of which would execute a different solver on it, and then race until
some algorithm found a solution. Given the number of plausible solver configurations, this approach is not realistic. It is, however, possible to learn to schedule a set
of solvers on a set of processors. Our approach combines case-based reasoning
(*CBR*), a greedy algorithm, and a set of heuristics. Although CBR [6] and greedy
algorithms [7] have been applied to construct portfolios for CSPs before, this work is,

Y. Hamadi and M. Schoenauer (Eds.): LION 6, LNCS 7219, pp. 323–338, 2012.
© Springer-Verlag Berlin Heidelberg 2012

to the best of our knowledge, the first to combine them in a single framework. Given a CSP, our method uses CBR to identify a small set of similar training problems, and then greedily generates an effective portfolio without the complete search necessary to find an optimal one. In addition, we introduce three heuristics that transform algorithm portfolios intended for a single processor into ones intended for parallel execution. Extensive experiments show that portfolios produced by our method would solve more problems, not only when they are designed for one processor, but also consistently improve performance when they are designed for as many as 16 processors.

The next two sections provide background on CSPs and algorithm portfolios. Section 4 formulates algorithm portfolio construction as a machine learning task and reviews related work. Section 5 discusses a general framework that combines CBR with a greedy algorithm to construct algorithm portfolios, and Section 6 generalizes that framework to parallel algorithms. Subsequent sections detail and discuss the experimental design and results, and offer some conclusions.

2   Constraint Satisfaction Problems

A CSP here is a triple $<X, D, C>$, where X is a set of variables, D is a set of finite domains associated with those variables, and C is a set of constraints that those variables must satisfy. A constraint defined on two variables is *binary*, and one defined on $n > 2$ variables is *n-ary*. An *extensional* constraint explicitly represents a set of tuples; an *intensional* constraint implicitly describes tuples with a predicate.

An *instantiation* of a CSP assigns values to its variables from their respective domains. A *consistent* instantiation violates no constraint. An instantiation of all the variables is a *complete* instantiation, and a complete and consistent instantiation is a *solution*. A CSP is *solvable* if it has at least one solution; otherwise it is *unsolvable*.

Many constraint solvers search for a solution to a CSP with *systematic backtracking*, which assigns values to variables one at a time and checks consistency after each assignment. After an assignment, any inconsistent value for an as-yet-unassigned variable is temporarily removed from that variable's domain. A *wipeout* occurs when a domain becomes empty. At that point, search backtracks to an earlier variable with an alternative value, restores removed values along the way, and assigns another value to the earlier variable. Search returns a solution when one is found, or halts when the domain of the variable at the root of the search tree becomes empty.

A CSP solver is typically a complex combination of fundamental search algorithms, along with a set of techniques, heuristics, and policies to realize and support them. To improve overall search performance, *preprocessing* techniques manipulate the problem before a full search, *variable-ordering* heuristics choose the next variable to be assigned a value, and *value-ordering* heuristics choose a value for it. Once a heuristic orders the possible variables or values, *randomization* chooses one at random, usually from a small set of the top-ranked candidates [8]. A *restart* policy is a sequence of termination conditions that trigger the re-initiation of the search. Combined with randomization, a restart policy may improve search performance. Although the many ways to assemble a solver's components and then set their parameters yield a broad spectrum of search performance, they also provide fertile raw material for effective algorithm portfolio construction.

3   Algorithm Portfolios

An algorithm portfolio for CSP solution was originally defined as a method that combined different algorithms to improve search performance while it lowered *search risk*, the standard deviation of a performance metric (e.g., expected CPU time or number of backtracks to solve a problem) [9, 10]. In other words, an algorithm portfolio searched for a Pareto frontier in the two-dimensional space defined by a given performance metric and its standard deviation. Later, an algorithm portfolio was generalized to denote a combination of different algorithms intended to outperform the search performance of any of its constituent algorithms [3, 6, 11-14]. Here we extend that formulation, so that an algorithm portfolio schedules its constituent algorithms to run concurrently on a set of processors.

Let an *algorithm* be any CSP solver, as described in the previous section. Given a set $A = \{a_1, a_2, ..., a_m\}$ of m algorithms, a set $P = \{x_1, x_2, ..., x_n\}$ of n problems, and a set of B consecutive time intervals $T = \{t_1, t_2, ..., t_B\}$, a *simple schedule* S_k for a problem on a single processor specifies which algorithm addresses the problem in each time interval, that is, $S_k: T \rightarrow A$. (At most one algorithm executes in any time interval in a simple schedule.) A *schedule* for K processors is a set of K simple schedules, one for each processor. (Here, a schedule addresses only one problem at a time.) An *algorithm portfolio* is then a quintuple $<P, A, K, S, B>$ where S is a set of schedules that deploy algorithms A on K processors to solve problems from P within B. Note that our definition includes both *simple* ($K = 1$) and *parallel* ($K > 1$) algorithm portfolios. Without loss of generality, we also simplify T to $\{1, 2, ..., B\}$. Of course, neither a simple nor a parallel schedule can outperform an oracle's perfect algorithm selection.

Clearly, on one processor at most B time can be allotted to any algorithm on any problem. Thus the performance of A on P can be represented as an $n \times m$ *performance matrix* τ. If the entry $\tau_{ij} \in \{1, 2, ..., B\}$ then a_j solves x_i in time τ_{ij}; otherwise x_i goes unsolved by a_j in time B. A *deterministic* algorithm consistently produces the same output given the same problem and time cutoff; that is, for a deterministic algorithm each τ_{ij} is fixed. In contrast, the output of a randomized algorithm may change from one run to the next (i.e., τ_{ij} is a random number).

Given a problem, a *sequential* algorithm portfolio executes algorithms on it in a specific order, but does not preserve any intermediate search data for an algorithm when the portfolio leaves it. Thus, a sequential portfolio must restart on the problem if it later reapplies a previous algorithm to it. In contrast, a *switching* algorithm portfolio interleaves algorithms, and preserves intermediate search data, so that search can continue from a previous state when it returns to an earlier algorithm. *Algorithm selection* is an algorithm portfolio that schedules only one algorithm [13, 15].

The schedule for a *static algorithm portfolio* is constructed in advance, and goes unchanged during search. In contrast, a *dynamic algorithm portfolio* can profit from feedback as it executes, and adjust its schedule accordingly. For example, the dynamic algorithm portfolios in [2, 16] iteratively share a (possibly varying-length) time slice among all available algorithms, but modify the algorithms' relative priorities based on their progress. Adjustments for a dynamic portfolio can be triggered by unsatisfactory performance during execution [17, 18]. Most of the work referenced thus far is for simple schedules, which interleave algorithms on a single processor.

There are other ways to exploit parallel processing beyond the scope of this paper. These include *search space splitting* to partition the search space of a CSP into subspaces and uses different processors to explore difference subspaces [19], and *structural decomposition* to separate a CSP into simpler, smaller-size subproblems based on the structure of its constraint hypergraph [20, 21]. Moreover, a parallel SAT solver can share clauses learnt on different processors, where each processor executes a manually pre-determined algorithm [22].

The current algorithm portfolio performance metric is runtime, which may be used to optimize different objective functions. For example, a portfolio may be required to minimize its expected runtime on a problem generated at random from some problem distribution. (Alternatives are introduced in [14].) Recent CSP solver competitions evaluated solvers on how many problems they solved under a fixed, per-problem time limit, and broke ties on average solution time across solved problems [23, 24]. We compare algorithm portfolio construction methods (henceforward, *constructors*) with the same standard. (In contrast, SAT solver competitions have compared solvers with a complex scoring function that includes the performance of all competitors [25].)

As formulated here, the differences between two solvers may be simply in their choice of even a single technique, heuristic, or policy that sustains performance diversity. Thus an algorithm portfolio can be thought of as a mixture of experts [26], including variable-ordering and value-ordering heuristics, restart policies, and nogood learning methods. In particular, even if only one heuristic is available, the portfolio could consist of the heuristic and its opposite, or the heuristic and random selection.

4   Learning an Effective Algorithm Portfolio

Algorithm portfolio constructors that learn are classified as online or offline based on the way they use their training problems. An *offline* constructor observes the performance of algorithms on a set of training problems and then builds a portfolio of those algorithms to optimize its performance on an entire testing set [3, 6, 13]. An *online* constructor solves one problem at a time, and the knowledge it relies on for that problem comes only from the problems that preceded it [2, 7, 16]. This paper focuses on offline algorithm constructors.

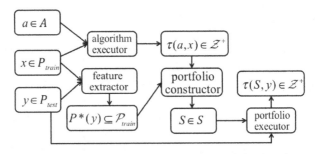

Fig. 1. Algorithm portfolio construction as offline learning

Our case-based approach to algorithm portfolio construction relies on feature extraction. Figure 1 represents offline algorithm portfolio construction with feature extraction as a machine-learning task. Given a set P_{train} of training problems, a set P_{test} of testing problems, and a performance matrix $\tau(a,x)$ that stores the time required by each algorithm $a \in A$ to solve each problem $x \in P_{train}$, the constructor's task is to find a schedule S with optimal performance that uses A to solve P_{test}. Here, all entries in τ are discrete, fixed positive integers, that is, all algorithms are assumed to be deterministic. $P^*(y)$ is a set of CSPs similar to testing problem y. (Portfolios of randomized algorithms are discussed in [3, 27].)

The two portfolio constructors most relevant here are CPHYDRA [6] and GASS [7]. Let $P(a_j, S)$ be the problems in P solved by a_j under schedule S. CPHYDRA defines the optimal schedule as one that maximizes the number of problems solved within B:

$$\underset{S}{\operatorname{argmax}} \left| \bigcup_j P(a_j, S) \right| \text{ such that length}(S) \leq B$$

Because it uses relatively few algorithms in competition, CPHYDRA can address optimality with exhaustive search, in time $O(2^m)$ where m is the number of algorithms. CPHYDRA had two entries in the 2008 competition, both with $m = 3$: CPHYDRA_k_10 used 10 similar training examples (i.e., $|P^*(y)| = 10$), and CPHYDRA_k_40 used 40. Among 24 competitiors, both versions finished in the top two solvers, except in the category for global constraints. CPHYDRA also weights training problems by their Euclidean distance from the testing problem. Its approach was later exploited and tailored for SAT problems [28] as well.

GASS' greedy algorithm bases its optimal schedule on $c_i(S)$, the expected time to solve x_i under schedule S. Its optimal schedule minimizes the overall runtime (equivalent to the average runtime under fixed n) to solve all problems in P_{train}:

$$\underset{S}{\operatorname{argmax}} \sum_{i=1}^{n} c_i(S)$$

At each step, GASS greedily maximizes the number of problems solved per unit of time, and counts only problems solved for the first time during the current time step. In time $O(nm \log n \cdot \min\{n, Bm\})$, GASS returns an approximate schedule that is at most four times worse (a *4-approximation*) than the optimal switching schedule. The computation of any better approximation is NP-hard [7].

5   WG, a New Constructor for Switching Algorithm Portfolios

Our Weighted Greedy (*WG*) algorithm is a new constructor for switching algorithm portfolios that exploits the perspectives of both GASS and CPHYDRA. For a single processor, CPHYDRA uses CBR to select a small set of similar training problems for each testing problem. It then does a complete search, exponential in the number of algorithms m, to find an optimal schedule for the new problem. In contrast, the impact of m on GASS is at worst quadratic; GASS' greedy approach is heavily dependent on the number of training problems n instead. WG exploits the fact that some problems are far more similar to a given testing problem than others, so that a properly selected subset of problems can estimate the runtime of the testing problem more precisely.

Input: training set $P = \{x_1, x_2, ..., x_n\}$, algorithms $A = \{a_1, a_2, ..., a_m\}$, time limit B,
testing problem y, weight function w: $\Re^d \rightarrow \Re^d$, neighbor set ratio r
Output: schedule S for a non-parallel switching algorithm portfolio

---

For $i = 1$ to n, compute Euclidean distance between x_i and y
$P^* \leftarrow \{100r\%$ of problems in P closest to $y\}$
For each x_i in P^*, compute weight $w_i = w(x_i)$
Initialize time step $z \leftarrow 1$, overall time $T \leftarrow 0$, and time spent $t_j \leftarrow 0$ for algorithm a_j
While $P^* \neq \emptyset$ and $T < B$
    Select a_j with execution time Δ_z to maximize $N_j^z(t_j + \Delta_z)/\Delta_z$
    Remove from P^* problems solved by a_j during step z
    Schedule a_j with execution time Δ_z in S
    Update times: $t_j \leftarrow t_j + \Delta_z$, $T \leftarrow T + \Delta_z$, and $z \leftarrow z + 1$
Return S

Fig. 2. High-level pseudocode for WG, a weighted greedy constructor for one processor

On one processor, to schedule within time limit B algorithms from A for a problem y given prior experience on a set of problems P, WG combines GASS and CPHYDRA into a single framework for switching scheduling. (See Figure 2.) WG is similar to GASS, except that it represents problems by numeric feature vectors, and restricts its attention to similar problems (i.e., reasons based only on similar cases). WG initially selects a *neighbor set* P^* that is the $100r\%$ of the most similar training problems (i.e., have feature vectors closest in Euclidean distance to that of y), $0 < r \leq 1$. The influence of these problems in the selection of an algorithm may be uniform, or be weighted in proportion to their distance d_i from y. The weight functions investigated here are shown in Table 1, where d_{min} denotes the smallest distance from a neighbor set problem to y, and d_{max} denotes the largest distance.

During each new interval Δ_z, WG counts (from the performance matrix τ) and weights how many training problems in the current neighbor set P^* it could solve within time $t + \Delta_z$ if it assigned problem x_i to algorithm a_j during that interval:

$$N_j^z(t) = \sum_{x_i \in P^*} w_i \zeta_{ij}(t) \text{ where } \zeta_{ij}(t) = \begin{cases} 1 & \text{if } \tau_{ij} \leq t \\ 0 & \text{otherwise} \end{cases} \tag{1}$$

WG then greedily maximizes (1) per unit of time expended, that is, it calculates

$$\underset{a_j, \Delta_z}{\text{argmax}} \frac{N_j^z(t + \Delta_z)}{\Delta_z}$$

and removes those now-solved similar problems from P^*. The time complexity of WG is $O(rnm \log rn \cdot \min\{rn, Bm\})$ because it considers every algorithm a_j with every interval length Δ_z.

Table 1. Three weight functions that measure problem similarity, where d_i denotes the Euclidean distance of problem y from the ith neighbor set problem x_i. Here, $\varepsilon = 0.001$.

Reciprocal weighting	Normalized weighting	Normalized-fixed weighting
$w_i = \dfrac{1}{1 + d_i}$	$w_i = 1 - \dfrac{(n-1)(d_i - d_{min})}{n(d_{max} - d_{min})}$	$w_i = 1 - \dfrac{(1-\varepsilon)(d_i - d_{min})}{d_{max} - d_{min}}$

6   Creation of Portfolios for Parallel Processing

An intuitive way to parallelize WG for K identical processors π_1, π_2, ..., π_K is to partition the similar training problems P^* into K subsets P^1, P^2, ..., P^K at random, and then use WG to construct a schedule for processor π_k based its corresponding subset P^k. We call this RPWG (randomized parallel WG). With uniform weights $w_i = 1$, RPWG is a naïve parallel version of GASS. (To reduce the impact of randomness, RPWG could construct such a partition v times, although to conserve time $v = 1$ here.) Thus the overall complexity of RPWG is $O(vrnm \log(rn/K) \cdot \min\{rn/K, Bm\})$. Similarly, RP-CPHYDRA, the naïve parallel version of CPHYDRA, randomly partitions the similar training problems into K subsets and then uses CPHYDRA on each subset to construct a schedule for each processor. Section 7 investigates both these naïve parallel constructors as baselines. (Other recent work relevant to parallel algorithm portfolios includes online learning [2, 16] and methods that split problems [29, 30].)

Effectively, the construction of a parallel algorithm portfolio to solve as many training problems as possible on K processors is an integer-programming (IP) problem. The goal is to find the schedule S that specifies the time allotments to all algorithms on all processors, such that no problem can receive more than B time from all the processors together, and the total number of problems solved is a maximum. The expression $(1 - \zeta_{ij}(t_{kj}))$ is 1 if problem x_i is unsolved by algorithm a_j after time t_{kj} allocated to it on π_k, and 0 otherwise. The product of $(1 - \zeta_{ij}(t_{kj}))$ over all j and k is 1 if problem x_i is not solved by any algorithm on any processor in schedule S_k, and 0 otherwise. Thus the best schedule is

$$\underset{S=\{S_1,...,S_K\}}{\arg\max} \sum_{i=1}^{n}[1 - \prod_{k=1}^{K}\prod_{j=1}^{m}(1 - \zeta_{ij}(t_{kj}))] \text{ such that } \sum_{j=1}^{m} t_{kj} \leq B \text{ and } t_{kj} \geq 0 \qquad (2)$$

Intuitively, when two schedules solve the same number of training problems, we would prefer the one that consumes less total time. Thus (2) becomes:

$$\underset{S=\{S_1,...,S_K\}}{\arg\min} \left\{ (KB+1)\sum_{i=1}^{n}\prod_{k=1}^{K}\prod_{j=1}^{m}(1 - \zeta_{ij}(t_{kj})) + \sum_{k=1}^{K}\sum_{j=1}^{m} t_{kj} \right\} \text{ such that } \sum_{j=1}^{m} t_{kj} \leq B \text{ and } t_{kj} \geq 0 \quad (3)$$

Expression (3) seeks to minimize the cost of schedule S, as measured by a penalty for unsolved problems (counted in the first sum) and the resources t_{kj} allocated to all processors. Each unsolved problem incurs cost $KB + 1$, which is greater than all available time on all processors. This guarantees that any benefit introduced by reduction in overall runtime will be overshadowed by the penalty for solving one less problem. The optimization in (3) is NP-hard; others have proposed the use of column generation to solve a simpler IP problem for algorithm scheduling for non-parallel algorithm portfolios [28]. Instead here we adopt heuristics to generalize WG for this IP problem.

We argue that the optimal solution to (3) can occur only when there exists at most one processor k for each algorithm a_j such that $t_{kj} > 0$. For example, consider a schedule that allocates time t_{1j} and t_{2j} ($0 < t_{1j} < t_{2j}$) to the same algorithm on processors 1 and 2, respectively. These times are resources only, and are not directed to any particular problem or algorithm. Any problem solved by some algorithm on processor

Input: training set $P = \{x_1, x_2, ..., x_n\}$, algorithms $A = \{a_1, a_2, ..., a_m\}$, time limit B,
testing problem y, weight function $w: \mathfrak{R}^d \to \mathfrak{R}^d$, neighbor set ratio r,
processors $\{\pi_1, \pi_2, ..., \pi_K\}$
Output: schedule $S = \{S_1, S_2, ..., S_K\}$ for a parallel switching algorithm portfolio

1 For $i = 1$ to n, compute Euclidean distance between x_i and y
2 $P^* \leftarrow \{100r\%$ of problems in P closest to $y\}$
3 Compute weight w_i for each x_i in P^* with w
4 Initialize time step $z \leftarrow 1$, overall time $T^u \leftarrow 0$ on processor π_u,
   time $t_{uj} \leftarrow 0$ for a_j on π_u
5 While $P^* \neq \emptyset$ and $T^u < B$ for at least one u
6   Select a_j on π_u with time Δ_z to maximize $N_j^z (t_j + \Delta_z)/ \Delta_z$ ** *Retain* **
7   Remove from P^* problems solved by a_j during step z
8   Schedule a_j with execution time Δ_z on π_u
9   Update times: $t_{uj} \leftarrow t_{uj} + \Delta_z$, $T^u \leftarrow T^u + \Delta_z$, and $z \leftarrow z + 1$
10 For each π_u where $T^u < B$
11   If $T^u = 0$ ** *Spread* **
12   then assign a_j to π_u for B, where a_j solves the most problems in P and $a_j \notin S$
13   update times: $t_{uj} \leftarrow B$, $T^u \leftarrow B$, and $z \leftarrow z + 1$
14   else π_u executes the first algorithm placed on π_u until B ** *Return* **
15   update times: $t_{uj} \leftarrow t_{uj} + (B - T^u)$, $T^u \leftarrow B$, and $z \leftarrow z + 1$
16 Return S

Fig. 3. High-level pseudocode for RSR-WG, a weighted greedy algorithm that constructs a parallel switching schedule with heuristics Retain, Spread, and Return

1 in t_{1j} can be solved by the same algorithm on processor 2 in t_{2j}. Removing the algorithm from processor 1 does not increase the number of unsolved training problems because the same problems will be solved on processor 2, but it does reduce the total runtime, and produces a better schedule.

Inspired by this argument, Figure 3 introduces *RSR-WG* for parallel algorithm portfolios, where RSR stands for three heuristics: Retain, Spread, and Return. Like WG, RSR-WG selects an initial set of similar training problems and tries to schedule greedily, but with modifications from our three heuristics. *Retain* (line 6) places algorithm a_j on processor π_u if that placement will maximize equation (1) per unit of expended time and π_u still has time available ($T^u < B$). Among such processors, Retain prefers one that has already hosted y before ($t_{uj} \neq 0$), and otherwise selects one that has thus far been used the least (i.e., has minimum T^u). If a parallel schedule S solves all training problems without making full use of all the processors, *Spread* (line 11) places the algorithm a_j that solves the most problems in P but does not appear in S on a processor that was idle throughout S (if one exists), breaking ties at random. (The rationale here is that a_j may be generally effective but not outstanding on y.) Finally, if a processor is not fully used in S (i.e., $T^u < B$), *Return* (line 14) places the first algorithm it executed on that processor until the time limit. Obviously, RSR-WG achieves the performance of an oracle when $K = m$, but it is also effective when K is relatively small compared to m, as demonstrated in the next section.

7   Experimental Design and Results

We compared the performance of parallel algorithm portfolios from three constructors to that of four non-parallel solvers on problems from the Third International CSP solver competition (*CPAI'08*). To extract the 36 features values (e.g., number of variables, maximum domain size) used by CPHYDRA and RSR-WG, we ran the CSP solver Mistral 1.550 ([31]). For feature extraction we allotted 1 second on an 8 GB Mac Pro with a 2.93 GHz Quad-Core Intel Xeon processor.

CPAI'08 included 3307 problems in 5 categories. Some solvers could not address problems in every category; we merged the 2-ARY-INT and N-ARY-INT ($N > 2$) categories because the same solvers addressed both. Because our experiments count solved problems (those where a solver finds a solution or proves that none exists), we excluded any problem that was not solved by any solver within the CPAI'08 time limit of 1800 seconds. If CPHYDRA does not extract features quickly enough, it simply splits its schedule evenly among its three algorithms. Rather than test portfolios' luck with an algorithm this way (and penalize a portfolio with more algorithms at its disposal), we chose to exclude such problems. Table 2 summarizes the remaining 2865 problems in 4 categories.

Stratified partitioning was used in all runs, to maintain the proportions of problems from different categories in each subset. Table 3 reports the performance, in number of problems solved within 1800 seconds each, of an oracle and three non-parallel algorithm portfolio constructors as baselines: CPHYDRA_k_10, CPHYDRA_k_40, and GASS. The data for GASS was obtained by 10-fold cross-validation with stratified partitioning on the 2865 problems.

All portfolio construction experiments ran under 10-fold cross-validation on a Dell PowerEdge 1850 cluster with one head node and 86 compute nodes, each with four Intel 2.80 GHz Woodcrest dual-core processors. RSR-WG results reported here are for portfolio construction (i.e., scheduling) time plus runtime. The runtimes of RPWG and RP-CPHYDRA did not include portfolio construction time, which gave them a slight advantage. In extensive testing, uniform weighting and the three weight functions in Table 1 produced slightly different performance improvements in RSR-WG, but no one statistically significantly outperformed the others consistently. Thus this paper reports only on the normalized-fixed weight function.

Table 2. Competition problems by category. Experiment problems were those for which at least one solver found a solution or showed that none existed, and also had features extractable within one second. Solvable problems had at least one solution.

Applicable solvers	Category	Competition problems	Experiment problems	Experiment solvable problems
17	GLOBAL	556	493	256
22	k-ARY-INT ($k \geq 2$)	1412	1303	739
23	2-ARY-EXT	635	620	301
24	N-ARY-EXT ($N > 2$)	704	449	156

Table 3. Benchmark results for the 3rd International CSP solver competition

Solver	Oracle	GASS	CPHYDRA_k_10	CPHYDRA_k_40
Number solved	2865	2773	2577	2573
% solved	100%	96.79%	89.95%	89.81%

In CPAI'08, CPHYDRA chose 10 or 40 similar problems from which to learn, so here RP-CPHYDRA selects $10*K$ neighbors, randomly distributes them to K processors, and executes a complete search for the optimal schedule on each processor. RP-CPHYDRA's portfolio construction time was limited to 180 seconds. If it did not produce the optimal schedule in that time, the best schedule found so far was used. To reduce search time, any algorithm dominated by another algorithm (i.e., always outperformed by it on all 2865 problems) was also eliminated from RP-CPHYDRA's consideration. RP-CPHYDRA also scaled all schedules (as discussed in Section 8) to exploit the full time limit B.

Table 4. Performance of 3 parallel portfolio constructors on 2865 problems, with best value for K processors in boldface. * means RSR-WG outperformed RPWG; † means RSR-WG outperformed RP-CPHYDRA.

K	RP-CPHYDRA	Neighbor set ratio					
		0.005		0.01		0.02	
		RPWG	RSR-WG	RPWG	RSR-WG	RPWG	RSR-WG
1	2779	2771	2773	2778	2779	**2787**	2786†
2	2807	2801	**2826***	2799	2821*	2802	2823*†
3	2817	2808	**2841***†	2810	2836*†	2808	2839*†
4	2827	2810	**2850***†	2812	2847*†	2811	2847*†
5	2830	2817	**2855***†	2819	2851*†	2816	2852*†
6	2831	2821	**2857***†	2818	2855*†	2819	2856*†
7	2834	2823	**2858***†	2823	**2858***†	2824	2857*†
8	2834	2825	2859*†	2825	**2860***†	2825	2858*†

Table 5. Mean and standard deviation for the number of problems solved by RSR-WG out of 2865, with normalized-fixed weight function over 10 runs with K processors. Best value for K processors is in boldface.

K	Neighbor set ratio											
	0.005		0.01		0.02		0.04		0.08		0.16	
1	2773	3.65	2779	3.20	2786	2.30	**2789**	3.17	2788	3.09	**2789**	2.51
2	**2826**	3.51	2821	2.49	2823	3.16	2816	2.97	2810	2.99	2809	2.87
3	**2841**	2.12	2836	1.93	2839	2.56	2832	2.07	2827	2.27	2819	2.07
4	**2850**	2.15	2847	1.57	2847	2.63	2843	2.06	2838	2.22	2832	2.50
5	**2855**	1.37	2851	2.35	2852	0.88	2850	1.78	2845	2.72	2843	3.26
6	**2857**	0.95	2855	1.07	2856	1.26	2853	1.64	2851	1.03	2850	1.07
7	**2858**	0.79	**2858**	0.57	2857	0.82	2855	1.83	2854	2.35	2854	1.14
8	2859	1.18	**2860**	1.34	2858	1.06	2858	1.18	2856	0.74	2855	1.43
16	**2864**	0.42	**2864**	0.00	**2864**	0.00	2863	0.00	2861	0.42	2861	0.47

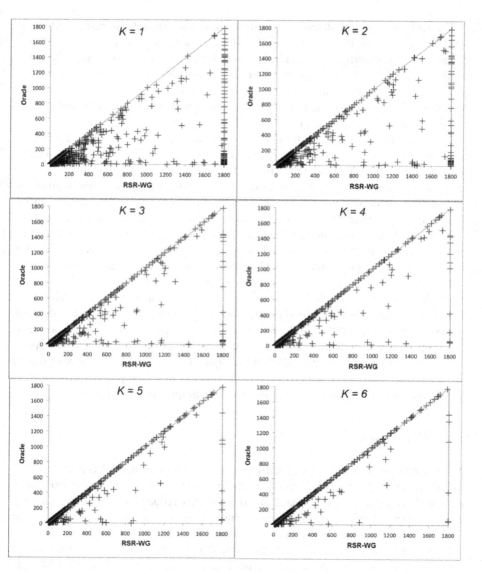

Fig. 4. Comparison of (ideal) oracle runtime (*y-axis*) to RSR-WG's time (*x-axis*) for 1 run with weight function normalized-fixed and neighbor set ratio 0.005. Each + denotes a result on one of the 2865 problems. Number of processors K ranges from 1 to 6.

Table 4 compares the performance of parallel portfolios from three constructors: RP-CPHYDRA (the parallel version of CPHYDRA), RPWG (the naïve parallel version of GASS), and RSR-WG. It lists the total number of problems (out of 2865) solved by each constructor's portfolios, and flags experiments where RSR-WG portfolios were statistically significantly better ($p < 0.005$) than those of a naïve parallel constructor.

For RSR-WG we simulated all 24 solvers from the original competition [23]. For RSR-WG only, we tested as many as $K = 16$ processors. Both $K = 8$ and $K = 16$ produced near-oracle performance; indeed, 2 out of 10 runs for $K = 16$ were perfect. Execution of RSR-WG on $K = 16$ processors is a reasonable approach for modern computers, where it would produce portfolios able to solve only one fewer problem than an oracle. (Execution of RSR-WG on one computer with multiple cores could degrade performance, for example, due to overhead introduced by memory sharing.)

One important question is the number of training problems to use for CBR, as measured by the *neighbor set ratio* (# problems / # training problems). For K from 1 to 16 we tested neighbor set ratios of 0.005 to 0.16, which yield neighbor sets that range in size from 14 to 458, respectively. Table 5 reports on how many problems (out of 2865) RSR-WG solved, and shows how the neighbor set ratio impacts performance under different numbers of processors K. Boldface entries in Tables 4 and 5 indicate the best performance for each K. Clearly RSR-WG efficiently generates effective algorithm portfolios, and does best with small neighbor set ratios for $K > 1$. On $K = 1$, RSR-WG outperforms GASS, CPHYDRA_k_10, and CPHYDRA_k_40.

Finally, Figure 4 compares the runtimes of an oracle solver and RSR-WG in one run with neighbor set ratio 0.005 and weight function normalized-fixed. (Again, RSR-WG's time includes both portfolio construction and search.) As in [23], each plus sign represents one of the 2865 problems. Those at the far right correspond to problems that went unsolved by RSR-WG in 1800 seconds. Those on the diagonal correspond to problems that were solved by RSR-WG as quickly as an oracle would have solved them. Clearly, more processors reduced the number of unsolved problems (from 90 to 6 in this particular run) and solved more problems as quickly as an oracle.

8   Discussion

As indicated above, the 1800-second runtime per problem for RSR-WG in these experiments includes the time to extract features, construct the schedule, and to execute it. RSR-WG adopts a greedy approach that dramatically reduces its scheduling time but still generates effective portfolios. For example, over 10 runs the average scheduling time of RSR-WG for $K = 8$ processors ranged from 14.56 to 14.96 seconds (σ in [6.05, 6.35]) with normalized-fixed weights and a neighbor set ratio of 0.16. For $K = 1$ processor under the same conditions, average scheduling time ranged from 14.30 to 14.80 seconds (σ in [5.75, 6.15]). These are small but statistically significant differences. In contrast, RP-CPHYDRA sometimes failed to compute an optimal schedule within 180 seconds. When $K = 1$, CPHYDRA failed to compute an optimal schedule 4.81% of the time. When $K > 1$, CPHYDRA must construct a schedule for each processor, on training sets that may be considerably more diverse. This can increase the search effort; indeed, for $K = 8$, CPHYDRA failed to compute an optimal schedule 14.39% of the time. As for GASS, because it learns on all the training problems, it required more than 5 days of execution time for its single entry in Table 2.

Instead of Spread, one might *scale* S to extend it to the entire time limit B, that is, allocate B to algorithms proportionally to their runtimes in S. CPHYDRA adopted scaling, and so did RP-CPHYDRA in our experiments. Scaling, however, would be unwise in RSR-WG because the earliest designated algorithms might be both most promising and quick, in which case they would only be allotted relatively short time intervals Δ_z. Whether or not scaling is appropriate, we believe, is probably determined by the problem set. The Return heuristic succeeds, we suspect, because as K approaches m it is better to allot larger time intervals to an algorithm on a single processor.

We temper the results on $K = 16$ with the observation that it is very nearly a race, when the problems in the neighbor set are sufficiently descriptive to eliminate the poorest performers on y. We prefer to consider the near-optimal performance for $K = 8$, and even $K = 4$, and to remember that RSR-WG was charged for scheduling time, while its competitor constructors were not.

Coarser granularity (indicated by a smaller B, which allocates longer intervals) impacts the scheduling efficiency of RSR-WG, but the effectiveness of the resultant portfolio depends on the performance matrix entries for the neighbors of the testing problem. A smaller B does not necessarily reduce the effectiveness of the resultant algorithm portfolio; if that were the case, a switching (or scheduling) portfolio would always be superior to algorithm selection. In addition to Table 5, where $B = 1800$, we tested RSR-WG with $B = 20, 10, 5, 4, 3, 2,$ and 1. (This is equivalent to time allocations that, instead of 1 second on a processor, are 90, 180, 360, 450, 600, 900, or 1800 seconds. Note that $B = 1$ is equivalent to racing one algorithm on each processor to address a problem.) In these granularity experiments, for $K = 1$ the number of solved problems peaked at $B = 10$. For $1 < K \leq 8$, no coarser granularity ever showed a significant improvement; indeed, performance degraded slightly as B decreased. Both improvement on $K = 1$ and failure to improve when $K > 1$ were consistent across all neighbor set ratios reported here, with peaks at either $B = 5$ or $B = 10$ when $K = 1$.

The success of RSR-WG algorithm portfolios relies heavily on the diversity of the performance of its constituent algorithms and the relevance of the extracted features. Typically, algorithm portfolio constructors select their algorithms and features based upon domain-specific knowledge. The reader may, for example, wonder how RSR-WG would perform if it relied on the three solvers CPHYDRA used in CPAI'08. The difficulty here is that CPHYDRA included solvers from the 2006 competition, solvers that did not enter CPAI'08, and whose performance was therefore unavailable on the 2008 problems. Although algorithm choice based on domain knowledge and feature selection can further enhance a portfolio's performance, it could also make it vulnerable to overfitting. When the number of features is larger, feature selection can be of considerable benefit to an algorithm portfolio constructor [13, 14], and we intend to explore it in future work.

Current work is proceeding in several directions. In practice, many algorithms may perform differently on the same problem in different runs, but still exhibit a certain level of consistency [3]. Indeed, in (sequential) CSP solver competitions, solvers typically fix their parameter values and introduce relatively little randomness to achieve stable performance. In that case, with coarse granularity (e.g., $B = 10$), a solver's performance is nearly deterministic. Greater randomness, however, could change solvers' performance dramatically, and thereby potentially benefit parallel constraint solving. A generalization of RSR-WG is in process to handle such behavior. On the other hand, automatic parameter tuning could introduce much diversity, and should fare well in algorithm portfolios [32]. Specifically, one may view different configurations of an algorithm as different algorithms, and thereby combine parameter tuning and an algorithm portfolio in the same framework. We are pursuing this avenue as well.

The performance of any algorithm portfolio is, of course, bounded by that of an oracle. The combination of algorithms as black boxes eliminates any opportunity to improve an individual algorithm. In contrast, parallelism can be achieved by a variety of problem decomposition methods (e.g., search space splitting), as discussed in Section 3. Although the results of recent SAT solver competitions suggest that a well-designed algorithm portfolio outperforms decomposition methods on a small number of processors [22], decomposition methods have shown their potential on many more processors (e.g., 64 cores or more in [19]). We will explore this in future work.

9   Conclusions

This paper presents WG, a constructor for non-parallel algorithm portfolios based on case-based reasoning and a greedy algorithm. It formulates parallel algorithm portfolio construction as an integer-programming problem, and generalizes WG to RSR-WG, a constructor for parallel algorithm portfolios based on a property of the optimal solution to the inherent integer-programming problem. To address a set of problems one at a time, RSR-WG creates portfolios of deterministic algorithms offline. Experiments show that the parallel algorithm portfolios produced by RSR-WG are statistically significantly better than those produced by naïve parallel versions of popular portfolio constructors. Moreover, with only a few additional processors, RSR-WG portfolios are competitive with an oracle solver on a single processor.

Acknowledgements. This research was supported in part by the National Science Foundation under grants IIS-0811437, CNS-0958379 and CNS-0855217, and the City University of New York High Performance Computing Center.

References

1. Gebruers, C., Hnich, B., Bridge, D.G., Freuder, E.C.: Using CBR to Select Solution Strategies in Constraint Programming. In: Muñoz-Ávila, H., Ricci, F. (eds.) ICCBR 2005. LNCS (LNAI), vol. 3620, pp. 222–236. Springer, Heidelberg (2005)
2. Gagliolo, M., Schmidhuber, J.: Learning Dynamic Algorithm Portfolios. *Annals of Mathematics and Artificial Intelligence 47(3), 295–328 (2006)*
3. Silverthorn, B., Miikkulainen, R.: Latent Class Models for Algorithm Portfolio Methods. In: Twenty-Fourth AAAI Conference on Artificial Intelligence, pp. 167–172 (2010)
4. Stern, D., Herbrich, R., Graepel, T., Samulowitz, H., Pulina, L., Tacchella, A.: Collaborative Expert Portfolio Management. In: Twenty-Fourth AAAI Conference on Artificial Intelligence, pp. 179–184 (2010)

5. Xu, L., Hutter, F., Hoos, H.H., Leyton-Brown, K.: SATzilla-07: The Design and Analysis of an Algorithm Portfolio for SAT. In: Bessière, C. (ed.) CP 2007. LNCS, vol. 4741, pp. 712–727. Springer, Heidelberg (2007)

6. O'Mahony, E., Hebrard, E., Holland, A., Nugent, C., O'Sullivan, B.: Using Case-Based Reasoning in an Algorithm Portfolio for Constraint Solving. In: Nineteenth Irish Conference on Artificial Intelligence and Cognitive Science (2008)

7. Streeter, M., Golovin, D., Smith, S.F.: Combing Multiple Heuristics Online. In: the Twentysecond National Conference on Artificial Intelligence, pp. 1197-1203 (2007)

8. Gomes, C., Selman, B., Crato, N.: Heavy-Tail Distributions in Combinatorial Search. In: Smolka, G. (ed.) CP 1997. LNCS, vol. 1330, pp. 121–135. Springer, Heidelberg (1997)

9. Huberman, B., Lukose, R., Hogg, T.: An Economics Approach to Hard Computational Problems. Science 256, 51–54 (1997)

10. Gomes, C., Selman, B.: Algorithm Portfolio Design: Theory vs. Practice. In: Thirteenth Conference On Uncertainty in Artificial Intelligence, pp. 190–197. Morgan Kaufmann (1997)

11. Guerri, A., Milano, M.: Learning Techniques for Automatic Algorithm Portfolio Selection. In: Sixteenth European Conference on Artificial Intelligence, pp. 475–479 (2004)

12. Xu, L., Hoos, H.H., Leyton-Brown, K.: Hydra: Automatically Configuring Algorithms for Portfolio-Based Selection. In: Twenty-Fourth AAAI Conference on Artificial Intelligence, pp. 179–184 (2010)

13. Xu, L., Hutter, F., Hoos, H.H., Leyton-Brown, K.: SATzilla: Portfolio-Based Algorithm Selection for SAT. Journal of Artificial Intelligence Research 32, 565–606 (2008)

14. Horvitz, E., Ruan, Y., Gomes, C.P., Kautz, H.A., Selman, B., Chickering, D.M.: A Bayesian Approach to Tackling Hard Computational Problems. In: Seventeenth Conference in Uncertainty in Artificial Intelligence, pp. 235-244. Morgan Kaufmann Publishers Inc. (2001) 720234

15. Rice, J.R.: The Algorithm Selection Algorithm. Advances in Computers 15, 65–118 (1976)

16. Gagliolo, M., Schmidhuber, J.: Towards Distributed Algorithm Portoflios. In: International Symposium on Distributed Computing and Artificial Intelligence, pp. 634–643 (2008)

17. Carchrae, T., Beck, J.C.: Low-Knowledge Algorithm Control. In: Nineteenth National Conference on Artificial Intelligence, Sixteenth Conference on Innovative Applications of Artificial Intelligence, pp. 49–54. AAAI Press / The MIT Press (2004) 1597158

18. Carchrae, T., Beck, J.C.: Applying Machine Learning to Low-Knowledge Control of Optimization Algorithms. Computational Intelligence 21(4), 372–387 (2005)

19. Bordeaux, L., Hamadi, Y., Samulowitz, H.: Experiments with Massively Parallel Constraint Solving. In: Twenty-First International Joint Conference on Artificial Intelligence, pp. 443–448. Morgan Kaufmann Publishers Inc. (2009) 1661516

20. Singer, D., Monnet, A.: JaCk-SAT: A New Parallel Scheme to Solve the Satisfiability Problem (SAT) Based on Join-and-Check. In: Wyrzykowski, R., Dongarra, J., Karczewski, K., Wasniewski, J. (eds.) PPAM 2007. LNCS, vol. 4967, pp. 249–258. Springer, Heidelberg (2008)

21. Li, W., van Beek, P.: Guiding Real-World SAT Solving with Dynamic Hypergraph Separator Decomposition. In: Sixteenth IEEE International Conference on Tools with Artificial Intelligence, pp. 542–548 (2004)

22. Hamadi, Y., Sais, L.: ManySAT: A Parallel SAT Solver. Journal on Satisfiability, Boolean Modeling and Computation 6, 245–262 (2009)

23. CPAI (2008), http://www.cril.univ-artois.fr/CPAI08/

24. Fourth International CSP Solver Competition, `http://www.cril.univ-artois.fr/CSC09/`
25. The SAT 2007 Competition, `http://satcompetition.org/2007/rules07.html`
26. Dietterich, T.G.: Ensemble Methods in Machine Learning. In: The First International Workshop on Multiple Classifier Systems, pp. 1–15 (2000)
27. Streeter, M., Golovin, D., Smith, S.F.: Restart Schedules for Ensembles of Problem Instances. In: The Twenty Second National Conference on Artificial Intelligence, pp. 1204–1210 (2007)
28. Kadioglu, S., Malitsky, Y., Sabharwal, A., Samulowitz, H., Sellmann, M.: Algorithm Selection and Scheduling. In: Lee, J. (ed.) CP 2011. LNCS, vol. 6876, pp. 454–469. Springer, Heidelberg (2011)
29. Segre, A.M., Forman, S., Resta, G., Wildenberg, A.: Nagging: A Scalable Fault-Tolerant Paradigm for Distributed Search. Artificial Intelligence 140, 71–106 (2002)
30. Vander-Swalmen, P., Dequen, G., Krajecki, M.: A Collaborative Approach for Multi-Threaded SAT Solving. International Journal of Parallel Programming 37, 324–342 (2009)
31. Mistral, `http://4c.ucc.ie/~ehebrard/Software.html`
32. Hutter, F., Hoos, H.H., Leyton-Brown, K., Stützle, T.: Paramils: An Automatic Algorithm Configuration Framework. Journal of Artificial Intelligence Research 36, 267–306 (2009)

Bayesian Optimization
Using Sequential Monte Carlo

Romain Benassi, Julien Bect, and Emmanuel Vazquez

SUPELEC, Gif-sur-Yvette, France

Abstract. We consider the problem of optimizing a real-valued continuous function f using a Bayesian approach, where the evaluations of f are chosen sequentially by combining prior information about f, which is described by a random process model, and past evaluation results. The main difficulty with this approach is to be able to compute the posterior distributions of quantities of interest which are used to choose evaluation points. In this article, we decide to use a Sequential Monte Carlo (SMC) approach.

1 Overview of the Contribution Proposed

We consider the problem of finding the global maxima of a function $f : \mathbb{X} \to \mathbb{R}$, where $\mathbb{X} \subset \mathbb{R}^d$ is assumed bounded, using the *expected improvement* (EI) criterion [1, 3]. Many examples in the literature show that the EI algorithm is particularly interesting for dealing with the optimization of functions which are expensive to evaluate, as is often the case in design and analysis of computer experiments [2]. However, going from the general framework expressed in [1] to an actual computer implementation is a difficult issue.

The main idea of an EI-based algorithm is a Bayesian one: f is viewed as a sample path of a random process ξ defined on $\mathbb{R}^d$. For the sake of tractability, it is generally assumed that ξ has a Gaussian process distribution conditionally to a parameter $\theta \in \Theta \subseteq \mathbb{R}^s$, which tunes the mean and covariance functions of the process. Then, given a prior distribution π_0 on θ and some initial evaluation results $\xi(X_1), \ldots, \xi(X_{n_0})$ at $X_1, \ldots, X_{n_0}$, an (idealized) EI algorithm constructs a sequence of evaluations points $X_{n_0+1}, X_{n_0+2}, \ldots$ such that, for each $n \geq n_0$,

$$X_{n+1} = \operatorname*{argmax}_{x \in \mathbb{X}} \bar{\rho}_n := \int_{\theta \in \Theta} \rho_n(x; \theta) \mathrm{d}\pi_n(\theta), \tag{1}$$

where π_n stands for the posterior distribution of θ, conditional on the σ-algebra $\mathcal{F}_n$ generated by $X_1, \xi(X_1), \ldots, X_n, \xi(X_n)$, and

$$\rho_n(x; \theta) := \mathsf{E}_{n,\theta}((\xi(X_{n+1}) - M_n)_+ \mid X_{n+1} = x)$$

is the EI at x given θ, with $M_n = \xi(X_0) \vee \cdots \vee \xi(X_n)$ and $\mathsf{E}_{n,\theta}$ the conditional expectation given $\mathcal{F}_n$ and θ. In practice, the computation of ρ_n is easily carried out (see [3]) but the answers to the following two questions will probably

Y. Hamadi and M. Schoenauer (Eds.): LION 6, LNCS 7219, pp. 339–342, 2012.
© Springer-Verlag Berlin Heidelberg 2012

have a direct impact on the performance and applicability of a particular implementation: a) How to deal with the integral in $\bar{\rho}_n$? b) How to deal with the maximization of $\bar{\rho}_n$ at each step?

We can safely say that most implementations—including the popular EGO algorithm [3]—deal with the first issue by using an *empirical Bayes* (or *plug-in*) approach, which consists in approximating π_n by a Dirac mass at the maximum likelihood estimate of θ. A plug-in approach using maximum a posteriori estimation has been used in [6]; *fully Bayesian* methods are more difficult to implement (see [4] and references therein). Regarding the optimization of $\bar{\rho}_n$ at each step, several strategies have been proposed (see, e.g., [3, 5, 7, 10]).

This article addresses both questions simultaneously, using a sequential Monte Carlo (SMC) approach [8, 9] and taking particular care to control the numerical complexity of the algorithm. The main ideas are the following. First, as in [5], a weighted sample $\mathfrak{T}_n = \{(\theta_{n,i}, w_{n,i}) \in \Theta \times \mathbb{R}, 1 \leq i \leq I\}$ from π_n is used to approximate $\bar{\rho}_n$; that is, $\sum_{i=1}^{I} w_{n,i}\, \rho_n(x; \theta_{n,i}) \to_I \bar{\rho}_n(x)$. Besides, at each step n, we attach to each $\theta_{n,i}$ a (small) population of candidate evaluation points $\{x_{n,i,j}, 1 \leq j \leq J\}$ which is expected to cover promising regions for that particular value of θ and such that $\max_{i,j} \bar{\rho}_n(x_{n,i,j}) \approx \max_x \bar{\rho}_n(x)$.

2   Algorithm and Results

At each step $n \geq n_0$ of the algorithm, our objective is to construct a set of weighted particles

$$\mathfrak{G}_n = \left\{\, (\gamma_{n,i,j}, w'_{n,i,j})\, , \right.$$
$$\left. \gamma_{n,i,j} = (\theta_{n,i}, x_{n,i,j}) \in \Theta \times \mathbb{X}, w'_{n,i,j} \in \mathbb{R}\, ,\; 1 \leq i \leq I, 1 \leq j \leq J \,\right\} \quad (2)$$

so that $\sum_{i,j} w'_{n,i,j}\delta_{\gamma_{n,i,j}} \to_{I,J} \pi'_n$, with

$$\mathrm{d}\pi'_n(\gamma) = \tilde{g}_n(x \mid \theta)\, \mathrm{d}\lambda(x)\, \mathrm{d}\pi_n(\theta)\, , \quad x \in \mathbb{X},\; \theta \in \Theta,\; \gamma = (\theta, x),$$

where λ denotes the Lebesgue measure, $\tilde{g}_n(x \mid \theta) = g_n(x \mid \theta)/c_n(\theta)$, $g_n(x \mid \theta)$ is a criterion that reflects the interest of evaluating at x (given θ and past evaluation results), and $c_n(\theta) = \int_{\mathbb{X}} g_n(x \mid \theta)\mathrm{d}x$ is a normalizing term. For instance, a relevant choice for g_n is to consider the probability that ξ exceeds M_n at x, at step n. (Note that we consider less θs than xs in $\mathfrak{G}_n$ to keep the numerical complexity of the algorithm low.)

To initialize the algorithm, generate a weighted sample $\mathfrak{T}_{n_0} = \{(\theta_{n_0,i}, w_{n_0,i}), 1 \leq i \leq I\}$ from the distribution π_{n_0}, using for instance importance sampling with π_0 as the instrumental distribution, and pick a density q_{n_0} over $\mathbb{X}$ (the uniform density, for example). Then, for each $n \geq n_0$:

Step 1: demarginalize — Using $\mathfrak{T}_n$ and q_n, construct a weighted sample $\mathfrak{G}_n$ of the form (2), with $x_{n,i,j} \overset{\text{iid}}{\sim} q_n$, $w'_{n,i,j} = w_{n,i}\dfrac{g_n(x_{n,i,j}|\theta_{n,i})}{q_n(x_{n,i,j})c_{n,i}}$, and $c_{n,i} = \dfrac{1}{J}\sum_{j'=1}^{J} \dfrac{g_n(x_{n,i,j'}|\theta_{n,i})}{q_n(x_{n,i,j'})}$.

Step 2: evaluate — Evaluate ξ at $X_{n+1} = \operatorname{argmax}_{i,j} \sum_{i'=1}^{I} w_{n,i'} \, \rho_n(x_{n,i,j}; \theta_{n,i'})$.

Step 3: reweight/resample/move — Construct $\mathfrak{T}_{n+1}$ from $\mathfrak{T}_n$ as in [8]: reweight the $\theta_{n,i}$s using $w_{n+1,i} \propto \frac{\pi_{n+1}(\theta_{n,i})}{\pi_n(\theta_{n,i})} w_{n,i}$, resample (e.g., by multinomial resampling), and move the $\theta_{n,i}$s to get $\theta_{n+1,i}$s using an independant Metropolis-Hastings kernel.

Step 4: forge q_{n+1} — Form an estimate q_{n+1} of the second marginal of π'_n from the weighted sample $\mathfrak{X}_n = \{(x_{n,i,j}, w'_{n,i,j}), 1 \le i \le I, 1 \le j \le J\}$. Hopefully, such a choice of q_{n+1} will provide a good instrumental density for the next demarginalization step. Any (parametric or non-parametric) density estimator can be used, as long as it is easy to sample from; in this paper, a tree-based histogram estimator is used.

Nota bene: when possible, some components of θ are integrated out analytically in (1) instead of being sampled from; see [4].

Experiments. Preliminary numerical results, showing the relevance of a fully Bayesian approach with respect to empirical Bayes approach, have been provided in [4]. The scope of these results, however, was limited by a rather simplistic implementation (involving a quadrature approximation for $\bar{\rho}_n$ and a non-adaptive grid-based optimization for the choice of X_{n+1}). We present here some results that demonstrate the capability of our new SMC-based algorithm to overcome these limitations.

The experimental setup is as follows. We compare our SMC-based algorithm, with $I = J = 100$, to an EI algorithm in which: 1) we fix θ (at a "good" value obtained using maximum likelihood estimation on a large dataset); 2) X_{n+1} is obtained by exhaustive search on a fixed LHS of size $I \times J$. In both cases, we consider a Gaussian process ξ with a constant but unknown mean function (with a uniform distribution on $\mathbb{R}$) and an anisotropic Matérn covariance function with regularity parameter $\nu = 5/2$. Moreover, for the SMC approach, the variance parameter of the Matérn covariance function is integrated out using a Jeffreys prior and the range parameters are endowed with independent lognormal priors.

Results. Figures 1(a) and 1(b) show the average error over 100 runs of both algorithms, for the Branin function ($d = 2$) and the log-transformed Hartmann 6 function ($d = 6$). For the Branin function, the reference algorithm performs better on the first iterations, probably thanks to the "hand-tuned" parameters, but soon stalls due to its non-adaptive search strategy. Our SMC-based algorithm, however, quickly catches up and eventually overtakes the reference algorithm. On the Hartmann 6 function, we observe that the reference algorithm always lags behind our new algorithm.

We have been able to find results of this kind for other test functions. These findings are promising and need to be further investigated in a more systematic large-scale benchmark study.

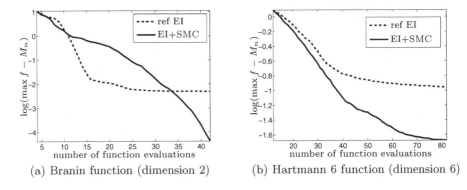

(a) Branin function (dimension 2)   (b) Hartmann 6 function (dimension 6)

Fig. 1. A comparison of the average error to the maximum (100 runs)

References

1. Mockus, J., Tiesis, V., Zilinskas, A.: The application of Bayesian methods for seeking the extremum. In: Dixon, L., Szego, G. (eds.) Towards Global Optimization, vol. 2, pp. 117–129. Elsevier (1978)
2. Santner, T.J., Williams, B.J., Notz, W.I.: The design and analysis of computer experiments. Springer (2003)
3. Jones, D.R., Schonlau, M., Welch, W.J.: Efficient global optimization of expensive black-box functions. J. Global Optim. 13(4), 455–492 (1998)
4. Benassi, R., Bect, J., Vazquez, E.: Robust Gaussian Process-Based Global Optimization Using a Fully Bayesian Expected Improvement Criterion. In: Coello Coello, C.A. (ed.) LION 5. LNCS, vol. 6683, pp. 176–190. Springer, Heidelberg (2011)
5. Gramacy, R., Polson, N.: Particle learning of Gaussian process models for sequential design and optimization. J. Comput. Graph. Stat. 20(1), 102–118 (2011)
6. Lizotte, D.J., Greiner, R., Schuurmans, D.: An experimental methodology for response surface optimization methods. J. Global Optim., 38 pages (2011)
7. Bardenet, R., Kégl, B.: Surrogating the surrogate: accelerating Gaussian-process-based global optimization with a mixture cross-entropy algorithm. In: ICML 2010, Proceedings, Haifa, Israel (2010)
8. Chopin, N.: A sequential particle filter method for static models. Biometrika 89(3), 539–552 (2002)
9. Del Moral, P., Doucet, A., Jasra, A.: Sequential Monte Carlo samplers. J. R. Stat. Soc. B 68(3), 411–436 (2006)
10. Ginsbourger, D., Roustant, O.: DiceOptim: Kriging-based optimization for computer experiments, R package version 1.2 (2011)

Influence of the Migration Period in Parallel Distributed GAs for Dynamic Optimization

Yesnier Bravo[1], Gabriel Luque[2], and Enrique Alba[2]

[1] Universidad de las Ciencias informática, Cuba
yesnier@uci.cu
[2] Universidad de Málaga, España
{gabriel,eat}@lcc.uma.es

Abstract. Dynamic optimization problems (DOP) challenge the performance of the standard Genetic Algorithm (GA) due to its *panmictic* population strategy. Several approaches have been proposed to tackle this limitation. However, one of the barely studied domains has been the parallel distributed GA (dGA), characterized by decentralizing the population in islands communicating through *migrations* of individuals. In this article, we analyze the influence of the migration period in dGAs for DOPs. Results show how to adjust this parameter for addressing different change severities in a comprehensive set of dynamic test-bed functions.

1 Introduction

Solving dynamic optimization problems (DOPs) means pursuing an optimal value that changes over time. Job shop scheduling (dealing with new arrivals), semaphores (adapting to traffic), and elevator systems (minimizing customer waiting time while receiving new calls), are some of these scenarios. These systems raise big challenges for researchers in Genetic Algorithms (GAs) [1, 2]. The reason is that GAs hardly can, once converged, to escape from old optima and adapt to the new environment.

An important weak part of the standard GA model lies in its panmictic population strategy, consisting on a single pool of individuals where any two of them can potentially mate. Consequently, a few authors have used multiple populations for specializing and tracking promising regions of the search space [1–3]. Most of these approaches perform periodical migrations of individuals among the populations. However, there is no unified and comprehensive study of the influence of the migration period in the literature.

In this article, we adopt the parallel distributed GA (dGA), which has been barely studied in this domain [3]. Many dGAs have proven effective in scenarios of high diversity requirements and computing resources, so their use should be valuable to DOPs. Our contributions are twofold: (1) we analyze the influence of the migration period in the performance of the dGA for a comprehensive set of DOP benchmarks (Section 4) and (2) we illustrate and discuss the diversity enhancement and speciation-like features of dGA models for DOPs (Section 5). Let us start by providing a brief background on DOP and dGA model.

Y. Hamadi and M. Schoenauer (Eds.): LION 6, LNCS 7219, pp. 343–348, 2012.
© Springer-Verlag Berlin Heidelberg 2012

2 Background

A dynamic optimization problem (DOP) is a real-world problems that change on time, where the fitness function is deterministic over time intervals. The goal here is to find the optimal solution for each time interval quickly and accurately, and do it by reusing information from previous time intervals rather than restarting the search from scratch. Two of the most important features of DOPs are the *change frequency* (how often the changes occur) and *change severity* (how different the environment is after a change) [1].

The dGA model [5] structures the population in demes named *islands*. Each island independently evolves, usually in parallel, and communicates with the other ones through migration of individuals. The migration period (ζ), amount of migrants to exchange (m), criteria for selecting (ω_s) or accepting (ω_r) migrants, neighborhood among islands and synchronization, form the *migration policy*.

Two main reasons drove us to use dGAs in DOPs. First, different islands can naturally evolve to different solutions (speciation), which is useful to track multiple peaks at the same time, i.e., potential optima after an environmental change. Second, the coarse grained distribution and migrations among islands, improves the population diversity due to the recombination of different genetic material. This last can be seen as a mechanism to adapt to the changes in a DOP, since both behaviors depend on the coupling degree among the islands, which is highly influenced by the migration period.

3 Experimental Setup

The behavior of algorithms is tested using two well-known benchmarks for binary and real encoding GAs, thus addressing both discrete and continuous DOPs. The first one is the technique introduced in [6] to build DOPs from a given binary-encoded stationary function $f(x)(x \in \{0,1\}^l)$. We use that technique on three different functions: Onemax, Royal Road, and Deceptive [6]. We vary the change severity ($\rho \in \{0.05, 0.1, 0.2, 0.5, 0.7\}$) to provide a wide set of difficulty degrees. The second type of generator is the moving peaks benchmark (MPB) with the parameter setting of the first standard scenario[1] and vary the *number of peaks* ($n = \{5, 50, 200\}$) and the *step severity* ($\rho = \{0.0, 0.5, 1.0, 2.0, 3.0\}$). Since we are interested in studying the adaptation ability of the dGA, we set the same change frequency of $\tau = 50$ generations for all problem instances tested.

Our dGA consists of eight islands evolving homogenously. In every island, we use a sequential GA with generational replacement. Migrations occur synchronously on a unidirectional ring topology and the migration policies used are defined in Table 1. The migration periods used are set in number of generations and proportional to the change frequency ($\tau = 50$). Thus, we test the influence of migrations at each generation ($\zeta = 1$), four times at each stationary interval ($\zeta = \frac{\tau}{4}$), one time in the half and other after a change ($\frac{\tau}{2}$), only after a change ($\zeta = \tau$), plus other at alternating intervals ($\zeta = \frac{2\tau}{3}$), respectively.

---

[1] Online available at http://people.aifb.kit.edu/jbr/MovPeaks

Table 1. Parameter settings for GAs and migration policy

Population Size	512 (64 × 8 islands)	ζ	$\in \{1, 12, 25, 50, 75\}$
Parent's Selection	(Binary tournament,	m	One copy
	Binary tournament)	ω_s	*Random* selection
Crossover	SPX, pc=0.6,	ω_r	Replace *if-better* than *least-fit*
	(BLX$_{\alpha=0.5}$ for MPB)		
Bit Mutation	pm=1/L, L=string length		Synchronous migrations
	(Polynomial for MPB)		Unidirectional ring topology

Algorithms and benchmarks were implemented in C++, using the MALLBA library[2]. All experiments were performed in a PC with an Intel Core i7-720QM processor at 1.60GHz, 4GB of RAM, and running GNU/Linux Ubuntu 10.10. To describe the behavior of algorithms we compute the *accuracy* ($\overline{acc}$) metric, also known as *relative error*. Then, we use the area below the curve (ABC) tool [7] to compare the results. Finally, we average the results over 100 independent runs and evaluate the statistical significance with a level of confidence of 95 %.

4   Influence of the Migration Period on the Performance

Lets us first analyze the influence of the migration period in dGAs for DOPs. Fig. 1. summarizes the $ABC_{\overline{Acc}}$ achieved with several migration periods and change severities. High values of this metric indicate a better adaptation of the algorithm to the changing optimum throughout all the run.

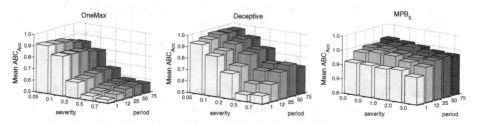

Fig. 1. Influence of the migration period in the performance of the dGA model for DOPs with different severity degrees

As a first conclusion, you can notice that the effect of the migration period is dependent on the severity of change. The lowest migration period ($\zeta = 1$) is notably better for Onemax with low severity. This instance consists of a fitness landscape with a single optimal solution drifting slowly. Therefore, a high coupling among the islands produces an accumulation of visited solutions around the optimum which is useful to pursue small variations of it, but at the expense of the global diversity. In fact, if the severity degree is higher ($\rho > 0.1$) then the algorithm hardly react and adapt to the changes in the environment (see

---

[2] Online available at http://neo.lcc.uma.es/mallba/easy-mallba

Fig. 1a). Conversely, a high migration period ($\zeta = \tau = 50$) results beneficial for unimodal DOPs with high severity, since a loose coupling improves the global diversity of the population.

Multimodal DOPs (Deceptive or MPB) add an additional behavior due to the large number of suboptimal solutions that arise. In these scenarios, a small change in the problem can produce abrupt and discontinuous shifts of the optimum in the search space. Then, a high migration period ($\zeta = 50$ or $\zeta = 75$) produces better performance, even when the step severity is low, since in addition to the diversity enhancement it allows *speciation* for tracking several optima candidate at the same time (see next section). We can see in Table 2 the numerical results with all instances tested. For each severity value (columns in the table), the best result is marked with a star (*) character and the bold type is applied to those which are not significantly different from this one.

Table 2. Mean $ABC_{\overline{Acc}}$ computed for dGA with different migration periods for DOPs with several change severities

ζ	$\rho = 0.05$	$\rho = 0.1$	$\rho = 0.2$	$\rho = 0.5$	$\rho = 0.7$	$\rho = 0.0$	$\rho = 0.5$	$\rho = 1.0$	$\rho = 2.0$	$\rho = 3.0$
	Onemax					*MPB₅*				
1	0.919*	0.845*	0.639	0.539	0.528	0.912	0.911	0.908	0.912	0.896
12	0.912	0.843	0.649	0.561	0.554	0.939	0.946	0.945	0.932	0.923
25	0.905	0.842	0.671	0.585	0.579	0.948	0.952	0.945	0.937	0.937
50	0.879	0.824	0.681*	0.599*	0.592*	0.946	0.956*	0.959*	0.947*	0.942*
75	0.852	0.800	0.664	0.588	0.579	0.958*	0.953	0.947	0.938	0.935
	RoyalRoad					*MPB₅₀*				
1	0.734*	0.601*	0.306	0.0758	0.0777	0.874	0.860	0.870	0.860	0.850
12	0.720	0.599	0.313	0.0944	0.098	0.906	0.898	0.907	0.903	0.887
25	0.711	0.579	0.311	0.110	0.114	0.915	0.903	0.911	0.899	0.897
50	0.661	0.562	0.319*	0.118*	0.122*	0.913	0.916	0.912	0.908*	0.903*
75	0.589	0.491	0.286	0.111	0.114	0.923*	0.916*	0.912*	0.908	0.896
	Deceptive					*MPB₂₀₀*				
1	0.936	0.851	0.722	0.559	0.570	0.856	0.852	0.856	0.848	0.851
12	0.957	0.877	0.765	0.631	0.648	0.898	0.900	0.894	0.887	0.879
25	0.977*	0.917	0.846	0.723	0.732	0.901	0.901	0.900	0.894	0.882
50	0.976	0.937*	0.867*	0.764*	0.772*	0.897	0.910	0.908*	0.900*	0.894*
75	0.969	0.921	0.846	0.723	0.728	0.913*	0.914*	0.897	0.886	0.888

Results in Table 2 corroborate the previous observations statistically. Another finding is that migrating after a change produces the best overall performance. Since it insufflates diversity into the population, through the crossbreeding between individuals with different genotypes. In addition, we note that it can only be effective if islands have had enough isolation time as to promote the speciation of individuals, as we will illustrate in the next section.

5   Benefits of Speciation for DOPs

With the aim at illustrating the speciation feature of a dGA, we use only two migration periods: a low one ($\zeta = 1$) and a high value ($\zeta = 50$), and the MPB_5 instance with change severity of $\rho = 3.0$, ensuring the same dynamic behavior throughout all the runs. Fig. 2 shows the best fitness evolution and the peak being exploited by each deme. The grey line depicts the optimum trajectory.

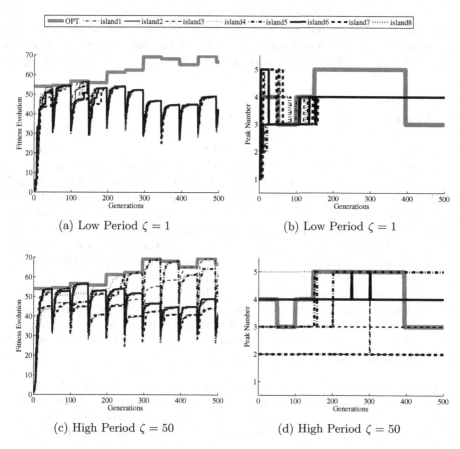

Fig. 2. Fitness evolution (left) and peak tracking (right) by each island of a dGA with low (upper half) and high (bottom half) migration periods for the MPB_5

On the one hand (low period), the dGA losses the evolutionary potential in the Fig. 2a (curves join in a straight line), which is due to islands exploit in parallel a reduced number of promising areas converging to a single solution. Such behavior is also depicted in Fig. 2b, during the first 150 generations (peak number 4). As noted in previous section, this behavior could be useful for unimodal DOPs with continuous and drifting landscapes. However, it raises the convergence problem in the long term, since it resembles the panmictic population strategy. On the other hand, a high migration period improves the population diversity and promotes speciation by the isolated evolution of islands. Speciation process consists of the natural grouping of individuals with similar traits (species), because of the constrained mating induced by structuring the population in several demes. Note in Fig. 2c that the curves are more widely spaced than the ones obtained above with a low migration period. This behavior corresponds to the ability of the algorithm to track several peaks at the same time. This is more clear in Fig 2c from generation 150 up to 400, where the problem changes but the optimal peak remains the same, and the two islands exploiting this peak dynamically adapt to its movement. If the optimal peak alternates, as can be seen in the remainder time intervals, a new specie is able to adapt to the new environment (track island number 4 after the third change and island number 2 after the eighth change in Fig. 2d).

6 Conclusions

In this paper we analyzed the influence of the migration period, an important parameter for dGA models, for DOPs. We used a comprehensive test environment based on unimodal and multimodal DOP benchmarks. On the one hand, results showed the benefits of a low migration period to address unimodal DOPs with small changes. On the other hand, a high migration period showed more robust to tackle a wide range of change severities in all DOP instances tested, enhancing the diversity and speciation features of the population. In particular, migrating as response to a change in the environment shown effective as a mechanism to adapt to dynamic environments.

In future works, we aim at developing adaptive or self-adaptive dGA models that exploit the main findings of this work with respect to the migration period in function of the severity of change, a unified study of all parameters governing the migration policy, and enhancing the basic behavior with other DOP techniques like memory, hypermutation, etc.

Acknowledgments. Authors acknowledge funds from the Spanish Ministry of Sciences and Innovation European FEDER under contract TIN2008-06491-C04-01 (M* project), CICE Junta de Andalucía under contract P07-TIC-03044 (DIRICOM project) and AUIP as sponsors of the Scholarship Program Academic Mobility.

References

1. Jin, Y., Branke, J.: Evolutionary optimization in uncertain environments-a survey. IEEE Trans. Evolutionary Computation 9(3), 303–317 (2005)
2. Yang, S., Ong, Y.-S., Jin, Y.: Evolutionary Computation in Dynamic and Uncertain Environments. Springer (2007)
3. Homayounfar, H., Areibi, S., Wang, F.: An advanced island based ga for optimization (2003)
4. Ayvaz, D., Topcuoglu, H., Gurgen, F.S.: Hybrid Techniques for Dynamic Optimization Problems. In: Levi, A., Savaş, E., Yenigün, H., Balcısoy, S., Saygın, Y. (eds.) ISCIS 2006. LNCS, vol. 4263, pp. 95–104. Springer, Heidelberg (2006)
5. Alba, E., Troya, J.M.: A survey of parallel distributed genetic algorithms. Complex 4, 31–52 (1999)
6. Yang, S.: Non-stationary problem optimization using the primal-dual genetic algorithm. In: IEEE CEC, pp. 2246–2253 (2003)
7. Alba, E., Sarasola, B.: ABC, a new performance tool for algorithms solving dynamic optimization problems. In: IEEE CEC, pp. 1–7 (2010)

A Hyper-Heuristic Inspired by Pearl Hunting

C.Y. Chan, Fan Xue, W.H. Ip, and C.F. Cheung

Department of Industrial and Systems Engineering,
The Hong Kong Polytechnic University, Hunghom,
Kowloon, Hong Kong
{mfcychan,mffxue,mfwhip,mfbenny}@inet.polyu.edu.hk

1   Pearl Hunter: An Inspired Hyper-Heuristic

Pearl hunting is a traditional way of diving to retrieve pearl from pearl oysters or to hunt some other sea creatures. In some areas, hunters need to dive and search seafloor repeatedly at several meters depth for pearl oysters. In a search perspective, pearl hunting consists of repeated diversification (to surface and change target area) and intensification (to dive and find pearl oysters). A Pearl Hunter (PHunter) hyper-heuristic is inspired by the pearl hunting, as shown in Fig. 1. Given a problem domain and some low-level heuristics (LLHs), PHunter can group, test, select and organize LLHs for the domain by imitating a rational diver.

PHunter, which executes a repeated "move-dive-move-dive" sequence in the main phase in Fig. 1, is in the Iterated Local Search (ILS) scheme [1] in general. In PHunter, a "surface move" (or move) action involves usually one diversification LLH which is not hill climbing. However, PHunter can try more moves if the new position (solution) is trapped around a "buoy". In other words, a diversification will not be accepted if the new objective value does not meet a low threshold "buoy". In practice, the buoy can be set to the best result of the first iteration.

A "dive" action refers to a sequential execution of hill climbing LLHs. There are two kinds of dives: "snorkeling" and "deep dive" (scuba). Snorkeling involves a short sequence of hill climbing algorithms with a low "depth of search" and stops once an improvement is found. Deep dive iteratively carries out a long sequence with a high "depth of search" until no further improvement can be found. Experience showed that there were different positive coefficients between snorkeling and deep dive in different domains. In a typical "move-dive" iteration, PHunter generates a number ($Num\_of\_snorkeling$) of new solutions and ranks them by snorkeling. Only a few promising (best ranked) solutions can be further processed by deep dives.

PHunter decides a "mode" consisting of a portfolio of grouped moves and a way of diving for a given problem. In fact, the idea of portfolio was proven successful in SAT (Boolean satisfiability) competitions [2]. In the rehearsal run, PHunter employs counters to record how many suboptimal solutions are found by different groups of moves and different dives. The final mode is determined according to the rules obtained by off-line learning. The diving environment is also discovered in the rehearsal run. For example, if the snorkeling and the deep

Y. Hamadi and M. Schoenauer (Eds.): LION 6, LNCS 7219, pp. 349–353, 2012.
© Springer-Verlag Berlin Heidelberg 2012

```
 1 procedure PHunter()
 2    test_and_order_dives_and_moves();
 3    mode ←rehearsal_run();
 4    loop while (terminate_condition_not_met())
 5       for each move m in mode.portfolio
 6          P ←∅;
 7          loop for Num_of_snorkeling times
 8             p ←apply_move_to_pool(m);
 9             loop while (trapped_around_buoy(p))
10                p ←apply_more_moves(p);
11             end loop
12             p' ←snorkeling(p, mode.env);
13             P ← P ∪ p';
14          end loop
15          p* ←select_promising_positions(P);
16          deep_dive(p*, mode.env);
17       end for
18       if (mission_restart_condition_is_met())
19          clear_pool();
20       end if
21    end loop;
22    return BestEverFound;
23 end procedure
```

Fig. 1. Pseudo code of Pearl Hunter

dives always find the same result for every move, the environment is flagged as "Shallow Water". In Shallow Water, PHunter simplifies the sequence in snorkeling and disables deep dives. Another environment is "Sea Trench", where at least one hill climbing heuristic consumes too much time (e.g., 3% of overall time) in a single execution. In this case, the depths of search are tuned to lower values and the sequences in snorkeling and deep dives are also simplified.

In practice, a hashed cache can be employed to record courses of deep dives and it is also used as an unwanted (inferior to tabu) list for surface moves at the same time. A restart mechanism can reset the search procedure when the suboptimal solution pool is over-converged or no better solutions are found for a certain amount of time.

2   Implementation and Experiments

PHunter was implemented on a Java cross-domain platform named HyFlex[1] (Hyper-heuristics Flexible framework) [3]. HyFlex provides a random initialization, a set of LLHs in 4 groups (Crossover, Mutation, Ruin-recreate and Local search), two parameters (the "intensity" of mutation and "depth of local

---

[1] See http://www.asap.cs.nott.ac.uk/chesc2011/hyflex_description.html

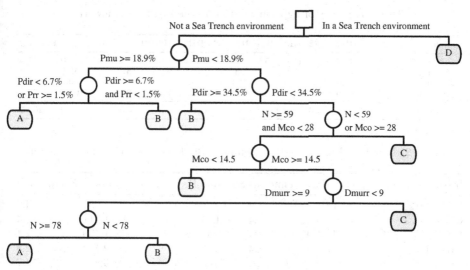

D_{murr}: Depth of the mission in the Mutation and Ruin-recreate test,

M_{co}: Number of missions completed in the Crossover test,

N: Number of suboptimal solutions found in total,

P_{dir}: Percentage of suboptimal solutions found right after some moves (before any dive),

P_{mu}: Percentage of suboptimal solutions found in iterations started with Mutation moves,

P_{rr}: Percentage of suboptimal solutions found in iterations started with Ruin-recreate moves,

Fig. 2. Decision tree on modes obtained by off-line learning

search"), and a list of easily accessible (but functionally limited) solutions for each problem in each problem domain.

Five portfolios of moves were defined on the 3 groups (Crossover, Mutation and Ruin-recreate) of moves: average calls (A), Crossover emphasized (B), Crossover only (C), average calls with an online pruning (D), Mutation and Ruin-recreate only (E). The portfolio A chooses a move from the 3 groups with the same probability. The portfolio D selects in the same way and eventually prunes some moves according to the history. An off-line classification procedure was carried out to identify the best mode. The decision attributes (counters) were gathered from a 1-minute test on mode C followed by a 1-minute test on mode E. The latter test inherits the solution pool. A decision tree was discovered by the Best-first tree classifier provided by WEKA[2] with default parameters, as shown in Fig. 2, where the mode E was dominated.

Given a set of hill climbing heuristics $\{A, B, C\}$ (ordered by performances) and an initial solution, the result of applying heuristics in order "ABC" is usually different from that in "CBA" in practice. Possible reasons include complex shape of solution space and occasionally inconsistency of local search algorithms. In PHunter, deep dives exploit parallel sequences (such as "$A\text{-}BA\text{-}CBA$" and "$CBA\text{-}BA\text{-}A$"),

---

[2] Version 3.5.6, see http://www.cs.waikato.ac.nz/ml/weka/

Table 1. Scores of hyper-heuristics on the HyFlex framework

Domain	HH1	HH2	HH3	HH4	HH5	HH6	HH7	HH8	PHunter	PH$_{w/o\_Snor}$	PH$_{Hsiao}$
MAX-SAT	41.3	54.8	19.5	5.0	0.0	25.0	38.3	2.0	**91.3**	46.0	67.0
1D Bin-packing	29.0	30.0	53.0	35.0	0.0	23.0	12.0	0.0	**73.0**	63.0	72.0
Personnel Scheduling	**53.0**	50.5	8.0	39.0	42.0	0.0	36.0	18.0	52.0	42.0	49.5
Flow Shop	12.0	3.0	11.0	54.5	5.0	42.5	3.0	37.0	**82.5**	65.0	74.0
Overall	135.3	138.3	91.5	133.5	47.5	90.5	89.3	57.0	**298.8**	216.0	262.5

Table 2. New best-known solutions found in the Personnel Scheduling domain

Instance	Size (Men * days)	Time (h)	Solution	Previous best-known[†]	% improved
BCV-A.12.2	12 * 31	24	1,875	1,953	4.0
CHILD-A2	41 * 42	24	1,095	1,111	1.4
ERMGH-B	41 * 42	24	1,355	1,459	7.1
ERRVH-A	51 * 42	24	2,135	2,197	2.8
ERRVH-B	51 * 42	24	3,105	6,859	54.7
MER-A	54 * 42	24	8,814	9,917	11.1

†: Collected from http://www.cs.nott.ac.uk/~tec/NRP/misc/NRP_Results.xls

swap and repeat until no improvements can be further found. The complex sequences introduce potential redundancy but return better results generally.

Tests have been conducted on 4 problem domains: MAX-SAT, 1D Bin-packing, Personnel Scheduling and Flow Shop, each with 10 difficult instances. The results are shown in Table 1, where PH$_{w/o\_S}$ was the PHunter without snorkeling and PH$_{Hsiao}$ was the PHunter that used Hsiao et al.'s local search scheme [4] in deep dive. The results were on average of 10 independent trials. HH1 to HH8 were 8 default hyper-heuristics and their results were provided in HyFlex. The scoring system was the Formula 1 point system provided by HyFlex, greater number meant better. The computation time was benchmarked to be equal to 10 CPU minutes on an Intel P4 3.0GHz CPU.

As shown in Table 1 scores of PHunter were competitive. PHunter won the 4th place overall and the 1st place in the hidden domains out of 20 competitors in the CHeSC 2011 competition[3]. The score of PH$_{w/o\_Snor}$ was significantly lower than PHunter's in Table 1. It can be concluded that the snorkeling trial is one of the keys to the success of PHunter. Another key should be the effectiveness of the ILS scheme. The parallel sequences of local search tests in deep dive might also be a reason by comparing the scores of PHunter and PH$_{Hsiao}$. However, the complex sequence in deep dive is a specified compromise to the LLHs implemented in HyFlex and may not work in other practice.

In fact some results of the tests approximated or had broken the best-known solutions. One exception is the Personnel Scheduling domain. PHunter classified most of the environments as Sea Trench in the domain. So further tests were made, where the computation time was 24 CPU hours and more benchmark

---

[3] See http://www.asap.cs.nott.ac.uk/chesc2011/results.html.

problems were included. The results were much satisfied. Especially, PHunter discovered 6 new best-known records, as shown in Table 2. One possible reason was the new "vertical" swap local search was first implemented in an LLH on the HyFlex.

Acknowledgements. The work described in this paper was partially supported by a grant from the Hong Kong Polytechnic University (POLYU5110/10E) and partially supported by a grant from the Department of Industrial and Systems Engineering of The Hong Kong Polytechnic University (No. RP1Z).

References

1. Lourenço, H., Martin, O., Stützle, T.: Iterated Local Search. In: Glover, F., Kochenberger, G. (eds.) Handbook of Metaheuristics, pp. 320–353. Springer, New York (2003)
2. Xu, L., Hutter, F., Hoos, H., Leyton-Brown, K.: SATzilla: Portfolio-based Algorithm Selection for SAT. Journal of Artificial Intelligence Research 32, 565–606 (2008)
3. Ochoa, G., Hyde, M., Curtois, T., Vazquez-Rodriguez, J.A., Walker, J., Gendreau, M., Kendall, G., McCollum, B., Parkes, A.J., Petrovic, S., Burke, E.K.: HyFlex: A Benchmark Framework for Cross-Domain Heuristic Search. In: Hao, J.-K., Middendorf, M. (eds.) EvoCOP 2012. LNCS, vol. 7245, pp. 136–147. Springer, Heidelberg (2012)
4. Hsiao, P.-C., Chiang, T.-C., Fu, L.-C.: A variable neighborhood search-based hyperheuristic for cross-domain optimization problems in CHeSC 2011 competition. In: Fifty-Third Conference of OR Society (OR53), Nottingham, UK, September 6-8 (2011)

Five Phase and Genetic Hive Hyper-Heuristics for the Cross-Domain Search

Tomasz Cichowicz, Maciej Drozdowski, Michał Frankiewicz, Grzegorz Pawlak, Filip Rytwiński, and Jacek Wasilewski

Institute of Computing Science, Poznan University of Technology, Poland
{Tomasz.Cichowicz,Michal.Frankiewicz,Filip.Rytwinski,
Jacek.Wasilewski}@student.put.poznan.pl,
{Maciej.Drozdowski,Grzegorz.Pawlak}@cs.put.poznan.pl

Abstract. In this paper we present two hyper-heuristics: Five Phase Approach (5Ph) and Genetic Hive (GH), developed for the Cross-Domain Heuristic Search Challenge held in 2011. Performance of both methods is studied. Experience gained in construction of the hyper-heuristics is presented. Conclusions and recommendations for the future advancement of hyper-heuristic methodologies are discussed.

Keywords: hyper-heuristics, cross-domain heuristic search, HyFlex.

1 Introduction

Hyper-heuristics (HH) are supposed to bring a new quality to solving hard combinatorial problems. Instead of directly searching the space of various combinatorial optimization problems, hyper-heuristics explore the space of low level heuristics (LLHs). The LLHs perform moves in the space of solutions of a ground combinatorial optimization problem similarly to the classic local search methods. Thus, LLHs serve as an interface between the problem *domain* and the guiding algorithm of a hyper-heuristic. This approach has a potential advantage of automating construction and tuning of algorithms. That allows solution of a broad range of combinatorial problems (domains). Still, this general concept to be fruitful needs considering of at least two issues: Which LLHs concepts are general enough to be implemented in every domain, and how can they be controlled?

The first issue has been tackled in the HyFlex framework [1]. HyFlex is a Java library implementing four LLH types on six domains. The LLH types are: local search heuristics, mutational heuristics, ruin-recreate heuristics, crossover heuristics. The domains were: maximum satisfiability (Max-SAT), bin packing, flowshop (FS), personnel scheduling (PS), and later also traveling salesman problem (TSP), vehicle routing problem (VRP) [4]. HyFlex maintains a population of solutions initialized by randomized constructive heuristics. The objective functions are uniformly minimized in all domains. Some of the LLHs have additional parameters controlling, i.e. depth of search, intensity of mutation.

Y. Hamadi and M. Schoenauer (Eds.): LION 6, LNCS 7219, pp. 354–359, 2012.
© Springer-Verlag Berlin Heidelberg 2012

In this paper we report on two hyper-heuristics developed by a CS-PUT team of students and researchers of the Institute of Computing Science, Poznan University of Technology. Two methods were independently developed: Five Phase (5Ph) and Genetic Hive (GH).

2   Five Phase Approach

The idea was to build an algorithm which iteratively goes through three main phases: *intensification*, *stagnation*, and *diversification*. Moreover, to avoid getting stuck in some bad solution, the algorithm should work on a number of solutions in parallel applying also the *mutation* and *crossingover* phases (see Figure 1).

Solution Streams Initialization. This step was applied once for each thread to scatter the search paths into different areas of the solutions space. Random LLHs were applied for five seconds on every thread.

LLH classification. The classification algorithm ran an LLH in its thread (solution stream) a predetermined number of times to collect the statistics. A linear regression was used to calculate the slope a_i of the linear approximation of the objective function in the repetition count, for each LLH i. The duration Δ_i of the classification period for LLH i was also recorded. The score of LLH i was $stat_i = -a_i/\Delta_i$. The LLH with $stat_i > 0$ was labeled as an *improver*, with $stat_i < -0.2$ LLH was classified as *masher*.

Intensification. A random LLH from the triplet(triple cluster of LLHs instead of just singleton LLHs) was applied in the solution. Probability of selecting the LLH was proportional to its score. The scores were calculated on the basis of statistics collected while running the thread. For each LLH i recent improvement of the objective function value ϕ_i ($\phi_i > 0$ means improvement), and execution time δ_i were recorded. The score of LLH i was $score_i = score_i * e^{\phi_i/\delta_i}$, where initial $score_i = 1$ - not selected firstly, might have been dominated and eliminated due to quickly growing score of just one LLH. To counter such effect, the LLHs that were not applied so far, had their score increased by 5% with each execution of any LLH in the thread. The selection of LLH and its run to stagnation was repeated $NoIt = \lceil gs/3 + 3 \rceil$ times, where gs is the number of global stagnations.

Stagnation. In the stagnation state, the improvement in the objective function stalled. It had been defined as a situation in which the solution did not improve in a number of consecutive iterations of LLH. Global stagnation phase occurred when, after 3 global iterations, the best objective function value was not changed.

Diversification. In this phase masher LLHs were chosen randomly and applied for a predetermined time period. It was proposed to use short clusters - triplet- of LLHs on the solutions instead of just singleton LLHs. The architecture of 5Ph is depicted in Figure 1.

Triplet Mutation. After the intensification phase LLH triples were mutated. For each thread the algorithm of LLH mutation proceeded in two steps: 1. Randomly selecting a triplet from some other thread. Probability of drawing a triplet was

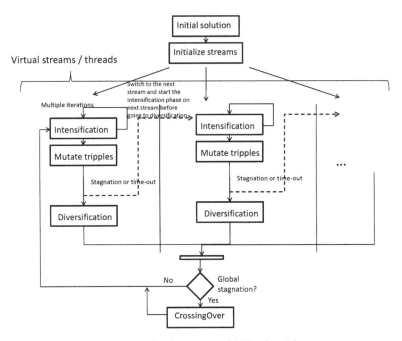

Fig. 1. Architecture of 5Ph algorithm

proportional to the mean value of the triplets LLH scores; 2. Replacing the worst LLH in the current thread with the best LLH from the drawn one.

Solution Crossingover. If the state of global stagnation was reached, 5Ph applied a random LLH of the crossover class on the solutions from the threads. The two solutions to the crossover were selected randomly - the first solution with the probability proportional to the quality of the solution and the second with the probability inversely proportional to the solution quality. At the end of this phase, the global stagnation counter gs was reset to 0, and 5Ph restarted in the intensification phase.

3   Genetic Hive Algorithm

This algorithm consisted in parallel search of the solution space using evolving sequences of low level heuristics. It was inspired by the Bees Algorithm presented in [2] and genetic algorithms mentioned in [3] imitating the behavior of bees searching for food.

In the algorithm, bees correspond with the LLH sequences. They will be labeled as agents. Searching for locations correspond with searching for current problem solutions. The agents (bees) in the hive remain passive, while the agents outside attempt to improve the current solutions. Thus, this algorithm is a combination of evolutionary approach and simulation of agent colony searching for resources. Let us denote by H the set of all agents, by L the set of active agents (outside the hive), and by $B \subseteq L$ the set of agents continuing their search in their locations.

GHHH-SEARCH()

```
 1   INITSEARCH()
 2   while timeSpent < timeLimit
 3       do
 4           for every agent a ∈ L
 5               do EVALUATE-AGENT(a)
 6           ▷ Select agents with the best scores that stay in search locations
 7           B ← bSize best agents from set L
 8           ▷ Select additional agents that stay in hive to simulate partial extinction
 9           S ← sSize random agents from set H \ B
10           ▷ Evolve new agents and update hive with them
11           O ← EVOLVE(B, oSize)
12           H ← B ∪ S ∪ O
13           ▷ Assign random agents from hive to free search locations
14           L ← B ∪((lSize − bSize) random agents from set H − B);
15           timeSpent ← currentTime − startTime
16       End.
```

where: $hSize$ - population size - the total number of agents, $lSize$ - number of search locations, $bSize$ - number of agents staying in their search locations, $sSize$ - number of additional agents that don't evolve, $oSize$ - number of agents replaced by the offspring, size of the agent - number of LLHs in agent definition, probability of mutation.

4   Computational Experiments

The aforementioned GH and 5Ph hyper-heuristics, as well as some other hyper-heuristics created by the team were subject to a number of tests comparing their performance. The results of that comparisons are summarized in Table 1. The first two are ad hoc methods. *RND LLH* - for each of 10 parallel solutions

Table 1. Pairwise comparison of CS-Put HHs

Hyper Heuristics	1	2	3	4	5	6	7	8	9	10	11	12	13
1 RND LLH	x												
2 Each LLH	22/17	x											
3 4Ph-LS	36/2	35/5	x										
4 4Ph-RND LLH	25/14	21/17	6/32	x									
5 5Ph-LS 3pl	37/3	35/5	19/17	36/3	x								
6 5Ph-RND LLH 3pl	25/15	28/12	6/33	18/15	4/34	x							
7 5Ph-154	37/2	35/5	17/20	33/4	20/17	37/2	x						
8 5Ph-155	32/7	31/8	17/20	31/8	16/18	31/8	16/22	x					
9 5Ph-160-40	35/5	27/10	19/18	28/12	18/21	25/15	17/22	19/20	x				
10 5Ph-160-46	30/10	27/12	19/20	24/15	18/21	24/14	16/20	15/24	14/24	x			
11 5Ph-160-63	32/6	35/5	17/21	29/11	20/19	30/10	15/23	18/20	20/19	22/17	x		
12 GenHiv-35	32/8	28/10	14/24	30/9	17/21	28/12	16/22	17/20	17/22	19/21	14/23	x	
13 GenHiv-65	36/4	27/13	19/19	29/9	19/18	26/13	18/21	21/18	21/17	21/16	19/21	20/19	x
14 GenHiv-68	35/5	29/11	21/17	30/10	24/15	30/10	19/19	26/14	22/18	24/15	23/15	26/11	23/15

Table 2. Comparison results for 5Ph and GH - No. of wins out of 10 instances

SAT	FS	PS	BP	SAT	FS	PS	BP
5Ph				**GH**			
Minimum							
1	1	1	2	2	8	9	6
Median							
8	0	4	7	2	10	4	3
Average							
10	10	9	10	0	0	0	0

randomly choose LLH and apply it to a solution. *Each LLH* - means executing each LLH and choosing the best solution. Further suffixes denote: LS means using *local search*, LLHs in the intensification phase *3pl* means that the triplets of LLHs were introduced. The particular method can be executed with the different parameters settings. The values presented in the table are the numbers of wins against each other. The total number of evaluated instances was 40. For some entries, the total number of wins is smaller than 40. It means that there were ties and that both methods gave the same solution. In Table 2, the two final HH of CS-PUT team are compared on all the HyFlex instances. Minima, medians and averages of the objective function are presented. A value presents number of wins for particular method and instance out of 10 evaluated instances. From the results gathered in the Table 2 one can conclude that GH method is generally better, according to the competition rules, than 5Ph. On contrary, 5Ph is better in averages and there is a tie in medians. Hence, distributions of the results are different. This demonstrates that a single performance index may be insufficient to show the complexity of the results.

5   Conclusions

Controlling the search process is a complex problem. One of the difficulties was to decide when to stop applying the current LLH. If the current LLH exhausts its potential for improving the current solution the search arrives in the state of "stagnation". However, simple statistical methods of detecting stagnation, we applied, were insufficient to quickly discover that the search had already stalled. Similar difficulty is guiding the diversification process to move the solution away from the local minimum. On one hand, it is necessary to leave quickly the current part of the solution space, on the other hand, the good parts of the current solution should be preserved to avoid rebuilding the solution from the scratch. We attempted to improve the performance of 5Ph by diversifying the population of the solutions supplied to the crossover: some of the solutions were the local minima, some were random, and some were the worst solutions visited so far. Still, the results were not satisfactory and in the tuning the number of crossovers gradually decreased to 1. It seems that convergence of crossover was too slow for the rules of the competition.

A HH, applying certain types of LLH, iteratively performed well in one domain, but was unsuccessful in other domains. This suggests that the sequence of the LLH types should be broken, and the type of LLH currently applied should be varied. To avoid being trapped in a bad solution path, we introduced "parallel" search threads in 5Ph. In the course of tunning, the number of threads decreased from over 100 to 3-7. It can be noted that the chance of avoiding being stucked up by parallelism is not bigger than by the use of other diversification tools. Thus, massive parallelism does not provide real advantage in HH search.

Performance of LLHs is domain-, instance-, and solution-dependent. This has consequences in reasoning about LLHs and HHs: Classifying LLHs by their average behavior makes no sense, and such classifications cannot be applied to guide HHs. Since, a HH does not "know" what instance in what domain is solved and the LLHs can perform so unpredictably, each instance becomes a unitary combinatorial optimization problem. Consequently, unless the domain is fixed and the instances are similar, HHs cannot be preconditioned for efficient solving of any instance in any domain. All the information fed to the AI of HH must be collected while solving the actual instance.

The performance of hyper-heuristic search has at least two criteria: time and quality of solutions. However, the issue of time was more involved. The HH should be perceived as combination of three elements: architecture, control parameters, and guiding algorithm. The architecture dictates the mode of LLH usage. The architecture can be, i.e., Tabu Search, Memetic Algorithm, 5Ph, GH, etc. In the classic meta-heuristics, the control parameters are set in the tuning process. Consequently, HH with sophisticated architecture and a lot of control parameters (as in 5Ph) need more time and more data to tune to the solved instances. This leads to a conclusion that for better understanding of HH search, it would be more advantageous to start with a rudimentary HH.

Acknowledgments. Research partially supported by Polish National Science Center (No. 519 643340)

References

1. Burke, E., Curtois, T., Hyde, M., Ochoa, G., Vazquez-Rodriguez, J.A.: HyFlex: A Benchmark Framework for Cross-domain Heuristic Search. ArXiv e-prints (2011), http://adsabs.harvard.edu/abs/2011arXiv1107.5462B
2. Pham, D.T., Ghanbarzadeh, A., Koc, E., Otri, S., Rahim, S., Zaidi, M.: The Bees Algorithm. Manufacturing Engineering Centre, Cardiff University, UK (2005)
3. Chakhlevitch, K., Cowling, P.: Hyperheuristics: Recent Developments. In: Cotta, C., et al. (eds.) Adaptive and Multilevel Metaheuristics. SCI, vol. 136, pp. 3–29. Springer, Heidelberg (2008)
4. Hyde, M., Ochoa, G., Parkes, A.: Cross-domain Heuristic Search Challenge (2011) http://www.asap.cs.nott.ac.uk/chesc2011/
5. Hyde, M., Ochoa, G.: ASAP Default Hyper-heuristics (2011) http://www.asap.cs.nott.ac.uk/chesc2011/defaulthh.html

Implicit Model Selection Based on Variable Transformations in Estimation of Distribution

Emanuele Corsano, Davide Cucci, Luigi Malagò, and Matteo Matteucci

Department of Electronics and Information, Politecnico di Milano
Via Ponzio, 34/5, 20133 Milano, Italy
emanuele.corsano@mail.polimi.it, {cucci,malago,matteucci}@elet.polimi.it

Abstract. In this paper we address the problem of model selection in Estimation of Distribution Algorithms from a novel perspective. We perform an implicit model selection by transforming the variables and choosing a low dimensional model in the new variable space. We apply such paradigm in EDAs and we introduce a novel algorithm called I-FCA, which makes use of the independence model in the transformed space, yet being able to recover higher order interactions among the original variables. We evaluated the performance of the algorithm on well known benchmarks functions in a black-box context and compared with other popular EDAs.

Keywords: Estimation of Distribution Algorithms, Transformation of Variables, Implicit Model Selection, Minimization of Mutual Information.

1 Introduction

Estimation of Distribution Algorithms (EDAs) belong to the class of meta-heuristics for optimization where the search is guided by a statistical model able to capture the interactions among the variables in the problem. The choice of the model is crucial, indeed much of the literature in the EDAs community is focused on applying machine learning techniques for model selection, able to identify the correct interactions among the variables from a sample of observations. Some examples are the algorithms which learn the structure of a Bayesian Network, as in the Bayesian Optimization Algorithms (BOA) [4], clustering algorithms for the variables that appear to be correlated, extended Compact Genetic Algorithm (eCGA) [2] or model selection for Markov Random Field, as in DEUM [5]. Although very powerful, these techniques have their main drawback in the computational complexity of the model selection and sampling phases [1].

In this paper we propose a novel approach to the problem of model selection based on the idea of applying a transformation of variables and then employing fixed, low dimensional model in the new transformed space. This corresponds to implicitly identify a different statistical model in the original space which depends on the particular transformation applied. Obviously we moved much of the computational complexity from model selection to the choice of a good

Y. Hamadi and M. Schoenauer (Eds.): LION 6, LNCS 7219, pp. 360–365, 2012.
© Springer-Verlag Berlin Heidelberg 2012

transformation of variables; on the other side it becomes easier to select models able to capture higher order interactions among the variables. Instead of limiting the search up to a given order of interactions, due to the family of transformations we introduced we are able to identify non hierarchical models that can be efficiently employed in an EDAs.

This paper is organized as follows: we first introduce how transformation of variables can be employed in EDAs, then we present I-FCA, a novel algorithm which employs this technique. Finally we compare the performances with other popular EDAs.

2   Variable Transformations in EDAs: Function Composition Algorithms

In this section we apply the idea of choosing a transformation of variables and then considering low-dimensional statistical models in the transformed space to introduce a novel family of EDAs called Function Composition Algorithms (FCAs). In the following we address the maximization of $f(x) : \Omega^n \to \mathbb{R}$, $\Omega = \{\pm 1\}$.

We introduce a new vector of variables $y = (y_1, \ldots, y_n)$ in Ω and a one-to-one map $h : \Omega \to \Omega$ such that $y = h(x)$. We can thus express f as the composition of a function $g(y) : \Omega \to \mathbb{R}$ with h, i.e., $f = g \circ h$ and $g = f \circ h^{-1}$. Since h defines a permutation of the points in Ω, follows that $\max g = \max f$.

Recall the basic iteration of an EDA:

$$\mathcal{P}^t \xrightarrow{\text{selection}} \mathcal{P}^t_s \xrightarrow{\text{estimation}} p(x; \theta^t) \in \mathcal{M} \xrightarrow{\text{sampling}} \mathcal{P}^{t+1}$$

At each iteration, EDAs start with a population $\mathcal{P}^t$, chose a subset of individuals according to a selection policy and use this sample to estimate the parameters of a distribution $p(x; \theta)$ belonging to a model $\mathcal{M}$. For instance this can be done by means of statistical techniques such as max-likelihood estimation. A new population $\mathcal{P}^{t+1}$ is finally generated sampling individuals from $p(x; \theta)$. In the estimation phase, some algorithms, such as UMDA [3], employ a fixed model while more powerful EDAs, such as BOA [4], DEUM [5] perform a model selection step using machine learning techniques in order to chose a good model able to express the interactions among variables in the selected population $\mathcal{P}_s$.

We introduce the following variation of an EDA, where estimation and sampling are preceded and followed by two transformation steps: first a one-to-one map $y = h(x)$ is applied to each individual in the selected sample, obtaining $\tilde{\mathcal{P}}_s$, then the new sample $\tilde{\mathcal{P}}^{t+1}$ is mapped back in the original space with h^{-1}:

$$\mathcal{P}^t_s \xrightarrow{y=h(x)} \tilde{\mathcal{P}}_s \xrightarrow{\text{estimation}} q(y; \xi^t) \in \mathcal{N} \xrightarrow{\text{sampling}} \tilde{\mathcal{P}}^{t+1} \xrightarrow{x=h^{-1}(y)} \mathcal{P}^{t+1}$$

Here $\mathcal{N}$ identifies a model for the transformed variables y which corresponds to a model $\mathcal{M}$ for x which depends on the particular map h applied. Both models are characterized by the same dimension of the parameter space.

In the following we give the details of Independence-FCA (I-FCA), a novel EDA which fixes $\mathcal{N}$ to be the independence model for Y and performs an implicit model selection step among a wide family of n-variate models by means of the choice of the one-to-one map h. At each iteration a map h is chosen among a subset of all the possible one-to-one maps by means of a greedy strategy which maximizes the likelihood of $q(y; \xi^t)$ with respect to $\tilde{\mathcal{P}}_s$. The resulting low-dimensional model $\mathcal{M}$ for X achieves a better approximation of the sample $\mathcal{P}_s$ with respect to the independence model for X.

The subset of the class of the one-to-one maps employed by I-FCA, indexed by j, $k \in \{1, \ldots, n\}$, with $j \neq k$, is defined such that

$$h_i^{(j,k)} : \begin{cases} y_i = x_i x_k & \text{if } i = j \\ y_i = x_i & \text{otherwise.} \end{cases}$$

Obviously we have $n(n-1)$ different $h^{(j,k)}$ transformations. It is easy to see that they are one-to-one and that $h^{-1} = h$, since $x_i^2 = 1$ and $\Omega = \{\pm 1\}$. Next we extend the class of transformations we consider by allowing elements h to be the composition of a finite number $\overline{m}$ of maps of the form $h^{(j,k)}$:

$$h = h^{(j_1, k_1)} \circ \ldots \circ h^{(j_m, k_m)} \circ \ldots \circ h^{(j_{\overline{m}}, k_{\overline{m}})}.$$

Since the inverse of each transformation in the sequence of compositions is the element itself, it is easy to see that h^{-1} is the compositions of all the $h^{(j_m, k_m)}$ in the inverse order.

In I-FCA we propose a strategy for the choice of map h based on the maximization of the likelihood of the transformed selected sample $\tilde{\mathcal{P}}_s$ with respect to the estimated distribution $q(y, \hat{\xi}) \in \mathcal{N}$, where $\mathcal{N}$ is the independence model for Y. This is equivalent to minimize the Kullback-Leibler divergence between the empirical distribution representing the selected population and its projection on the independence model, which gives a measure of the loss of information which occurs when $\tilde{\mathcal{P}}_s$ is approximated with $q(y, \xi)$. In order to make the search for h feasible, we chose a greedy approach: we initialize h to be the identity map $y = x$, then we iteratively examine all the $n(n-1)$ maps $h^{(j,k)}$ and compose the h map obtained at the previous step with the map $h^{(j,k)}$ which better improves the likelihood of $(h \circ h^{(j,k)})(\mathcal{P}_s)$ with respect to the independence model. The iteration stops when no improvement in the likelihood is achievable composing further maps of the form $h^{(j,k)}$ or when the maximum number $\overline{m}$ of transformations in h has been reached. See Algorithm 1.

Since the chosen encoding for h is redundant, the procedure IsAllowed() is needed to avoid the evaluation of maps which lead to configurations already appeared in previous stages. The worst case time complexity of the search strategy for h is $\mathcal{O}(n^2 \overline{m} N)$, where N is the population size, even though it is possible to take advantage of the likelihood decomposition to cut most of the complexity which comes from iterations over the selected population.

3   Experimental Results

In this section we present the results of a preliminary performance evaluation for the novel I-FCA algorithm on a set of well known benchmarks functions:

Algorithm 1. I-FCA - Choice of the map h

```
1:  m ← 0;
2:  maxL ← L_ind;                    ▷ The likelihood of P_s w.r.t the independence model
3:  repeat
4:      h[m] ← NULL;                 ▷ The m-th element of the composition sequence
5:      for all j, k ∈ {1, ..., n}, j ≠ k do
6:          if IsAllowed(h^(j,k)) then
7:              P̃_s ← h^(j,k)(P_s);
8:              θ̂ ← MaxLikelihoodEstimation(P̃_s);
9:              L ← Likelihood(P̃_s, q(y; θ̂));
10:             if L > maxL then
11:                 h[m] ← h^(j,k);
12:             end if
13:         end if
14:     end for
15:     P_s ← h[m](P_s);
16:     m ← m + 1;
17: until m ≥ m̄ ∨ h[m − 1] = NULL;
18: return h;
```

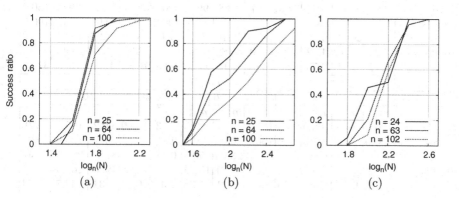

Fig. 1. Probability of success for three problems as a function of the population size for different problem sizes n. (a) Alternated Bits, $n \in \{25, 64, 100\}$, 50 runs (b) 2D Ising spin glass, $n \in \{25, 64, 100\}$, 5 instances, 16×5 runs (c) Trap3, $n \in \{24, 63, 102\}$, 50 runs.

Alternated Bits, 2D Ising spin glass and Trap3. The first two functions are quadratic while Trap3 includes hierarchical interactions up to order three.

After a preliminary tuning the I-FCA parameters were chosen as follows. As the selection policy we perform truncation selection and keep the S highest fitness individuals, where S is a function of the problem size n but it is independent with respect to N. We found that $S = 5n$ is a good choice for all the benchmark functions considered. Moreover we set $\overline{m} = n$. This result is also supported by an analysis on the set of models $\mathcal{M}$ obtainable mapping the independence model for Y into a model for X by means of h. The success ratio as a function of the

Table 1. Statistics of the best solutions averaged over 48 runs for Alternated Bits and Trap3 and 8 runs × 5 instances for Sping glass. UMDA: truncation selection 50%, DEUMce: truncation selection 30%, Cross Entropy min significance 2.0. CPU: AMD Opteron[TM] 6176, 2.3 GHz

	I-FCA	UMDA	hBOA	DEUMce	I-FCA	UMDA	hBOA	DEUMce	I-FCA	UMDA	hBOA	DEUMce
	Alternated Bits											
n	25				64				100			
N	$n^{2.0}$	$n^{2.2}$	$n^{2.0}$	$n^{2.2}$	$n^{2.2}$	$n^{1.4}$	$n^{2.2}$	$n^{2.2}$	$n^{2.2}$	$n^{1.4}$	$n^{2.2}$	$n^{2.0}$
$f_{best}[\%]$	100.0	94.4	100.0	97.7	100.0	88.8	100.0	100.0	100.0	85.4	100.0	100.0
success [%]	100.0	14.6	100.0	83.3	100.0	0.0	100.0	100.0	97.9	0.0	100.0	100.0
$f_{evals} \times 10^3$	2.43	13.14	3.82	1.12	51.08	7.42	119.0	9.42	179.3	15.69	444.9	10.01
iteration	3.60	10.62	4.71	–	5.19	21.56	11.25	–	6.98	24.50	16.35	–
t [s]	0.26	0.22	0.48	0.98	8.77	0.18	79.19	1.72	23.07	0.50	1649.7	8.04
	Ising spin glass 2D											
n	25				64				100			
N	$n^{2.4}$	$n^{2.4}$	$n^{2.2}$	$n^{2.4}$	$n^{2.4}$	$n^{1.8}$	$n^{2.2}$	$n^{2.4}$	$n^{2.4}$	$n^{1.6}$	$n^{2.2}$	$n^{2.4}$
$f_{best}[\%]$	99.5	97.3	100.0	94.9	99.7	92.8	99.7	100.0	99.6	85.2	99.9	100.0
success [%]	92.5	55.0	100.0	50.0	87.5	2.5	87.5	100.0	70.0	0.0	95.0	100.0
$f_{evals} \times 10^3$	5.55	26.78	6.89	2.03	130.4	43.16	155.5	21.62	488.0	39.27	498.4	63.10
iteration	2.23	11.40	4.42	–	5.80	23.82	15.15	–	7.53	24.32	18.48	–
t [s]	0.17	0.24	0.73	2.88	10.67	0.91	95.26	2.54	76.83	1.18	1707.6	37.53
	Trap3											
n	24				63				102			
N	$n^{2.4}$	$n^{2.6}$	$n^{2.4}$	$n^{1.8}$	$n^{2.6}$	$n^{1.8}$	$n^{2.2}$	$n^{1.8}$	$n^{2.4}$	$n^{1.8}$	$n^{2.0}$	$n^{1.8}$
$f_{best}[\%]$	100.0	95.5	100.0	90.9	100.0	90.3	100.0	82.3	100.0	90.2	100.0	77.9
success [%]	100.0	4.2	100.0	0.0	100.0	0.0	100.0	0.0	95.8	0.0	100.0	0.0
$f_{evals} \times 10^3$	4.81	6.78	13.68	0.32	159.8	23.65	150.8	0.91	348.3	82.15	287.5	2.46
iteration	2.12	1.29	5.35	–	3.15	13.21	15.29	–	5.19	19.48	26.29	–
t [s]	0.19	0.10	1.18	4.26	8.69	0.62	97.63	2.05	20.43	2.31	992.7	15.74

population size N is shown in Figure 1, for different problem sizes n. We next compare the performances of I-FCA with three well known EDAs: UMDA [3], hBOA and DEUMce [5]. UMDA employs the independence model and it is identical to I-FCA once h is fixed to be the identity map $y = x$. hBOA make use of densities which factorize according to the structure of a Bayesian Network which is learned at every iteration from the selected sample. DEUMce employs a Cross Entropy criterion to learn the structure of a Markov random field where interactions up to order two are considered.

Straightforward implementations of these algorithms have been implemented in Evoptool [6] and all code is available at[1]. We run experiments with different parameter settings and population sizes and we computed the normalized value of f, the success ratio, the number of fitness evaluations, time and algorithm

---

[1] http://airlab.elet.polimi.it/index.php/Evoptool

iterations when the best found individual appeared in the population, averaged over multiple runs. The results for the parameter settings which gave highest success ratio and least number of fitness function evaluations are presented in Table 1. It is possible to see that I-FCA outperforms UMDA: this proves the viability of the variables transformation approach. Moreover, the performances of I-FCA are comparable with hBOA, although the latter is able to achieve reliable convergence to the global optimum with smaller populations. Part of this comes from the more advanced selection scheme employed. DEUMce fails to find any good solution on Trap3, because of the third interaction present in the function, which instead are correctly handled by I-FCA and hBOA.

4   Conclusions

In this work we have introduced a novel EDA called I-FCA and we have tested out algorithm on three well known benchmark function. I-FCA has a low number of parameters for which we were able to give problem independent settings. Although a wider set of benchmark functions has to be analyzed, our preliminary experiments have shown that I-FCA can challange algorithms which learn expressive models, such as hBOA, employing only a low dimensional model, once a proper variable transformation has been learnt.

References

1. Echegoyen, C., Zhang, Q., Mendiburu, A., Santana, R., Lozano, J.: On the limits of effectiveness in estimation of distribution algorithms. In: 2011 IEEE Congress on Evolutionary Computation (CEC), pp. 1573–1580 (June 2011)
2. Harik, G.: Linkage Learning via Probabilistic Modeling in the ECGA, No. 99010, Illinois Genetic Algorithms Lab, University of Illinois at Urbana-Champaign (1999)
3. Mühlenbein, H., Mahnig, T.: Mathematical analysis of evolutionary algorithms. In: Essays and Surveys in Metaheuristics, Operations Research/Computer Science Interface Series, pp. 525–556. Kluwer Academic Publishers (2002)
4. Pelikan, M., Goldberg, D.: Hierarchical Bayesian Optimization Algorithm. In: Pelikan, M., Sastry, K., CantúPaz, E. (eds.) Scalable Optimization via Probabilistic Modeling. SCI, vol. 33, pp. 63–90. Springer, Heidelberg (2006)
5. Shakya, S., Brownlee, A., McCall, J., Fournier, F., Owusu, G.: A fully multivariate DEUM algorithm. In: IEEE Congress on Evolutionary Computation (2009)
6. Valentini, G., Malagò, L., Matteucci, M.: Evoptool: an extensible toolkit for evolutionary optimization algorithms comparison. In: Proceedings of IEEE World Congress on Computational Intelligence, pp. 2475–2482 (July 2010)

Improving the Exploration in Upper Confidence Trees

Adrien Couëtoux[1,2,3], Hassen Doghmen[1], and Olivier Teytaud[1,2]

[1] TAO-INRIA, LRI, CNRS UMR 8623,
Université Paris-Sud, Orsay, France
[2] OASE Lab, National University of Tainan, Taiwan
[3] Artelys, 12 rue du Quatre Septembre Paris, France

Abstract. In the standard version of the UCT algorithm, in the case of a continuous set of decisions, the exploration of new decisions is done through blind search. This can lead to very inefficient exploration, particularly in the case of large dimension problems, which often happens in energy management problems, for instance. In an attempt to use the information gathered through past simulations to better explore new decisions, we propose a method named Blind Value (BV). It only requires the access to a function that randomly draws feasible decisions. We also implement it and compare it to the original version of continuous UCT. Our results show that it gives a significant increase in convergence speed, in dimensions 12 and 80.

1 Introduction and Motivation

We consider a high dimensional continuous and stochastic sequential decision making problem. Both the decision space and state space are continuous. For the sake of simplicity, both have the same dimension N. There is a finite time horizon H, after which no further decisions are made. At each time step $t < H$, given a current state s_t, the optimizer has to make a decision d. In this paper, we will denote the set of feasible decisions from state s as $X(s)$. We will also denote the set of explored decisions from state s after the nth iteration, $n \geq 1$, as $D_n(s)$.

We consider that the optimizer has at its disposal a model with a transition function f and a sampling function φ. The transition function takes as inputs a state s_t and a decision $x \in X(s_t)$. Its outputs are a state s_{t+1} and a reward $r_{t+1} \in \mathbb{R}$. The sampling function takes as input a state s_t, and its output is a decision $x \in X(s_t)$. The optimizer has no other knowledge of the model than these two functions. Both functions can be (and are, in our experiments) stochastic: $f(s_t, d)$ and $\varphi(s_t)$ are two multidimensional real random variables, their probability distributions being unknown to the optimizer.

The optimizer is given an initial state s_0, and its objective is to maximize the accumulated reward $\sum_{1 \leq i \leq H} r(i)$.

In this context, UCT like algorithms have been some of the most efficient methods, like MCTS in the game of Go [6,4,7,8], or continuous MCTS on energy

Y. Hamadi and M. Schoenauer (Eds.): LION 6, LNCS 7219, pp. 366–371, 2012.
© Springer-Verlag Berlin Heidelberg 2012

management problems [5]. This is due, mostly, to the fact that without any dimension reduction technique (that is application specific), all other well known methods, like dynamic programming, fail [1,2].

However, in its current form, continuous MCTS does not use any information when it explores new decisions. This is not such a dramatic issue in discrete cases, or in low dimensional continuous cases, as one can just blindly cover most of the search space. But, in the case of a high dimensional continuous decision space, we think that the convergence speed could be greatly increased by biasing the way new decisions are explored. As mentioned before, we consider the case where the sampling of feasible decisions is "black box". This means that the set of feasible decision, as well as the inside of the sampling function φ, are unknown to the optimizer. This keeps biased sampling methods out of our options. We chose to consider this case because, to have access to the inside of the sampling function, one needs to know precisely the constraints of the model, and to implement them. Not only can this work take enourmous amounts of time (i.e.,hundreds of man hours), but the resulting feasible space may also be highly non convex, resulting in even more work to develop sampling functions on these spaces. In short, biased sampling, in our opinion, should be relevant for very problem-specific methods.

One could also count on an approximate knowledge of the feasible space, sample on this approximation, and apply a large penality to infeasible decisions. This is a valid option for many problems, but, in energy management problems (our main current application of interest), the feasible decisions are extremely sparse in the convex envelop, making this approach less interesting.

Our approach is focused on using the very limited information of the transition and sampling functions, in combination with the information progressively made available during the course of the simulations.

This is inspired by a work on continuous and stochastic bandits problems [3], which provides a method that ensures that no area of the feasible decisions set is left unexplored, while focusing on promising areas of this set. However, it focuses on the theoretical aspect of the problem, and requires some assumptions that often do not hold in our case. In particular, it requires the knowledge of the set of feasible decisions. Still, our approach follows the same ideas: explore the empty areas first, and then to focus on areas where promising decisions are.

We first quickly review the state of the art form of MCTS in a continuous setting. Then, we introduce its new variant termed MCTS with Blind Value (MCTS-BV). Then, we show some experimental results on an energy management problem, where we compare the two versions, first in a small dimension setting, then in a larger dimension setting ($N = 80$). Finally, we present some experimental results on the tuning of one parameter of MCTS-BV.

2   State of the Art of Continuous Upper Confidence Trees

The UCT algorithm builds a tree where the nodes represent the reachable states, and the arcs the feasible decisions. By progressively adding arcs and nodes from its root (that represents the initial state given to the optimizer), more information is gathered. When the algorithm runs out of time, it selects the most promising decision reachable from the root (usually, the one that has been simulated the most). Among the crucial mechanisms in this algorithm, there are: when to add a new decision to the tree or when to exploit a known decision, and how to select a known decision once we have chosen not to add a new one. The first part is usually dealt with by using Progressive Widening [5], while the second is usually dealt with by using an Upper Confidence Bound formula (UCB).

For more detailed information about the current state of the art form of MCTS with Double Progressive Widening (MCTS-DPW), please read [5].

3   Blind Value

The principle of Blind Value is to help the exploration of new decisions. One can want to explore a new decision from any state already in the tree. Although in our case, the optimizer cannot bias the sampling of new decisions, we propose a method that does use the information available in the tree. More precisely, we use the information about the children of the current node. In terms of states and decisions, it means: when we want to explore a new decision from state s, we use information about all the decisions explored from this state s in the past simulations to select a new decision $x \in X(s)$ to be explored from state s.

Note that one could use any other information in the tree: brother nodes, grand children nodes, father node, etc. However, even the exploitation of the direct children of a node only is computationally costly. And, the more distant in the tree some information is, the more likely it is to be irrelevant to the node we are currently in (states might be very different, and this type of problem is also highly time step dependant). [8] has proposed the use of Rapid Action Value Estimates (RAVE), which are an interesting other possibility; we will consider the mixing of blind value with RAVE values in a further work. [8] also proposed the use of information from related nodes; after preliminary positive results, this was later removed from the corresponding implementations (for the game of Go) for correctly tuned implementations.

The idea of BV is to try to explore decisions that are far away from known decisions during the first simulations, and then to focus on areas that have a lot of decisions with high UCB values. This is done by sampling a number of new decisions, and by selecting one of them according to a combination of these two criterions (explore unknown regions and explore regions with many decisions with high UCB values in it).

More precisely, we sample a number $M \geq 1$ of random decisions, and we pick the one that is the most interesting to explore. The way we measure the interest of a decision is through a function from the decision space to $\mathbb{R}$, denoted $BV(.)$ (Blind Value). This function can be defined in many different ways. In this paper, at an iteration n, given a state s and a decision $x \in X(s)$, we chose to define $BV(x)$ as the minimum over $D_n(s)$ of the sum of two parts. The first part is the UCB value of $d \in D_n(s)$, the second is the distance between $x \in X(s)$ and $d \in D_n(s)$, multiplied by an adaptation coefficient. We use the standard euclidian distance, but any other distance could be used instead.

More precisely, the Blind Value of x in state s with decisions $d \in D_n(s)$ already explored is

$$BV(x) = \min_{d \in D_n(s)} UCB(d) + \rho dist(d, y).$$

What follows is the detailed blind value algorithm:

Exploration of new decisions
Input: a state s, a set D of already explored decisions, an integer M, and a distance function over the decision space, *dist*
Output: an unexplored decision x.
Generate M random decisions. Let X be the set composed of these decisions.
Compute $a = UnbiasedStandardDeviation_{d \in D}(UCB(d))$
Compute $b = UnbiasedStandardDeviation_{x \in X}(dist(x, 0))$, 0 being the center of the domain
Compute $\rho = \frac{a}{b}$
return $x = argmax_{y \in X} BV(y, \rho, D)$

Computing BV (Blind Value)
Input: an unexplored decision y, a real number ρ, and D the set of explored decisions

Output: a real number $BV(y, \rho, D)$.
return $min_{d \in D}(\rho \times dist(d, y) + UCB(d))$

---

4   Experimental Comparison

Our test case is an energy management problem. There are N energy stocks, H time steps, and a thermal power plant with a given maximum capacity and production cost function. In our experiments, we used a quadratic cost function. At each time step, each stock also receives an inflow. Each inflow follows its own independent random distribution.

At each time step, the decision maker has to decide how much to produce from each stock, and how much to produce from the thermal plant. His goal is to satisfy a time varying demand at the lowest possible cost.

We ran two algorithms on this problem: the continuous version of MCTS, as introduced in [5], and the same algorithm with the addition of Blind Value (MCTS-BV), with the sample size parameter set to 20. This experiment was run with 12 stocks and 16 time steps. The results are shown in fig 1 (left). In this experiment, as in the following ones, each point is computed from 10000 runs of the algorithm on one problem instance. The 95% confidence intervals are plotted as blue segments around the points, even though their small size can make them very hard to see in some cases.

This figure shows that even in dimension 12, BV already gives an edge of magnitude 10 to MCTS, in terms of computation time (to reach a certain level

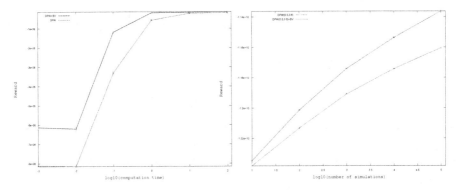

Fig. 1. Left: reward, as a function of the computation time. Problem settings: 12 stocks, 16 time steps. MTCS with BV is 10 times faster than MCTS, for budgets up to 10 seconds per decision. Right: reward, as a function of the number of simulations per decision. Problem settings: 80 stocks, 6 time steps. MTCS with BV is 10 times faster than MCTS, for all budgets.

of performance, MCTS requires 10 times as many simulations as MCTS-BV). The problem being reasonably easy, we also see that this edge decreases when the computation time increases (but only for a computation time of about 3 minutes). This is due to the fact that, as the budget gets bigger, both algorithms get very close to the optimum, in terms of reward.

This is why we made a second experiment, on the same problem, with a much higher dimension. In this experiment, there are 80 stocks, 6 time steps, and $M = 640$. With the information given to the algorithms (just the transition and the sampling functions), this problem is incredibly difficult, and naturally has very low reward (the highest average possible reward being around -2.5×10^7). Given its dimension, there is no way of exploring the entire decision space, even with a very low density. The results are shown in Fig. 1 (right).

This figure shows that BV still gives an edge of magnitude 10 to MCTS in terms of computation time, even though we were not able to approach the optimum with our computing capacities. One can also note that in this setting, the difference seems to be increasing as the budget increases. This leads to think that on very difficult problems, BV can divide by ten, or even more, the computing time necessary to reach a certain level of performance.

5 Conclusion and Future Work

We introduced a new variant of continuous MCTS to better solve high dimensional problems. Our experimental results show that this variant, MCTS-BV, improves the original algorithm by a factor 10 (it reaches the same level of performance with 10 times less simulations than the original). This result holds for even relatively small dimension problems (dimension 12). The edge given by BV seems to be even bigger in very high dimensions, but this could be confirmed by longer experiments.

We believe that, especially in its simplest form, Blind Value is very easy to implement, and can significantly increase the convergence speed of MCTS on continuous and stochastic planning problems.

It could also be coupled with other ways of exploiting information throughout the tree. One of the promising leads is the use of RAVE values, first introduced in the game of Go [8], and currently extended to continuous domains. While Blind Value only uses the decisions at the current level in the tree (horizontal propagation), RAVE propagates information through the path of each simulation, from child to father node (vertical propagation) enabling the algorithm to attribute a value to decisions not yet explored from one specific state.

Our future work will focus on setting an adaptive form for the sampling size parameter, to adapt it to the time budget. We also plan on trying different variants of the formula used to rank the decisions in the pool, by changing the distance, for example.

Acknowledgments. We thank Grid5000, that made our experiments possible (www.grid5000.fr). We also thank the National Science Council of Taiwan for grants NSC97-2221-E-024-011-MY2 and NSC 99-2923-E-024-003-MY3.

References

1. Bellman, R.: Dynamic Programming. Princeton Univ. Press (1957)
2. Bertsekas, D., Tsitsiklis, J.: Neuro-dynamic Programming. Athena Scientific (1996)
3. Bubeck, S., Munos, R., Stoltz, G., Szepesvári, C.: Online optimization in x-armed bandits. In: Advances in Neural Information Processing Systems 22 (2008)
4. Chaslot, G., Winands, M., Uiterwijk, J., van den Herik, H., Bouzy, B.: Progressive Strategies for Monte-Carlo Tree Search. In: Wang, P., et al. (eds.) Proceedings of the 10th Joint Conference on Information Sciences (JCIS 2007), pp. 655–661. World Scientific Publishing Co. Pte. Ltd. (2007)
5. Couëtoux, A., Hoock, J.-B., Sokolovska, N., Teytaud, O., Bonnard, N.: Continuous Upper Confidence Trees. In: Coello, C.A.C. (ed.) LION 5. LNCS, vol. 6683, pp. 433–445. Springer, Heidelberg (2011)
6. Coulom, R.: Efficient Selectivity and Backup Operators in Monte-Carlo Tree Search. In: van den Herik, H.J., Ciancarini, P., Donkers, H.H.L.M(J.) (eds.) CG 2006. LNCS, vol. 4630, pp. 72–83. Springer, Heidelberg (2007)
7. Coulom, R.: Computing elo ratings of move patterns in the game of go. In: Computer Games Workshop, Amsterdam, The Netherlands (2007)
8. Gelly, S., Silver, D.: Combining online and offline knowledge in UCT. In: ICML 2007: Proceedings of the 24th International Conference on Machine Learning, pp. 273–280. ACM Press, New York (2007)

Parallel GPU Implementation of Iterated Local Search for the Travelling Salesman Problem

Audrey Delévacq, Pierre Delisle, and Michaël Krajecki

CReSTIC, Université de Reims Champagne-Ardenne, Reims, France
{audrey.delevacq,pierre.delisle,michael.krajecki}@univ-reims.fr

Abstract. The purpose of this paper is to propose effective parallelization strategies for the Iterated Local Search (ILS) metaheuristic on
Graphics Processing Units (GPU). We consider the decomposition of
the 3-opt Local Search procedure on the GPU processing hardware and
memory structure. Two resulting algorithms are evaluated and compared
on both speedup and solution quality on a state-of-the-art Fermi GPU architecture. We report speedups of up to 6.02 with solution quality similar
to the original sequential implementation on instances of the Travelling
Salesman Problem ranging from 100 to 3038 cities.

Keywords: TSP, ILS, Parallel Metaheuristics, 3-opt, GPU, CUDA.

1 Introduction

Iterated Local Search (ILS) is a metaheuristic that successively applies a Local
Search (LS) procedure to an initial solution and incorporates mechanisms to
climb out of local optima. It finds good solutions to many optimization problems in a reasonable time which may remain too high in practice. Even though
this time can be reduced by parallel computing, most approaches are dedicated
to CPU-based architectures. As research on computer architectures is rapidly
evolving, new types of hardware have recently become available. Among them
are Graphics Processing Units (GPU) which provide great and affordable computing power but also require new algorithmic paradigms to be used efficiently.

The purpose of this paper is to propose parallelization strategies for ILS to
efficiently solve the Travelling Salesman Problem (TSP) in a GPU computing
environment. We first present k-opt LS algorithms and the ILS metaheuristic.
Then, after a literature review on parallel LS and ILS, the proposed GPU parallelization strategies are explained and experimented.

2 Iterated Local Search for the TSP

The Travelling Salesman Problem (TSP) may be defined as a complete directed
graph $G = (V, A, d)$ where $V = \{1, 2, ..., n\}$ is a set of vertices, $A = \{(i, j) \mid
(i, j) \in V \times V\}$ is a set of arcs and $d : A \to \mathbb{N}$ is a function assigning a weight d_{ij}
to every arc. The objective is to find a minimum weight Hamilton cycle in G.

Y. Hamadi and M. Schoenauer (Eds.): LION 6, LNCS 7219, pp. 372–377, 2012.
© Springer-Verlag Berlin Heidelberg 2012

Local Search (LS) aims to iteratively improve a solution by local transformations, replacing it by a better neighbor until no more improving moves are possible. Most known LS algorithms for the TSP are based on k-opt exchanges which delete k arcs of a solution and reconnect partial tours with k other arcs. Figure 1 describes the 3-opt method [4]. As a LS procedure may become trapped in a local optimum, it is often embedded in a guiding construction such as Iterated Local Search (ILS) [5]. Figure 2 shows the main steps of this metaheuristic.

Compute length L of solution S
while S is improved **do**
for all $a, b, c \in [0; n]$ **do**
Delete arcs $(a,a+1)$, $(b,b+1)$ and $(c,c+1)$
Produce S' by reconnecting partial tours
with other arcs
Compute length L' of solution S'
if $L' < L$ **then**
$S = S'$ and $L = L'$
Return best solution S

Generate solution S
Apply LS procedure on S
Compute length L of solution S
while end criterion is not reached **do**
Transform S into S' by a perturbation move
Apply LS procedure on S'
Compute length L' of solution S'
if $L' < L$ **then** //*acceptance criterion*
$S = S'$ and $L = L'$
Return best solution S

Fig. 1. 3-opt LS pseudo-code   **Fig. 2.** ILS pseudo-code

The works of Stützle and Hoos [10] and Lourenço *et al.* [5] show the efficiency of ILS in solving TSP problems ranging from 100 to 5915 cities. However, faced to large and hard optimization problems, it may need a considerable amount of computing time and memory space to be effective in the exploration of the search space. A way to accelerate this exploration is to use parallel computing.

3   Literature Review on Parallel LS and ILS

Verhoeven and Aarts [11] proposed a classification that distinguishes *single-walk* and *multiple-walk* parallelization approaches for LS algorithms. In the first category, one search process goes through the search space and its steps are decomposed for parallel execution. In that case, neighbors of a solution may be evaluated in parallel (*single-step*) or several exchanges may be performed on different parts of that solution (*multiple-step*). In the second category, many search processes are distributed over processing elements and designed either as *multiple independent walks* or *multiple interacting walks*.

Johnson and McGeoch [3] defined three parallelization strategies for k-opt algorithms. The first one uses *geometric partitioning* to divide the set of cities into subgroups that are sent to different processors to be improved by a constructive algorithm and a LS procedure. As this partitioning has the drawback of isolating subgroups without reconnecting subtours intelligently, the second strategy favors *tour-based partitioning* to divide tours into partial solutions that includes a part of the edges of the current solution. The third approach is a simple parallelization of neighborhood construction and exploration.

Works on parallelization of ILS for the TSP mainly follow the population-based, multiple-walk approach where many solutions are built concurrently. Hong *et al.* [2] designed a parallel ILS which executes a total of m iterations using a pool of p solutions. Martin and Otto [8] proposed an implementation in which several solutions are computed simultaneously on different processors and the best solution replaces all solutions at irregular intervals.

Luong *et al.* [7] proposed a methodology for implementing large neighborhood LS on GPU. The CPU runs the LS processes while the GPU generates neighbor solutions which are associated to CUDA threads. It is experimented with Tabu Search on the Permuted Perceptron Problem and maximal speedup of 25.8 is reported. In Luong *et al.* [6], Tabu Search is embedded with ILS to solve the Quadratic 3-dimensional Assignement Problem with maximal speedup of 6.1.

Most works related to parallel LS and ILS are based on traditional CPU architectures. As LS algorithms are key underlying components of most high-performing metaheuristics, a natural fit would be to run a guiding metaheuristic on CPU while the GPU, acting as a co-processor, takes charge of running the LS procedure. However, there is still much conceptual and technical work to achieve in order to design such hybrid parallel combinatorial optimization methods. This paper aims to partially fill this gap by proposing and evaluating GPU implementations of an ILS algorithm for the TSP based on 3-opt parallel LS.

4   Parallel GPU Strategies for ILS

We present two GPU strategies for ILS which mainly differ by the way they distribute solutions to processing elements and by their use of GPU shared memory. Beforehand, we provide a brief description of the GPU computing environment.

The NVIDIA GPU [9] architecture includes many *Streaming Multiprocessors* (SM), each one of them being composed of *Streaming Processors* (SP). Each SM allows the execution of many threads in a data-parallel fashion. On this special hardware, the *global* memory is a specific region of the *device* memory that can be accessed in read and write modes by all SPs of the GPU. It is relatively large in size but slow in access time. *Constant* and *texture memory caches* provide faster access to device memory but are read-only. All SPs can read and write in their *shared memory*, which is fast in access time but small in size and local to a SM. In the CUDA programming model [9], the GPU works as a co-processor of a conventional CPU. It allows the parallel execution of many CUDA threads that are grouped into *blocks* to be executed by the SMs. However, the number of blocks that a SM can process at the same time (*active blocks*) is restricted by the available shared memory and registers. Special care must also be taken to avoid flow control instructions (if, switch, do, for, while) that may force threads of a same block to take different paths in the program and serialize the execution.

The proposed ILS implementations are inspired by the multiple independent walks strategy described in Section 3. However, only the LS is parallelized on GPU instead of entire walks. In this scheme, illustrated in Figure 3(a), LS is applied on each solution on different processing elements.

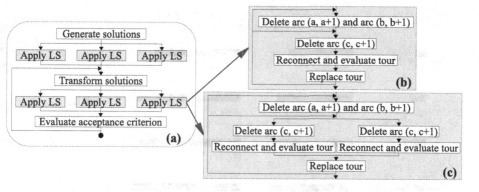

Fig. 3. Parallelization models for ILS : general (a), ILS_{thread} (b) and ILS_{block} (c)

On a conventional CPU architecture, the concept of processing element is usually associated to a single-core processor or to one of the cores of a multi-core processor. On a GPU, the obvious choice is to associate this concept to a single SP. In that case, a first strategy that may be defined is to associate each LS to a CUDA thread. Each thread then improves its solution in a SIMD fashion. This strategy, called ILS_{thread}, is illustrated in Figure 3(b) and has been proposed for the parallelization of Ant Colony Optimization in previous work by the authors [1]. It has the advantage of allowing the execution of a great number of LSs on each SM and the drawback of limiting the use of fast GPU memory.

The second strategy, called ILS_{block} and illustrated in Figure 3(c), is based on associating the concept of processing element to a whole SM. In that case, each solution is associated to a CUDA block and parallelism is preserved for the LS phase. A single thread of a given block is still in charge of applying LS to a solution, but another level of parallelism is exploited by sharing the multiple neighbors between many threads of a block. Following the idea of ILS_{thread}, a simple implementation would then imply keeping the private data structures of a solution in the global memory. However, as only one solution is assigned to a block and so to a SM, it becomes possible to store the data structures needed to improve the solution in the shared-memory. Two variants of the ILS_{block} strategy are then distinguished and experimented : ILS_{block}^{global} and ILS_{block}^{shared}.

5   Experimental Results

GPU strategies are experimented on TSP problems ranging in size from 100 to 3038 cities. Speedups are computed by dividing the sequential CPU time with the parallel time obtained with the GPU acting as a co-processor. Experiments were made on a NVIDIA Fermi C2050 GPU (14 SMs, 32 SPs per SM and 48 KB of shared memory). Code was written in the "C for CUDA V4.0" [9] programming environment. As a premiminary step, we validated our sequential ILS with a comparative study with Stützle and Hoos [10] and Lourenço et al. [5] works.

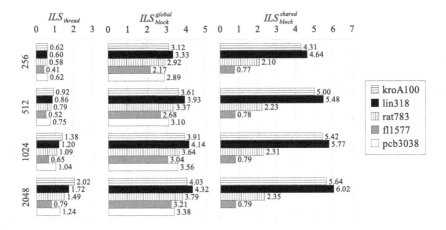

Fig. 4. Speedups for ILS_{thread}, ILS_{block}^{global} and ILS_{block}^{shared} strategies for each nb_{sol}

The parallel ILS is configured with a population of nb_{sol} solutions, a total number of 1048576 iterations and an ILS procedure limited to $it_{lim} = \frac{1048576}{nb_{sol}}$ iterations for each solution. Speedups are computed from 20 trials for problems with less than 1000 cities and from 10 trials for larger instances.

Figure 4 illustrates the speedups obtained for each problem and each strategy. It shows that increasing nb_{sol} and so, the total number of threads, generally leads to increasing speedups for all strategies. Moreover, speedups obtained with ILS_{thread} are limited to 2.02 and are always lower than with ILS_{block}. This strategy does not execute enough threads to efficiently hide memory latency. Furthermore, code divergence induced by computing the neighbors of many solutions/threads on the same block involves significant algorithm serialization.

The greater speedups and the maximal value of 4.32 obtained with ILS_{block}^{global} show that sharing the work associated to each solution between several threads is more efficient. However, speedups increase from 100 to 318 cities and then slightly decrease. In that case, the larger data structures and frequent memory accesses imply memory latencies that grow faster than the benefits of parallelizing available computations. Further improvements are brought by the use of shared memory of ILS_{block}^{shared}, which provides a maximal speedup of 6.02. However, results for the three biggest problems show that the limits of this kind of memory are quickly reached. In fact, since this memory is very limited in size, either speedup is not achieved or the problem can not be solved at all.

An analysis of the average percentage deviation Δ from the optimum showed that the optimal solution is always found ($\Delta = 0.000$) by the parallel implementations for small problems. For medium-sized problems, the more nb_{sol} increases, the less frequently the optimal solution is found ($\Delta = 0.002$ to $\Delta = 0.137$). As the number of iterations becomes too low to provide a thorough search, the optimal solution is never found for the bigger problem ($\Delta = 0.520$ to $\Delta = 1.112$). This indicates that when choosing appropriate parameters for the parallel algorithms, a compromise must be achieved between speedup and solution quality.

6   Conclusion

The aim of this paper was to design efficient parallelization strategies for the implementation of Iterated Local Search on Graphics Processing Units to solve the Travelling Salesman Problem. The ILS_{thread} and ILS_{block} strategies associated the Local Search phase to the execution of streaming processors and multiprocessors respectively. Experimental results showed significant speedups of up to 6.02 with solution quality often equal or close to optima, but also considerable limitations on large problems. Moreover, they highlighted that maximal exploitation of GPU ressources often requires algorithmic configurations that do not let ILS perform an effective exploration and exploitation of the search space.

In future work, we plan to study the GPU performance of other decomposition approaches like tour-based partitioning. We would also like to design k-opt based parallel algorithms that provide a better compromise between GPU efficiency and search robustness for the TSP and related problems.

Acknowledgments. This work has been supported by the Agence Nationale de la Recherche (ANR) under grant no. ANR-2010-COSI-003-03. The authors would also like to thank the Centre de Calcul Régional Champagne-Ardenne for the availability of the computational resources used for experiments.

References

1. Delévacq, A., Delisle, P., Gravel, M., Krajecki, M.: Parallel ant colony optimization on graphics processing units. In: PDPTA 2010, pp. 196–202. CSREA Press (2010)
2. Hong, I., Kahng, A., Moon, B.: Improved large-step markov chain variants for the symmetric tsp. Journal of Heuristics 3, 63–81 (1997)
3. Johnson, D., McGeoch, L.: The Travelling Salesman Problem: A Case Study in Local Optimization. In: Aarts, E.H.L., Lenstra, J.K. (eds.) Local Search in Combinatorial Optimization, pp. 215–310. John Wiley & Sons (1997)
4. Lin, S.: Computer solutions of the traveling salesman problem. Bell System Technical Journal 44, 2245–2269 (1965)
5. Lourenço, H., Martin, O., Stützle, T.: Iterated local search: framework and applications. In: Handbook of Metaheuristics, pp. 363–397. Springer (2010)
6. Luong, T., Loukil, L., Melab, N., Talbi, E.: A gpu-based iterated tabu search for solving the quadratic 3-dimensional assignment problem. In: AICCSA, pp. 1–8 (2010)
7. Luong, T., Melab, N., Talbi, E.: Neighborhood structures for gpu-based local search algorithms. Parallel Processing Letters 20(4), 307–324 (2010)
8. Martin, O., Otto, S.: Combining simulated annealing with local search heuristics. Annals of Operations Research 63, 57–75 (1996)
9. NVIDIA Corporation: CUDA : Computer Unified Device Architecture Programming Guide 4.0 (2011), http://www.nvidia.com
10. Stützle, T., Hoos, H.: Analysing the run-time behaviour of iterated local search for the traveling salesman problem. In: Essays and Surveys in Metaheuristics, pp. 21–43. Springer (2001)
11. Verhoeven, M., Aarts, E.: Parallel local search. J. Heuristics 1, 43–65 (1995)

Constraint-Based Local Search
for the Costas Array Problem

Daniel Diaz[1], Florian Richoux[2], Philippe Codognet[2],
Yves Caniou[3], and Salvador Abreu[4]

[1] University of Paris 1-Sorbonne, France
Daniel.Diaz@univ-paris1.fr
[2] JFLI, CNRS / UPMC / University of Tokyo, Japan
{richoux,codognet}@is.s.u-tokyo.ac.jp
[3] JFLI, CNRS / NII, Japan
Yves.Caniou@ens-lyon.fr
[4] Universidade de Évora and CENTRIA FCT/UNL, Portugal
spa@di.uevora.pt

Abstract. The Costas Array Problem is a highly combinatorial problem
linked to radar applications. We present in this paper its detailed modeling and solving by Adaptive Search, a constraint-based local search
method. Experiments have been done on both sequential and parallel
hardware up to several hundreds of cores. Performance evaluation of
the sequential version shows results outperforming previous implementations, while the parallel version shows nearly linear speedups w.r.t. the
sequential one, for instance 120 for 128 cores and 230 for 256 cores.

1 Introduction

During the last decade, the family of Local Search methods and Metaheuristics
has been quite successful in solving large real-life problems.

A generic Constraint-based Local Search method named Adaptive Search was
proposed in [4,5]. It is a metaheuristic that takes advantage of the structure
of the problem to guide the search and that can be applied to a large class
of constraints (*e.g.*, linear and non-linear arithmetic constraints and symbolic
constraints). Moreover, it intrinsically copes with over-constrained problems.

A parallel version with a multi-start approach requiring no communication
between processes has been defined in [7,3]. On classical CSP benchmarks from
CSPLib, this simple parallelization scheme gives good results, with a factor 50-70
speedup for 256 cores, but this is far from ideal speedup (*e.g.*, factor 256 speedup
for 256 processors), even for large problem instances. It is thus an open question
to know whether this is due to the classical (structured) CSP benchmarks used
or if this is a limitation of the method. In this paper we address the problem
of modeling a very combinatorial problem, with a low density of solutions the
Costas Array Problem (CAP) in the sequential version and we further investigate
if it scales up to a large number of processors and exhibits better speedups.
The CAP is an abstract problem that was motivated by a sonar application in

Y. Hamadi and M. Schoenauer (Eds.): LION 6, LNCS 7219, pp. 378–383, 2012.
© Springer-Verlag Berlin Heidelberg 2012

the 1960's but still has practical interest in radar and software-defined radio applications [2]. Experiments in solving the CAP by Adaptive Search on two parallel platforms, the Hitachi HA8000 supercomputer at University of Tokyo and GRID'5000, the French national Grid for the research, show nearly linear speedups w.r.t. the sequential version, for instance 120 for 128 cores and 230 for 256 cores.

The rest of this paper is organized as follows. Section 2 presents the Costas Array Problem. Section 3 presents the modeling of the Costas Array Problem within the Adaptive Search formalism. Section 4 details the experiments up to 256 cores on the HA8000 supercomputer and the GRID'5000 platform. Section 5 concludes the paper and briefly discusses about future work.

2   The Costas Array Problem

A Costas array is an $n \times n$ grid containing n marks such that there is exactly one mark per row and per column and the $n(n-1)/2$ vectors joining the marks are all different. We give here an example of Costas array of size 5. It is convenient to see the Costas Array Problem (CAP) as a permutation problem by considering an array of n variables $(V_1, \ldots, V_n)$ which forms a permutation of $\{1, 2, \ldots, n\}$. The Costas array above can thus be represented by the array $[3, 4, 2, 1, 5]$.

Historically, these arrays have been developed in the 1960's to compute a set of sonar and radar frequencies avoiding noise [6]. The problem to find a Costas array of size n is very complex since the required time grows exponentially with n. In the 1980's, several algorithms have been proposed to build a Costas array given n, such as the Welch construction and the Golomb construction [9], but these methods cannot built Costas arrays of size 32 and some higher non-prime sizes. Nowadays, after many decades of research, it remains unknown if there exist any Costas arrays of size 32 or 33. Another difficult problem is to enumerate all Costas arrays for a given size. Using the Golomb and Welch constructions, Drakakis *et. al* present in [8] all Costas arrays for $n = 29$. They show that among the 29! permutations, there are only 164 Costas arrays, and 23 unique Costas arrays up to rotation and reflection. There are constructive methods known to produce Costas arrays of order 24 to 29.

The Costas array problem has been proposed as a challenging combinatorial problem by Kadioglu and Sellmann in [10]. They propose a local search metaheuristic, *Dialectic Search*, for constraint satisfaction and optimization, and show its performance for several problems. Clearly this problem is too difficult for propagation-based solvers, even for medium size instances (*i.e.*, with n around $18-20$). Let us finally note that we do not pretend that using local search is better than constructive methods in order to solve the CAP. We rather consider the CAP as a very good benchmark for testing local search and constraint-based systems and to investigate how they scale up for large instances and parallel execution.

In [12], Rickard and Healy studied a stochastic search method for CAP and concluded that such methods are unlikely to succeed for $n > 26$. Although their conclusion is true for their stochastic method, it cannot be extended to all stochastic searches: their method uses a restart policy which is too simple and they also used an approximation of the Hamming distance between configurations in order to guide the search which they recognized themselves not be be a very good indicator. However, they studied in this paper the distribution of solutions in the search space and shown that clusters of solutions tend to spread out from $n > 17$, which justify our multi-walk approach presented in Section 4 to reach linear speedup for high values of n.

3 Solving the CAP with Adaptive Search

The CAP can be modeled as a permutation problem by considering an array of n variables $(V_1, \ldots, V_n)$ which forms a permutation of $\{1, 2, \ldots, n\}$ (*i.e.*, implicit alldifferent constraint on variables V_i). A variable $V_i = j$ iff there is a mark at column i and row j. To take into account constraints on vectors between marks (which must be different) it is convenient to use the so-called *difference triangle*.

This triangle contains $n - 1$ rows, each row corresponding to a distance d. The dth row of the triangle contains the differences $V_{i+d} - V_i$ for all $i = 1, \ldots, n - d$ (*i.e.*, the difference of values at a distance d). Ensuring all vectors are different comes down to ensure the triangle contains no repeated values on any given row (*i.e.*, alldifferent constraint on each row).

	3	4	2	1	5
$d = 1$	1	-2	-1	4	
$d = 2$		-1	-3	3	
$d = 3$			-2	1	
$d = 4$				2	

Here is the difference triangle for the Costas array given as example in Section 2.

In the Adaptive Search (AS) method, the way to define a constraint is done via error functions [4]. At each new configuration, the difference triangle is checked to compute the global cost and the cost of each variable V_i. Each row d of the triangle is checked one by one. Inside a row d, if a pair (V_i, V_{i+d}) presents a difference which has been already encountered in the row, the error is reported as follows: increment the global cost and the cost of both variables V_i and V_{i+d} by $ERR(d)$ (a strictly positive function). For a basic model we can use $ERR(d) = 1$ (to simply count the number of errors). Obviously a solution is found when the global cost equals 0. Otherwise AS selects the most erroneous[1] variable and will try to improve it.

Our AS sequential version has been tested over the CAP and compared to Dialectic Search (DS). AS outperforms DS on the CAP: for small instances AS is five times faster but the speedup seems to grow with the size of the problem, reaching a factor 8.3 for $n = 18$. [10] does not provide results for instances with $n > 18$.

CAP has also been used as a benchmark in the Constraint Programming community and we can compare with a CP Comet program made by Laurent Michel and based on the modeling in MiniZinc by Barry O'Sullivan[2]. As could

---

[1] i.e. the variable with the highest total error.

[2] http://www.g12.cs.mu.oz.au/mzn/costas_array/CostasArray.mzn

be expected, CP is much less efficient than local search, and this Comet program is about 400 times slower than AS for CAP19.

4 Parallel Implementation and Performance Analysis

We implemented a parallel version of AS using OpenMPI, an implementation of the MPI standard. Experiments and performance results on classical CSP benchmarks are described in [3]. The parallelization is straightforward and based on the idea of multi-starts and independent multiple-walks: fork a sequential AS method on every available cores. But on the opposite of the classical fork-join paradigm, parallel AS shall terminate as soon as a solution is found, not wait until all the processes have finished (since some searches initialized with "bad" initial configurations can take some time). Thus, some non-blocking tests are involved every c iterations to check if there is a message indicating that some other processes has found a solution; in which case it terminates the execution properly. This results in a high number of independent work units, a high CPU to I/O ratio, and no inter-process communication Three different testbeds were used on two platforms: The supercomputer HA8000 at the University of Tokyo (with a maximum of nearly 16000 cores) and the French national grid for research GRID'5000 (on two nodes at Sophia-Antipolis: Suno with 360 cores and Helios with 224 cores). Tables 1 & 2 show the execution times of the parallel executions on the HA8000 supercomputer and GRID'5000. Timings are given in seconds and are the average of 50 executions for each benchmark; they do not include the deployment time, negligible on big benchmarks.

Table 1. Speedups on HA8000, Suno and Helios for small instances of CAP

Platform	Problem	Time on 1 core	Speedup on k cores			
			32	64	128	256
HA8000	CAP 18	6.76	27.0	29.4	28.2	26.0
	CAP 19	54.54	29.6	54.5	75.7	99.2
	CAP 20	367.2	26.6	42.4	98.2	168
Suno	CAP 18	5.28	33.0	63.6	94.3	139
	CAP 19	49.5	36.1	83.9	121	226
	CAP 20	372	30.5	63.5	139	208
	CAP 21	3743	21.9	72.8	107	218
Helios	CAP 18	8.16	34.0	74.2	136	-
	CAP 19	52.0	22.6	59.8	130	-
	CAP 20	444	31.0	58.2	98.2	-

Behaviors on all three platforms are similar and exhibit very good speedups for larger instances. For $n = 21$ on Suno we have **a 218 times speedup on 256 cores** w.r.t. sequential execution. For the bigger instances CAP21 and CAP22, we present in Table 2 results for executions from 32 to 256 cores only, because the sequential time becomes prohibitive (*e.g.,* more than one hour on average

Table 2. Speedups on HA8000, Suno and Helios for large instances of CAP

Platform	Problem	Time on 32 cores	Speedup on k cores 64	128	256
HA8000	CAP21	160.4	1.96	4.16	10.0
	CAP22	501.2	2.01	3.90	8.24
Suno	CAP21	171	3.32	4.90	9.94
	CAP22	731	1.92	3.66	7.09
Helios	CAP21	153	1.51	4.17	-
	CAP22	1218	2.34	5.53	-

for CAP21 and more than 10 hours for CAP22 on HA8000). We can see that on all platforms, **execution times are halved when the number of cores is doubled**, thus achieving ideal speedup. As a final result, we can now solve $n = 22$ in about one minute on average with 256 cores on HA8000.

Up to now, we focused on the *average* execution time in order to measure the performance of the method, but a more detailed analysis could be done. In [1,11], a method is introduced to represent and compare execution times of stochastic optimization methods by using so-called *time-to-target plots*. Observe that, for the CAP, the target value to achieve is obviously *zero*, meaning that a solution is found. It is then easy to check if runtime distributions could be approximated by a (shifted) exponential distribution of the form: $1 - e^{-(x-\mu)/\lambda}$. Then, according to [13], it is possible to achieve linear speedups by multiple independent walks if we have an exponential runtime distribution.

The following figure presents time-to-target plots for CAP 21 in order to compare runtime distributions over 32, 64, 128 and 256 cores.

Points represent execution times obtained over 200 runs and lines correspond to the best approximation by an exponential distribution. It can be seen that the actual runtime distributions are very close to exponential distributions. Time-to-target plots also give a clear visual comparison between instances of the same method running on a different number of cores. For instance it can be seen that we have around 50% chance to find a solu-

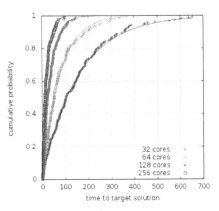

tion within 100 seconds using 32 cores, and around 75%, 95% and 100% chance respectively with 64, 128 and 256 cores.

5 Conclusion and Future Work

The CAP is a hard combinatorial problem for medium and large instances, too difficult to solve with classical propagation-based solver and we thus used a constraint-based local search solver. We proposed a parallel version based on the idea of multi-starts and independent multiple-walks which naturally provides *Pleasantly Parallel* computations and appears viable as it exhibits a nearly linear speedup behavior. We are currently continuing our experiments by tackling larger instances and using more cores.

Future work will focus on more complex parallel execution methods with inter-processes communication, *i.e.*, in the dependent multiple-walk scheme, in order to further improve performance. The communication mechanism will be designed with the goals of (1) minimizing data transfers as much as possible, as we aim at massively parallel machines with no hierarchical memory, and (2) re-using some common computations and/or recording previous interesting crossroads in the resolution, from which a restart can be operated.

References

1. Aiex, R., Resende, M., Ribeiro, C.: TTT Plots: a perl program to create time-to-target plots. Optimization Letters 1, 355–366 (2007)
2. Beard, J., Russo, J., Erickson, K., Monteleone, M., Wright, M.: Combinatoric collaboration on costas arrays and radar applications. In: Proceedings of the IEEE Radar Conference, Philadelphia, USA, pp. 260–265 (2004)
3. Caniou, Y., Codognet, P., Diaz, D., Abreu, S.: Experiments in parallel constraint-based local search. In: EvoCOP 2011, 11th European Conference on Evolutionary Computation in Combinatorial Optimisation, Italy. Springer (2011)
4. Codognet, P., Diaz, D.: Yet Another Local Search Method for Constraint Solving. In: Steinhöfel, K. (ed.) SAGA 2001. LNCS, vol. 2264, pp. 73–90. Springer, Heidelberg (2001)
5. Codognet, P., Diaz, D.: An efficient library for solving CSP with local search. In: Ibaraki, T. (ed.) MIC 2003, 5th International Conference on Metaheuristics (2003)
6. Costas, J.: A study of detection waveforms having nearly ideal range-doppler ambiguity properties. Proceedings of the IEEE 72(8), 996–1009 (1984)
7. Diaz, D., Abreu, S., Codognet, P.: Parallel Constraint-Based Local Search on the Cell/BE Multicore Architecture. In: Essaaidi, M., Malgeri, M., Badica, C. (eds.) Intelligent Distributed Computing IV. SCI, vol. 315, pp. 265–274. Springer, Heidelberg (2010)
8. Drakakis, K., Iorio, F., Rickard, S., Walsh, J.: Results of the enumeration of costas arrays of order 29. Advances in Mathematics of Communications 5(3), 547–553 (2011)
9. Golomb, S., Taylor, H.: Constructions and properties of Costas arrays. Proceedings of the IEEE 72(9), 1143–1163 (1984)
10. Kadioglu, S., Sellmann, M.: Dialectic Search. In: Gent, I.P. (ed.) CP 2009. LNCS, vol. 5732, pp. 486–500. Springer, Heidelberg (2009)
11. Ribeiro, C., Rosseti, I., Vallejos, R.: Exploiting run time distributions to compare sequential and parallel stochastic local search algorithms. Journal of Global Optimization 17, 1–25 (2011) (published online August 17, 2011)
12. Rickard, S., Healy, J.: Stochastic search for costas arrays. In: Proceedings of the 40th Annual Conference on Information Sciences and Systems, Princeton, NJ, USA (March 2006)
13. Verhoeven, M., Aarts, E.: Parallel local search. Journal of Heuristics 1(1), 43–65 (1995)

Evaluation of a Family of Reinforcement Learning Cross-Domain Optimization Heuristics

Luca Di Gaspero and Tommaso Urli

DIEGM, Università degli Studi di Udine
via delle Scienze 208 – I-33100, Udine, Italy
{luca.digaspero,tommaso.urli}@uniud.it

Abstract. In our participation to the *Cross-Domain Heuristic Search Challenge (ChesC 2011)* [1] we developed an approach based on Reinforcement Learning for the automatic, on-line selection of low-level heuristics across different problem domains. We tested different memory models and learning techniques to improve the results of the algorithm. In this paper we report our design choices and a comparison of the different algorithms we developed.

1 Introduction

ChesC 2011 aims at fostering the development of methods for the automatic design of heuristic search methodologies, which are applicable to multiple problem domains (see [1] for more details). The competition sits on the underlying framework of hyper-heuristics, i.e., automatic methods to generate or select heuristics for tackling hard combinatorial problems. In order to ease the implementation of hyper-heuristics and to let the participant compete on a common ground, the competition organizers released a software framework, called HyFlex [3], to be used as a basis for their algorithms. HyFlex is a Java API that provides the basic functionalities for loading problem instances, generating solutions and applying low-level heuristics. Low-level heuristics are black-boxes, and only information about their *family* is known (i.e., *ruin-recreate*, *mutation*, *cross-over* and *local search* steps). The six problem domains considered in the competition are: *MAX-SAT, (1D) Bin Packing, Permutation Flow Shop Scheduling, Personnel Scheduling, TSP* and *VRP*. We refer the reader to the ChesC 2011 website [1] for further details.

2 Reinforcement Learning Basics

The algorithmic alternatives that we have considered for ChesC 2011 are all based on Reinforcement Learning (RL) [6]. In order to use RL, one needs to specify at least three components: an *environment* (whose observable features are encoded in *states*), a set of *actions* which can be pursued by the learning agent and a *reward* function which is a numeric feedback about the agent's actions.

Y. Hamadi and M. Schoenauer (Eds.): LION 6, LNCS 7219, pp. 384–389, 2012.
© Springer-Verlag Berlin Heidelberg 2012

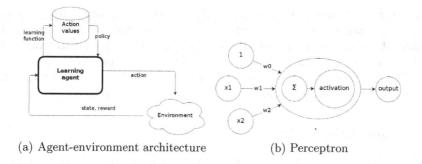

(a) Agent-environment architecture   (b) Perceptron

The various choices for these elements, together with the action selection *policy* and the *learning function*, determine a range of different behaviors.

At each iteration, the agent selects which action to take in the current state according to its policy. The policy is a function for selecting an action based on its *value*. Together with the set of learned action values, a policy determines completely (barring stochastic effects) the agent's behavior at each decision step. The execution of an action updates the state of the agent and yields a *reward* value. The *learning function* uses this reward to update one (or mode) action values. Since the policy is usually fixed, this concretely changes the behavior of the agent for the future iterations. The state represents the agent's beliefs about the environment at a specific time; representing the state in a rich, yet complete, manner is key to the success of RL.

2.1 Function Approximation with MLPs

When the states and actions are discrete and finite, a simple way to store action values is to keep them in a table (*tabular RL*). However, when the states or the actions are continuous or the number of states is just too large, this solution is no more viable. In these cases, the only option is to consider action values as a continuous function and to use function approximation techniques to model it. Multi-Layer Perceptrons (MLP) are a function approximation mechanism which belongs to the class of *supervised* Machine Learning algorithms. We are going to briefly revise MLPs in this section, by starting from the simpler concept of Perceptron. A (Single-Layer) Perceptron is a processing unit with a number of weighted inputs (one of which has always a value of 1) and an output. Upon activation, the Perceptron computes the weighted sum of its inputs and outputs a function of this sum. The algorithm implemented by the perceptron in Figure 1b can be summarized with the following formula:

$$h(\boldsymbol{x}) = activation(\boldsymbol{w}^T\boldsymbol{x}) \tag{1}$$

By varying *activation* one can use perceptrons to approximate different functions, however the complexity of these is very limited. MLPs are layered networks of Perceptrons in which outputs of nodes in a layer are connected to inputs of nodes in the following. Since Perceptrons are actually inspired to neurons, these networks are commonly known as Artificial Neural Networks (ANN). Layers other than input or output are called *hidden*. There are no constraints on the number of hidden layers or nodes, however it has been demonstrated [4] that a MLP with a single, large-enough, hidden layer can approximate any nonlinear function of the input. Unfortunately, there is no rule of thumb on the right number of hidden neurons, which must be worked out with parameter tuning.

2.2 Eligibility Traces

Eligibility Traces (ET) are a RL mechanism for *temporal credit assignment*. The idea in temporal credit assignment is that each action on the trajectory to a reward, and not just the last one, must take some credit for that reward. To accomplish this, one a chance is to keep an $e_{s,a}$ value for each visited (s, a) pair. This value tells how long before the pair was visited and is updated as follows

$$e_{s,a} = \begin{cases} 1 & \text{if state is } s \text{ and action is } a \\ \lambda e_{s,a} & \text{otherwise} \end{cases} \qquad (2)$$

where λ is a decay factor in $[0, 1)$. In RL, $e_{s,a}$ is called *eligibility trace* of (s, a) and is used by the learning function to weight the update to (s, a). Intuitively, recently visited (s, a) pairs are more likely to be responsible for a reward with respect to older ones, and should benefit (or suffer more) from the last obtained reward. In practice, one can implement ETs efficiently by keeping a queue of the last $\lceil \log(threshold)/\log(\lambda) \rceil$ visited pairs where *threshold* is the value of e under which an update is considered neglectable. Then $e_{s,a}$ is computed as $\lambda^{position}$ where *position* is the pair's position in the queue.

3   Reinforcement Learning for Heuristic Selection

In order to describe our RL hyper-heuristics we must identify the following elements: (i) the *environment states*, (ii) the set of *actions*, (iii) the *reward*, (iv) the *policy* and (v) the *learning function*.

Environment: HyFlex is designed to support the construction of cross-domain hyper-heuristics. For this reason, all the information about a solution (except its cost) domain is hidden to the user. This makes things difficult because, in principle, RL states must be Markov (i.e. enough informative to allow choosing the right action). After attempting some variants, we resorted to a state representation which tries to capture the concept of reward trend. In particular, when a reward is obtained, the new state is computed as $s_{i+1} = \lfloor (s_i + reactivity * (r_i - s_i)) \rfloor$, where r_i is a normalized cross-domain reward measure and *reactivity* defines the attitude of the agent to switch state.

Actions: We defined a possible *action* a in a given state as the choice of the heuristic family to be used, plus an intensity (or depth of search) value in

Table 1. Common hyper-heuristic parameters

parameter name	values	explanation
agents	3, 4, 5, 6, 8, 10	number of concurrent agents
crossoverWith	bestAgent, incumbentOptimum	secondary solution for cross-over
reactivity	0.05, 0.1, 0.25, 0.5, 0.9	readiness to change state
learningRate	0.1, 0.2, 0.3	readiness to update action values
epsilon	0.01, 0.05, 0.1	probability to pick random actions

the quantized set of values $0.2, 0.4, 0.6, 0.8, 1.0$. Once the family has been determined, a random heuristic belonging to that family is chosen and applied to the current solution with the specified intensity (or depth of search).

Reward: The reward r is computed as the solution's Δ_{cost} before and after action application. Variants which also take into account the time spent applying an action have been tried, but with poor results.

Policy: The ϵ-greedy policy (which chooses $\arg\max_a \pi_{s,a}$ with probability $1 - \epsilon$ and a random action otherwise) proved to perform better than a number of alternatives and is currently our policy of choice. A note on the *move acceptance criterion*: in our approach we decided to always trust the policy hence we always apply the action chosen even if it deteriorates the solution.

Learning function: The learning function is based on a very simple update $\pi(s, a)_{k+1} = \pi(s, a)_k + learningRate * (r_k - \pi(s, a)_k)$ which always moves the estimated reward for a (s, a) pair towards the last reward. The discount factor *learningRate* is needed to tackle non-stationarity (i.e. updates are constant, the policy never converges).

4 Experimental Analysis

In order to tune the variants' parameters and to understand their relationships we carried out an experimental analysis based on the tools commonly employed in the statistical analysis of algorithms. To collect the required data we ran all configurations on three different Intel machines equipped with Quad Core processors (resp. at 2.40, 2.40 and 3.00 GHz) and running Ubuntu 11.04. Differences were leveled through a benchmarking tool provided by CHeSC organizers.

We compare three hyper-heuristics inspired to the RL variants in section 2: tabular reinforcement learning (RLHyperHeuristic), tabular reinforcement learning with ETs (RLHyperHeuristic-ET) and reinforcement learning with MLP function approximation (RLHyperHeuristic-MLP). Although each approach has its own parameters, some of them (Table 1) are common to all hyper-heuristics.

RLHyperHeuristic-MLP requires a number of extra parameters (see Table 2a) related to MLPs. The parameters *hiddenLayers* and *hiddenNeurons* determine the complexity of the function that the MLP is able to approximate. *inputScale* is a parameter to control input normalization. Similarly RLHyperHeuristic-ET introduces two parameters (see Table 2b): *threshold* and *traceDecay* (λ), which are used to compute the length of the eligibility queue.

Table 2. Specific parameters for RLHyperHeuristic variants

(a) RLHyperHeuristic-MLP

parameter name	values
hiddenLayers	1, 2
hiddenNeurons	20, 30, 40
inputScale	1, 3

(b) RLHyperHeuristic-ET

parameter name	values
threshold	0.01
traceDecay	0.5, 0.9

Since the evaluation has to be performed across different domains and on instances with different scales of cost functions we decided to consider as the response variable of our statistical tests the normalized cost function value. That is, the cost value y is transformed by means of the following transformation, which aggregates the results on the same problem instance π:

$$e(y, \pi) = \frac{y(\pi) - y^*(\pi)}{y_*(\pi) - y^*(\pi)} \tag{3}$$

where $y^*(\pi)$ and $y_*(\pi)$ denote the best known value and the worst known value of cost on instance π. This information has been computed by integrating the data gathered by our experiments with the information made public by CHeSC organizers. In all the following analyses we employ the R system [5].

Parameter influence. The first analysis aims at clarifying the influence of parameters on the outcome of the algorithms, in order to fix some of the parameters to *reasonable* values and to perform further tuning of the relevant ones. For this purpose we perform an *analysis of variance* on a comprehensive dataset including all configurations run throughout all the problem domains. Each variant has been run on each instance for 5 repetitions. We perform separate analysis for each variant and we set the significance level of the tests to 0.95.

RLHyperHeuristic: The most relevant parameters are the selection of the cross-over solution and the number of agents, but there seems to be no detectable interaction among them. As for this variant, ϵ is also significant.
RLHyperHeuristic-ET: The relevant parameter is the *traceDecay*, apart of the selection of the agent for the crossover that was relevant also in the previous variant. The number of agents doesn't seem to be relevant.
RLHyperHeuristic-MLP: Apart *crossOver*, the *inputScale* and *hiddenNeurons* are relevant in explaining the different performances of the algorithm. We do not found any significant second-order interaction among parameters.

Parameter tuning. For tuning the parameters identified in the previous analysis we employed the F-Race technique [2]. As for the selection of the best candidates, we took the ones that had the lowest median value of the normalized cost function across all instances and all domains. The setting of the parameters for the three different variants of the algorithm are reported in Table 3.

Table 3. **CW** = crossoverWith, **LR** = learningRate, **R** = reactivity, **TH** = treshold, **TD** = traceDecay, **IS** = inputScale, **HL** = hiddenLayers and **HN** = hiddenNeurons

variant	agents	CW	ϵ	LR	R	TH	TD	IS	HL	HN
RL	5	incumbentOptimum	0.05	0.2	0.5					
RL-ET	4	incumbentOptimum	0.1	0.1	0.1	0.01	0.5			
RL-MLP	4	incumbentOptimum	0.05	0.1	0.5			1	1	20

Comparison with the other participants. We have compared our variants with the other participants in the CHeSC competition by using the median value of the normalized cost function for ranking. Overall the variant using function approximation improves over our original algorithm (13th place, against 16th), while the one which uses eligibility traces doesn't. A proper tuning of the algorithm we sent to CHeSC has determined a relevant improvement but overall we are still far from the first positions. This is likely to be caused by the state representation, which seems to be insufficiently informative.

5 Conclusions and Future Work

This work is part of our investigation about the use of Machine Learning techniques for driving combinatorial optimization algorithms. In our opinion the results are interesting given the fact that there is no move acceptance criteria and the whole control is in the hands of a learning algorithm. However the intrinsic limitations imposed by the competition are too tight to allow a proper RL integration. For this reason, we are currently investigating these approaches outside the HyFlex framework.

References

[1] ASAP Research Group, Nottingham: CHeSC: the cross-domain heuristic search challenge (2011)
[2] Birattari, M., Stutzle, T., Paquete, L., Varrentrapp, K.: A racing algorithm for configuring metaheuristics. In: GECCO 2002: Proceedings of the Genetic and Evolutionary Computation Conference, vol. 2, pp. 11–18 (2002)
[3] Burke, E., Curtois, T., Hyde, M., Ochoa, G., Vazquez-Rodriguez, J.: HyFlex: A Benchmark Framework for Cross-domain Heuristic Search. Arxiv preprint arXiv:1107.5462, pp. 1–27 (2011), http://arxiv.org/abs/1107.5462
[4] Hornik Maxwell, K., White, H.: Multilayer feedforward networks are universal approximators. Neural networks 2(5), 359–366 (1989)
[5] R Development Core Team: R: A Language and Environment for Statistical Computing. R Foundation for Statistical Computing, Vienna, Austria (2011) ISBN 3-900051-07-0, http://www.R-project.org
[6] Sutton, R.S., Barto, A.G.: Reinforcement Learning: An Introduction (Adaptive Computation and Machine Learning). The MIT Press (1998)

Autonomous Local Search Algorithms
with Island Representation

Adrien Goëffon and Frédéric Lardeux

LERIA, University of Angers (France)
name@info.univ-angers.fr

Abstract. The aim of this work is to use this dynamic island model to autonomously select local search operators within a classical evolutionary algorithm. In order to assess the relevance of this approach, we will use the model considering a population-based local search algorithm, with no crossover and where each island is associated to a particular local search operator. Here, contrary to recent works [6], the goal is not to forecast the most promising crossovers between individuals like in classical island models, but to detect at each time of the search the most relevant LS operators. This application constitutes an original approach in defining autonomous algorithms.

1 Introduction

Island Models [9] are simultaneously considering a set of populations clustered in islands which are evolving independently during some search stages while interacting periodically. This model, which constitutes an additional abstraction level in comparison to classical genetic and memetic algorithms, allows to propose several diversification levels and to simplify its parallelization.

Most of the time, island models are used in a static way, where individuals are migrating from population to population following a determinate scheme [7], or are specifically chosen in order to reinforce the populations diversities [8,4,1]. Nevertheless, it is possible to dynamically regulate migrations between islands in considering a transition matrix [5]. Such a model can reinforce or reduce the migration probabilities during the evolutionary process in function of the impact of previous analog migrations. The aim is to auto-adapt migration without any given scheme, to dynamically regulate the gathering or isolation of individuals in function of the search progress, and consequently to adapt the population sizes.

In classical uniform island models, islands are following the same evolutionary rules, so they differ only by their individuals. The dynamic model allows to regulate interactions between individuals or group of individuals. We propose to extend this model in assigning to each island different local search operators. An appropriate and autonomous regulation of migration flows will affect dynamically the resources to the most pertinent operators in function of the search progress. In experimenting this model without crossovers but with a proper local search operator for each island, the objective is not only to regulate the interactions between individuals, but to simulate a reactive controller which assigns individuals to the most promising islands.

Y. Hamadi and M. Schoenauer (Eds.): LION 6, LNCS 7219, pp. 390–395, 2012.
© Springer-Verlag Berlin Heidelberg 2012

2   Island Models Framework

2.1   Island Models as a Complete Digraph

In [5] we proposed an island model framework which dynamically supervises the commonly-used specification parameters [1] like the number of individuals undergoing migration, the policy for selecting immigrants or the topology of the communication among subpopulations. An island model topology is represented by a complete labeled digraph $G = (X, X^2)$.

Migration policies are given by a transition (stochastic) matrix T, where $T(i, j)$ represents the probability for an individual to migrate from island i to island j (or to stay at the same island if $i = j$).

One can denote T_t the matrix at time (or generation) t.

An application of this dynamic evolution of the model topology is to determine pertinent migration probabilities at each time of the search, considering a classical multi-population based genetic algorithm. The dynamic regulation of migration policies can produce different size islands, which prevents poor-quality subpopulations or islands to require as many computational effort as promising ones. However, if different islands represent different mutation or local search operators, then the aim is to dynamically provide a well-adapted repartition of individuals in function of these operators and considering the search progression, which can be assimilated to an operator selection process.

2.2   Migration Policy

Algorithm 1 is the generic algorithm we used for the autonomous operator selection within an island model context. In order to allow a maximum of adaptability, we chose to update the migration process after each local search iteration (for the whole population). Ideational, less frequent mutations process do not minimize the effective number of mutations (individuals moving to other islands) but only provide a less reactive search. As a dynamic algorithm, transition values are expected to be regulate accordingly.

Initialize population;
repeat
    foreach *population* **do**
        foreach *individual* **do**
            ∟ One local search iteration;
    Update the Transition Matrix T;
    Migration Process;
until *stop condition*;

Algorithm 1. Generic Dynamic Island Model (DIM) Algorithm

The crucial point concerns the update of the transition matrix T, which follows a learning process:

$$T_t = (1 - \beta)(\alpha.T_{t-1} + (1 - \alpha).R_t) + \beta.N_t$$

R_t is a *reward matrix*, computed after migration process $t - 1$ and LS step t, and which takes into account the comparative pertinence of the last migrations. Using an *intensive* strategy, for each island, the migration which have brought the best average accuracy score acc of individuals (typically their fitness improvement) receives the maximum reward. More formally, if M_{ijt} is the set of individuals which have migrated from island i to island j in migration process $t - 1$ ($\cup_i M_{ijt}$ is the set of individuals in island j during iteration t):

$$R_t(i, j) = \begin{cases} 1/|B| & \text{if } j \in B, \\ 0 & \text{otherwise,} \end{cases}$$

$$\text{with } B = \underset{j'}{\text{argmax}} \frac{\sum_{x \in M_{ij't}} acc(x)}{|M_{ij't}|}$$

N_t is a noise stochastic matrix with random values.

The two parameters α and β allow to manage the update of the transition matrix. α represents the importance of the knowledge accumulated during the last migrations and β the amount of noise which is necessary to explore alternative ways and to keep the model reactive.

3   Experimentations

In this section we show that the behavior of our population-based local search algorithm is very close of the theoretical results. Moreover, we remark that it is not very dependent of the parameter tuning.

3.1   One-Max Problem

The One-Max problem is a simple and well-known problem, commonly used to assess the performance of Adaptive Operator Selection algorithms [3,2]. The n-bits One-Max problem considers n-length bit strings; starting from 0^n individuals (*i.e.* strings made up of n *zeros*), the aim is to maximize the number of *ones*, that is to reach the 1^n bit string. The *score* of a bit string x, noted $|x|_1$, corresponds to its number of *ones*.

Recent works cited above use four mutation (or local search) operators: *bit-flip*, which flips every bit with probability $1/n$, and *k-flip* (with $k = \{1, 3, 5\}$), which flips exactly k bits. In the following and depending on the context, bit-flip and k-flip can denote the mutation operator as well as the corresponding neighborhood relation. k-flip can easily be modelized as a neighborhood relation $\mathcal{N}_k : \{0, 1\}^n \to 2^{\{0,1\}^n}$ such as $x' \in \mathcal{N}_k(x)$ if and only if $|h(x, x')| = k$ (hamming distance). It is more difficult to exprim the bit-flip operator with a neighborhood relation, since it corresponds to a complete neighborhood with a non-uniform move probability. However, one bit-flip move can be reduced in one k-flip move with a determined probability of chosing k.

Intuitively, the 5-flip operator mutation will be more efficient when applied on weak individuals (with a majority of *zeros*), while 1-flip will improve with a higher probability individuals with a high proportion of *ones*.The domination rates evolution of the four considered operators in function of the score of an individual is shown in Figure 1 (with $M = \{$ 1-flip, 3-flip, 5-flip, bit-flip $\}$).

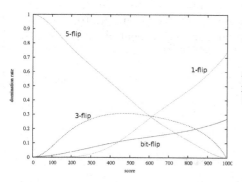
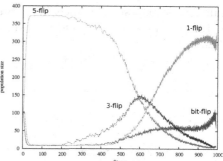

Fig. 1. Domination rates evolution for the 1000-bits One-Max problem

Fig. 2. Evolution of the population size in each island with respect to the average fitness of the population

3.2   Theoretical vs Empirical Results

The expected behavior during the search is to use the 5-flip operator when the population quality is weak (at the beginning), then the 3-flip operator and finally the 1-flip operator when the population quality is sufficiently high. In our experiments, this can be observed by the evolution of the population size in each island with respect to the migrations. The more an island attracts individuals, the more its assigned operator is applied.

Parameters for this experiment are:

- number of islands: 4 (one for each LS operator)
- population size: 400
- initial probabilities of migrations: 1 to stay in the same island
- (α, β): $(0.8, 0.01)$

To compare the experimental results with the theoretical values, we represent in Figure 2 the population size in each island with respect to the average fitness of the population. The fact that this evolution of population sizes, *i.e.* the computational effort of each operators, match with the theoretical domination rates, show the accuracy of the proposed model and its pertinence to simulate an operator selection mechanism.

3.3 Dynamic Model Parameters

Default used values for α and β are respectively 0.8 and 0.01. An increasing value of α makes the search slower since informations obtained by recent migrations are less considered for the update. On the contrary, decreasing value of α minimizes the impact of the knowledge (learning process) and overestimates the last migration effects, so the search can be wrong oriented by a migration which provides exceptionally a good result.

The influence of β is important, but its exact setting is not crucial to the smooth-running of the algorithm, even if a too high value of β make the search slower. On the other hand, it must make sure that $\beta \neq 0$, otherwise some islands can become and stay unreachable (transition probability equal to 0).

Effect of parameters α and β on the model are experimentally shown Figures 3 and 4.

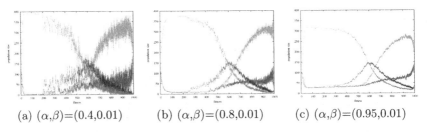

(a) $(\alpha,\beta)=(0.4,0.01)$ (b) $(\alpha,\beta)=(0.8,0.01)$ (c) $(\alpha,\beta)=(0.95,0.01)$

Fig. 3. Changing the value of α: less or more inertness makes the model more stable but dos not modify the global repartition of individuals

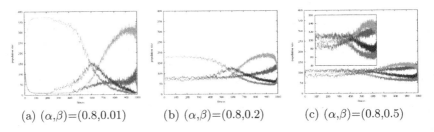

(a) $(\alpha,\beta)=(0.8,0.01)$ (b) $(\alpha,\beta)=(0.8,0.2)$ (c) $(\alpha,\beta)=(0.8,0.5)$

Fig. 4. Changing the value of β: more noise makes the repartition of individuals more uniform

4 Conclusion

This paper presents an original and efficient approach to design an autonomous local search algorithm with an accurate selection of operators. The proposed mechanism use a dynamic island model, where each island represents an operator. A learning process regulates and adapts migration policies during the search depending to the impact of previous migrations. At each stage of the search, the more efficient operators receive dynamically the great majority of computational resources. In other words, the model is able to auto-adapt the attractive power of each islands.

This application is an extension of the dynamic island model approach. In previous work, we focus on the capacity for the model to dynamically regulate the interaction between individuals in an evolutionary context, with crossovers and the same configuration on each island, with promising results. Here, we dissociate the exploitation / exploration dilemma to focus on the capacity to allocate with relevance the resources to the most suitable operators. For that, we used an experimental protocol which makes possible to assess the real efficiency of the model (One-Max problem and comparison with theoretical values). The next step is to apply this operator selection strategy to difficult problems, and then to assemble this heterogeneous model within a more general evolutionary context.

References

1. Araujo, L., Guervós, J.J.M., Mora, A., Cotta, C.: Genotypic differences and migration policies in an island model. In: GECCO, pp. 1331–1338 (2009)
2. Derbel, B., Verel, S.: DAMS: Distributed Adaptive Metaheuristic Selection. In: Proceedings of the Annual Conference on Genetic and Evolutionary Computation (GECCO 20111), United Kingdom, pp. 1–18 (July 2011)
3. Fialho, Á., Da Costa, L., Schoenauer, M., Sebag, M.: Extreme Value Based Adaptive Operator Selection. In: Rudolph, G., Jansen, T., Lucas, S., Poloni, C., Beume, N. (eds.) PPSN X. LNCS, vol. 5199, pp. 175–184. Springer, Heidelberg (2008)
4. Gustafson, S., Burke, E.K.: The speciating island model: An alternative parallel evolutionary algorithm. J. Parallel Distrib. Comput. 66(8), 1025–1036 (2006)
5. Lardeux, F., Goëffon, A.: A Dynamic Island-Based Genetic Algorithms Framework. In: Deb, K., Bhattacharya, A., Chakraborti, N., Chakroborty, P., Das, S., Dutta, J., Gupta, S.K., Jain, A., Aggarwal, V., Branke, J., Louis, S.J., Tan, K.C. (eds.) SEAL 2010. LNCS, vol. 6457, pp. 156–165. Springer, Heidelberg (2010)
6. Maturana, J., Lardeux, F., Saubion, F.: Autonomous operator management for evolutionary algorithms. J. Heuristics 16(6), 881–909 (2010)
7. Rucinski, M., Izzo, D., Biscani, F.: On the impact of the migration topology on the island model. CoRR, abs/1004.4541 (2010)
8. Skolicki, Z., De Jong, K.A.: The influence of migration sizes and intervals on island models. In: GECCO, pp. 1295–1302 (2005)
9. Whitley, D., Rana, S., Heckendorn, R.B.: The island model genetic algorithm: On separability, population size and convergence. Journal of Computing and Information Technology 7, 33–47 (1998)

An Approach to Instantly Use Single-Objective Results for Multi-objective Evolutionary Combinatorial Optimization

Christian Grimme[1] and Joachim Lepping[2]

[1] Robotics Research Institute, TU Dortmund University, 44221 Dortmund, Germany
christian.grimme@udo.edu
[2] INRIA Rhône-Alpes, Grenoble University, 38330 Montbonnot-Saint-Martin, France
joachim.lepping@inria.fr

Abstract. Standard dominance-based multi-objective evolutionary algorithms hardly allow to integrate problem knowledge without redesigning the approach as a whole. We present a flexible alternative approach based on an abstraction from predator-prey interplay. For parallel machine scheduling problems, we find that the combination of problem knowledge principally leads to better trade-off approximations compared to standard class of algorithms, especially NSGA-2. Further, we show that the incremental integration of existing problem knowledge gradually improves the algorithm's performance.

Keywords: Predator-Prey Model, Evolutionary Multi-Objective Optimization, Multi-objective Scheduling, Knowledge Integration.

1 Introduction

In multi-objective evolutionary optimization, *dominance-based methods* are currently used as quasi-standard. They extend the concept of original single-objective evolutionary algorithms to the multi-objective domain introducing mechanisms for selecting solutions regarding multiple objectives. For the NSGA-2 [2] algorithm, the particular fitness assignment is based on sorting the population into different fronts using the non-domination order relation. To form the next generation of candidate solutions, NSGA-2 combines the current population and its offspring generated by standard variation operators. Such a strong focus on selection may devalue variation operators to a subordinate influence. That means, for expertise integration advanced variation operators can unfold their full benefit only along with an alternative and more dynamic selection scheme which replaces the monolithic algorithmic architecture of dominance-based approaches. Such an alternative appears in this paper.

Our approach uses the predator-prey model (PPM) proposed by Laumanns et al. [4] which adapts the well-known predation paradigm from biology: a population of prey is arbitrarily distributed on a spatial structure which is represented by a toroidal grid. Predators pursue only *one objective* and favor only *one special*

Y. Hamadi and M. Schoenauer (Eds.): LION 6, LNCS 7219, pp. 396–401, 2012.
© Springer-Verlag Berlin Heidelberg 2012

variation operator each. They randomly roam the population to chase prey which are weak regarding their specific objective. Multiple predators are expected to force the prey likewise to adapt to the threats and thus result in suitable trade-off solutions for the complete optimization problem. In our approach the coupling of special heuristics (which realize the actual variation) to predators allows to integrate expert knowledge from single-objective problems.

2 Multi-objective Optimization and Scheduling Problems

In multi-objective optimization, a problem instance comprises multiple and (at least partly) contradicting goals that should be fulfilled simultaneously. Usually it is impossible to find a single optimal solution but only a set of good trade-offs among those goals. This solution set is called Pareto-optimal set and forms the Pareto-front in solution space.

A scheduling problem—denoted by $\alpha|\beta|\gamma$—is commonly concerned with allocating n jobs to a machine environment α with m machines such that all constraints β are met [3]. The resulting schedule should be optimal for one or more given objective(s) γ. For example, a check-in counter queue problem is denoted as $P_m|r_j, d_j|\sum U_j$. Here, P_m denotes an environment of m identical counters (machines). Passenger j arrives at time r_j and needs to catch a flight at time d_j. The objective is to minimize the total number of passengers $\sum U_j$, $U_j \in \{0,1\}$ missing their flight ($U_j = 1$). Under multiple objectives, the γ-field contains all objectives that have to be optimized simultaneously. Regarding our example the problem $P_m|r_j, d_j|\sum C_j, \sum U_j$ states that not only the flight misses should be minimized but also all customers should be served as fast as possible. This is expressed as minimizing the sum of all completion times, while the condition $C_j \geq r_j + p_j$ holds. There, p_j is the processing time of job j. While for the $\sum U_j$ objective ascending ordering of due dates d_j is reasonable the $\sum C_j$ can be solved optimally sequencing jobs in ascending order of p_j. However, as due dates and processing times might be unrelated (except that all due dates can be always met, thus $d_j > p_j$ holds.) both objectives are fundamentally conflicting.

As almost all multi-objective scheduling problems are NP-hard, practitioners have to use general techniques or randomized heuristic approaches like multi-objective evolutionary algorithms (MOEAs), see Coello et al. [1] for a detailed overview. Today, the practitioner finds a huge amount of standard algorithms that either apply non-dominated ranking, sorting and archiving techniques (e.g., NSGA-2, SPEA2, PAES) or indicator-based selection mechanisms (e.g., SMS-EMOA, HypE) to generate a precise and diverse Pareto-front approximation. As these methods are rather general, the practitioner still faces the problem of integrating his already available expert knowledge into the algorithm. That is more complicated than expected: Due to the monolithic and integrated structure of most approaches it is usually not sufficient—in contrast to single-objective problems—to only change variation operators. He rather has to redesign many parts of the original algorithmic scheme in order to bring in expertise. We address this problem by revisiting the predator-prey idea and show that it offers the property to integrate expertise seamlessly.

3   Predator-Prey Model for Multi-objective Optimization

The nature-inspired principle of predator and prey interaction proposed by Laumanns et al. [4] considers *prey* as solutions for multi-objective problems which are placed at vertices of a two-dimensional toroidal grid representing the spatially distributed population. *Predators* move across the spatial structure according to a random walk scheme (usually a uniformly distributed movement) and chase the prey only within their current neighborhood on the torus. This "hunting" process consists of evaluating all prey in the direct neighborhood of a predator's position according to a *single objective* assigned to it. The *worst* prey within this neighborhood is "eaten" and replaced by an offspring prey, which is created out of neighboring prey using variation operators. In our realization, the replacement approach follows an elitist philosophy: the worst prey is only replaced, if the offspring is better regarding the predators objective. The process is repeated until a termination criterion is reached. As the described action is restricted to each predator and completely self-contained, multiple predators can act in parallel and bring their influence to the distributed population.

For transferring the PPM to scheduling problems we encode schedules in prey using a permutation encoding of length n which represents the sequence of jobs. To map the permutation into schedules, we use a non-delay First-Come-First-Served (FCFS) approach. The main goal of expertise integration is to foster convergence to the Pareto-front. This expertise is here provided by simple sequencing heuristics: The *shortest processing time first* (SPT) rule is known to be optimal for the total completion time objective $(1|| \sum C_j, P_m || \sum C_j)$. Further, for the number of late jobs problem on parallel machines $(P_m || \sum U_j)$ the *earliest due date first* (EDD) rule is reasonable[1]. We designed a variation operator which allows to bring the effects of SPT, EDD, or any other sorting scheme randomly and well-dosed into the population. Figure 1 exemplary depicts the application of this operator to a given sequence with processing times p_j. A position is selected randomly in the permutation representation of the genotype. Then, a subsequence of $2\delta + 1$ genes is sorted according to the heuristic. Here, we show the application of SPT sorting. The size of this δ-neighborhood is determined by a always positive normal distributed value with adjustable step-size σ. A larger δ leads to a higher probability of a completely ordered genome, $\delta = 0$ leads to no reordering at all.

4   Experiments and Results

To evaluate our approach, we generated 50 synthetic job sets, $(\mathcal{J}_1^{50} \dots \mathcal{J}_{50}^{50})$ containing 50 jobs each. We sampled all sets with characteristics of processing time $p_j = \lfloor \mathcal{U}(1,50) \rfloor$, $\forall j = 1 \dots n$ and due dates $d_j = p_j + \lfloor \mathcal{U}(1,100) \rfloor$, $\forall j = 1 \dots n$. Release dates are generated depending on p_j and d_j according to $r_j = \mathcal{U}(0, d_j - p_j)$ in 90 % of the cases and $r_j = 0$ otherwise. As we consider a parallel machine

---

[1] Certainly, a better way is to apply SBC3 by Süer et al. which however incorporates aspects of EDD and SPT [5].

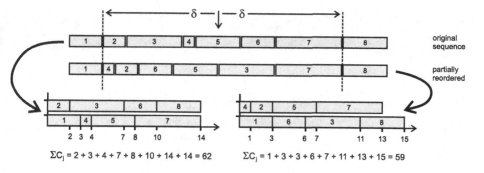

Fig. 1. Schematic depiction of the mutation operator concept with $\delta = 2$ and SPT-mutation and FCFS schedule generation

setup, we fix the machine size for this benchmark to $m = 8$ identical machines. For statistical soundness, we simulated every instance 30 times.

The PPM is applied a standard configuration consisting of a 10×10 toroidal grid with 100 immobile prey individuals and a uniformly distributed predator step size of 1. Overall, we allow a maximum of 12,000 function evaluations (and fix this maximum number as termination criterion).

For comparison reasons, we also apply NSGA-2 to the considered problems to acquire some landmark results. There, we also chose a population size of 100 individuals and allow also a maximum of 12,000 function evaluations. Based on extensive pre-experimental testing, we used NSGA-2 with a random swap mutation operator and step size setting $\delta = 8$. This mutation randomly swaps δ jobs in the sequence and is applied to each individual (variation probability of 1.0). With this setting we achieved the best NSGA-2 results. Mutation with expertise integration as well as (interestingly) recombination have shown negative effects on the solution quality and are thus excluded.

For the qualitative evaluation, we apply two well-known metrics: the hypervolume metric as well as the ε-Indicator, see Zitzler at al. [6]. For all evaluations, we use the reference point $r = (\sum C_j, \sum U_j) = (5500, 50)$ that is beyond all maximum solution values in function space. We further apply the ε-dominance metric $I_\varepsilon(A, B)$ [6] which determines, whether a solution set A dominates another solution set B entirely. Only if $I_\varepsilon(A, B) > 0$ and the inverse comparison $I_\varepsilon(B, A) \leq 0$ holds, the set A dominates set B completely. Otherwise, the intersections of the determined solution fronts do not allow a domination statement.

First, we compare the PPM with NSGA-2 on a parallel machine problem with two criteria: $P_8 || \sum C_j, \sum U_j$. We apply an SPT-based mutation which sorts a randomly selected part of the genome according to SPT ($\delta = 2$), an EDD-based mutation ($\delta = 2$) which orders according to EDD, and a Gaussian swap mutation ($\delta = 4$). These operators are each attached to two predators of which one predator selects regarding $\sum C_j$ and the other on $\sum U_j$ (resulting in overall six predators).

Evaluation results are shown in Figure 2(a). Compared to the application of NSGA-2, the expertise-guided PPM generates a better Pareto-front approximation. In this figure, gray shaded areas depict the hypervolumes enclosed by

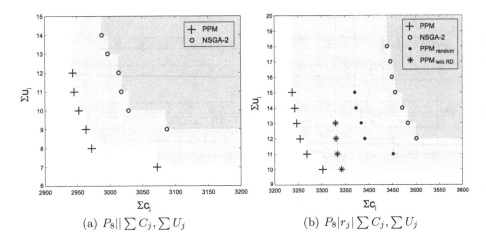

(a) $P_8 || \sum C_j, \sum U_j$ (b) $P_8 | r_j | \sum C_j, \sum U_j$

Fig. 2. Result of two examined scheduling problems

the generated solution fronts. Over all 50 examined instances, the comparison to NSGA-2 revealed a significant dominance of the PPM approximations (Wilcoxon rank-sum test with $p \leq 0.05$ regarding the enclosed hypervolume). Further, Table 1 summarizes the ε-Indicator point of view on the acquired solutions. There, columns show how often (in %) the results of PPM dominate the results of NSGA-2 and vice versa compared pair-wise over all instances. The operator $A \triangleright_\varepsilon B$ denotes the percentage domination count of A over B with respect to ε-dominance. The first line shows that about 65 % of all PPM solutions dominate NSGA-2 solutions completely. However, NSGA-2 often finds some solutions that are missed by PPM and also not dominated by any PPM solution. Still, none of the NSGA-2 solutions dominates the PPM results.

Table 1. Results of the ε-Indicator evaluation, • for pure random, * for only SPT and EDD and + for SPT, EDD, and RD configurations

Problem	PPM $\triangleright_\varepsilon$ NSGA-2		NSGA-2 $\triangleright_\varepsilon$ PPM
	% (mean)	std.	%
$P_8 || \sum C_j, \sum U_j$	64.54	25.25	0.00
	96.42•	3,70•	0.01•
$P_8 | r_j | \sum C_j, \sum U_j$	92.55*	9.47*	0.03*
	98.94+	2.16+	0.00+

Adding release dates to the previous problem results in $P_8 | r_j | \sum C_j, \sum U_j$ which is far more difficult to optimize, even for each objective separately. SPT is not optimal anymore for $P_8 | r_j | \sum C_j$ and EDD is far from being optimal for $P_8 | r_j | \sum U_j$. For our experiments, we use the same settings as before but extend the set of variation operators by a release date related operator. The RD-mutation operator orders a subsequence according to release dates ($\delta = 3$).

Some exemplary results are shown in Figure 2(b). The statistically more detailed evaluation (Wilcoxon rank sum test, $p \leq 0.05$) proves that the PPM with solely mutation significantly dominates NSGA-2 on the hypervolume. The remaining two extensions also proof their benefit significantly. Integrating the expertise from our previous case study strongly improves convergence while adding RD-mutation increases the solution quality once more. Further, the ε-Indicator results from Table 1 show for all three setups a strict domination of PPM over NSGA-2 in more that 90 % of the cases. Only in the completely random setup, NSGA-2 can by chance dominate a solution in very few cases. We were able to show that expertise integration with the PPM is highly beneficial for improving solution quality.

5 Conclusion

We presented the predator-prey model that allows to effectively support the optimizer by integrating available problem specific knowledge. The expertise is expressed by variation operators which can be seamlessly used in the algorithm. Our results show that this is a great advantage over traditional dominance-based methods. In our case studies, we investigated multi-objective combinatorial scheduling problems and found that we can reliably achieve better trade-off approximations. Furthermore, the incremental integration of existing problem knowledge gradually improves the algorithms performance.

References

1. Coello Coello, C., Lamont, G.B., Veldhuizen, D.v.: Evolutionary Algorithms for Solving Multi-Objective Problems, 2nd edn. Springer, New York (2007)
2. Deb, K., Pratap, A., Agarwal, S., Meyarivan, T.: A fast and elitist multiobjective genetic algorithm: NSGA-II. IEEE Transactions on Evolutionary Computation 6(2), 182–197 (2002)
3. Graham, R.L., Lawer, E.L., Lenstra, J.K., Kan, A.H.G.R.: Optimization and Approximation in Deterministic Sequencing and Scheduling: A Survey. Annals of Discrete Mathematics 5, 287–326 (1979)
4. Laumanns, M., Rudolph, G., Schwefel, H.P.: A Spatial Predator-Prey Approach to Multi-objective Optimization: A Preliminary Study. In: Eiben, A.E., Bäck, T., Schoenauer, M., Schwefel, H.-P. (eds.) PPSN 1998. LNCS, vol. 1498, pp. 241–249. Springer, Heidelberg (1998)
5. Süer, G.A., Báez, E., Czajkiewicz, Z.: Minimizing the number of tardy jobs in identical machine scheduling. Computers and Industrial Engineering 25(1–4), 243–246 (1993)
6. Zitzler, E., Thiele, L., Laumanns, M., Fonseca, C.M., Grunert da Fonseca, V.: Performance Assessment of Multiobjective Optimizers: An Analysis and Review. IEEE Transactions on Evolutionary Computation 7(2), 117–132 (2003)

Lower Bounds and Upper Bounds for MaxSAT[*]

Federico Heras, Antonio Morgado, and Joao Marques-Silva

CASL, University College Dublin, Ireland

Abstract. This paper presents several ways to compute *lower* and *upper bounds* for MaxSAT based on calling a complete SAT solver. Preliminary results indicate that (i) the bounds are of high quality, (ii) the bounds can boost the search of MaxSAT solvers on some benchmarks, and (iii) the upper bounds computed by a *Stochastic Local Search* procedure (*SLS*) can be substantially improved when its search is initialized with an assignment provided by a complete SAT solver.

1 Introduction

Weighted Partial MaxSAT (WPMS) [3] is a well-known optimization variant of Boolean Satisfiability (SAT) that finds a wide range of practical applications [3]. WPMS divides the formula in two sets of clauses: The *hard* clauses that must be satisfied and the *soft* clauses that can be unsatisfied with a penalty of their associated *weight*.

Early complete algorithms for MaxSAT solving were based on branch-and-bound search [3]. These algorithms perform very well on *crafted* and *random* instances, but are in general inefficient for industrial instances. An alternative approach is based on iteratively calling a SAT solver. The most widely used approach consists on relaxing the *soft* clauses and then iteratively refining upper bounds on the optimum solution (e.g. [2]). Recent work proposed to guide the search with *unsatisfiable subformulas* [3] (or *cores*) and is most often based on refining lower bounds (e.g. [4,11,1]). Other approaches refine both an upper bound and a lower bound [12]. Finally, a more recent approach based on combining *binary search* and *core-guided search* [7] computes the middle value between both bounds. Observe that all the above approaches could benefit from higher quality initial lower bounds and upper bounds to boost the search.

An alternative way to solve MaxSAT is *stochastic local search* (*SLS*). Such methods are incomplete but they can find approximate solutions for problem instances. However, SLS algorithms have a number of drawbacks. First, they are known to provide low quality solutions (ie. upper bounds) for industrial instances. Second, they are unable to take advantage of *partial MaxSAT* instances with hard and soft clauses.

This paper studies existing lower bounds and upper bounds based on calling a SAT solver, presents some improvements and relate them with recent work in the field. The empirical study shows that (i) SLS can improve its performance when initializing its search with an assignment computed by a complete SAT solver, (ii) the new bounds are tighter than the previous ones and finally that (iii) *core-guided MaxSAT algorithms* boost their performance when enhanced with the new bounds in some benchmarks.

---

[*] This work was partially supported by SFI PI grant BEACON (09/IN.1/I2618).

Y. Hamadi and M. Schoenauer (Eds.): LION 6, LNCS 7219, pp. 402–407, 2012.
© Springer-Verlag Berlin Heidelberg 2012

2 Computing Lower and Upper Bounds

In this section lower bounds (LB) and upper bounds (UB) for MaxSAT are introduced. In what follows, standard SAT and MaxSAT definitions are introduced (e.g. [3]).

Let $X = \{x_1, x_2, \ldots, x_n\}$ be a set of Boolean variables. A *literal* is either a variable x_i or its negation $\bar{x}_i$. A *clause* C is a disjunction of literals. An *assignment* is a set of literals $\mathcal{A} = \{l_1, l_2, \ldots, l_k\}$. If variable x_i is assigned to *true* (*false*), literal x_i ($\bar{x}_i$) is *satisfied* and literal $\bar{x}_i$ (x_i) is *falsified*. An assignment *satisfies* a literal iff it belongs to the assignment, it satisfies a clause iff it satisfies one or more of its literals and it *falsifies* a clause iff it contains the negation of all its literals. A *model* is a complete assignment that satisfies all the clauses in a CNF formula φ. SAT is the problem of deciding whether there exists a model for a given propositional formula. Given an unsatisfiable SAT formula φ, a subset of clauses φ_C whose conjunction is still unsatisfiable is called an *unsatisfiable core* (or *core*) of the original formula. Modern SAT solvers can be instructed to generate an unsatisfiable core [17].

A *weighted* clause is a pair (C, w), where C is a clause and the *weight* w is the cost of its falsification. Weighted clauses that must be satisfied are called *mandatory* (or *hard*) and are associated with a special weight $\top$. Non-mandatory clauses are called *soft* clauses and have a weight $w < \top$. A *weighted* formula in *conjunctive normal form* (WCNF) φ is a set of weighted clauses. A *model* is a complete assignment $\mathcal{A}$ that satisfies all hard clauses. Given a WCNF formula, the *Weighted Partial* MaxSAT is the problem of finding a model of minimum cost.

The remainder of this section introduces the notation used to describe bound computation algorithms. φ_W is the current working formula. Soft clauses may be extended with additional variables called *relaxation variables*. The bounds may use these functions: $\mathtt{Soft}(\varphi)$ returns the set of all *soft* clauses in φ and $SAT(\varphi)$ makes a call to the SAT solver which returns whether φ (ignoring weights) is satisfiable (SAT or UNSAT). Without loss of generality, this paper assumes that the input formula has a model.

2.1 Lower Bounds

Consider Algorithm 1. Let λ be the lower bound, initially $\lambda = 0$. A SAT solver is iteratively called while the formula is unsatisfiable. For each core φ_C, the minimum weight $\min(\varphi_C)$ among the soft clauses is computed, the lower bound is updated as $\lambda = \lambda + \min(\varphi_C)$ and the weight of the soft clauses in φ_C is decreased by $\min(\varphi_C)$. Besides, each soft clause that reaches a weight of 0 is removed from the working formula. This lower bound will be referred to as sat-lb-s. The lower bound in [7] is similar to the described one but all the soft clauses in φ_C are removed from the formula which provides a weaker lower bound (for weighted MaxSAT but equivalent for unweighted MaxSAT) and will be referred to as sat-lb. In [10], cores are detected by unit propagation (UP), whereas the LB in [11,7] additionally detects cores that cannot be identified by UP. Given that a SAT solver always detects first all the cores by solely applying UP and then the remaining ones, it is straightforward that the LB sat-lb is stronger that the one in [10], but sat-lb makes calls to a SAT solver which can require exponential time. sat-lb-s is an extension of [10] for weighted MaxSAT that provides a stronger LB because it is not restricted to UP.

Algorithm 1. Lower Bound

Input: φ
1 $(\varphi_W, \lambda, \varphi_R) \leftarrow (\varphi, 0, \emptyset)$
2 while true do
3       $(st, \varphi_C, \mathcal{A}) \leftarrow$ SAT (φ_W)
4   if $st = SAT$ then return (λ, φ_R)
5       $(\lambda, \varphi_R) \leftarrow (\lambda + \min(\varphi_C), \varphi_R \cup$ Soft $(\varphi_C))$
6   foreach $(C, w) \in$ Soft (φ_C) do
7           $w \leftarrow w - \min(\varphi_C)$
8   if $w = 0$ then $\varphi_W \leftarrow \varphi_W \setminus \{(C, w)\}$
9   end
10 end

Algorithm 2. Upper Bound

Input: φ
1 $(\mu, last\mathcal{A}) \leftarrow (\sum_{i=1}^{m} w_i + 1, \emptyset)$
2 $(R, \varphi_W) \leftarrow$ Relax $(\emptyset, \varphi,$ Soft $(\varphi))$
3 $(st, \varphi_C, \mathcal{A}) \leftarrow$ SAT (φ_W)
4 if $st = true$ then $(last\mathcal{A}, \mu) \leftarrow (\mathcal{A}, \sum_{i=1}^{m} w_i \times (1 - \mathcal{A}\langle C_i \setminus \{r_i\}\rangle))$
5 return $SLS(last\mathcal{A}, \varphi)$

2.2   Upper Bounds

Consider Algorithm 2. Let μ be an UB. Initially, each soft clause is extended with a relaxation variable in function Relax. Then, the SAT solver is called and it returns a satisfying assignment $\mathcal{A}$. Then, the sum of weights of the soft clauses for which the relaxation variable has been assigned to true provides an UB in $\mathcal{A}$ [7]. Note that, for non-optimal assignments $\mathcal{A}$, a relaxation variable assigned to true does not mean necessarily that the soft clause associated to such variable must be unsatisfied. As a result, a slight improvement is to sum the weights of unsatisfied soft clauses by $\mathcal{A}$ disregarding the relaxation variables in the soft clauses. Such UB will be referred to as sat-ub and is inspired in [5]. Additionally, a stochastic local search (SLS) solver is called providing the previous computed assignment restricted to original variables. Recall that such assignment satisfies all hard clauses. The SLS solver may return an improved solution (or the given one, in the worse case). This UB will be referred to as sat-ub+s.

Using *non-random* initial assignments to improve the performance of a local search procedure was first studied in [13] for partial MaxSAT. The work in [9] executes in parallel a SAT solver and an SLS procedure. The variables to be *flipped* by the SLS depend on the current *partial assignment* of the SAT solver. However, such approach is (i) unable to take advantage of hard and soft clauses and (ii) cannot improve the SLS solver in the instances from MaxSAT Evaluations, essentially because the SAT solver proves the unsatisfiability very quickly and cannot guide the SLS procedure. Differently, sat-ub+s provides an assignment that satisfies (i) all hard clauses and (ii) its performance only depends on the ability of the SAT solver to find such an assignment. As a result, it can be applied on the benchmarks of MaxSAT Evaluations and still obtain significant improvements as shown in the empirical section.

Table 1. Quality of the upper bounds and lower bounds

Benchmark	#Inst.	sls	sat-ub	sat-ub+s	sat-lb	sat-lb-s
circ	9	94892	99	35	4	4
sean	112	69595	265	171	16	16
fir	59	4570	36	27	22	22
simp	138	31	41	28	25	25
msp	148	20787	375	350	227	227
mtg	215	515	18	16	6	6
haplo	6	3690	1151	1068	352	352
frb	25	447	449	446	233	233
mo3sat	80	46	55	37	26	26
mostr	60	39244	246	239	139	139
plan	56	294881	2171	2169	760	1371
spot	21	146940	159734	146739	63408	68743
rnet	78	156099	296800	156230	113019	143922
upgrade	100	-	10849700000	-	251240000	416861000
time	32	19354800	742	704	13	18
pedi	100	216139000	110344	91520	13792	15391
Aborted	-	0	27	27	30	33
AverageTime	-	34.61	3.65	5.73	17.16	21.11

3   Experimental Evaluation

Experiments were conducted on a HPC cluster (3GHz) with linux. For each run, the time limit was set to 1200 seconds and a memory limit of 4GB. The bounds were implemented in the MSUNCORE [14] system. All benchmarks from 2009-2011 MaxSAT Evaluations (2067 instances) were considered.

3.1   Analysis of the Bounds

Table 1 summarizes the quality of the computed bounds only for some benchmark sets, but similar improvements are observed in the remaining ones. The first column shows the name of the set of benchmarks, the second column shows the number of instances in the set. The three following columns show three different upper bounds. The two final columns show two different lower bounds. All five columns present the average value of the bound for all instances in the benchmark set. Column sls refers to an upper bound computed by the SLS procedure ADAPTNOVELTY+ [8] included in the UBC-SAT (with default parameters) [16] solver but any other SLS algorithm could be used. sat-ub+s uses ADAPTNOVELTY+ as the SLS algorithm. Regarding the upper bounds, the solutions provided by the SLS algorithm are of very low quality. Differently, sat-ub provides a solution orders of magnitude better than the previous one. Finally, sat-ub+s is more accurate than the previous one. One of the reasons why sat-ub and sat-ub+s are better than sls is because calling a SAT solver with the additional relaxation variables provides a good initial assignment that *satisfies all hard clauses*. Note that the benchmark set *upgrade* contains very large weights and the sls algorithm cannot handle such weights. For this reason they are omitted from the average for 2 upper bounds.

Recall that the approach [9] is unable to improve the upper bound provided by a SLS procedure in the MaxSAT Evaluation instances. Regarding the lower bounds, both sat-lb and sat-lb-s provide the same value for unweighted MaxSAT as expected given that in such case they are equivalent. Differently, for weighted MaxSAT sat-lb-s provides substantially higher lower bounds.

Note the last two rows in the Table 1 that show summarized results over the 2067 instances. One shows the number of *aborted* instances within the time limit while computing the bounds. The other one shows the average time in seconds to compute the bounds. The upper bounds based on calling a SAT solver can be aborted for some very hard instances, but they usually require much less time than SLS.

3.2   Improving Core-Guided MaxSAT Algorithms with the Bounds

In what follows, the performance of several *core-guided* MaxSAT algorithms [7] is studied. Each sub-table in Table 2 shows the results for msu3 [11] (left), msu4 [12] (mid), and core-guided binary search [7] (right), respectively. All three algorithms use exactly *one relaxation variable per soft clause*. Once the LBs are computed, the algorithms will add one relaxation variable to each soft clause returned in φ_R (See Algorithm 1). For each sub-table in Table 2, the first and second columns show the benchmark set and its number of instances, respectively. The remaining three columns show the performance of an algorithm with different bounds in terms of solved instances within the time limit. Note that the necessary time to compute the bounds *is included* in the time limit for each execution. For each algorithm some sets of instances are shown where significant differences in the performance are reported.

msu3 [11] iteratively refines a LB. Table 2 (left) shows the performance of msu3 without LB (3rd column), with sat-lb (4th col.) and with sat-lb-s (5th col.). Clearly, the use of lower bounds improve the performance of msu3. For unweighted problem sets (msp and frb), both lower bounds provide the same improvement as expected. For weighted problem sets (planning, upgrade and pedigree), sat-lb-s is noticeable better than sat-lb.

msu4 [12] refines both a LB and a UB but empirical observation shows that in most of its iterations, msu4 refines an UB. Hence, msu4 may benefit from both bounds but specially from a good initial UB. Table 2 (mid) shows the performance of msu4 where the LB is fixed to sat-lb-s, while the UBs considered are none (3rd col.), sat-ub (4th col.) and sat-ub+s (5th col.). Clearly, the use of UBs improve the performance of msu4, being sat-ub+s the one that provide the best results.

Core-guided binary search [7] refines both a lower bound and upper bound, and at each iteration it asks for the middle value between them. Table 2 (right) shows the performance of core-guided binary search without bounds (3rd col.), with both sat-lb and sat-ub as in [7] (4th col.) and with the two new bounds sat-lb-s and sat-ub+s (5th col.). The additional sixth column shows the results for sat-lb and sat-ub+s. The performance of core-guided binary search is quite good without the bounds and their use improves the performance in 4 of 5 sets. Note that the efficiency for the *upgrade* set of problems is slightly worsened. While the use of bounds can save calls to the SAT solver in binary search, they may move the search to *harder* calls of the SAT solver [15].

Table 2. Bounds on msu3 (left), msu4 (mid) and core-guided binary search (right)

Set	#I.	None	sat-lb	sat-lb-s
msp	148	89	92	92
frb	25	0	14	14
plan.	56	38	40	44
upgr.	100	0	0	10
pedi.	100	24	40	44
total	370	151	186	**204**

Set	#I.	None	sat-ub	sat-ub+s
sean	112	51	77	78
fir	59	46	53	53
mostr	60	44	44	59
msp	148	75	86	108
plan.	56	21	35	50
total	435	237	295	**348**

Set	#I.	None	sat-lb sat-ub	sat-lb-s sat-ub+s	sat-lb sat-ub+s
sean	112	72	77	78	78
frb	25	0	15	15	15
msp	148	98	107	107	107
upgr.	100	63	59	52	59
pedi.	100	32	34	34	33
total	485	265	**292**	286	**292**

4   Conclusions and Future Work

This paper introduces new LB and UB based on calling a SAT solver and studies their effect on the performance core-guided MaxSAT solvers. The bounds presented in this paper can be integrated in *branch and bound* MaxSAT solvers and MaxSAT solvers based on computing unsatisfiable cores that exploit *disjoint cores* [1,7], and which add more than one relaxation variable per soft clause [4]. Additionally, the bounds can be extended to other boolean optimization frameworks [6].

References

1. Ansótegui, C., Bonet, M.L., Levy, J.: A new algorithm for weighted partial MaxSAT. In: AAAI (2010)
2. Le Berre, D., Parrain, A.: The Sat4j library, release 2.2. JSAT 7, 59–64 (2010)
3. Biere, A., Heule, M., van Maaren, H., Walsh, T. (eds.): Handbook of Satisfiability (2009)
4. Fu, Z., Malik, S.: On Solving the Partial MAX-SAT Problem. In: Biere, A., Gomes, C.P. (eds.) SAT 2006. LNCS, vol. 4121, pp. 252–265. Springer, Heidelberg (2006)
5. Giunchiglia, E., Maratea, M.: Solving optimization problems with DLL. In: ECAI, pp. 377–381 (August 2006)
6. Heras, F., Manquinho, V.M., Marques-Silva, J.: On applying unit propagation-based lower bounds in pseudo-boolean optimization. In: FLAIRS Conference, pp. 71–76 (2008)
7. Heras, F., Morgado, A., Marques-Silva, J.: Core-guided binary search algorithms for maximum satisfiability. In: AAAI (2011)
8. Hoos, H.H.: An adaptive noise mechanism for WalkSAT. In: AAAI, pp. 655–660 (2002)
9. Kroc, L., Sabharwal, A., Gomes, C.P., Selman, B.: Integrating systematic and local search paradigms: A new strategy for MaxSAT. In: IJCAI, pp. 544–551 (2009)
10. Li, C.-M., Manyà, F., Planes, J.: Exploiting Unit Propagation to Compute Lower Bounds in Branch and Bound Max-SAT Solvers. In: van Beek, P. (ed.) CP 2005. LNCS, vol. 3709, pp. 403–414. Springer, Heidelberg (2005)
11. Marques-Silva, J., Planes, J.: On using unsatisfiability for solving maximum satisfiability. Computing Research Repository, abs/0712.0097 (December 2007)
12. Marques-Silva, J., Planes, J.: Algorithms for maximum satisfiability using unsatisfiable cores. In: DATE, pp. 408–413 (2008)
13. Menai, M.E., Batouche, M.: An effective heuristic algorithm for the maximum satisfiability problem. Appl. Intell. 24(3), 227–239 (2006)
14. Morgado, A., Heras, F., Marques-Silva, J.: The MSUnCore MaxSAT solver. In: POS (2011)
15. Sellmann, M., Kadioglu, S.: Dichotomic Search Protocols for Constrained Optimization. In: Stuckey, P.J. (ed.) CP 2008. LNCS, vol. 5202, pp. 251–265. Springer, Heidelberg (2008)
16. Tompkins, D.A.D., Hoos, H.H.: UBCSAT: An Implementation and Experimentation Environment for SLS Algorithms for SAT and MAX-SAT. In: Hoos, H.H., Mitchell, D.G. (eds.) SAT 2004. LNCS, vol. 3542, pp. 306–320. Springer, Heidelberg (2005)
17. Zhang, L., Malik, S.: Validating sat solvers using an independent resolution-based checker: Practical implementations and other applications. In: DATE, pp. 10880–10885 (2003)

Determining the Characteristic of Difficult Job Shop Scheduling Instances for a Heuristic Solution Method

Helga Ingimundardottir and Thomas Philip Runarsson

School of Engineering and Natural Sciences, University of Iceland
{hei2,tpr}@hi.is

Abstract. Many heuristic methods have been proposed for the job-shop scheduling problem. Different solution methodologies outperform other depending on the particular problem instance under consideration. Therefore, one is interested in knowing how the instances differ in structure and determine when a particular heuristic solution is likely to fail and explore in further detail the causes. In order to achieve this, we seek to characterise features for different difficulties. Preliminary experiments show there are different significant features that distinguish between easy and hard JSSP problem, and that they vary throughout the scheduling process. The insight attained by investigating the relationship between problem structure and heuristic performance can undoubtedly lead to better heuristic design that is tailored to the data distribution under consideration.

1 Introduction

Hand crafting heuristics for NP-hard problems is a time-consuming trial and error process, requiring inductive reasoning or problem specific insights from their human designers. Furthermore, within a problems class, such as job-shop scheduling, it is possible to construct problem instances where one heuristic would outperform another. Depending on the underlying data distribution, different heuristics perform differently, commonly known as the *no free lunch* theorem [1]. The success of a heuristic is how it manages to deal with and manipulate the characteristics of its given problem instance. So in order to understand more fully how a heuristic will eventually perform, one needs to look into what kind of problem instances are being introduced to the system. What defines a problem instance, e.g. what are its key features? And how can they help with designing better heuristics?

In investigating the relationship between problem structure and heuristic effectiveness one can research what [2] calls *footprints* in instance space, which is an indicator how an algorithm generalises over the instance space. This sort of investigation has also been referred to as *landmarking* [3]. It is evident from experiments performed in [2] that one-algorithm-for-all problem instances is not ideal. An algorithm may be favoured for its best overall performance, however

Y. Hamadi and M. Schoenauer (Eds.): LION 6, LNCS 7219, pp. 408–412, 2012.
© Springer-Verlag Berlin Heidelberg 2012

it was rarely the best algorithm available over various subspaces of the instance space. Thus when comparing different algorithms one needs to explore how they perform w.r.t. the instance space, i.e. their footprint.

In this study, the same problem generator is used to create 1,500 problem instances, however the experimental study in section 3 shows that MWRM works well/poorly on a subset of the instances. Since the problem instances are only defined by processing times and its permutation, the interaction between the two is important, because it introduces hidden properties in the data structure making it easy or hard to schedule with for the given algorithm. These underlying characteristics or features define its data structure. So a sophisticated way of discretising the instance space is grouping together problem instances that show the same kind of feature behaviour, in order to infer what is the feature behaviour between *good* and *bad* schedules.

It is interesting to know if the difference in the structure of the schedule is time dependent, is there a clear time of divergence within the scheduling process? Moreover, investigation of how sensitive is the difference between two sets of features, e.g. can two schedules with similar feature values yield completely contradictory outcomes, i.e. one poor and one good schedule? Or will they more or less follow the same path? This essentially answers the question of whether is is in fact feasible to discriminate between *good* and *bad* schedules using the currently selected features as a measure. If results are contradictory, it is an indicator the features selected are not robust enough to capture the essence of the data structure. Additionally, there is also the question of how can one define 'similar' schedules, what measures should be used? This paper describes some preliminary experiments with the aim of investigating the feasibility of finding distinguishing features corresponding to *good* and *bad* schedules in JSSP.

Instead of searching through a large set of algorithms (creating an algorithm portfolio) and determining which algorithm is the most suitable for a given subset of the instance space, as is generally the focus in the current literature [4,5,2], our focus is rather on a single algorithm and understanding *how* it works on the instance space – in the hopes of being able to extrapolate where it excels in order to aid its failing aspects.

The outline of the paper is as follows, in section 2 priority dispatch rules for the JSSP problem are discussed, what features are of interest and how data is generated. A preliminary experimental study is presented in section 3. The paper concludes with a summary of main findings and points to future work.

2   Job-Shop scheduling

The job-shop scheduling task considered here is where n jobs are scheduled on a set of m machines, subject to the constraint that each job must follow a predefined machine order and that a machine can handle at most one job at a time. The objective is to schedule the jobs so as to minimize the maximum completion times, also known as the makespan. For a mathematical formulation of JSSP the reader is recommended [6].

Table 1. Feature space $\mathcal{F}$ for JSSP. Features 1–13 can vary throughout the scheduling process w.r.t. tasks that can be dispatched next, however features 14–16 are static

ϕ	Feature description
ϕ_1	processing time for job on machine
ϕ_2	start-time
ϕ_3	end-time
ϕ_4	when machine is next free
ϕ_5	current makespan
ϕ_6	work remaining
ϕ_7	most work remaining
ϕ_8	slack time for this particular machine
ϕ_9	slack time for all machines
ϕ_{10}	slack time weighted w.r.t. number of operations already assigned
ϕ_{11}	time job had to wait
ϕ_{12}	size of slot created by assignment
ϕ_{13}	total processing time for job
ϕ_{14}	total processing time for all jobs
ϕ_{15}	mean processing time for all jobs
ϕ_{16}	range of processing times over all jobs

2.1 Single-Priority Dispatching Heuristic

Dispatching rules are of a construction heuristics, where one starts with an empty schedule and adds on one job at a time. When a machine is free the dispatching rule inspects the waiting jobs and selects the job with the highest priority. A survey of more than 100 of such priority rules was given in 1977 by [7]. In this paper however, only most work remaining (MWRM) dispatching rule will be investigated.

In order to apply a dispatching rule a number of features of the schedule being built must be computed. The features of particular interest were obtained from inspecting the aforementioned single priority-based dispatching rules. The temporal scheduling features applied in this paper are given in Table 1. These are not the only possible set of features, they are however built on the work published in [6,4] and deemed successful in capturing the essence of a JSSP data structure.

2.2 Data Generation

Problem instances were generated stochastically by fixing the number of jobs and machines and sampling a discrete processing time from the uniform distribution $U(1, 200)$. The machine order is a random permutation of $\{1, ..., m\}$. A total of 1,500 instances were generated for a six job and six machine job-shop problem.

In the experimental study the performance of the MWRM, μ_{MWRM}, and compared with its optimal makespan, μ_{opt}. Since the optimal makespan varies between problem instances the following performance measure is used:

$$\rho = \frac{\mu_{\mathrm{MWRM}}}{\mu_{\mathrm{opt}}}. \tag{1}$$

3   Experimental Study

In order to differentiate between problems, a threshold of a $\rho < 1.1$ and $\rho > 1.3$ was used to classify *easy* and *hard* problems. Of the 1500 instances created, 271 and 161 problems were classified *easy* and *hard*, respectively.

Table 2. Features for *easy* and *hard* problems are drawn from the same data distribution (denoted by ·)

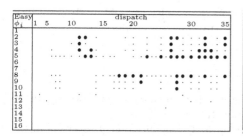

Table 3. Significant correlation (denoted by ·) for *easy* (left) and *hard* (right) problems and resulting ratio from optimality, ρ defined by (1). Commonly significant features across the tables are denoted by •.

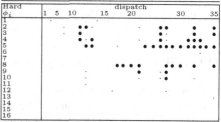

Table 2 reports where data distributions are the same (denoted by ·). From the table one can see that distribution for ϕ_1, ϕ_6, ϕ_{12} and ϕ_{16}) are (more or less) the same throughout the scheduling process. However there is a clear time of divergence for distribution of slacks; step 6 for ϕ_8 and step 12 for ϕ_9 and ϕ_{10}.

In order to find defining characteristics for *easy* and *hard* problems, a (linear) correlation was computed between features (on a step-by-step basis) to the resulting ratio from optimality. Significant features are reported in Table 3 for *easy* and *hard* problems, (denoted by ·). As one can see from the tables, the significant features for the different difficulties are varying. Some are commonly significant features across the tables (denoted by •).

4   Discussion and Conclusion

From the experimental study it is apparent that features have different correlation with the resulting schedule depending in what stage it is in the scheduling process, implying that their influence varies throughout the scheduling process. And features constant throughout the scheduling process are not correlated with the end-result. There are some common features for both difficulties considered which define JSSP on a whole. However the significant features are quite different across the two difficulties, implying there is a clear difference in their data structure. The amount of significant features were considerably more for easy problems, indicating their key elements had been found. However, the features distinguishing hard problems were scarce. Most likely due to their more complex data structure their key features are of a more composite nature.

The feature attributes need to be based on statistical or theoretical grounds. Thus scrutiny in understanding the nature of problem instances is of paramount importance in feature engineering for learning. Which yields feedback into what features are important to devout more attention to, i.e. features that result in a failing algorithm. In general, this sort of investigation can undoubtedly be used in better algorithm design which is more equipped to deal with varying problem instances and tailor to individual problem instance's needs, i.e. a footprint-oriented algorithm.

Although this methodology was only implemented on a simple single-priority dispatching rule heuristic, the methodology is easily adaptable for more complex algorithms. The main objective of this work is to illustrate the interaction of a specific algorithm on a given problem structure and its properties.

References

1. Wolpert, D.H., Macready, W.G.: No free lunch theorems for optimization. IEEE Transactions on Evolutionary Computation 1(1), 67–82 (1997)
2. Corne, D.W., Reynolds, A.P.: Optimisation and Generalisation: Footprints in Instance Space. In: Schaefer, R., Cotta, C., Kołodziej, J., Rudolph, G. (eds.) PPSN XI, Part I. LNCS, vol. 6238, pp. 22–31. Springer, Heidelberg (2010)
3. Pfahringer, B., Bensusan, H.: Meta-learning by landmarking various learning algorithms. In: Machine Learning (2000)
4. Smith-Miles, K.A., James, R.J.W., Giffin, J.W., Tu, Y.: A Knowledge Discovery Approach to Understanding Relationships between Scheduling Problem Structure and Heuristic Performance. In: Stützle, T. (ed.) LION 3. LNCS, vol. 5851, pp. 89–103. Springer, Heidelberg (2009)
5. Smith-Miles, K., Lopes, L.: Generalising Algorithm Performance in Instance Space: A Timetabling Case Study. In: Coello Coello, C.A. (ed.) LION 5. LNCS, vol. 6683, pp. 524–538. Springer, Heidelberg (2011)
6. Ingimundardottir, H., Runarsson, T.P.: Supervised Learning Linear Priority Dispatch Rules for Job-Shop Scheduling. In: Coello Coello, C.A. (ed.) LION 5. LNCS, vol. 6683, pp. 263–277. Springer, Heidelberg (2011)
7. Panwalkar, S., Iskander, W.: A Survey of Scheduling Rules. Operations Research 25(1), 45–61 (1977)

Expected Improvements for the Asynchronous Parallel Global Optimization of Expensive Functions: Potentials and Challenges

Janis Janusevskis[1], Rodolphe Le Riche[1,2], David Ginsbourger[3], and Ramunas Girdziusas[1]

[1] Ecole des Mines de Saint-Etienne, Saint-Etienne, France
[2] CNRS, UMR 5146 Cl. Goux, France
[3] Departement of Mathematics and Statistics, University of Bern, Switzerland

Abstract. Sequential sampling strategies based on Gaussian processes are now widely used for the optimization of problems involving costly simulations. But Gaussian processes can also generate parallel optimization strategies. We focus here on a new, parameter free, parallel expected improvement criterion for asynchronous optimization. An estimation of the criterion, which mixes Monte Carlo sampling and analytical bounds, is proposed. Logarithmic speed-ups are measured on 1 and 9 dimensional functions.

1 Introduction to Parallel Expected Improvements

The current technological solution for optimizing functions of numerically costly simulators is to rely on an increasing number of processing units (processors, cores, GPUs). Demanded features of new parallel optimization algorithms are not only high speed-ups but also the ability to work with heterogeneous processing units (e.g., computing grids) and fault tolerance. Evolutionary algorithms offer many opportunities for parallelization, including in heterogeneous computing networks, and in master-worker or island computing structures [2]. Master-worker parallel optimization algorithms are more common as they fit the costly simulator case: the optimizer is the master node, the workers evaluate the objective functions and constraints. In [6] for example, a parallel and asynchronous version of the local pattern search algorithm is described. [9] presents a deterministic global parallel algorithms which, contrarily to the forthcoming method is based on radial basis functions and has a synchronization step. [10] and [1] describe methods where Gaussian processes are used through expected improvement maximization with restarts and infill sampling, respectively, to provide the set of points to evaluate in parallel. Boths methods have a synchronisation step when the master optimizer iterates.

This article also proposes a sampling criterion based on Gaussian processes. Its originality relies on an asynchronous extension of the multi-points expected improvement of [3] which, itself, was a parallel extension of the expected improvement of [8]. In the sequel, the parallel asynchronous expected improvement will be denoted $EI^{(\mu,\lambda)}$.

Y. Hamadi and M. Schoenauer (Eds.): LION 6, LNCS 7219, pp. 413–418, 2012.
© Springer-Verlag Berlin Heidelberg 2012

A distinctive feature of $EI^{(\mu,\lambda)}$ is to provide a unified mathematical treatment of the already evaluated and the currently running points for optimization, therefore no additional parameter is introduced to parallelize the search. Initial empirical results on 1D and 9D functions show that logarithmic speed-ups are obtained. This work also illustrates that $EI^{(\mu,\lambda)}$ is difficult to compute.

2 A Unified Presentation of Parallel Expected Improvements

Let us consider the optimization problem $\min_{x \in \mathbb{R}^d} f(x)$ where each evaluation of f implies a call to a numerically intensive simulation program, and assume that a set of past observations $(\mathbb{X}, f(\mathbb{X}))$ has been gathered. In the Bayesian Global Optimization settings considered here, the unknown f is represented *a priori* by a Gaussian Process $(Y(x))_{x \in \mathbb{R}^d}$, and is being approximated relying on the conditional distribution of Y knowing that $Y(\mathbb{X}) = f(\mathbb{X})$ (*Kriging metamodel*). We focus here on cases where λ computing nodes are available for starting new simulations while μ computing nodes are currently evaluating f at a set of $\mu \geq 0$ "busy" points $\mathbb{X}_{\text{busy}} := \{x_{\text{b}}^1, \ldots, x_{\text{b}}^\mu\}$. We define the **asynchronous multi-points Expected Improvement** of $\mathbb{X}_{\text{asy}} := \{x_a^1, \ldots, x_a^\lambda\}$ as

$$EI^{(\mu,\lambda)}(\mathbb{X}_{\text{asy}}) := \mathbb{E}\left[(\min(Y(\mathbb{X} \cup \mathbb{X}_{\text{busy}})) - \min(Y(\mathbb{X}_{\text{asy})}))^+ \, | Y(\mathbb{X}) = f(\mathbb{X}) \right],$$

where $[\bullet]^+ \equiv \max(0, \bullet)$. This criterion, which was initially introduced in [4], measures how much progress with respect to already calculated or currently calculating points $(\mathbb{X} \cup \mathbb{X}_{\text{busy}})$ will be made on average at $\mathbb{X}_{\text{asy}}$. The case $\mu = 0$, $\lambda = 1$ is the usual EI defined in the EGO algorithm [8]. A parallel algorithm that works by maximizing $EI^{(\mu,\lambda)}$ can now be introduced:

Asynchronous Parallel EI algorithm
1. Generate $\mathbb{X}$ through a space-filling design. Calculate $f(\mathbb{X})$.
2. While calculation budget not exhausted do
   (a) [non blocking] Retrieve new x's and $f(x)$'s if any. Update the kriging model (optionally with parameter re-estimation). Update λ and μ.
   (b) Generate λ points by max $EI^{(\mu,\lambda)}(x_a^1, \ldots, x_a^\lambda)$ using a global optimizer (e.g. CMA-ES [5]). Send them to worker nodes for evaluation.

The classic EI criterion has the desirable property that its maximum lies away from already sampled points of $\mathbb{X}$ and strikes a comprise between the exploration of unknown regions of the design space and the intensification of the search in known highly performing regions [8]. In addition, the multi-points asynchronous $EI^{(\mu,\lambda)}$ has its maximum away from any subset of already sampled and currently running points, $\mathbb{X} \cup \mathbb{X}_{\text{busy}}$ (it is null there while I is a positive variable, [7]). Another advantage of $EI^{(\mu,\lambda)}$ is that it does not introduce extra parameters.

$EI^{(\mu,\lambda)}$ accounts for all ratios of optimizer over simulation computation times. If the simulations are much longer than any optimizer iteration (kriging update and EI maximization), points will be allocated to newly available nodes one at a time, in which case $EI^{(\mu,\lambda=1)}$ will be used. Vice versa, if the optimizer iteration takes longer than the simulations, no busy point occurs and $EI^{(\mu=0,\lambda)}$ is relevant. Intermediate cases call for general $EI^{(\mu,\lambda)}$'s.

3   Bounds and Estimation of $EI^{(\mu,\lambda)}$

While the calculation of $EI^{(0,1)}$ and $EI^{(0,2)}$ is analytical [3], no general expression for $EI^{(\mu,\lambda)}$ is known. We propose to estimate $EI^{(\mu,\lambda)}$ through a Monte Carlo strategy augmented by bounds knowledge. In particular, the following bounds are established in [7] where $EI^*(x_b^i, x_a^j) := \mathbb{E}\left[(Y(x_b^i) - Y(x_a^j))^+ | Y(\mathbb{X}) = f(\mathbb{X})\right]$:

$$\max_{i=1,\lambda} EI(x_a^i) \le EI^{(0,\lambda)}(\mathbb{X}_{\mathrm{asy}}) \le \sum_{i=1}^{\lambda} EI(x_a^i)$$

$$0 \le EI^{(\mu,\lambda)}(\mathbb{X}_{\mathrm{asy}}) \le \min\left(\sum_{i=j}^{\lambda} EI(x_a^j), \sum_{j=1}^{\lambda} EI^*(x_b^1, x_a^j), \ldots, \sum_{j=1}^{\lambda} EI^*(x_b^\mu, x_a^j)\right)$$

All the expressions in the bounds are analytical, including the EI^* terms, because they are instances of the usual EI formula [8].

The Monte Carlo estimator of the mean of $I^{(\mu,\lambda)}$ and its variance are calculated from N samples of the conditional Gaussian process Y (i.e., samples i_j of the improvements) as follows:

$$EI_{MC}^{(\mu,\lambda)}(\mathbf{x}) = \frac{1}{N}\sum_{j=1}^{N} i_j^{(\mu,\lambda)}(\mathbf{x}) \ , \ \sigma_{MC}^2(\mathbf{x}) = \frac{1}{N(N-1)}\sum_{j=1}^{N}(i_j^{(\mu,\lambda)}(\mathbf{x}) - EI^{(\mu,\lambda)})^2$$

We now introduce the above bounds through Bayes law. The a priori density of $EI^{(\mu,\lambda)}$ is uniform between the lower and upper bounds, $\mathcal{U}(L,U)$. Since the likelihood of the expectation estimator, $p\left(EI_{MC}^{(\mu,\lambda)} | EI^{(\mu,\lambda)}\right)$, is Gaussian, the a posteriori density is a known truncated Gaussian as can be seen from $p\left(EI^{(\mu,\lambda)} | EI_{MC}^{(\mu,\lambda)}\right) = p\left(EI_{MC}^{(\mu,\lambda)} | EI^{(\mu,\lambda)}\right)\mathcal{U}(L,U)/\mathrm{const}$. Calculation details are given in [7] and yield estimations of the mean and the variance of $EI^{(\mu,\lambda)}(\mathbf{x}) | EI_{MC}^{(\mu,\lambda)}(\mathbf{x})$, denoted $M(\mathbf{x})$ and $V^2(\mathbf{x})$, respectively, which account for both Monte Carlo simulations and the bounds.

The asynchronous parallel EI algorithm introduced earlier has a step that maximizes $EI^{(\mu,\lambda)}(\bullet)$ with the CMA-ES algorithm ([5]). In CMA-ES, the objective function values are used to rank the explored points. We propose to modify this comparison of points (say $\mathbf{x}^i$ and $\mathbf{x}^j$) in order to control the Monte Carlo simulations:

Pairwise ranking procedure

1. Set confidence k (e.g., over 60% confidence for $k = 1$), N, ΔN,
2. While not stop do
   (a) If $M(\mathbf{x^i}) - kV(\mathbf{x^i}) \geq M(\mathbf{x^j}) + kV(\mathbf{x^j})$ then $EI^{(\mu,\lambda)}(\mathbf{x^i}) \geq EI^{(\mu,\lambda)}(\mathbf{x^j})$, stop. (And vice versa.)
   (b) Decrease the Monte Carlo variances by allocating extra samples: $\Delta N_i = V(\mathbf{x^i})\Delta N/(V(\mathbf{x^i}) + V(\mathbf{x^j}))$, $\Delta N_j = \Delta N - \Delta N_i$.

---

4   Test Results

The tests reported in [7] are now summarized.

Firstly, 100 functions in 1D have been generated by sampling Gaussian process trajectories (see the 2 examples of Fig.1). For each function, a pair of points was randomly chosen and their $EI^{(\mu,\lambda)}$ compared, $\mu = 0, 1, 3$ and $\lambda = 1$ to 5. It is observed that, on average, accounting for the bounds divides by 7 the total number of Monte Carlo simulations necessary to discriminate the points. However, a first difficulty appeared. When $\mu > 0$, 25% of the pairs needed over $N = 100,000$ MC samples for the comparison, i.e., their points had very close $EI^{(\mu,\lambda)}$ values: the $EI^{(\mu,\lambda)}$ function has plateaus. This proportion increased when the pairs were generated by optimization because these plateaus correspond to high performance regions of the design space.

The second issue stems directly from $EI^{(\mu,\lambda)}$'s definition: its input is high dimensional (dim $\mathbb{X}_{\mathrm{asy}} = \lambda d$), making its maximization potentially costly.

Thirdly, on the positive side, the experiments summarized in Fig.1 and 2 show that logarithmic speed-ups are obtained: we have observed

$$\frac{\text{time to solve, } \lambda = 1}{\text{time to solve, } \lambda} \approx 1 + b\log(\lambda)$$

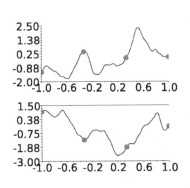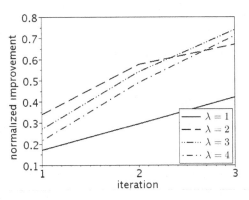

Fig. 1. Left: two of the ten sampled trajectories with Matern 5/2 kernel and scale $\theta = 0.15$. Right: mean normalized improvement as a function of iteration for $\lambda = 1$ to 4 ($\mu = 0$), averaged over 10 test functions (defined as 1D trajectories such as those on the left).

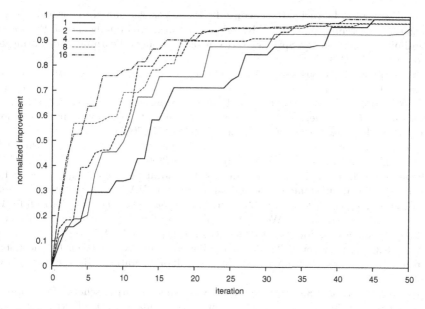

Fig. 2. Mean normalized improvement as a function of iteration for $\lambda = 1$ to 16 ($\mu = 0$), 9 dimensional test case

where $b = 0.85$ and 0.49 in 1 and 9 dimensions, respectively. The 9D test function is the approximation of a matrix by its rank 1 counter part:

$$(\mathbf{B}^*_{mr}, \mathbf{C}^*_{rn}) = \arg \min_{\mathbf{B}_{mr}, \mathbf{C}_{rn}} \|\mathbf{A}_{mn} - \mathbf{B}_{mr}\mathbf{C}_{rn}\|, \quad r < \mathrm{rank}(\mathbf{A}_{mn}), \qquad (1)$$

where $\| \cdot \|$ is the Frobenius norm, $\mathbf{A}$ is a 4×5 matrix of uniformly distributed elements in $[0, 1]$, $\mathbf{B}$ and $\mathbf{C}$ are a column and a row vector whose elements are constrained to be in $[-1, 1]$, respectively. This is a continuous nonconvex 9-dimensional box-constrained optimization problem whose solution is given by the first singular vectors. An instance of a typical run is shown for $\lambda = 1$ to 16, $\mu = 0$, in Fig. 2.

The observed speed-up is due to the kriging model which summarizes all of the information gathered by the worker nodes, including the currently running simulations. Further work addressing the aforementioned estimation and optimization difficulties of $EI^{(\mu,\lambda)}$ is needed.

References

1. Berbecea, A.C., Kreuawan, S., Gillon, F., Brochet, P.: A Parallel Multiobjective Efficient Global Optimization: The Finite Element Method in Optimal Design and Model Development. IEEE Transactions on Magnetics 46(8), 2868–2871 (2010)
2. Branke, J., Kamper, A., Schmeck, H.: Distribution of Evolutionary Algorithms in Heterogeneous Networks. In: Deb, K., Tari, Z. (eds.) GECCO 2004. LNCS, vol. 3102, pp. 923–934. Springer, Heidelberg (2004)

3. Ginsbourger, D., Le Riche, R., Carraro, L.: Kriging is well-suited to parallelize optimization. In: Tenne, Y., Goh, C.-K. (eds.) Computational Intelligence in Expensive Optimization Problems. Springer series in Evolutionary Learning and Optimization, pp. 131–162 (2009)

4. Ginsbourger, D., Janusevskis, J., Le Riche, R.: Dealing with asynchronicity in parallel Gaussian Process based global optimization. HAL technical report no. hal-00507632 (July 2010), http://hal.archives-ouvertes.fr/hal-00507632/

5. Hansen, N., Ostermeier, A.: Completely derandomized self-adaptation in evolution strategies. Evolutionary Computation 9(2), 159–195 (2001)

6. Kolda, T.G.: Revisiting asynchronous parallel pattern search for nonlinear optimization. SIAM J. Optimization 16(2), 563–586 (2005)

7. Janusevskis, J., Le Riche, R., Ginsbourger, D.: Parallel expected improvements for global optimization: summary, bounds and speed-up. HAL technical report no. hal-00613971 (August 2011), http://hal.archives-ouvertes.fr/hal-00613971_v1

8. Jones, D.R., Schonlau, M., Welch, W.J.: Efficient global optimization of expensive black-box functions. Journal of Global Optimization 13(4), 455–492 (1998)

9. Regis, R.G., Shoemaker, C.A.: Parallel radial basis function methods for the global optimization of expensive functions. European J. of Operational Research 182, 514–535 (2007)

10. Sobester, A., Leary, S.J., Keane, A.J.: A parallel updating scheme for approximating and optimizing high fidelity computer simulations. J. of Structural and Multidisciplinary Optimization 27, 371–383 (2004)

Effect of SMS-EMOA Parameterizations on Hypervolume Decreases

Leonard Judt[1], Olaf Mersmann[1], and Boris Naujoks[2]

[1] Faculty of Statistics, TU Dortmund University, 44221 Dortmund, Germany
[2] Cologne University of Applied Sciences, 51643 Gummersbach, Germany
{leonard.judt,olafm}@statistik.tu-dortmund.de,
boris.naujoks@fh-koeln.de

Abstract. It is possible for the $(\mu+1)$-SMS-EMOA to decrease in dominated hypervolume w.r.t. a global reference point. We study the influence of SMS-EMOA parameter settings on number and amount of the observed decreases. We show that the number of decreases drop and the number of increases rise with a higher population size. In addition, a positive correlation between mean increase and mean decrease can be observed. Our findings further indicate a substantial impact of the mutation operators on the number and amount of decreases.

Keywords: EMO, hypervolume decreases, parameter influence.

1 Introduction

The dominated hypervolume was defined by Zitzler and Thiele as the the size of the space covered by a Pareto front[1] with respect to a given reference point and special properties of this indicator have been proven [6]. As recent results show, unsuspected decreases in the hypervolume progression during an optimization run are possible [4]. These results are summarised in the following section. Section 3 provides our latest results on the dependency of such decreases on EMOA parameter settings as well as some initial insight into how these decreases influence the final dominated hypervolume of an optimization run. Finally, we summarise our latest findings and give an outlook to future work in section 4.

2 Decreases in Hypervolume Progressions

The progression of the hypervolume in a 1-greedy hypervolume selection based EMOA w.r.t. a fixed global reference point was thought to never decrease in the course of an optimization run. This belief arose from the design of the algorithms, in which the individual with the least hypervolume contribution is discarded in every generation. However, Judt et al. [4] showed that this intuition is wrong for an adaptive reference point.

---

[1] For details on EMO related definitions and vocabulary the reader is refered to Deb [3] or Coello Coello [2].

Y. Hamadi and M. Schoenauer (Eds.): LION 6, LNCS 7219, pp. 419–424, 2012.
© Springer-Verlag Berlin Heidelberg 2012

For the paper a total of 71 250 reproducible runs of the SMS-EMOA [1] on both 2-dim. and 3-dim. test cases were conducted. Different parameterizations have been considered, e.g. three different population sizes $\mu \in \{10, 20, 100\}$ were examined. Parameters of the Simulated Binary Crossover (SBX, [3]) and the Polynomial Mutation (PM, [3]) operator were chosen according to a Latin Hypercube Sample. For each combination of test function, population size and variation operator set, 50 independent runs were conducted. The hypervolume w.r.t. a fixed reference point was calculated for each generation and the number of times this hypervolume drops was stored. More details on exact parameterizations and reference points are provided in the supplementary material[2].

In both experiments, a considerable number of decreases in hypervolume were observed. In the 2-dim. case, an exceptional rule for the two boundary solutions is responsible for the drops. In the 3-dim. case, drops were explained by the hypervolume contributions being calculated w.r.t. an adaptive reference, which depends on the current population. While internally never attaining a decrease in hypervolume w.r.t. this adaptive reference point, drops may occur w.r.t. a global reference point. This 3-dim. effect can be generalized to a higher number of objectives. Nevertheless, the authors still believe that hypervolume is the most effective selection scheme for MOP known today.

3   Results

In the following analysis we focus primarily on the 3-dim. cases. More results and similar results for the 2-dim. cases are provided in the supplementary material.

3.1   Population Size Influence

It is well known that choosing a suitable population size μ is crucial for the performance of an evolutionary algorithm (EA). We therefore hypothesized that it could also have an influence on the number of decreases in hypervolume as well as the magnitude of these decreases. We find evidence for the latter hypothesis in figure 1 which looks at the absolute change of hypervolume in each iteration and depicts the trade-off between the average increase and decrease of a run.

There appears to be a barrier at 10^{-6} below which the mean decrease in hypervolume does not drop. This is a numerical artifact. The observed absolute hypervolume values are of the order of 10^9 and the double precision floating point values used for all calculations only have ≈ 16 significant digits. It is therefore not possible to observe smaller differences than $\approx 10^{-6}$ in the hypervolume directly. This problem of the hypervolume indicator seems to be largely ignored and will only worsen as the hypervolume indicator or approximations of it are used for problems with more than 3 objectives.

Another interesting thing in figure 1 is that the mean increase and decrease are positivly correlated. This implies that if we observe large increases in hypervolume during our optimization run, we can assume that we will also have large

---

[2] Supplementary Material is available at http://ptr.p-value.net/lion12a

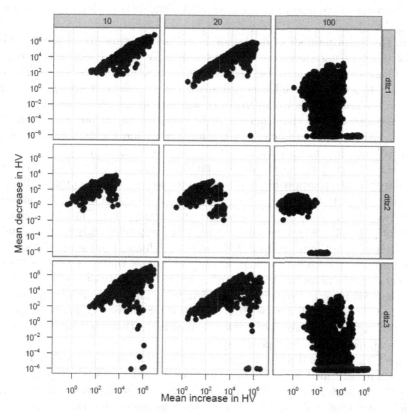

Fig. 1. Increases vs decreases in HV averaged over all iterations in each of the 22 500 3-dim. runs. The data is futher subdivided by population size μ along the top and test function along the right side.

decreases in hypervolume w.r.t. a fixed reference point. Less obvious is the effect of μ on the mean increase and decrease. Both tend to drop for larger values of μ, which is plausible given that with a larger population size the expected contribution of each individual is lower.

In addition, plots depicting the influence of the population size μ on the number of occurrences of a decrease in hypervolume can be seen in the supplementary material. It is shown there that for increasing μ the number of hypervolume decreases drops, whereas the number of increases rises.

3.2 Eliminating Parameters

The influence of variation operator parameterization on the final SMS-EMOA performance has been studied in [5]. Here, we wish to see if they also influence the number or the amount of decrease in hypervolume. The four parameters are sbx_n, sbx_p, pm_n, pm_p from SBX and PM. The first two control the distribution

and probability of crossover and the second two the distribution and probability
of mutation.

Both, the mean number of decreases as well as the variance of the decreases
over the different parameterizations vary widely for a fixed setting of sbx_n or
sbx_p. There is no similarity between the functions and only slight similarities
for a fixed function and different values of the populations size μ. The only
consistent phenomenon is the decrease in variance for larger values of μ.

We therefore believe that there is little incentive to further investigate the
influence of these two parameters on the number of and the amount of decrease
in hypervolume. For sbx_n this observation is constitent with [5].

For pm_n and pm_p the situation is a bit different. While there is still strong
non-linearity in the relationship between the parmeter and the number of hypervolume decreases, there are regularities in the plot. For pm_n there is a consistent
spike between 20 and 25 for each function and setting of μ. The magnitude decreases for $\mu = 100$ and the variance seems to be approximately proportional to
the mean number of hypervolume decreases. This hints at a linear relationsship
between μ and both pm_n and pm_p. From [5] we know that these values lead to
good overall performance of the algorithm on a wide varity of test functions.

The pm_p parameter also exibits some structure. There is a peak in the number
of decreases in hypervolume for small values of pm_p and the curve drops and
levels off as pm_p increases. The high variance is concentrated in the region of
low pm_p values which seems peculiar.

Figure 2 visualizes all different parameter settings in a single plot. It shows
the maximum attained hypervolume (rescaled to aid in visualizing the result)
against the number of decreases in hypervolume. Each different shade of grey
is a different parameter setting. There is obvious underlying structure. Similar
parameter settings tend to 'cluster' together. That is, a parameter setting tends
to produce similar number of hypervolume decreases and a similar attained hypervolume over many runs of the SMS-EMOA.

Even more interesting is the result that decreases seem to help the algorithm.
Note that due to rescaling, larger values of the dominated hypervolume gap are
worse than smaller values. So from figure 2 we see that by increasing the number
of decreases the worst case performance (measured in dominated hypervolume)
increases by 1 to 2 orders of magnitude. This would suggest that while largely
ignored so far, this form of non-elitism might be beneficial to the optimization
process.

However, decreases do ultimately hurt the best case performance, sometimes
by several orders of magnitude. But due to the non-linear scale the dominated
hypervolume is measured on, it is unclear how signigicant this result is.

4   Summary and Outlook

The paper at hand intended to find dependencies between EMO algorithm parameter settings and the decreases in hypervolume progression [4]. The common
SBX and polynomial mutation variation parameters as well as the population

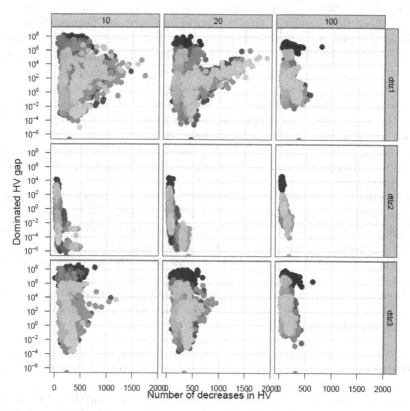

Fig. 2. Number of decreases in hypervolume in relation to the difference between the maximum hypervolume over all runs and the hypervolume of the run. The latter transformation was chosen for visualization purposes only. Note that smaller gap values correspond to a larger dominated hypervolume. The shading of the points marks different combinations of parameter settings of the algorithm.

size of the $(\mu + 1)$ SMS-EMOA were analyzed for possible influences on the number or amount of decrease in hypervolume.

The population size has a major effect on the number and amount of hypervolume decreases. The results show that for increasing μ the number of hypervolume decreases drops. This was traced back to the larger number of boundary solutions for larger population sizes.

We found no evidence that the recombination parameters sbx_n and sbx_p influence the number of decreases in hypervolume or the amount of decreases. This observation underpins the results of [5] that the sbx_n parameter has no influence on SMS-EMOA results in general. We therefore suggest to ignore this parameter in future experiments and thus, reduce the number of parameters of EMO algorithms and ultimately the complexity of the parameter tuning.

For pm_n and pm_p we found evidence that suggests a highly non-linear relationship between these parameters and the number and amount of hypervolume decreases. Interestingly, the standard settings from the literature for pm_n and pm_p are those that yield the largest number of decreases, suggesting that decreases might have a positive influence on the final result of the optimization process. Further proof for this was given by relating the number of decreases to the maximally attained hypervolume, w.r.t a fixed reference point, for each run. From this data, we could also deduce, that decreases help the algorithm to avoid stagnation. Runs with a large number of decreases tend to be orders of magnitude better than those with a lower number of decreases in the worst case. The price for this is a small decrease in maximally dominated hypervolume. It would be interesting to study this effect further, possibly by deliberately introducing hypervolume decreases into the optimization run and studying their influence on the final result.

Our results indicate that choosing "good" parameter settings has a measurable impact on the number and the amount of decreases in hypervolume. An effect on the overall quality of the results cannot be deduced due to missing data for comparison. There are currently no guards in the algorithm to discard points that lead to a decrease in hypervolume w.r.t. the fixed reference point. The simplest solution to this dilemma is to check each new point w.r.t a predefined fixed reference point and to forget those individuals that cause a decrease and continue with the previous population. This will form the foundation for upcoming experiments.

References

1. Beume, N., Naujoks, B., Emmerich, M.: SMS-EMOA: Multiobjective selection based on dominated hypervolume. European Journal of Operational Research 181(3), 1653–1669 (2007)
2. Coello Coello, C.A., Van Veldhuizen, D.A., Lamont, G.B.: Evolutionary Algorithms for Solving Multi-Objective Problems, 2nd edn. Springer, New York (2007)
3. Deb, K.: Multi-Objective Optimization using Evolutionary Algorithms. Wiley, Chichester (2001)
4. Judt, L., Mersmann, O., Naujoks, B.: Non-monotonicity of Obtained Hypervolume in 1-greedy S-Metric Selection. Journal of Multi-Criteria Decision Analysis (2011), preprint at http://maanvs03.gm.fh-koeln.de/webpub/CIOPReports.d/Judt11a.d/
5. Wessing, S., Beume, N., Rudolph, G., Naujoks, B.: Parameter Tuning Boosts Performance of Variation Operators in Multiobjective Optimization. In: Schaefer, R., Cotta, C., Kołodziej, J., Rudolph, G. (eds.) PPSN XI. LNCS, vol. 6238, pp. 728–737. Springer, Heidelberg (2010)
6. Zitzler, E., Thiele, L., Laumanns, M., Fonseca, C.M., da Fonseca, V.G.: Performance Assessment of Multiobjective Optimizers: An Analysis and Review. IEEE Transactions on Evolutionary Computation 7(2), 117–132 (2003)

Effects of Speciation on Evolution of Neural Networks in Highly Dynamic Environments

Peter Krčah*

Computer Center, Charles University
Ovocný Trh 5, 116 36 Prague 1, Czech Republic
peter.krcah@ruk.cuni.cz

Abstract. Using genetic algorithms for solving dynamic optimization problems is an important area of current research. In this work, we investigate effects of speciation in NeuroEvolution of Augmenting Topologies (NEAT), a well-known method for evolving neural network topologies, on problems with dynamic fitness function. NEAT uses speciation as a method of maintaining diversity in the population and protecting new solutions against competition. We show that NEAT outperforms non-speciated genetic algorithm (GA) not only on problems with static fitness function, but also on problems with gradually moving optimum. We also demonstrate that NEAT fails to achieve better performance on problems where the optimum moves rapidly. We propose a novel method called DynNEAT, which extends NEAT by changing the size of each species based on its historical performance. We demonstrate that Dyn-NEAT outperforms both NEAT and non-speciated GA on problems with rapidly moving optimum, while it achieves performance similar to NEAT on problems with static or slowly moving optimum.

Keywords: NEAT, speciation, neural networks, dynamic optimization.

1 Introduction

Real-world optimization problems often contain various sources of uncertainty. When evolutionary algorithms are used to solve such problems, this uncertainty translates into dynamic fitness function, often posing a challenge for standard EAs. One common approach to improve efficiency of the search is to use multiple concurrently evolving populations to track different optima (a survey of methods is given in [2]). In this work, we investigate behavior of NEAT [4], a method for optimizing structure and weights of a neural network, on dynamic problems. We show that NEAT performs poorly on problems where the optimum of the fitness function changes rapidly between generations. We propose a new method, called DynNEAT, which extends NEAT by taking into account multiple previous generations when choosing the size of the species in the next generation, thus stabilizing the speciation process and improving performance of the search.

---

* The research is supported by the Charles University Grant Agency under contract no. 9710/2011.

Y. Hamadi and M. Schoenauer (Eds.): LION 6, LNCS 7219, pp. 425–430, 2012.
© Springer-Verlag Berlin Heidelberg 2012

2 Methods

NeuroEvolution of Augmenting Topologies (NEAT). NEAT is a method
for evolving both structure and connection weights of artificial neural networks [4].
NEAT uses speciation to maintain diversity in the population by protecting new
solutions from direct competition with currently best individuals. Speciation in
NEAT works in the following way. In the first generation, each individual is assigned to a different species and its fitness function is evaluated. Each subsequent
generation is constructed by first dividing all slots in the population among all
species present in the previous generation. Species sizes in generation $i + 1$ are
allocated proportionally to $s_{NEAT}(i + 1)$, an average fitness of all individuals
belonging to the given species in the previous generation:

$$s_{NEAT}(i + 1) = \frac{\sum_{j=1}^{N_i} f_{ij}}{N_i},$$

where f_{ij} is the fitness value of jth individual of the given species in generation i
and N_i is the number of individuals in generation i in the given species. When the
new size of each species is known, each slot is populated by performing crossover
and mutation of individuals selected from the given species in the previous generation. Newly created individuals are assigned to species not based on their
ancestral species, but by comparing them one at a time to representatives of
each species from the previous generation and assigning them to the first species
whose representative is sufficiently similar (based on a defined threshold). If an
individual is not sufficiently similar to a representative of any existing species, a
new species is created for it. When all individuals are assigned to species, NEAT
continues by evaluating their fitness and repeating the same process for the new
generation. Two other major components of NEAT are *historical markings* of
neurons and growing neural networks incrementally. Comprehensive description
of NEAT is available in [4].

**NeuroEvolution of Augmenting Topologies for Dynamic Fitness Func-
tions (DynNEAT).** On highly dynamic problems, speciation scheme used by
NEAT can be disadvantageous. In NEAT, the size of the species is chosen based
solely on the average fitness value of individuals from the previous generation. In
highly dynamic problems, this value will change dramatically from generation to
generation, leading to dramatic changes in the size of the species. Such radical
changes in species size can be detrimental to the progress of the search by removing novel solutions from a species before they can be optimized. To improve
the behavior of speciation on such problems, we propose DynNEAT method.
In DynNEAT, decisions about the size of the species are based not just on the
previous generation, but on t previous generations. Species sizes in generation
$i + 1$ are allocated proportionally to $s_{DynNEAT}(i + 1)$, the maximum average
fitness of last t generations:

$$s_{DynNEAT}(i + 1) = \max_{j=i-t+1}^{i} \frac{\sum_{k=1}^{N_j} f_{jk}}{N_j}.$$

Such method of sizing the species ensures stability of the species across generations even when the optimum of a fitness function moves rapidly. Parameter t controls for how long can DynNEAT maintain the size of a species when the average fitness of individuals in the species decreases over generations. In this work, the parameter value was set to 5 in all experiments.

3   Experiments

To evaluate the performance of NEAT and DynNEAT on dynamic optimization problems, we perform three different experiments, using increasingly dynamic versions of the *function approximation* problem. Each fitness evaluation consists of 600 steps, during which a single input value increases linearly from -1 to $+1$. The resulting fitness value is computed as $1 - min(\mu, 1)$, where μ is the mean squared error of the differences between expected and real output value measured in each step. In all three experiments, non-speciated GA is compared to NEAT and DynNEAT. As a non-speciated GA, we use a standard GA with the addition of historical markings used for crossover of neural networks. Each configuration was tested in 50 runs. Each run was stopped after 200 generations (after which most runs achieved a plateau). Significance levels were computed using Student's t-test. Population size was set to 300, to allow more species to form concurrently and cover different optima of the changing fitness function. In the first experiment, with a static fitness function, the target function $f_A(x)$ consists of a linear part, a constant part and a sine-wave part and is defined in the following way (see solid line in fig. 1):

$$f_A(x) = \begin{cases} 8x - 1 & \text{if } 0 \le x < \frac{1}{4} \\ 1 & \text{if } \frac{1}{4} \le x < \frac{1}{2} \\ cos(4\pi n(x - \frac{1}{2})) & \text{if } \frac{1}{2} \le x \le 1 \end{cases},$$

where $x = (k - 1)/599$ is the current evaluation step scaled to interval $[0, 1]$.

In the second experiment, with a slowly moving optimum (SMO), the target function is defined in the following way (see fig. 1):

$$f_B(x, y) = \frac{1}{2}(1 + \sin \frac{2\pi y}{50}) f_A(x),$$

where x is defined as in $f_A(x)$ and y is the generation counter. The target function oscillates between $f_A(x)$ and 0 with a period of 50 generations.

In the third experiment, with rapidly moving optimum (RMO), the target function is defined in the following way:

$$f_C(x, y) = \begin{cases} 1 & \text{if } (x < 0.5) \text{ xor } (y \text{ is even}) \\ -1 & \text{otherwise} \end{cases},$$

where x and y are defined as in $f_B(x, y)$. For odd generations, $f_C(x, y)$ is a step function with value 1 in first half of the domain and -1 in the second half of the domain. For even generations the function values are reversed.

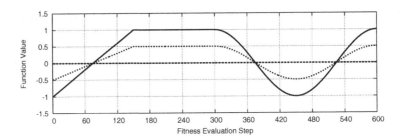

Fig. 1. Expected Output in Experiments with Static (solid line) and Slowly Moving Optimum (solid, dashed and dotted lines)

4   Results

In the experiment with the static fitness function, both DynNEAT and NEAT were able to consistently find good solutions (see fig. 2). The average maximum fitness achieved by NEAT in the 200th generation was 0.980 ($\sigma=0.009$), while DynNEAT achieved 0.988 ($\sigma=0.0067$). Non-speciated GA achieved average maximum fitness of only 0.741 ($\sigma=0.169$), with 30 of 50 runs failing to find a solution with fitness value above 0.603. Differences between any two methods are statistically significant ($p < 0.01$) in each generation since generation 60.

Results of experiments with slowly moving optimum (see fig. 2) reflect the periodicity of the problem. Since zero target output is trivial to solve compared to more complex outputs, the difficulty of the problem changes from generation to generation. Fig. 2 shows that all methods were able to find successful solutions when fitness function was close to zero, but their performance differed in generations where fitness function is furthest away from zero. In these generations, both DynNEAT and NEAT significantly outperformed non-speciated GA ($p < 0.01$ since generation 20, except 36-40 and 85-90). Differences between NEAT and DynNEAT were not statistically significant ($p > 0.2$).

In the experiment with rapidly moving optimum, DynNEAT was the only method capable of consistently finding good solutions to both target functions (see fig. 2). The average maximum fitness achieved by DynNEAT was 0.9547 ($\sigma=0.0295$), while neither NEAT nor non-speciated GA achieved fitness over 0.8. NEAT performed only marginally better than GA, with the average maximum fitness of 0.7794 ($\sigma=0.1044$) compared to 0.7376 ($\sigma=0.1280$) in GA ($p < 0.01$).

5   Discussion and Future Work

The experiment with the static fitness confirmed that the lack of speciation results in a significant drop in the performance of non-speciated GA. Moreover, the same effect occurs with a slowly moving optimum (SMO), which shows that advantages of speciation can also be utilized in dynamic problems. However, the experiment with rapidly moving optimum (RMO) demonstrates that when

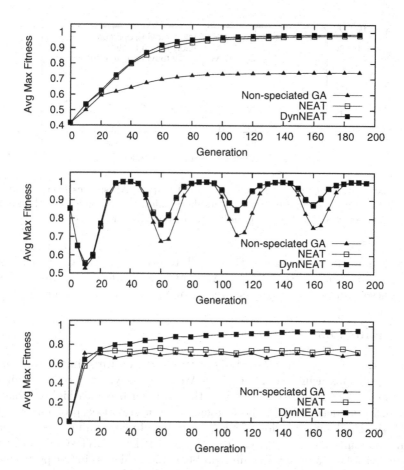

Fig. 2. Comparison of Average Maximum Fitness on a Fitness Function with Static, Slowly Moving and Rapidly Moving Optimum (from top to bottom)

the fitness is highly dynamic, NEAT fails to provide significant benefit over non-speciated GA (see fig. 2). The drop in performance can be explained by examining the dynamics of the speciation process. In order for the speciation to be effective, species must be long-lived to give their members enough time for adaptation. In RMO experiment, NEAT required 7.16 times more species than in SMO experiment (580 vs. 80.9), with an average lifespan shorter by a factor of 6.85 (3.64 vs. 24.94). DynNEAT, on the other hand, achieves high average lifespan of species even in RMO experiment (10.13 generations vs. 3.64 in NEAT) and smaller average number of species (202 vs. 580 in NEAT). This is further demonstrated by the distribution of species lifespans shown in fig. 3.

We applied DynNEAT to a problem where fitness alternates between two simple states. Such problem was chosen to clearly demonstrate the differences in speciation dynamic between NEAT and DynNEAT. In future works we would like to extend these results to study the influence of parameter t (which was

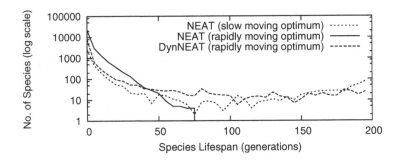

Fig. 3. Distribution of Species Based on Their Lifespan. All species from all 50 runs are included in the distribution for each method. Lifespan of a species is the number of generations in which at least one individual belonged to that species.

fixed in this work) and compare DynNEAT to other dynamic optimizers using benchmark functions (e.g. MPB [1]). Another direction for future research is to extend these results to methods derived from NEAT, such as HyperNEAT [3].

6   Conclusions

In this work, we investigate the effects of speciation on three increasingly more dynamic problems using NEAT method. We have shown that the dynamics of speciation in NEAT is disrupted when the problem becomes highly dynamic, significantly impacting NEAT performance. To address this problem, we proposed DynNEAT, an extension of NEAT capable of maintaining species even in a highly dynamic environment. We have shown that DynNEAT significantly outperforms NEAT on a highly dynamic problem and achieves similar performance on problems with static or slow-changing optimum. Analysis of the speciation has confirmed that DynNEAT achieves its performance by stabilizing speciation, which allows long-lived species to form even in cases when fitness of individuals dramatically changes between generations.

References

1. Branke, J.: Memory enhanced evolutionary algorithms for changing optimization problems. In: Congress on Evolutionary Computation, CEC 1999, pp. 1875–1882. IEEE (1999)
2. Jin, Y., Branke, J.: Evolutionary optimization in uncertain environments-a survey. IEEE Trans. Evolutionary Computation, 303–317 (2005)
3. Stanley, K.O., D'Ambrosio, D.B., Gauci, J.: A hypercube-based encoding for evolving large-scale neural networks. Artificial Life 15(2), 185–212 (2009)
4. Stanley, K.O., Miikkulainen, R.: Efficient reinforcement learning through evolving neural network topologies. In: Proceedings of the Genetic and Evolutionary Computation Conference. Morgan Kaufmann, San Francisco (2002)

Natural Max-SAT Encoding of Min-SAT

Adrian Kügel

Faculty of Engineering and Computer Sciences
Ulm University, Ulm, Germany
Adrian.Kuegel@uni-ulm.de

Abstract. We show that there exists a natural encoding which transforms Min-SAT instances into Max-SAT instances. Unlike previous encodings, this natural encoding keeps the same variables, and the optimal assignment for the Min-SAT instance is identical to the optimal assignment of the corresponding Max-SAT instance. In addition to that the encoding can be generalized to the Min-SAT variants with clause weights and hard clauses. We conducted experiments which give evidence that our encoding is practically relevant, as Min-2-SAT instances can be solved much faster by transforming them to Max-SAT and using a Max-SAT solver than by using the best Min-SAT solver directly.

Keywords: Min-SAT, Max-SAT.

1 Introduction

The Minimum Satisfiability Problem (Min-SAT) asks for an assignment of Boolean variables which satisfies the minimum number of clauses of a given formula, whereas the Maximum Satisfiability Problem (Max-SAT) seeks to maximize the number of satisfied clauses. Both problems can be seen as a generalization of the Satisfiability Problem (SAT).

Recently, Li et al. [8] presented a Min-SAT solver called MinSatz and showed that a Min-SAT encoding of the MaxClique problem (and related problems) makes MinSatz competitive with the best MaxClique solvers. MinSatz was also tested against Max-SAT solvers using the three different encodings presented in [7]. In these tests, MinSatz was faster than the Max-SAT solvers on the encodings. These experiments indicated that solving a Min-SAT instance can be done faster using MinSatz than encoding it to a Max-SAT instance and using one of the available Max-SAT solvers. We show that when using our *better* Max-SAT encoding of Min-SAT instances it is still possible to outperform MinSatz on several kind of Min-SAT instances by encoding them to Max-SAT and using the Max-SAT solver akmaxsat.

Our paper is structured as follows: in Section 2 we provide basic definitions, then in Section 3 we describe our encoding. In Section 4 we present experimental data and finally in Section 5 we draw our conclusions.

Y. Hamadi and M. Schoenauer (Eds.): LION 6, LNCS 7219, pp. 431–436, 2012.
© Springer-Verlag Berlin Heidelberg 2012

2   Definitions

A *CNF formula* $\mathcal{F}$ is a conjunction of clauses consisting of Boolean variables. A *clause* C is a disjunction of literals and is written as $(l_1 \vee l_2 \vee \cdots \vee l_k)$, where $l_1, \ldots, l_k$ are from the set of variables and their negations. A literal $\overline{x}_i$ is true if the variable x_i is false, and it is false otherwise. We call a clause *satisfied* if at least one of its literals is true, and we call it *unsatisfied* if all its literals are false.

A *hard clause* is a clause which needs to be satisfied, whereas a *soft clause* specifies a clause which may be unsatisfied by the optimal assignment. The *partial Min-SAT problem* and the *partial Max-SAT problem* deal with both soft and hard clauses. Another variant of the Min-SAT problem is the *weighted Min-SAT problem*; in this variant, each clause has a positive weight which indicates the relative importance of the clause, and the sum of the weights of satisfied clauses has to be minimized. Likewise, in the *weighted Max-SAT problem* the sum of the weights of satisfied clauses has to be maximized.

We define the size of a clause to be the number of literals it consists of. A CNF formula which consists only of clauses of size k is also called a k-SAT formula, the corresponding Max-SAT instance Max-k-SAT, and the corresponding Min-SAT instance Min-k-SAT.

3   A Natural Max-SAT Encoding of Min-SAT Instances

The idea of the natural Max-SAT encoding of a Min-SAT instance is quite simple: we replace each original clause C by its negation $\overline{C}$. However we need to transform $\overline{C}$ into conjunctive normal form in order to get a Max-SAT instance. We apply an idea based on Max-SAT resolution rules. Max-SAT resolution rules were developed by Bonet et al. ([2], [3]) and Larrosa et al. ([5]). It is shown that $\overline{C} = \overline{(l_1 \vee l_2 \vee \ldots \vee l_k)}$ (where $l_1, \ldots, l_k$ are literals) can be transformed into a set of k clauses. We have adjusted the more general recursive resolution rule in [5] to our special case:

$$\mathrm{CNF}_{\mathrm{linear}}(\overline{l_1 \vee \ldots \vee l_k}) = \overline{l_1} \wedge (l_1 \vee \overline{l_2}) \wedge (l_1 \vee l_2 \vee \overline{l_3}) \wedge \ldots \wedge (l_1 \vee l_2 \vee \ldots \vee \overline{l_k}) \quad (1)$$

We will illustrate the transformation rule with a small example. Let $C = (x_1 \vee \overline{x_2} \vee x_3)$. $\mathrm{CNF}_{\mathrm{linear}}(\overline{x_1 \vee \overline{x_2} \vee x_3}) = \overline{x_1} \wedge (x_1 \vee x_2) \wedge (x_1 \vee \overline{x_2} \vee \overline{x_3})$. It can be easily verified that only for the assignment for which C is unsatisfied, all new clauses are satisfied, and at most one of the new clauses will be unsatisfied by any assignment.

Using this rule, each clause C of the original Min-SAT instance is replaced by $\mathrm{CNF}_{\mathrm{linear}}(\overline{C})$. The set of new clauses has the property, that any assignment that does not satisfy the original clause C satisfies all clauses of the new clause set, and any assignment that satisfies the original clause satisfies all but one of the new clauses. In this case, trying to maximize the number of satisfied clauses in the encoded instance is equivalent to trying to minimize the number of satisfied clauses in the original Min-SAT instance. Also, for any assignment, the number of unsatisfied clauses in the encoded instance corresponds to the number of satisfied clauses in the original Min-SAT instance.

As a clause consisting of k literals is replaced by k clauses, for a Min-k-SAT instance with m clauses we get a Max-SAT instance with the same number of variables and with $k \cdot m$ soft clauses. Previous encodings presented in [7] used m variables and up to $\Theta(m^2)$ clauses (including some hard clauses).

Our encoding can be used in the reverse direction too, transforming a Max-SAT instance into a Min-SAT instance. In this case, the number of unsatisfied clauses in the Min-SAT instance corresponds to the number of satisfied clauses of the Max-SAT instance. The encoding can also be used for weighted Min-SAT instances; in that case, each replacement clause gets the weight of the original clause. If a Min-SAT instance contains hard clauses, the hard clauses are not replaced, but kept as they are. It is interesting to note that the transformation of a partial Min-SAT encoded Maximum Clique instance yields the commonly used partial Max-SAT encoding of the Maximum Clique instance (and vice versa). In the next section we will present experimental evidence that our encoding is especially useful for Min-2-SAT instances.

4   Experimental Results

In order to evaluate our natural Max-SAT encoding of Min-SAT, we selected our Max-SAT solver akmaxsat [4] which performed best in several categories of random and crafted Max-SAT instances in the Max-SAT evaluation 2011. Our solver akmaxsat can be found at http://www.uni-ulm.de/in/theo/m/kuegel. For comparison reasons we also used the Max-SAT solver MaxSatz in its publicly available version from 2009 ([6]). Also, we obtained the solver MinSatz from the authors of [8] (the same version that was used in their tests).

As benchmark instances we generated randomly unweighted Min-2-SAT and Min-3-SAT instances. For each selection of number of variables and clauses-to-variables ratio we generated 30 instances. Each Min-SAT instance was also encoded to the corresponding Max-SAT instance using our natural Max-SAT encoding. Also, we generated the corresponding partial Max-SAT instance using the best encoding E3 of [7]. We ran akmaxsat and MaxSatz on both Max-SAT encodings of each Min-SAT instance and compared its performance with the performance of MinSatz on the corresponding Min-SAT instance.

We ran the experiments on a node of the bwGRiD [1] which provides two Intel Harpertown quad-core CPUs with 2.83 Ghz and 8GB RAM each. The installed operating system was Scientific Linux. We used a timeout of 1 hour for each instance. Instances which were not solved within 1 hour are regarded as unsolved.

Table 1 shows for all three solvers the average runtime in seconds on the solved Min-2-SAT instances of each kind (showing in parentheses the number of instances solved). Our natural Max-SAT encoding is labeled with *NE*. The test results show that akmaxsat (using our encoding) clearly outperforms MinSatz on random Min-2-SAT instances. For clauses-to-variables ratios of at most 3, the

434   A. Kügel

Table 1. Average runtime in seconds on Min-2-SAT instances

C/V	#var	MinSatz	akmaxsat		maxsatz	
			NE	E3	NE	E3
2	160	0.05 (30)	**0.01 (30)**	0.04 (30)	**0.01 (30)**	0.02 (30)
2	180	0.08 (30)	**0.01 (30)**	0.04 (30)	**0.01 (30)**	0.03 (30)
2	200	0.13 (30)	**0.01 (30)**	0.05 (30)	**0.01 (30)**	0.04 (30)
3	160	0.19 (30)	**0.02 (30)**	0.21 (30)	0.07 (30)	0.14 (30)
3	180	0.31 (30)	**0.02 (30)**	0.26 (30)	0.08 (30)	0.17 (30)
3	200	0.52 (30)	**0.04 (30)**	0.48 (30)	0.18 (30)	0.34 (30)
4	160	0.96 (30)	**0.11 (30)**	40.06 (30)	2.59 (30)	52.58 (30)
4	180	1.47 (30)	**0.15 (30)**	44.80 (30)	5.18 (30)	44.94 (30)
4	200	4.09 (30)	**0.28 (30)**	87.73 (30)	25.80 (30)	142.62 (30)
5	160	16.85 (30)	**0.78 (30)**	605.81 (25)	41.41 (30)	1027.26 (21)
5	180	51.80 (30)	**1.29 (30)**	940.13 (19)	175.03 (30)	930.35 (13)
5	200	117.4 (30)	**2.45 (30)**	1474.66 (12)	722.35 (30)	1987.71 (4)
6	160	349.9 (30)	**3.67 (30)**	1533.23 (6)	321.30 (30)	2088.89 (2)
6	180	861.6 (26)	**11.26 (30)**	1617.81 (1)	1136.41 (26)	- (0)
6	200	1330 (16)	**31.96 (30)**	2485.65 (1)	1540.48 (8)	- (0)

Table 2. Average runtime in seconds on Min-3-SAT instances

C/V	#var	MinSatz	akmaxsat		maxsatz	
			NE	E3	NE	E3
2	80	**0.01 (30)**	0.09 (30)	0.04 (30)	1.01 (30)	0.05 (30)
2	90	**0.02 (30)**	0.22 (30)	0.10 (30)	2.77 (30)	0.10 (30)
2	100	**0.05 (30)**	0.68 (30)	0.31 (30)	16.65 (30)	0.28 (30)
3	80	**0.15 (30)**	0.71 (30)	1.55 (30)	3.93 (30)	1.93 (30)
3	90	**0.45 (30)**	2.89 (30)	8.87 (30)	20.43 (30)	10.70 (30)
3	100	**1.41 (30)**	9.34 (30)	22.85 (30)	77.29 (30)	29.16 (30)
4	80	**1.71 (30)**	4.67 (30)	5.51 (30)	18.07 (30)	84.11 (30)
4	90	**8.54 (30)**	24.17 (30)	395.7 (30)	115.44 (30)	582.39 (30)
4	100	**37.06 (30)**	80.61 (30)	961.9 (27)	460.72 (30)	1372.18 (25)
5	80	**14.43 (30)**	21.20 (30)	952.7 (29)	76.97 (30)	1226.06 (26)
5	90	**112.2 (30)**	134.2 (30)	1762 (13)	412.77 (29)	1684.11 (6)
5	100	**439.0 (30)**	587.4 (30)	1554 (1)	1824.86 (22)	3460.76 (1)
6	80	**68.77 (30)**	73.06 (30)	2139 (8)	264.24 (30)	3549.32 (1)
6	90	552.9 (30)	**537.0 (30)**	- (0)	1519.72 (27)	- (0)
6	100	1551 (24)	**1514 (25)**	- (0)	2983.00 (2)	- (0)

solver MaxSatz is faster than MinSatz, too. When comparing the two encodings, the new encoding always leads to a faster runtime for both Max-SAT solvers.

Table 2 shows the average runtime on the Min-3-SAT instances in the same format as in Table 1. On Min-3-SAT instances, the new encoding seems to work better than the encoding E3 for clauses-to-variables ratios above 3. A clauses-to-variables ratio of 3 leads to different results for the two Max-SAT solvers:

akmaxsat can handle the new encoding better, whereas MaxSatz is faster on encoding E3. Note that in this case, akmaxsat on the new encoding is faster than MaxSatz on the encoding E3. For small clauses-to-variables ratios of less than 3, the encoding E3 seems to be always superior. Comparing the results of akmaxsat on the new encoding to the results of MinSatz, we can see that in most cases, MinSatz is still faster, but for some instances with a clauses-to-variables ratio of 6, akmaxsat outperforms Minsatz.

5   Conclusions

We have presented a natural Max-SAT encoding of Min-SAT instances that has the following advantages:

1. The encoding keeps the same variables and just increases the number of clauses by a factor of number of variables per clause.
2. The optimal assignment of the encoded instance is identical to the optimal assignment of the Min-SAT instance.
3. The encoding works notably well on Min-2-SAT instances, and the Max-SAT solver akmaxsat on the encoded instance is much faster than the Min-SAT solver MinSatz on the original instance.
4. Our encoding can also be used to transform Max-SAT instances into Min-SAT instances.

As the tests in [8] have shown, the MinSatz solver performs much better than Max-SAT solvers on encoded Maximum Clique instances. There might be other optimization problems where this could be true, and our encoding could be used to automatically transform Max-SAT encoded optimization problems into Min-SAT encoded optimization problems.

Acknowledgments. We gratefully thank the bwGRiD project [1] for the computational resources. Also we thank Zhu Zhu for sending us an executable of the solver MinSatz, and we thank the reviewers for helpful comments.

References

1. bwGRiD, member of the German D-Grid initiative, funded by the Ministry for Education and Research and the Ministry for Science, Research and Arts Baden-Wuerttemberg, http://www.bw-grid.de
2. Bonet, M.L., Levy, J., Manyà, F.: A Complete Calculus for Max-SAT. In: Biere, A., Gomes, C.P. (eds.) SAT 2006. LNCS, vol. 4121, pp. 240–251. Springer, Heidelberg (2006)
3. Bonet, M.L., Levy, J., Manyà, F.: Resolution for Max-SAT. Artif. Intell. 171, 606–618 (2007)
4. Kuegel, A.: Improved exact solver for the weighted max-sat problem. In: Workshop Pragmatics of SAT (2010)

5. Larrosa, J., Heras, F., de Givry, S.: A logical approach to efficient Max-SAT solving. Artif. Intell. 172(2-3), 204–233 (2008)
6. Li, C.M., Manyà, F., Mohamedou, N., Planes, J.: Exploiting Cycle Structures in Max-SAT. In: Kullmann, O. (ed.) SAT 2009. LNCS, vol. 5584, pp. 467–480. Springer, Heidelberg (2009)
7. Li, C.M., Manyà, F., Quan, Z., Zhu, Z.: Exact MinSAT Solving. In: Strichman, O., Szeider, S. (eds.) SAT 2010. LNCS, vol. 6175, pp. 363–368. Springer, Heidelberg (2010)
8. Li, C.M., Zhu, Z., Manyà, F., Simon, L.: Minimum Satisfiability and Its Applications. In: Walsh, T. (ed.) IJCAI, pp. 605–610. IJCAI/AAAI (2011)

A New Hyperheuristic Algorithm for Cross-Domain Search Problems

Andreas Lehrbaum and Nysret Musliu

Vienna University of Technology, Database and Artificial Intelligence Group
{lehrbaum,musliu}@dbai.tuwien.ac.at

Abstract. This paper describes a new hyperheuristic algorithm that performs well over a variety of different problem classes. A novel method for switching between working on a single solution and a pool of solutions is proposed. This method is combined with an adaptive strategy that guides the selection of the underlying low-level heuristics throughout the search. The algorithm was implemented based on the HyFlex framework and was submitted as a candidate for the Cross-Domain Heuristic Search Challenge 2011.

1 Introduction

Hyperheuristics (as introduced in [1]), are a way to incorporate existing problem-class specific simple low-level heuristics into a higher-level search strategy, which schedules and guides their execution. The main idea is that an ensemble of heuristics orchestrated by a top-level strategy is able to perform better on average at solving a wide range of problems, than any of the underlying heuristics alone. A good survey of existing hyperheuristic techniques is given in [2]. The HyFlex [3] framework offers an intuitive interface to utilise a set of given low-level search and mutation heuristics, easing the task of working purely on a high-level search strategy without any a priori knowledge about the problem instance or the heuristics available.

In this paper, we propose a hyperheuristic algorithm that consists of two distinct phases and uses a quality-based selection strategy for the local-search heuristics using feedback from the search progress. The implementation is tailored towards the interface of the HyFlex framework and the specific requirements of the CHeSC competition rules[1]. Development and testing was performed on 4 different problem domains: Boolean Maximum Satisfiability, One-dimensional Bin Packing, Personnel Scheduling and Permutation Flow Shop. The provided low-level heuristics were used, implementing standard heuristic search and mutation operations for each of the domains.

---

[1] http://www.asap.cs.nott.ac.uk/chesc2011/rules.html

Y. Hamadi and M. Schoenauer (Eds.): LION 6, LNCS 7219, pp. 437–442, 2012.
© Springer-Verlag Berlin Heidelberg 2012

2   Algorithm Description

2.1   Overview

Following the classification of Burke et al. [2], our algorithm is an online learning hyperheuristic, working mainly on heuristic selection. The novelty of the proposed approach lies in the repeated switching between two search variants, namely a serial search phase working only on a single solution and a systematic parallel search phase working with a set of different solutions at the same time. We further propose a grading mechanism for the ordering of the low-level heuristics and a selection strategy for the available mutation heuristics.

After an initialisation phase, in which the preliminary scores of the available heuristics are determined, the search continues by systematically executing the local-search heuristics in order of their respective quality scores. The splitting in a serial and a parallel search phase balances the focus of the search between exploration of new parts of the search space and the exploitation of the quality of the currently best working solution. Throughout the search process, the performance and runtime characteristics of the low-level heuristics are measured and the scores responsible for their selection are updated accordingly. In addition to that, the overall search progress is monitored continuously and mechanisms such as the temporary blocking of ineffective heuristics or the restart of the algorithm from the last best solution are applied. A global tabu-list in form of a ringbuffer which always contains the last 40 visited distinct solutions is used to avoid cycles and to prune already explored branches of the search tree. If the tabu list already contains 40 elements, the next solution replaces the oldest entry. The fixed size of the tabu-list was chosen based on experimental data to balance memory and runtime requirements with the effectiveness of the tabu mechanism.

2.2   Detailed Description of the Algorithm

Initialisation Phase. The algorithm begins with calculating preliminary quality-scores of all the local-search heuristics available for the given problem instance. It does this by initialising the first solution and applying the heuristics in turn, recording their gain and their required runtime. The heuristics are called with the highest possible parameter settings for the depth of search and the intensity of the mutations (where applicable). The initial quality-score for each heuristic is calculated as gain per runtime. The best solution found so far is returned as the working solution for the subsequent phase.

Serial Search Phase. The available local-search heuristics $LS = \{ls_1, \ldots, ls_n\}$ are applied sequentially, in order of decreasing quality, to the current working solution s_{work}. After each application of a heuristic, the fitness function $f(s)$ of the resulting solution is evaluated. If the fitness value is better, the solution is accepted as the next working solution. If the local search heuristic resulted in a different solution with the same fitness value then it is accepted as well. Additionally, a global tabu-list T keeps track of the last 40 distinct solutions encountered and prohibits the acceptance of a solution that is already on the tabu-list. In case a new working solution was accepted, the search continues with the application of the best local search heuristic, otherwise the next best heuristic is chosen from the set LS. The parameters for the depth of the search and the intensity of the mutation are set to random values before the application of each heuristic. This serial search phase ends whenever no further improvement could be found with all the available heuristics, therefore resulting in a locally optimal solution. See algorithm 1 for a pseudocode implementation.

Algorithm 1. Serial search working on a single solution s_{work}

1: **for** $i = 1 \rightarrow |LS|$ **do**
2:     setDepthOfSearch(random$(0 \ldots 1)$)
3:     setIntensityOfMutation(random$(0 \ldots 1)$)
4:     $s_{temp} \leftarrow$ applyHeuristic(ls_i, s_{work})
5:     **if** $f(s_{temp}) < f(s_{work})$ **or** $(f(s_{temp}) = f(s_{work})$ **and** $s_{work} \neq s_{temp})$
        then
6:        **if** $s_{temp} \notin T$ **then**
7:            $s_{work} \leftarrow s_{temp}$
8:            $T \leftarrow T \cup s_{work}$
9:            $i \leftarrow 1$
10:       **end if**
11:    **end if**
12: **end for**
13: **return** s_{work}

Quality Updates. In fixed time intervals of 5 seconds (or after every call to a heuristic, in case it takes longer to complete), the qualities of the local-search heuristics are updated to reflect their performance during the whole search process. This can result in the re-ordering of the search sequence of the heuristics. The quality metric is calculated as the number of times the heuristic resulted in an accepted solution, divided by the total runtime of this heuristic so far. The time interval between updates was determined experimentally for the CHeSC setting to balance the cost of the updates and the possible gain of reordering.

Generation of Mutated Solutions. The available mutation and ruin-recreate heuristics are placed in a roulette-wheel reflecting their relative performance. Initially all mutation heuristics have the same chance of being selected. The performance of a mutation heuristic is judged based on the solution quality of the mutated solution after the subsequent serial search phase has finished. This method is used in order to assess the likelihood of a given mutation heuristic to result in an improvement during the further course of the search. Given the current working solution as input, 7 mutated offsprings are generated by applying a set of mutation heuristics chosen by the roulette-wheel process. The mutation intensity and search depth parameters are again set to a new random value each time before the mutation heuristics are applied. However, the probability of strong mutations is lowered towards the end of the search.

Algorithm 2. Parallel search working on a set of solutions $s_{1...7}$

1: **for** $i = 1 \rightarrow 7$ **do**
2:     $h_i \leftarrow 1$ // set working heuristic index h_i to the best LS heuristic
3: **end for**
4: **repeat**
5:     $candidatesLeft \leftarrow$ **false**
6:     **for** $i = 1 \rightarrow 7$ **do**
7:         **if** $h_i \leq |LS|$ **then**
8:             setDepthOfSearch(random($0 \ldots 1$))
9:             setIntensityOfMutation(random($0 \ldots 1$))
10:             $s_{temp} \leftarrow$ applyHeuristic(ls_{h_i}, s_i)
11:             **if** $f(s_{temp}) < f(s_{best})$ **then**
12:                 $s_{best} \leftarrow s_{temp}$
13:                 $T \leftarrow T \cup s_{temp}$
14:                 **return** s_{best}
15:             **else**
16:                 $candidatesLeft \leftarrow$ **true**
17:                 **if** $f(s_{temp}) < f(s_i)$ **then**
18:                     $s_i \leftarrow s_{temp}$
19:                     $T \leftarrow T \cup s_{temp}$
20:                     $h_i \leftarrow 1$ // continue with best LS heuristic
21:                 **else**
22:                     $h_i \leftarrow h_i + 1$ // continue with the next best LS heuristic
23:                 **end if**
24:             **end if**
25:         **end if**
26:     **end for**
27: **until** $candidatesLeft =$ **false**
28: **return** SELECTSOLUTION($s_{1...7}$)

Parallel Search Phase. Starting from the set of 7 mutated solutions, the parallel search phase begins to work on the candidate solutions one after another. It begins with applying the best local-search heuristic to the first candidate. If an improvement is found, the new solution is accepted, otherwise it is discarded. Afterwards the search continues with the next candidate solution and the local-search heuristic scheduled for this solution. Whenever a global improvement is found (i.e. the result is better than the currently best found solution so far), it is immediately accepted as the working solution for the next serial search phase and the parallel search is aborted. Otherwise the search goes on until all solutions have reached a local optimum with respect to all available local-search heuristics. See algorithm 2 for a pseudocode implementation.

Working Solution Selection. Assuming no global improvement was found during the parallel search phase, a solution is selected from the pool of solutions containing the locally optimal output from the serial search phase as well as the parallel search phase. Solutions with a better quality have higher probability of being selected. If all solutions in the set are contained in the global tabu-list, a random mutation will be applied to the best solution until the result is not contained in the buffer anymore.

Excluding Inefficient Heuristics. If a local-search heuristic is found to be ineffective (determined by a low count of successful applications) it is excluded for a single iteration from the search process in both the serial and the parallel phase with a 50% chance.

Restarting the Search. Whenever the search continues for a certain amount of time (10% of the available runtime), producing only solutions which are worse than a preset threshold (130% of the so far best solution), the search continues with the generation of mutated solutions from the currently best solution. This prevents the search process from slowly generating ever worse solutions and diverging too far from the best candidate found so far.

3 Results

Tables 1 and 2 show the final results of the 20 participating teams (with our algorithm named HAHA), as published by the organizers of the CHeSC competition[2]. The values represent the median resulting function value per instance of 31 subsequent runs with different random seeds and 600 seconds CPU runtime. All problem instances were minimisation tasks and selected by the organisers of the competition. The rank column denotes the final overall rank according to the official scoring system and the best value in each column is marked bold.

Table 1. Median results for the Max-SAT ($MS_{1...5}$), Bin Packing ($BP_{1...5}$) and Personnel Scheduling ($PS_{1...5}$) problem instances

Rank	Algorithm	MS_1	MS_2	MS_3	MS_4	MS_5	BP_1	BP_2	BP_3	BP_4	BP_5	PS_1	PS_2	PS_3	PS_4	PS_5
1	AdapHH	**3**	5	**2**	3	8	**0.01607**	0.00360	**0.00356**	**0.10828**	**0.00354**	24	9667	3289	1765	325
2	VNS-TW	**3**	3	**2**	3	10	0.03696	0.00715	0.01671	0.10878	0.02776	19	9628	**3223**	1590	320
3	ML	5	10	3	9	8	0.04214	0.00753	0.01456	0.10852	0.02182	**18**	9812	3228	1605	**315**
4	PHUNTER	5	11	4	9	8	0.04787	0.00360	0.02012	0.10908	0.03948	25	10136	3255	1595	320
5	EPH	7	11	6	14	13	0.05042	0.00360	0.01127	0.10866	0.02238	22	10074	3232	1615	345
6	HAHA	**3**	4	**2**	5	8	0.08829	0.00726	0.01450	0.11023	0.02790	21	9666	3236	**1558**	335
7	NAHH	8	10	4	9	**7**	0.05504	0.00347	0.00473	0.10878	0.00554	27	9827	3246	1644	345
8	ISEA	5	11	4	9	11	0.03422	**0.00328**	0.00365	0.10862	0.00640	20	9966	3308	1660	**315**
9	KSATS	4	7	**2**	4	9	0.01923	0.00780	0.01149	0.10892	0.02199	22	9681	3241	1640	355
10	HAEA	6	12	5	12	11	0.04522	0.00363	0.01379	0.10873	0.02400	25	9795	3266	1699	345
11	ACO-HH	11	35	9	17	13	0.04771	0.00320	0.00388	0.10986	0.01486	26	11212	3346	1760	355
12	GenHive	16	44	31	19	14	0.02994	0.00708	0.01037	0.10859	0.02286	21	12708	3274	1727	330
13	DynILS	23	56	37	31	19	0.04027	0.00767	0.01016	0.10872	0.01285	33	9893	3324	1870	465
14	SA-ILS	13	23	12	15	9	0.07873	0.01153	0.01458	0.11039	0.02958	20	9750	3228	1625	340
15	XCJ	6	8	5	9	10	0.02201	0.01145	0.01569	0.10856	0.02850	30	33390	3277	1658	380
16	AVEGNep	8	10	5	9	**7**	0.08737	0.00773	0.01807	0.11139	0.03750	26	10230	3283	1765	360
17	GISS	16	21	13	17	9	0.06917	0.00837	0.03218	0.11259	0.05922	25	**9625**	3294	1785	370
18	SelfS	13	36	14	14	10	0.06642	0.00736	0.01441	0.10968	0.02391	26	9803	3249	1635	350
19	MCHH-S	8	14	8	8	9	0.06225	0.00729	0.01459	0.10976	0.02861	32	13297	3344	1785	370
20	Ant-Q	23	52	38	27	14	0.04909	0.01650	0.02102	0.10990	0.03765	33	73535	3348	1970	425

4 Conclusion

Our algorithm was ranked 6th (out of 20 teams) in the final CHeSC competition. The good results at the completely different domains of Max-SAT and Personnel Scheduling indicate a rather good general problem solving capability. The

---

[2] http://www.asap.cs.nott.ac.uk/chesc2011/results.html

Table 2. Median results for the Flow Shop ($FS_{1...5}$), Travelling Salesman ($TSP_{1...5}$) and Vehicle Routing ($VRP_{1...5}$) problem instances

Rank	Algorithm	FS_1	FS_2	FS_3	FS_4	FS_5	TSP_1	TSP_2	TSP_3	TSP_4	TSP_5	VRP_1	VRP_2	VRP_3	VRP_4	VRP_5
1	AdapHH	**6240**	26814	6326	**11359**	26643	**48194**	**20822145**	**6810**	66879	53099	60900	13347	148516	20656	148689
2	VNS-TW	6251	26803	6328	11376	**26602**	48194	21042675	6819	67378	**54028**	76147	13367	148206	21642	149132
3	ML	6245	**26800**	**6323**	11384	26610	**48194**	21093828	6820	66893	54368	80671	13329	**145333**	20654	148975
4	PHUNTER	6253	26858	6350	11388	26677	**48194**	21246427	6813	67136	52934	64717	**12290**	146944	**20650**	148658
5	EPH	6250	26816	6347	11397	26640	**48194**	21064606	6811	**66756**	52925	74715	13335	162188	**20650**	155224
6	HAHA	6269	26850	6353	11419	26663	48414	21291914	6917	69324	56039	65498	13317	155941	20654	148655
7	NAHH	6245	26885	**6323**	11383	26671	**48194**	20971771	6841	67418	53097	65398	13358	157242	20654	152081
8	ISEA	6262	26844	6366	11419	26663	**48194**	20868203	6832	67282	54129	70471	13339	149149	20657	150474
9	KSATS	6292	26860	6366	11466	26683	48578	21557455	6947	72027	58738	64495	13296	156577	20655	**147124**
10	HAEA	6261	26826	6353	11408	26651	**48194**	20925949	6824	67488	54144	**60608**	13342	146951	20655	147283
11	ACO-HH	6249	26904	6353	11393	26724	48200	21137472	6851	67202	53428	73348	14371	149672	21663	151610
12	GenHive	6279	26835	6366	11434	26648	48271	21083157	6868	67236	56022	67475	13353	167297	20718	147960
13	DynILS	6269	26875	6365	11419	26670	**48194**	20987358	6823	67308	54100	69798	14359	149869	21654	150060
14	SA-ILS	6336	26886	6390	11514	26703	49046	21281226	6994	70614	57607	64185	13390	162642	20667	152271
15	XCJ	6271	26910	6366	11419	26710	48412	21162559	6884	68005	54967	63654	13354	152321	20658	153110
16	AVEGNep	6322	26952	6379	11507	26743	48639	21520601	6969	70194	57998	77884	12397	184710	20655	166742
17	GISS	6329	26979	6385	11516	26758	49010	21651052	7001	72630	58004	61580	13352	162266	20657	149590
18	SelfS	6287	26859	6369	11443	26678	49043	21040810	6984	69646	56647	73894	14386	203667	20687	153590
19	MCHH-S	6336	26937	6397	11527	26716	49412	21504030	6997	70685	57836	72005	13534	207891	20850	160303
20	Ant-Q	6358	26971	6407	11545	26792	49613	21277953	7016	69987	55314	76678	14382	193827	21656	160684

algorithm has however some shortcomings in domains where different fast local search heuristics all result in small improvements most of the time (like the tested Bin Packing instances). It seems that the rather strong bias for the local-search heuristic with the best quality score can be both a strength and a weakness, indicating that a more adaptive process could be advantageous. More extensive study is needed to assess the further potential of this algorithmic approach.

Acknowledgments. The research herein is partially conducted within the competence network Softnet Austria II (www.soft-net.at, COMET K-Projekt) and funded by the Austrian Federal Ministry of Economy, Family and Youth (bmwfj), the province of Styria, the Steirische Wirtschaftsförderungsgesellschaft mbH. (SFG), and the city of Vienna in terms of the center for innovation and technology (ZIT).

References

1. Cowling, P.I., Kendall, G., Soubeiga, E.: A Hyperheuristic Approach to Scheduling a Sales Summit. In: Burke, E., Erben, W. (eds.) PATAT 2000. LNCS, vol. 2079, pp. 176–190. Springer, Heidelberg (2001)
2. Burke, E., Hyde, M., Kendall, G., Ochoa, G., Ozcan, E., Qu, R.: A survey of hyper-heuristics. Computer Science Technical Report No. NOTTCS-TR-SUB-0906241418-2747, School of Computer Science and Information Technology, University of Nottingham (2009)
3. Burke, E., Curtois, T., Hyde, M., Kendall, G., Ochoa, G., Petrovic, S., Antonio, J.: HyFlex: A Flexible Framework for the Design and Analysis of Hyper-heuristics. In: Multidisciplinary International Scheduling Conference (MISTA 2009), Dublin, Ireland, p. 790 (2009)

Brain Cine-MRI Sequences Registration Using B-Spline Free-Form Deformations and MLSDO Dynamic Optimization Algorithm

Julien Lepagnot, Amir Nakib, Hamouche Oulhadj, and Patrick Siarry

Université Paris-Est Créteil (UPEC), LISSI, EA 3956
61 Avenue du Général de Gaulle, 94010 Créteil, France
siarry@u-pec.fr

Abstract. In this paper, a dynamic optimization algorithm is used to assess the deformations of the wall of the third cerebral ventricle in the case of a brain cine-MR imaging. In this method, a nonrigid registration process is applied to a 2D+t cine-MRI sequence of a region of interest. In this paper, we propose to use a B-spline Free-Form deformation model. The registration process consists of optimizing an objective function that can be considered as a dynamic function. Thus, a dynamic optimization algorithm, called MLSDO, is used to accomplish this task. The obtained results are compared to those of several well-known static optimization algorithms. This comparison shows the relevance of using a dynamic optimization algorithm to solve this kind of problems, and the efficiency of MLSDO.

Keywords: registration, image sequences, dynamic optimization, meta-heuristics, B-splines, MRI.

1 Introduction

Recently, optimization in dynamic environments has attracted a growing interest, due to its practical relevance. Almost all real-world problems are dynamic, *i.e.* their objective function changes over the time. Then, the goal is not only to find the global optimum, but also to track it as closely as possible over the time. In this paper, we propose to apply the *Multiple Local Search algorithm for Dynamic Optimization* (MLSDO) [4] to the registration of sequences of images.

We focus on a method based on cine-MRI sequences to facilitate the diagnosis, and to assist neurosurgeons in the characterization of the pathology called *hydrocephalus*. In order to characterize hydrocephalus, doctors need to estimate the amplitude and nature of the movements of the brain ventricles. Then, we need an image registration procedure to approximate it. In this work we consider the nonrigid (or elastic) registration to register regions containing non-rigid objects. We propose a method inspired from [5,4] to assess the movements of a region of interest (ROI), using a more accurate deformation model. Besides, another contribution of the present work is to show the importance of the use of dynamic optimization algorithms for brain cine-MRI registration.

Y. Hamadi and M. Schoenauer (Eds.): LION 6, LNCS 7219, pp. 443–448, 2012.
© Springer-Verlag Berlin Heidelberg 2012

The rest of this paper is organized as follows. In section 2, the method proposed to register sequences of images is described. In section 3, the MLSDO algorithm and its use for the problem at hand are presented. In section 4, a comparison of the results obtained by MLSDO on this problem to the ones of several well-known static optimization algorithms is performed. This comparison shows the relevance of using MLSDO on this problem. Finally, a conclusion and the works under progress are given in section 5.

2   The Registration Process

We propose a method inspired from [5,4] to evaluate the movement in sequences of cine-MR images. This operation is required in order to assess the movements in the ROI over time. In [5,4], a segmentation process is performed on each image of the sequence, to determine the contours (as a set of points) of the walls of the third cerebral ventricle. Then, a geometric registration of each successive contours is performed, based on an affine deformation model. In the present work, we propose to use an intensity based registration instead of a geometric registration process. This way, we do not have to use a segmentation process anymore. Moreover, to evaluate the pulsatile movements of the third cerebral ventricle more precisely, a nonrigid deformation model is used in this paper.

In order to accurately model the deformations in the ROI over time, we propose to use B-spline Free-Form Deformations (FFDs) [3]. An advantage of B-splines over other spline functions, such as thin-plate splines and elastic-body splines, is that B-splines are locally controlled, so they are easier to understand and to manipulate, and they can be computed in parallel [3].

A B-spline FFD is a nonrigid transformation based on the manipulation of a grid of control points overlaid on the image. Let Φ be a 2D grid of control points $\phi_{i,j}$, with uniform spacing d_x on the x-axis and d_y on the y-axis. Let Im_1 and Im_2 be two successive images of the sequence. Let the transpose of a matrix A be denoted by A^T, and $T_\Phi : o \mapsto o'$ be the transformation of any point $o = (x\ y)^T$ in image Im_2 to its corresponding point $o' = (x'\ y')^T$ in image Im_1. Then, the nonrigid transformation T_Φ by B-spline functions is defined by $T_\Phi(o) = \sum_{l=0}^{3} \sum_{m=0}^{3} B_l(u)\, B_m(v)\, \phi_{i+l,j+m}$, where $i = \left\lfloor \frac{x}{d_x} \right\rfloor - 1$, $j = \left\lfloor \frac{y}{d_y} \right\rfloor - 1$, $u = \frac{x}{d_x} - \left\lfloor \frac{x}{d_x} \right\rfloor$, $v = \frac{y}{d_y} - \left\lfloor \frac{y}{d_y} \right\rfloor$, and B_l is the lth basis function of cubic B-splines. The control points are the parameters of the B-spline FFD, so the number of degrees of freedom of the transformation depends on the resolution of the grid of control points. Denoting the cardinal function by *card*, the 2D grid Φ has $(2\,card(\Phi))$ degrees of freedom. Then, this set of parameters is estimated through the maximization of the following criterion:

$$C(\Phi) = \frac{NMI(\Phi)}{P(\Phi) + 1} \tag{1}$$

where $NMI(\Phi)$ computes the normalized mutual information [8] of Im_1 and Im_1', and Im_1' is the image that results from the transformation of Im_2 ; $P(\Phi)$

is part of a regularization term that penalizes large deformations of Im_2, as we are dealing with slight movements in the ROI. $P(\Phi)$ and $NMI(\Phi)$ are defined in (2) and (3), respectively.

$$P(\Phi) = \frac{1}{2 \, \text{card}(\Phi)} \sum_{\phi_{i,j} \in \Phi} \left(\phi_{i,j} - \tilde{\phi}_{i,j}\right)^{\text{T}} \left(\phi_{i,j} - \tilde{\phi}_{i,j}\right) \tag{2}$$

where $\tilde{\phi}_{i,j}$ is the position of a control point $\phi_{i,j}$, in the grid that corresponds to the identity transformation $(\tilde{\phi}_{i,j} = (d_x i \ \ d_y j)^{\text{T}})$.

$$NMI(\Phi) = \frac{H(Im_1) + H(Im_1')}{H(Im_1, Im_1')} \tag{3}$$

where $Im_1 \cap Im_1'$ is the overlapping area of Im_1 and Im_1' ; $H(Im_1)$ and $H(Im_1')$ compute the Shannon entropy of Im_1 and Im_1', respectively, in their overlapping area ; $H(Im_1, Im_1')$ computes the joint Shannon entropy [7] of Im_1 and Im_1', in their overlapping area.

The registration problem is formulated as an optimization problem by:

$$\Phi^* = \max \ C(\Phi) \tag{4}$$

For the problem at hand, a grid of 3×3 control points is used. It is sufficient to accurately model the deformations in the ROI. Then, the B-spline FFD has 18 degrees of freedom.

3 The MLSDO Algorithm

3.1 Description of the Algorithm

MLSDO uses several coordinated local searches, performed by two types of agents: the exploring agents (to explore the search space in order to discover the local optima), and the tracking agents (to track the found local optima over the changes in the objective function). An exclusion radius is attributed to each agent. This way, if several agents converge to a same local optimum, then only one of them can continue to converge to this optimum. Another important strategy is the use of two levels of precision in the stopping criterion of the local searches of the agents. This way, we prevent the fine-tuning of low quality solutions, which could lead to a waste of fitness function evaluations. Furthermore, the local optima found during the optimization process are archived, to accelerate the detection of the global optimum after a change in the objective function. These archived optima are used as initial solutions of the local searches performed by the tracking agents.

3.2 Cine-MRI Registration as a Dynamic Optimization Problem

The registration of a cine-MRI sequence can be seen as a dynamic optimization problem. Then, the dynamic objective function optimized by MLSDO changes according to the following rules:

- The criterion in (1) has to be maximized for each couple of successive images, as we are in the case of a sequence, then the optimization criterion becomes $C(\Phi(t)) = \frac{NMI(\Phi(t))}{P(\Phi(t))+1}$, where t is the index of the current couple in the sequence. $\Phi(t)$, $NMI(\Phi(t))$ and $P(\Phi(t))$ are the same as Φ, $NMI(\Phi)$ and $P(\Phi)$ defined before, respectively, but here depend on the couple of images.
- Then, the dynamic optimization problem is defined by: max $C(\Phi(t))$.
- If the current best solution (transformation) found for the couple t cannot be improved anymore (according to a stagnation criterion), the next couple $(t + 1)$ is treated.
- The stagnation criterion of the registration of a couple of images is satisfied if no significant improvement (higher than 1E-5) in the current best solution is observed during 5000 successive evaluations of the objective function.
- Thus, the end of the registration of a couple of images and the beginning of the registration of the next one constitute a change in the objective function.

3.3   Parameter Fitting of MLSDO

To perform the experiments reported in the following section, we use the following values for the six parameters of MLSDO. The initial step sizes of tracking and exploring agents are set to $r_l = 0.005$ and $r_e = 0.1$, respectively. The highest and the lowest precision parameters of the stopping criterion of the agents local searches are set to $\delta_{ph} = $ 1E-5 and $\delta_{pl} = $ 1E-4, respectively. The maximum numbers of tracking and exploring agents are set to $n_c = 2$ and $n_a = 1$, respectively. These values are suitable for the problem at hand, and they were fixed experimentally. Among several sets of values for the parameters, we selected the one that minimizes the number of evaluations performed. One can see that only one exploring agent is used to solve this problem. It is indeed sufficient for this problem, and using more than one exploring agent increases the number of evaluations required to register a sequence. However, using more than one exploring agent can improve the performance of MLSDO on other problems.

4   Experimental Results and Discussion

The registration of a couple of images is illustrated in Figures 1. As we can see, the movements in the ROI leave an important white trail in the difference images, as illustrated in Figures 1(e). Then, applying the found transformation (Figures 1(d)) eliminates the white trail and only noise remains in the difference images (Figures 1(f)).

A comparison between the results obtained by MLSDO and those obtained by several well-known static optimization algorithms is presented in this section. These algorithms, and their parameter setting, empirically fitted to the problem at hand, are defined below (see references for more details on these algorithms and their parameter fitting):

- CMA-ES (*Covariance Matrix Adaptation Evolution Strategy*) [2] using the recommended parameter setting, except for the initial step size σ, set to $\sigma =$

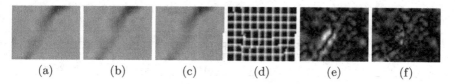

(a)   (b)   (c)   (d)   (e)   (f)

Fig. 1. Illustration of the registration of a couple of images of a sequence: (a) the first image of the couple, (b) the second image, (c) the second image after applying the found transformation to it, (d) illustration better showing this transformation, by applying it to the image of a grid, (e) illustration showing the difference, in the intensity of the pixels, between the two images of the couple: a black pixel indicates that the intensities of the corresponding pixels in the images are the same, and a white pixel indicates the highest difference between the images, (f) illustration showing the difference, in the intensity of the pixels, between the first image and the transformed second image

0.5. The population size λ of children and the number of selected individuals μ are set to $\lambda = 11$ and $\mu = 5$;
- SPSO-07 (*Standard Particle Swarm Optimization* in its 2007 version) [1] using the recommended parameter setting, except for the number S of particles ($S = 12$) and for the parameter K used to generate the particles neighborhood ($K = 8$) ;
- DE (*Differential Evolution*) [6] using the "DE/target-to-best/1/bin" strategy, a number of parents equal to $NP = 30$, a weighting factor $F = 0.8$, and a crossover constant $CR = 0.9$.

The image sequence used to fit their parameters is the same as the one used for MLSDO. However, it is not needed to fit the parameters of the algorithms for each sequence, and the same values are used for the other ones.

As these algorithms are static, we have to consider the registration of each couple of successive images as a new problem to optimize. Thus, these algorithms are restarted after the registration of each couple of images, using the stagnation criterion defined in section 3.2. Initializing these algorithms using the best solution found for the last registered couple of images cannot be used to improve their performance in our case. If we do so, algorithms perform a significant number of iterations without improving their current solution. Indeed, they progressively decrease the diversity of the population, before starting the intensification phase. In this comparison, the results obtained using MLSDO, as a static optimization algorithm, are also given.

In Table 1, the average number of evaluations among 20 runs of the algorithms are given. We can see that the number of evaluations of the objective function performed by MLSDO, used as a dynamic optimization algorithm, is significantly lower than the ones of the static optimization algorithms. A Jarque-Bera statistical test has been applied on the numbers of evaluations performed by the compared algorithms. This test indicates at a 95% confidence level that the numbers of evaluations follow a normal distribution. Then, we can perform a Welch's one-way ANOVA on these numbers of evaluations. This test confirms at a 95%

Table 1. Average number of evaluations to register a couple of images

Dynamic optimization	Static optimization			
MLSDO	CMA-ES	SPSO-07	DE	MLSDO
7655.16	9805.61	10155.35	10785.27	10880.14
± 584.30	± 669.32	± 733.00	± 850.99	± 820.49

confidence level that there is a significant difference between the performances of at least two of the compared algorithms. Then, the Tukey-Kramer multiple comparisons procedure has been used to determine which algorithms differ in terms of number of evaluations. It indicates that MLSDO performs significantly differently from all the other tested algorithms. Thus, this comparison shows the efficiency of MLSDO and the significance of using a dynamic optimization algorithm on this problem.

5   Conclusion

In this paper, a registration process, based on a B-spline FFD model and on a dynamic optimization algorithm, is proposed to register quickly all the images of a cine-MRI sequence. It takes profit from the effectiveness of the dynamic optimization paradigm. The process is sequentially applied on all the 2D images. The entire procedure is fully automated and provides an accurate assessment of the ROI deformation. Our work under progress consists of the parallelization of the MLSDO algorithm using Graphics Processing Units.

References

1. Clerc, M., et al.: Particle Swarm Central website, http://www.particleswarm.info
2. Hansen, N., Ostermeier, A.: Completely derandomized self-adaptation in evolution strategies. Evolutionary Computation 9(2), 159–195 (2001)
3. Ino, F., Ooyama, K., Hagihara, K.: A data distributed parallel algorithm for nonrigid image registration. Parallel Computing 31(1), 19–43 (2005)
4. Lepagnot, J., Nakib, A., Oulhadj, H., Siarry, P.: Brain cine-MRI registration using MLSDO dynamic optimization algorithm. In: Proceedings of the 9th Metaheuristics International Conference (MIC 2011), Udine, Italy, vol. 1, pp. 241–249 (2011)
5. Nakib, A., Aiboud, F., Hodel, J., Siarry, P., Decq, P.: Third brain ventricle deformation analysis using fractional differentiation and evolution strategy in brain cine-MRI. In: Medical Imaging 2010: Image Processing, vol. 7623, pp. 76232I-1–76232I-10. SPIE, San Diego (2010)
6. Price, K., Storn, R., Lampinen, J.: Differential Evolution - A Practical Approach to Global Optimization. Springer (2005)
7. Shannon, C.E.: The mathematical theory of communication. The Bell System Technical Journal 27, 379–423 and 623–656 (1948)
8. Studholme, C., Hill, D., Hawkes, D.: An overlap invariant entropy measure of 3D medical image alignment. Pattern Recognition 32(1), 71–86 (1999)

Global Optimization for Algebraic Geometry – Computing Runge–Kutta Methods

Ivan Martino[1] and Giuseppe Nicosia[2]

[1] Department of Mathematics, Stockholm University, Sweden
martino@math.su.se
[2] Department of Mathematics and Computer Science, University of Catania, Italy
nicosia@dmi.unict.it

Abstract. This research work presents a new evolutionary optimization algorithm, EVO-RUNGE-KUTTA in theoretical mathematics with applications in scientific computing. We illustrate the application of EVO-RUNGE-KUTTA, a two-phase optimization algorithm, to a problem of pure algebra, the study of the parameterization of an algebraic variety, an open problem in algebra. Results show the design and optimization of particular algebraic varieties, the Runge-Kutta methods of order q. The mapping between algebraic geometry and evolutionary optimization is direct, and we expect that many open problems in pure algebra will be modelled as constrained global optimization problems.

1 Introduction

In science and engineering, problems involving ordinary differential equations (ODEs) can be always reformulated a set of N coupled first-order differential equations for the functions y_i, $i = 1, 2, \ldots, N$, having the general form $\frac{dy_i(t)}{dt} = f_i(t, y_1, \ldots, y_N), i = 1, 2, \ldots, N$. A problem involving ODEs is completely specified by its equations and by boundary conditions of the given problem. Boundary conditions are algebraic conditions, they divide into two classes, *initial value problems* and *two-point boundary value problems*. In this work we will consider the initial value problem, where all the y_i are given at some starting value x_0, and it is desired to find the y_i's at some final point x_f. In general, it is the nature of the boundary conditions that determines which numerical methods to use. For instance, the basic idea of the *Euler's method* is to rewrite the dy's and dx's of the previous equation as finite steps δy and δx, and multiply the equations by δx. This produces algebraic formulas for the change in the functions when the independent variable x is increased by on *stepsize* δx; for very small stepsize a good approximation of the differential equation is achieved. The Runge-Kutta method is a practical numerical method for solving initial value problems for ODEs [1]. Runge-Kutta methods propagate a numerical solution over an $N + 1$-dimensional interval by combining the information from several Euler-style steps (each involving one evaluation of the right-hand f's), and then using the information obtained to match a Taylor series expansion up to some higher order. Runge-Kutta is usually the fastest method when evaluating f_i is cheap and the accuracy requirement is not ultra-stringent. In this research work, we want to design a methodology that allow us to find new Runge-Kutta methods of order q with minimal approximation error; such a question

Y. Hamadi and M. Schoenauer (Eds.): LION 6, LNCS 7219, pp. 449–454, 2012.
© Springer-Verlag Berlin Heidelberg 2012

can be tackled as a constrained optimization problem. At first, we define the problem from a geometrical point of view, using the theory of labelled trees by Hairer, Norsett and Wanner [2], and then the study of the parameterization of the algebraic variety $RK_s^q = \{s\text{-level Runge-Kutta methods of order } q\}$ suggests the use, with good results, of a new class of evolutionary optimization [3]. Algebraic geometry provides justification for why it is important to use evolutionary optimization algorithm to design effective new Runge-Kutta methods under several constrains.

2   The Algebraic Variety RK_s^q

Let $\Omega \subseteq \mathbb{R}^{n+1}$ be an open set and $f : \Omega \to \mathbb{R}^n$ a function such that the following Cauchy problem makes sense: $y' = f(t, y)$ under $y(t_0) = y_0$. We can always reduce that to the autonomous system

$$\begin{cases} y' = f(y) \\ y(t_0) = y_0 \end{cases} \tag{1}$$

where we abuse of the notation f, but the meaning is clear. The Runge-Kutta methods are a class of schemes to approximate the exact solution of (1). The structure of s-level *Implicit* Runge-Kutta method (RK method) is $k_i = f(y_0 + h \sum_{j=1}^{s} a_{i,j} k_j)$ with $i = 1, \ldots, s$ and h is the *step size*. The final numerical solution $y_1 \in \mathbb{R}^n$ of the problem (1) is given by $y_1 = y_0 + h \sum_{i=1}^{s} w_i k_i$. A RK method is called *explicit* if $a_{i,j} = 0$ if $i \geq j$. The approach will use all the parameters of the Butcher Tableau [4]. All c_i, w_i and $a_{i,j}$ are in $\mathbb{R}$ and characterize a given method with respect another one.

To understand the relation between the parameters of the Butcher Tableau and the order of the approximated solution we need to express the *local truncation error*, $\sigma_1 = y(t_0 + h) - y_1$, with respect to h and then we have to force that the coefficients of h^k must be zero for $k = 0, 1, \ldots, p$. Then the Runge Kutta method has order $p+1$. Forcing the coefficients of h^k to be zero produces the so called *order condition equations*. It is well known a combinatorial interpretation of the order conditions for a Runge Kutta method, involving rooted labelled trees and elementary differentials. This connection is carefully constructed in [2], and we refer you there for all the details. We denote T_p the set of the rooted labelled trees of order p. The reader should simply know that it possible state a bijection between the *structure* of the p-elementary differentials and each p-order conditions. The following Theorem explains also how to use this information to state the order condition equations.

Theorem 1. *If the Runge-Kutta method is of order p and if f is $(p + 1)$- times continuously differentiable, we have*

$$y^J(y_0 + h) - y_1^J = \frac{h^{p+1}}{(p+1)!} \sum_{t \in T_{p+1}} \alpha(t) e(t) F^J(t)(y_0) + \vartheta(h^{p+2})$$

where $e(t) = 1 - \gamma(t) \sum_{j=1}^{s} w_j \Phi_j(t)$ is called the error coefficient of the tree t.
Using this result it is possible to compute symbolically the system of equations [5]; and it is easy to see that the number of equations blow up like the factorial: this fact plays a key role in the choice of the evolutionary algorithm for the solution of the

problem. Moreover the integer number $\Phi_j(t)$ and $\gamma(t)$ are also used to construct the local truncation error (Theorem 1). We skip their combinatorial definition but we remark that $\Phi_j(t)$ depends by $\{a_{i,j}\}$ and using $e(t)$ depends by the $\{a_{i,j}\}$ and $\{w_i\}$. We note that $F^J(t)(y_0)$ is the J-component of the elementary differential of f corresponding to the tree t evaluated at the point y_0.

Now we call RK_s the set of all s-level Runge-Kutta methods and $RK_s^q \subseteq RK_s$ the set of all s-level Runge-Kutta methods with accuracy order q; moreover the subset of explicit s-level Runge Kutta will be denoted as ERK_s and similarly we define also ERK_s^q. The coefficients that control the methods are all in the Butcher Tableau. For this reason, a priori, the RK methods have $s + s^2$ free coefficients. Moreover, it is simple to prove that the order conditions in Theorem 1 are polynomials, so we can consider the affine algebraic variety V_s^q in the affine real space $\mathbb{A}^{s(s+1)}(\mathbb{R})$, minimally defined by the following polynomials in $s(s + 1)$ variables:

$$\sum_{j=1}^{s} w_j \Phi_j(t) = \frac{1}{\gamma(t)}, \ \forall t \in T_1 \cup T_2 \cup \cdots \cup T_q; \tag{2}$$

this set of polynomials has a particular algebraic structured, in fact it is an ideal of the ring of polynomials in $s(s + 1)$ variables. We will denote it with I_s^q. Thus, we can rewrite the *Theorem of the Local Error* as:

Theorem 2. Let $\mathbf{x} = ((a_{i,j})_{1 \leq j \leq i \leq s}, (w_1, w_2, \ldots, w_s))$, then $\mathbf{x} \in RK_s^q \Leftrightarrow \mathbf{x} \in V_s^q$. Similarly the algebraic variety EV_s^q in $\mathbb{A}^{\frac{s(s-1)}{2}}(\mathbb{R})$ minimally defined by the same polynomial equation and by $\{a_{i,j} = 0 \forall i \geq j\}$ is the variety of the explicit Runge Kutta method of s levels; EV_s^q is also a subvariety of V_s^q.

We remark that claiming $\mathbf{V_s^q}$ and $\mathbf{EV_s^q}$ being an algebraic variety has some underlying effect. One of the main difference concerns the topology: it is used the *Zariski topology* and in contrast to the standard topology, the Zariski topology is not Hausdorff (one can not separate two points with different open sets). We are going to show the hardness of studying the dimension and the parameterization of the varieties $\mathbf{V_s^q}$ and $\mathbf{EV_s^q}$. From now on, avoiding repetitions of EV_s^q, every result that we state for V_s^q is true also for EV_s^q, with the obvious opportune changes. Again, to simplify the reading, we put $V = V_s^q$ and $I = I_s^q$ (the ideal of the variety V). The goal of the present research work is to design particular features of the RK methods; hence, if we want to use symbolical or numerical methods, we need to parameterize V: in fact, we need to control the set RK_s^q using free parameters $\{p_1, \ldots, p_m\}$ where m is at least the dimension d of V. More explicitly given a connected component decomposition $V = \cup_{i \in I} U_i$ we need to know the function $\theta_{s,i}^q : \mathbb{R}^{d_i} \to U_i$ This problem translated in the algebraic geometry $\theta_{s,i}^q : \mathbb{A}^{d_i}(\mathbb{R}) \to U_i$ is the problem of *local parameterization of an real algebraic variety*; we remark, that, even if the meaning and the appearance of the two $\theta_{s,i}^q$ are exactly the same, the algebraic one carries all the differences in the topologies and in the map. If we narrow down our investigation, for a moment, and consider only curves (one dimension algebraic varieties) in the affine plane [6,7], the question is the following: *is the curve rational?* We know that non-rational curve exists. We can suppose that not all varieties X are birational equivalent (generalization of rational concept) to $\mathbb{A}^d(\mathbb{R})$. The theory of Gröebner basis [8] or the *Newton Polygon* [9] could be applied to tackle this problem, but, with a lot of generators, computational time blows up. The problem is hardly structured. It is extremely difficult to compute the connected component decomposition and their local dimension d_i's [10]: there are some methods in computational algebra where the complexity of computation depends on number of generators m, the number of variables $s(s + 1)$ and the degree of the polynomial q as $m^{o(1)}q^{o(s(s+1))}$ [11,12]; so if s and q increase (so d increase) the computational time blow up. For the same reason a symbolic approach of the problem is not feasible [5]. Hence excluding some particular cases, finding a global solution for the parameterization of an algebraic variety is an open problem [13]. Now we want to state clearly the results in the most general condition. Let $X \subset \mathbb{R}^m$ be the set of real solutions of a system of n polynomial equations $f_i = 0$. Let f be a positive real values function defined over X, and consider the optimization problem consisting of finding $\mathbf{x} \in X$ such that the value $f(\mathbf{x})$ is minimal in $f(X) \subset \mathbb{R}_+$. Then for bigger value of m and n it is an open problem to find the connected components $\{U_i\}$ of X, their local dimensions d_i and their local parameterization $\{\theta_i : \mathbb{R}^{d_i} \rightarrow U_i\}$. For this reason, we suggest the use of evolutionary algorithms to search a good solution of the corresponding optimization problem. Of course, any kind of optimization over the varieties $\mathbf{V_s^q}$ and $\mathbf{EV_s^q}$ are of this type; thus we are going to show how this optimization should be.

3   The Approach and the Results

In this section, we define the optimization problem that we want to solve. Even if a Runge-Kutta method of order q has many features that we want to control (for instance, the convergence region for implicit RK, S_a), the aim of this research work is to obtain new explicit or implicit Runge-Kutta methods of maximal order q that minimizes local errors of order $q + 1$. We denote, $\mathbf{x} = ((a_{i,j})_{1 \leq j, i \leq s}, (w_1, w_2, \ldots, w_s)) \in \mathbb{R}^{s(s+1)}$ where $\{a_{i,j}\}$ and $\{w_i\}$ are the coefficients of Butcher Tableau of an implicit, and respectively explicit, Runge-Kutta method. Using these results, $\mathbf{x}$ is a feasible solution, i.e., it is in V_s^q (or EV_s^q) if it respects the constrains in 2. The fitness function is so defined $fitness(\mathbf{x}) = \sum_{i=0}^{q+1} \sum_{t \in T_i} \alpha(t)|e(t)|$. Moreover, we want that $\mathbf{x}$ minimizes the local errors. The impossibility of analytically computing the functions $\theta_{s,i}^q$ and the difficulty of trying to numerically optimize local error gives us the motivation to use evolutionary algorithms to face this difficult global optimization problem [3]. The problem has been divided and tackled in two parts: I) to find solution for order condition, and II) to optimize the RK method provided by the solution of the system. Let us fix the number s of level and let us discuss the implicit RK methods case; what shall follow holds for the explicit case too. Since a priori we do not know the local dimension of the varieties $\mathbf{V_s^q}$, we need to fix $q = 2$ and exploring if the variety V_s^2 has points; if it is, we can also produce a suitable set of feasible solutions. Thus, we will consider the next order until we will not find any solutions, i.e. for the maximal order q the variety $V_s^{q+1} = \emptyset$. Without to modify the designed evolutionary algorithm, it is possible to explore the solution space computing the non-dominated solutions (the Pareto optimal solutions) of the given problem. The evolutionary algorithm for a fixed level s and order q has the structure shown in Appendix A. We want to produce new Runge-Kutta methods of high order approximation, but to verify our theory and methodology we have tested the evolutionary algorithm with a 3-level explicit Runge-Kutta methods of order 3 and with 4-level and 5-level explicit Runge-Kutta methods of order 4: we have find respectively 146, 364 and 932 new explicit Runge-Kutta methods. Moreover we have produced a remarkable set of feasible solution that can be used for different optimization problem over the algebraic varieties V_s^q and EV_s^q. We show the results in the following table.

Table 1. Feasible solutions in the explicit Runge-Kutta methods

Order/Level	3	4	5	6	7	8	9	10
2	46	108	34	19	3	8	13	16
3	146	197	140	3	24	57	45	28
4	0	364	932	53	20	110	0	34

4 Conclusion

The designed and implemented evolutionary algorithm, Evo-Runge-Kutta, optimizes the Butcher Tableaux an implicit or explicit Runge Kutta methods in order to find the maximal order of accuracy and to minimize theirs local errors in the next order. The results presented in this article suggest that further work in this research field will advance the designing of Runge-Kutta methods, in particular, and the use of the evolutionary algorithm for any kind of optimization over an algebraic variety. To our knowledge this is the first time that algebraic geometry is used to state correctly that evolutionary algorithms have to be used to face a particular optimization problem. Again we think this is the first time that algebraic geometry and evolutionary algorithms are used to tackle a numerical analysis problem. Further refinement of our evolutionary optimization algorithm will surely improve the solutions of these important numerical analysis problems.

Acknowledgements. Ivan Martino want to thank his advisor Professor Torsten Ekedahl that has recently passed away. It has been a privilege to learn mathematics from him. His genius and his generosity will always inspire me.

A The Algorithm: Evo-Runge-Kutta

Evo-Runge-Kutta()
1. t=0;

2. inizializePopulation($pop_d^{(t)}$);   /* random generation of RKs*/
3. initialize($newSolution$);   /* new array of feasible solutions*/
4. evaluationPopulation($pop_d^{(t)}$);   /* evaluation of RK systems */
5. **while** ((t< I_{max})&&(meanError< $accuracy$)&&(bestError< $accuracy$)) **do** {
6.   Copy ($pop_d^{(t)}$, $popMut_d^{(t)}$, p_c);
7.   mutationOperator($popMut_d^{(t)}$, p_m, σ);
8   isFeasible($popMut_d^{(t)}$, $oldSolutions$);

9.   evaluationPopulation($popMut_d^{(t)}$);

10. $pop_d^{(t+1)}$=Selection($pop_d^{(t)}$, $popMut_d^{(t)}$, r_s);

11.   computeStatistics($pop_d^{(t+1)}$);

12   saveSolutions($newSolution$);

13.   t=t+1;

14.   }

In the following table we show the parameters used.

Parameter	Value
order of the RK-methods	2
level of the RK-methods	3
order of the optimization of the error	3
d = population size	10^3
I_{max} = Max iterations of the first & second cycle	4×10^4
p_m = mutation probability of weight vector	0.5
part of element of Butcher Tableau that does not change during mutation	0.3
r_s = part of population selected for elitism in the first & second cycle	0.3
p_c = selection probability in the first & second cycle	0.5
σ = variance of Gaussian perturbation	0.1

References

1. Kaw, A., Kalu, E.E.: Numerical Methods with Applications. Autarkaw (2010)
2. Hairer, E., Norsett, S.P., Wanner, G.: Solving Ordinary Differential Equation I: Nonstiff Problems, 3rd edn. Springer (January 2010)
3. Goldberg, D.: Design of Innovation. Kluwar (2002)
4. Butcher, J.C.: Numerical methods for ordinary differential equations. John Wiley and Sons (2008)
5. Famelis, T., Papakostas, S.N., Tsitouras, C.: Symbolic derivation of runge kutta order conditions. J. Symbolic Comput. 37, 311–327 (2004)
6. Fulton, W.: Algebraic Curves. Addison Wesley Publishing Company (December 1974)
7. Kollar, J.: Rational Curves on Algebraic Varieties. Springer (1996)
8. Cox, D., Little, J., O'Shea, D.: Ideals, Varieties, and Algorithms, An Introduction to Computational Algebraic Geometry and Commutative Algebra, 3rd edn. Springer (2007)
9. D'Andrea, C., Sombra, M.: The newton polygon of a rational plane curve. arXiv:0710.1103
10. Castro, D., Giusti, M., Heintz, J., Matera, G., Pardo, L.M.: The hardness of polynomial equation solving. Found. Comput. Math. 3(4), 347–420 (2003)
11. Giusti, M., Hägele, K., Lecerf, G., Marchand, J., Salvy, B.: The projective Noether Maple package: computing the dimension of a projective variety. J. Symbolic Comput. 30(3), 291–307 (2000)
12. Giusti, M., Lecerf, G., Salvy, B.: A Gröbner free alternative for polynomial system solving. J. Complexity 17(1), 154–211 (2001)
13. Greene, R.E., Yau, S.T. (eds.): Open Problems in Geometry. Proceedings of Symposia in Pure Mathematics, vol. 54 (1993)

Clause Sharing in Parallel MaxSAT

Ruben Martins, Vasco Manquinho, and Inês Lynce

IST/INESC-ID, Technical University of Lisbon, Portugal
{ruben,vmm,ines}@sat.inesc-id.pt

Abstract. In parallel MaxSAT solving, sharing learned clauses is expected to help to further prune the search space and boost the performance of a parallel solver. However, not all learned clauses should be shared since it could lead to an exponential blow up in memory and to sharing many irrelevant clauses. The main question is which learned clauses should be shared among the different threads. This paper reviews the existing heuristics for sharing learned clauses, namely, static and dynamic heuristics. Moreover, a new heuristic for clause sharing is presented based on *freezing* shared clauses. Shared clauses are only incorporated into the solver when they are expected to be useful in the near future. Experimental results show the importance of clause sharing and that the freezing heuristic outperforms other clause sharing heuristics.

1 Introduction

Nowadays multicore processors are becoming the dominant platform. As a result, parallel Maximum Satisfiability (MaxSAT) solvers have been recently presented to exploit this new architecture [10,9]. These parallel solvers simultaneously search on the lower and upper bound values of the optimal solution. Searching in both directions and sharing learned clauses between these two orthogonal approaches makes the search more efficient. However, it is not clear which clauses should be shared among the different threads. The problem of determining if a shared clause will be useful in the future remains challenging, and in practice heuristics are used to select which learned clauses should be shared. This paper sheds some light on the impact of different clause sharing heuristics in parallel MaxSAT solving. The main contribution of this paper is twofold: (1) a new heuristic for clause sharing that freezes shared clauses until they are expected to be useful and (2) an empirical evaluation of static, dynamic and freezing heuristics for clause sharing.

The paper is organized as follows. In the next section the MaxSAT problem is defined and MaxSAT solvers are briefly referred. Section 3 describes different clause sharing heuristics that will be analyzed in the paper. Afterwards, section 4 presents an experimental evaluation of the different clause sharing heuristics. Finally, the paper concludes and suggests future work.

2 Preliminaries

A Boolean formula in conjunctive normal form (CNF) is a conjunction of clauses, where a clause is a disjunction of literals and a literal is a Boolean variable x or its

Y. Hamadi and M. Schoenauer (Eds.): LION 6, LNCS 7219, pp. 455–460, 2012.

© Springer-Verlag Berlin Heidelberg 2012

negation $\neg x$. A Boolean variable may be assigned truth values true or false. A positive (negative) literal x ($\neg x$) is said to be satisfied if the respective variable is assigned value true (false). A positive (negative) literal x ($\neg x$) is said to be unsatisfied if the respective variable is assigned value false (true). A variable (and respective literals) not assigned is said to be unassigned. A clause is said to be satisfied if at least one of its literals is satisfied. A clause is said to be unsatisfied if all of its literals are unsatisfied. A clause is said to be unit if all literals but one are unsatisfied and the remaining literal is unassigned. Otherwise, a clause is said to be unresolved. A formula is satisfied is all of its clauses are satisfied. The Boolean satisfiability (SAT) problem is to decide whether there exists an assignment that makes the formula satisfied. Such assignment is called a solution.

Maximum Satisfiability (MaxSAT) is an optimization version of Boolean Satisfiability (SAT) which consists in finding an assignment that minimizes (maximizes) the number of unsatisfied (satisfied) clauses. MaxSAT has several variants such as partial MaxSAT, weighted MaxSAT and weighted partial MaxSAT. In the partial MaxSAT problem, some clauses are declared as hard, while the rest are declared as soft. The objective in partial MaxSAT is to find an assignment to problem variables such that all hard clauses are satisfied, while minimizing the number of unsatisfied soft clauses. Finally, in the weighted versions of MaxSAT, soft clauses can have weights greater than 1 and the objective is to satisfy all hard clauses while minimizing the total weight of unsatisfied soft clauses.

The parallel MaxSAT solver PWBO [9] used in this paper is based on having several threads running a portfolio of two orthogonal algorithms: (i) an unsatisfiability-based algorithm that searches on the lower bound of the optimal solution [8] and (ii) a classical linear search algorithm that searches on the upper bound [7]. Notice that PWBO is not limited to the best performing algorithm in the portfolio, since threads can cooperate by exchanging information on the lower and upper bounds found during the search, as well as exchanging learned clauses that can prune the search on the other threads. In this paper we focus on strategies for the sharing of learned clauses between threads that can be used for improving parallel MaxSAT solvers. It is assumed that the reader is familiar with algorithms for MaxSAT, and we refer to the literature [4,7,8,1,9] for details.

3   Clause Sharing Heuristics

Clause sharing heuristics can be divided into three categories: (1) static, (2) dynamic and (3) freezing. The static heuristics share clauses within a given cutoff, whereas the dynamic heuristics adjust this cutoff during the search. Alternatively, the freezing heuristics temporarily freeze shared clauses until they are expected to be useful.

3.1   Static

The static heuristics are the most used for clause sharing since they are simple but still efficient in practice. The following measures are used in these heuristics:

- *Size*: the clause size is given by the number of literals. Small clauses are expected to be more useful than larger clauses.

- *Literal Block Distance* (LBD) [3]: the literal block distance corresponds to the number of different decision levels involved in a clause. The decision level of a literal denotes the depth of the decision tree at which the corresponding variable was assigned a value. Clauses with small LBD are considered as more relevant.
- *Random*: randomly decide whether to share each learned clause with a given probability. This heuristic was designed to evaluate the other heuristics which are expected to be more effective than a random one.

3.2 Dynamic

The size of learned clauses tends to increase over time. Consequently, in parallel solving, any static limit may lead to halting the clause sharing process. Therefore, to continue sharing learned clauses it is necessary to dynamically increase the limit during search. Hamadi et al. [6] proposed the following dynamic heuristic. At every k conflicts (corresponding to a period α) the throughput of shared clauses is evaluated between each pair of threads $(t_i \rightarrow t_j)$ according to the following heuristic:

$$\text{limit}_{t_i \rightarrow t_j}^{\alpha+1} = \begin{cases} \text{limit}_{t_i \rightarrow t_j}^{\alpha} + \text{quality}_{t_i \rightarrow t_j}^{\alpha} \times \frac{b}{\text{limit}_{t_i \rightarrow t_j}^{\alpha}} & \text{if sharing is small} \\ \text{limit}_{t_i \rightarrow t_j}^{\alpha} - (1 - \text{quality}_{t_i \rightarrow t_j}^{\alpha}) \times a \times \text{limit}_{t_i \rightarrow t_j}^{\alpha} & \text{if sharing is large} \end{cases},$$

where a and b are positive constants and the value of $\text{quality}_{t_i \rightarrow t_j}^{\alpha}$ corresponds to the quality of shared clauses that were send from t_i to t_j.

A shared clause is said to have *quality* [6] if at least half of its literals are active. A literal is *active* if its VSIDS heuristic [11] score is high, i.e. it is likely to be chosen as a decision variable in the near future. Hence, $\text{quality}_{t_i \rightarrow t_j}^{\alpha}$ gives the ratio between quality shared clauses and the total number of shared clauses in the period α. If the quality is high then the increase (decrease) in the size limit of shared clauses will be larger (smaller). The reasoning behind this heuristic is that the information recently received from a thread t_i is qualitatively linked to the information which could be received from the same thread t_i in the near future. In our experimental setting, we have selected $a = 0.125, b = 8$ and $\alpha = 3000$ conflicts. The throughput at each period is set to 750, i.e. if a thread t_j receives less than 750 shared learned clauses in the period α, it increases the limit of the size of shared clauses. Otherwise, this limit is decreased. These parameters are similar to the ones used by Hamadi et al. [6].

3.3 Freezing

Shared learned clauses may not be useful when they are imported and can actually deviate the search from the correct path. Our motivation for the freezing heuristic is to only import shared clauses when they are expected to be useful in the near future. Figure 1 illustrates the freezing procedure. Each shared clause ω is evaluated to determine if it will be frozen or imported. If ω is frozen then it will be reevaluated later. However, if ω is assigned the frozen state more than k times it is permanently deleted. When evaluating ω, our goal is to import clauses that are unsatisfied or that will become unit clauses in the near future. Next, the freezing heuristic is presented. According to the *status* of ω (satisfied, unsatisfied, unit or unresolved), whether ω should be frozen is decided:

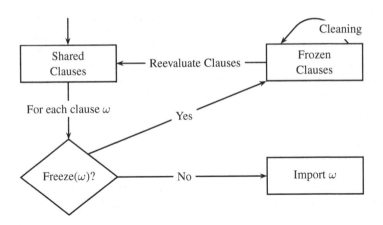

Fig. 1. Freezing procedure for sharing learned clauses

- ω is *satisfied*: Let *level* denote the current decision level, $level_h(\omega)$ the highest decision level of the satisfied literals in ω, *unassignedLits*(ω) the number of unassigned literals in ω and *activeLits*(ω) the number of active literals in ω. If $(level - level_h(\omega) \leq a)$ and $(unassignedLits(\omega) - activeLits(\omega) \leq b)$ then ω is imported, otherwise it is frozen. A satisfied clause is expected to be useful in the near future if it is not necessary to backtrack significantly to make the clause unit. It is also important that the number of unassigned literals is small or else the clause may not become unit in the near future. Active literals are also taken into consideration since they will be assigned in the near future.
- ω is *unsatisfied* or *unit*: ω is always imported;
- ω is *unresolved*: if $(unassignedLits(\omega) - activeLits(\omega) \leq b)$ then the clause is imported. Otherwise, it is frozen. Similarly to the satisfied case, if the number of unassigned literals is small then ω is likely to be unit in the near future.

In our experimental setting, we have selected $a = 31, b = 5$ and $k = 7$. In addition, the frozen clauses are reevaluated every 300 conflicts. These parameters were experimentally tuned. Freezing learned clauses has been recently proposed in the context of deletion strategies for learned clauses in SAT solving [2]. However, to the best of our knowledge, freezing shared clauses in a parallel solving context is a novel approach.

4   Experimental Results and Discussion

All experiments were run on the partial MaxSAT instances from the industrial category of the MaxSAT Evaluation 2011[1]. Instances that are easily solved will have similar solving times with and without sharing learned clauses. Hence, if an instance takes less than 60 seconds to be solved it is not considered. The evaluation was performed on two AMD Opteron 6172 processors (2.1 GHz with 64 GB of RAM) running Fedora

---

[1] http://www.maxsat.udl.cat/11/

Table 1. Comparison of the different heuristics for sharing learned clauses

	Heuristic	#Solved	Avg. #Clauses	Avg. Size	Time	Speedup
	No sharing	137	–	–	32,188.57	1.00
Static	Random 30	134	10,140.22	128.21	27,394.46	1.18
	LBD 5	137	8,947.36	9.94	25,346.69	1.27
	Size 8	137	7,529.18	5.30	25,098.85	1.28
	Size 32	138	18,027.48	11.76	25,174.29	1.28
	Dynamic	138	13,296.28	7.33	24,218.84	1.33
	Freezing	140	16,228.53	11.01	21,611.21	1.49

Core 13 with a timeout of 1,800 seconds (wall clock time). The different clause sharing heuristics were implemented on top of the portfolio version of PWBO [9] and were run with 4 threads. To have a better understanding of the impact of each heuristic, we have built a deterministic version of PWBO that is based on exchanging only information between the different threads at synchronization points (at every 100 conflicts). This is similar to what has been done in the deterministic version of MANYSAT [5].

Table 1 compares the different heuristics regarding the number of solved instances, average number of imported clauses by each thread, average size of imported clauses, solving time and speedup. Despite the number of solved instances not changing significantly, randomly sharing clauses can deteriorate the performance of the solver. Note that the solving time presented in table 1 only considers instances that were solved by all heuristics. LBD and size heuristics have similar speedups. Other sizes were also evaluated. It was observed that if the limit is too small then the speedup is reduced since not many clauses are shared. On the other hand, if the limit is too large then the speedup is also reduced since many irrelevant clauses are shared. Nevertheless, a size limit of 32 is comparable to a limit of 8 since there are some instances where learning larger clauses can be more useful than just learning smaller clauses. The dynamic heuristic outperforms the static heuristics but is outperformed by the freezing heuristic.

To summarize, although sharing learned clauses does not improve the number of solved instances significantly, it does reduce the solving time considerably. The freezing heuristic clearly outperforms all other heuristics in terms of solving time and number of instances solved and provides a strong stimulus for further research.

5   Conclusions

Recently, new parallel algorithms have been proposed for SAT as well as for MaxSAT. One of the main goals of parallel algorithms is to be able to take advantage of new multicore computer architectures by running several threads at the same time. One way to speedup these algorithms is to be able to share learned clauses between the different threads, thus allowing the pruning of the search space already explored in other threads.

In this paper different sharing heuristic procedures already proposed for SAT are described and used in a MaxSAT parallel solver. Moreover, a new heuristic based on the notion of freezing is proposed. This heuristic delays the import of shared clauses by a given thread until it is considered relevant in the context of its own search.

Experimental results show that sharing learned clauses in a portfolio-based parallel MaxSAT solver does not increase significantly the number of solved instances. However, it does allow a considerable reduction of the solving time. Finally, our proposed freezing heuristic outperforms all other heuristics both in solving time and number of solved instances.

As future work one might consider the aggregation of several heuristic criteria. Variations of the freezing heuristic can also be devised that take into consideration other information from the context of the search space being explored in the importing thread.

Acknowledgements. This work was partially supported by FCT under research project iExplain (PTDC/EIA-CCO/102077/2008), and INESC-ID multiannual funding through the PIDDAC program funds.

References

1. Ansótegui, C., Bonet, M.L., Levy, J.: Solving (Weighted) Partial MaxSAT through Satisfiability Testing. In: Kullmann, O. (ed.) SAT 2009. LNCS, vol. 5584, pp. 427–440. Springer, Heidelberg (2009)
2. Audemard, G., Lagniez, J.-M., Mazure, B., Saïs, L.: On Freezing and Reactivating Learnt Clauses. In: Sakallah, K.A., Simon, L. (eds.) SAT 2011. LNCS, vol. 6695, pp. 188–200. Springer, Heidelberg (2011)
3. Audemard, G., Simon, L.: Predicting Learnt Clauses Quality in Modern SAT Solvers. In: International Joint Conferences on Artificial Intelligence, pp. 399–404 (2009)
4. Fu, Z., Malik, S.: On Solving the Partial MAX-SAT Problem. In: Biere, A., Gomes, C.P. (eds.) SAT 2006. LNCS, vol. 4121, pp. 252–265. Springer, Heidelberg (2006)
5. Hamadi, Y., Jabbour, S., Piette, C., Sais, L.: Deterministic Parallel DPLL: System Description. In: Pragmatics of SAT Workshop (2011)
6. Hamadi, Y., Jabbour, S., Sais, L.: Control-Based Clause Sharing in Parallel SAT Solving. In: International Joint Conferences on Artificial Intelligence, pp. 499–504 (2009)
7. Li, C.M., Manyà, F.: MaxSAT, Hard and Soft Constraints. In: Handbook of Satisfiability, pp. 613–631. IOS Press (2009)
8. Manquinho, V., Marques-Silva, J., Planes, J.: Algorithms for Weighted Boolean Optimization. In: Kullmann, O. (ed.) SAT 2009. LNCS, vol. 5584, pp. 495–508. Springer, Heidelberg (2009)
9. Martins, R., Manquinho, V., Lynce, I.: Exploiting Cardinality Encodings in Parallel Maximum Satisfiability. In: International Conference on Tools with Artificial Intelligence, pp. 313–320 (2011)
10. Martins, R., Manquinho, V., Lynce, I.: Parallel Search for Boolean Optimization. In: RCRA International Workshop on Experimental Evaluation of Algorithms for Solving Problems with Combinatorial Explosion (2011)
11. Zhang, L., Madigan, C.F., Moskewicz, M.W., Malik, S.: Efficient Conflict Driven Learning in Boolean Satisfiability Solver. In: International Conference on Computer-Aided Design, pp. 279–285 (2001)

An Intelligent Hyper-Heuristic Framework for CHeSC 2011

Mustafa Mısır[1,2], Katja Verbeeck[1,2], Patrick De Causmaecker[2], and Greet Vanden Berghe[1,2]

[1] CODeS, KAHO Sint-Lieven
{mustafa.misir,katja.verbeeck,greet.vandenberghe}@kahosl.be
[2] CODeS, Department of Computer Science, K.U.Leuven Campus Kortrijk
patrick.decausmaecker@kuleuven-kortrijk.be

Abstract. The present study proposes a new selection hyper-heuristic providing several adaptive features to cope with the requirements of managing different heuristic sets. The approach suggested provides an intelligent way of selecting heuristics, determines effective heuristic pairs and adapts the parameters of certain heuristics online. In addition, an adaptive list-based threshold accepting mechanism has been developed. It enables deciding whether to accept or not the solutions generated by the selected heuristics. The resulting approach won the first Cross Domain Heuristic Search Challenge against 19 high-level algorithms.

1 Introduction

Selection hyper-heuristics are high-level search and optimisation strategies operating on a set of low-level heuristics in a problem-independent manner [1,2]. In this study, we developed an intelligent selection hyper-heuristic to deal with different heuristic sets for six problem domains provided by HyFlex [3]. The development phase consists of determining the generality requirements, designing effective components with online adaptation skills for these requirements, and combining them using certain decision mechanisms. The empirical results showed that the proposed approach is capable of delivering high performance over the tested problems.

2 Methodology

2.1 Adaptive Dynamic Heuristic Set Strategy

The adaptive dynamic heuristic set (ADHS) strategy [4,5] is responsible for monitoring the performance of each heuristic to determine elite heuristic subsets for consecutive iteration blocks, each referring to one particular phase. The underlying motivation is to specify the best performing heuristics that will be used during a number of phases to make the heuristic selection process easier. A performance metric (pm_i) based on simple quality indicators such as improvement

Y. Hamadi and M. Schoenauer (Eds.): LION 6, LNCS 7219, pp. 461–466, 2012.
© Springer-Verlag Berlin Heidelberg 2012

capability and speed, is used to decide upon exclusion of a heuristic. Equation 1 illustrates how the performance of heuristic i is measured. In this equation, $C_{p,best}(i)$ is the number of new best solutions explored. $f_{imp}(i)$ and $f_{wrs}(i)$ show the total improvement and worsening during the whole run. $f_{p,imp}(i)$ and $f_{p,wrs}(i)$ indicate the improvement and worsening provided during the current phase. t_{remain} is the remaining execution time. $t_{spent}(i)$ refers to the total time spent until that moment and $t_{p,spent}(i)$ demonstrates the execution time spent during the current phase. For each contributing performance element, a weight w_i is utilised. The values of these weights are set in a decreasing manner. The weights are sufficiently different to manage them in order of importance.

$$
pm_i = w_1\left[\left(C_{p,best}(i)+1\right)^2\left(t_{remain}/t_{p,spent}(i)\right)\right] \times b +
$$
$$
w_2\left(f_{p,imp}(i)/t_{p,spent}(i)\right) - w_3\left(f_{p,wrs}(i)/t_{p,spent}(i)\right) +
$$
$$
w_4\left(f_{imp}(i)/t_{spent}(i)\right) - w_5\left(f_{wrs}(i)/t_{spent}(i)\right) \tag{1}
$$
$$
b = \begin{cases} 1, & \sum_{i=0}^{n} C_{p,best}(i) > 0 \\ 0, & otw. \end{cases}
$$

The corresponding p_i values are ranked and a quality index $(QI \in [1,n])$ value is determined for each heuristic based on this ranking as a normalisation of the p_i values. The best performing heuristic gets the highest QI that is the number of heuristics (n) currently available. The QI values of the remainder of the heuristics decrease by 1 for each ranking level and the heuristic with the lowest p_i value has $QI = 1$. The heuristics with a QI less than the average of QIs are excluded, which means that it will not be called upon for a number of phases. These excluded heuristics have also $QI = 1$. If a heuristic is consecutively excluded, its tabu duration is incremented by 1. Alternatively, if a heuristic is not excluded after performing a phase, its tabu duration is set back to the initial value. This incrementation continues until the corresponding tabu duration reaches its upper bound, which is set to $2\sqrt{2n}$. Whenever the tabu duration is equal to its upper bound, ADHS permanently excludes this heuristic.

The phase length (pl) is set to $(d \times ph_{factor})$ iterations. ph_{factor} is a predetermined constant and it is set to 500. For instance, if the number of heuristics in the heuristic set is 10, then the tabu duration is set as $d = 4$ and pl is 2000 iterations. Whenever the heuristic subset is updated, pl is adjusted with respect to the average time required for performing a move by a non-tabu heuristic. This adjustment was performed based on the number of phases requested $(ph_{requested} = 100)$ which is a predefined value, as illustrated in Equation 2. $C_{moves}(i)$ shows the number of times heuristic i is called and t_{total} indicates the total execution time. The calculated value is constantly checked to keep it within its bounds, $pl \in [d \times 50, d \times ph_{factor}]$.

$$
pl = \left(t_{total}/ph_{requested}\right)/\sum_{i=0}^{n}\left(t_{spent}(i)/C_{moves}(i)\right).isTabu(i) \tag{2}
$$

Extreme Heuristic Exclusion. Some of the heuristics which did not find new best solutions during a phase are additionally excluded based on Equation 3 at the end of each phase. The idea behind this extra exclusion procedure is to fasten the search process by eliminating slow heuristics compared to the speed of the other heuristics in the heuristic set. The standard deviation (σ) and the average (ϖ) of the $exc(i)$ values together with the number of new best solutions (nb) found by the heuristics in the heuristic set are used for this additional exclusion as shown in Equation 4.

$$exc(i) = \Big(t_{spent}(i)/C_{moves}(i)\Big)/\Big(t_{spent}(fastest)/C_{moves}(fastest)\Big) \tag{3}$$

$$\sigma > 2.0 \;;\; exc(i) > 2\varpi \;;\; nb > 1 \tag{4}$$

Heuristic Selection. In order to choose a heuristic from the heuristic subset, a probability vector is maintained. The selection probabilities of the heuristics are the normalisation of the calculated values based on Equation 5.

$$pr_i = \big((C_{best}(i) + 1)/t_{spent}\big)^{(1+3tf^3)} \tag{5}$$

$$tf = (t_{total} - t_{elapsed})/t_{total}$$

2.2 Relay Hybridisation

The hyper-heuristic also investigates a simple relay hybridisation approach to determine effective pairs of heuristics that are applied consecutively. The details of this approach are presented in Algorithm 1. C_{phase} denotes the number of iterations that have been executed during the current phase. $C_{best,s}$ is a counter regarding the number of new best solutions found by the single heuristic selection method. $C_{best,r}$ is another counter for the number of new best solutions found by the relay hybridisation. p is a random variable to decide upon using relay hybridisation. p' is another random variable for choosing the second heuristic. $list_i$ indicates the list of heuristics to be applied after heuristic i. The size of the list is set to 10 for each heuristic. The choice of the first heuristic is made by a learning automaton (LA) that keeps a probability list with the selection probabilities of the first heuristics [6]. A *linear reward-inaction* update scheme is used for updating the probabilities as indicated in Equation 6 and 7. In these equations, the learning rates are set as $\lambda_1 = 0.5$ and $\lambda_2 = 0$. This update scheme increases the probability of a heuristic that has found new best solutions.

In addition, the tabu approach used for ADHS is applied to disable relay hybridisation if it could not deliver a new best solution after a phase.

$$p_i(t+1) = p_i(t) + \lambda_1\, \beta(t)(1 - p_i(t))$$

$$-\lambda_2(1 - \beta(t))p_i(t) \tag{6}$$

$$\text{if } a_i \text{ is the action taken at time step } t$$

Algorithm 1. Relay hybridisation

Input: $list_{size} = 10; \gamma \in (0.02, 50); p, p' \in [0:1]$
1 $\gamma = (C_{best,s} + 1)/(C_{best,r} + 1)$
2 **if** $p \leq (C_{phase}/pl)^{\gamma}$ **then**
3 | select LLH using a LA and apply to $S \rightarrow S'$
4 | **if** $size(list_i) > 0$ *and* $p' <= 0.25$ **then**
5 | | select a LLH from $list_i$ and apply to $S' \rightarrow S''$
6 | **else**
7 | | select a LLH and apply to $S' \rightarrow S''$
   | **end**
   end

$$p_j(t+1) = p_j(t) - \lambda_1 \, \beta(t) p_j(t)$$
$$+ \lambda_2 (1 - \beta(t))[(r-1)^{-1} - p_j(t)] \qquad (7)$$
$$\text{if } a_j \neq a_i$$

2.3 Heuristic Parameter Adaptation

Certain heuristics have a parameter called *"intensity of mutation"* representing the perturbation level. The other heuristics concentrating on improvement only have a parameter called *"depth of search"* related to the number of steps to be applied. A reward-penalty strategy is used to dynamically adapt these parameters.

2.4 Adaptive Iteration Limited List-Based Threshold Accepting

Adaptive iteration limited list-based threshold accepting (AILLA) is a move acceptance mechanism providing an adaptive diversification strategy in connection with the quality of the explored new best solutions earlier [4,7,5]. Its details are presented in Algorithm 2.

The iteration limit (k) is updated as shown in Equation 8. For the list length (l), the update rule presented in Equation 9 is utilised ($l_{base} = 5$, $l_{initial} = 10$).

$$k = \begin{cases} ((l-1) \times k + iter_{elapsed})/l, & \text{if } cw = 0 \\ ((l-1) \times k + \sum_{i=0}^{cw} k \times 0.5^i \times tf)/l, & \text{otherwise} \end{cases} \qquad (8)$$
$$cw = iter_{elapsed}/k$$
$$l = l_{base} + (l_{initial} - l_{base} + 1)tf^3 \qquad (9)$$

Re-initialisation. The threshold level ($best_{list}(i)$) starts from the lowest value and increases to the value placed in the l th location of the list. Each time the threshold level reaches value l, a new initial solution is randomly generated to find new best solutions in a faster way. Re-initialisation is disabled depending on the remaining execution time, its cost and the possibility of finding a new best solution afterwards.

Algorithm 2. AILLA move acceptance

Input: $i = 1, K \geq k \geq 0, l > 0$
    for $i=0$ **to** $l\text{-}1$ **do** $best_{list}(i) = f(S_{initial})$
1 **if** $adapt\_iterations \geq K$ **then**
2 | **if** $i < l - 1$ **then**
3 | | $i + +$
     | **end**
  end
4 **if** $f(S') < f(S)$ **then**
5 | $S \leftarrow S'$
6 | $w\_iterations = 0$
7 | **if** $f(S') < f(S_b)$ **then**
8 | | $i = 1$
9 | | $S_b \leftarrow S'$
10 | | $w\_iterations = adapt\_iterations = 0$
11 | | $best_{list}.remove(last)$
12 | | $best_{list}.add(0, f(S_b))$
     | **end**
13 **else if** $f(S') = f(S)$ **then**
14 | $S \leftarrow S'$
15 **else**
16 | $w\_iterations + +$
17 | $adapt\_iterations + +$
18 | **if** $w\_iterations \geq k$ and $f(S') \leq best_{list}(i)$ **then**
19 | | $S \leftarrow S'$ and $w\_iterations = 0$
     | **end**
  end

Table 1. CHeSC 2011 competition ranking and scores

Algorithm	Overall Score
ADAPHH (Our method)	181
VNS-TW	134
ML	131.5
PHUNTER	93.25
EPH	89.75
HAHA	75.75
NAHH	75
ISEA	71
KSATS-HH	66.5
HAEA	53.5
ACO-HH	39
GenHive	36.5
DynILS	27
SA-ILS	24.25
XCJ	22.5
AVEG-Nep	21
GISS	16.75
SelfSearch	7
MCHH-S	4.75
Ant-Q	0

3   Results and Conclusion

This study is about designing an intelligent hyper-heuristic to provide high quality performance across different optimisation problems. The hyper-heuristic presented here was submitted to the first international Cross-domain Heuristic Search Challenge (CHeSC 2011) to show its generality and robustness across multiple problem domains. It ended up as the competition winner out of 20 submissions. The performance of the competing algorithms were compared for five instances from six problem domains, i.e. max SAT, 1D bin packing, permutation flowshop scheduling, personnel scheduling, travelling salesman, vehicle routing. The last two domains were added to the problem set as hidden domains. The ranking and scores[1] of the corresponding algorithms are shown in Table 1. A detailed experimental analysis is available in [8].

References

1. Ozcan, E., Misir, M., Ochoa, G., Burke, E.: A reinforcement learning - great-deluge hyper-heuristic for examination timetabling. International Journal of Applied Meta-heuristic Computing 1(1), 39–59 (2010)
2. Burke, E., Gendreau, M., Hyde, M., Kendall, G., Ochoa, G., Ozcan, E., Qu, R.: Hyper-heuristics: A survey of the state of the art. Journal of the Operational Research Society (to appear)
3. Ochoa, G., Hyde, M., Curtois, T., Vazquez-Rodriguez, J.A., Walker, J., Gendreau, M., Kendall, G., McCollum, B., Parkes, A.J., Petrovic, S., Burke, E.K.: HyFlex: A Benchmark Framework for Cross-Domain Heuristic Search. In: Hao, J.-K., Middendorf, M. (eds.) EvoCOP 2012. LNCS, vol. 7245, pp. 136–147. Springer, Heidelberg (2012)
4. Misir, M., Verbeeck, K., De Causmaecker, P., Vanden Berghe, G.: A new hyper-heuristic implementation in HyFlex: a study on generality. In: Fowler, J., Kendall, G., McCollum, B. (eds.) The 5th Multidisciplinary International Scheduling Conference: Theory & Applications (MISTA 2011), Phoenix/Arizona, USA, pp. 374–393 (2011)
5. Misir, M., Smet, P., Verbeeck, K., Vanden Berghe, G.: Security personnel routing and rostering: a hyper-heuristic approach. In: Gunalay, Y., Kadipasaoglu, S. (eds.) Proceedings of the 3rd International Conference on Applied Operational Research (ICAOR 2011), Istanbul, Turkey. LNMS, vol. 3, pp. 193–205 (2011)
6. Misir, M., Wauters, T., Verbeeck, K., Vanden Berghe, G.: A Hyper-heuristic with Learning Automata for the Traveling Tournament Problem. In: Metaheuristics: Intelligent Decision Making, the 8th Metaheuristics International Conference - Post Conference Volume. Springer (to appear)
7. Misir, M., Vancroonenburg, W., Vanden Berghe, G.: A selection hyper-heuristic for scheduling deliveries of ready-mixed concrete. In: Proceedings of the 9th Metaheuristic International Conference (MIC 2011), Udine, Italy (2011)
8. Misir, M., Verbeeck, K., De Causmaecker, P., Vanden Berghe, G.: Design and analysis of an evolutionary selection hyper-heuristic framework. Tech. report, KAHO Sint-Lieven (2011)

---

[1] http://www.asap.cs.nott.ac.uk/chesc2011/results.html

An Efficient Meta-heuristic Based on Self-control Dominance Concept for a Bi-objective Re-entrant Scheduling Problem with Outsourcing

Atefeh Moghaddam*, Farouk Yalaoui, and Lionel Amodeo

Charles Delaunay Institute (ICD-LOSI), University of Technology of Troyes,
STMR, UMR CNRS 6279,
12 rue Marie Curie, 10010 Troyes, France
{atefeh.moghaddam,farouk.yalaoui,lionel.amodeo}@utt.fr

Abstract. We study a two-machine re-entrant flowshop scheduling problem in which the jobs have strict due dates. In order to be able to satisfy all customers and avoid any tardiness, scheduler decides which job shall be outsourced and find the best sequence for in-house jobs. Two objective functions are considered: minimizing total completion time for in-house jobs and minimizing outsource cost for others. Since the problem is NP-hard, an efficient genetic algorithm based on modified self-control dominance concept with adaptive generation size is proposed. Non-dominated solutions are compared with classical NSGA-II regarding different metrics. The results indicate the ability of our proposed algorithm to find a good approximation of the middle part of the Pareto front.

Keywords: Scheduling, re-entrant, bi-objective, outsourcing, genetic algorithm, dominance area.

1   Introduction

In today's competitive market, one of the most important survival factor for a company is the achievement of customer satisfaction which guarantees its long-run financial performance. Due to the resource constraints and clients' requirements, manufacturers are not always able to meet customers' due dates so tardiness is occurred. Outsourcing is an alternative to avoid loosing clients. In this paper, we study a bi-objective two-machine re-entrant permutation flowshop scheduling problem in which completing an order after its due date is not allowable so that order will be outsourced. Recent literature surveys on re-entrant scheduling problems and outsourcing can be found in [1] and [2] respectively.

The system that we study in this paper is illustrated in Fig. 1. For each job, its processing time on both machines on both cycles, due date and outsourcing cost are known in advance and the processing route for in-house jobs

---

* Corresponding author.

Y. Hamadi and M. Schoenauer (Eds.): LION 6, LNCS 7219, pp. 467–471, 2012.
© Springer-Verlag Berlin Heidelberg 2012

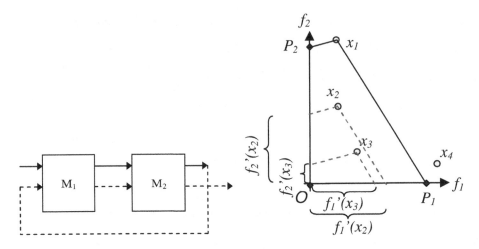

Fig. 1. Re-entrant flowshop scheme   **Fig. 2.** Dominance area regarding m-SCD

is (M_1, M_2, M_1, M_2). Other assumptions are the same as classical scheduling problem (i.e. no preemption, no break-down, ...). We are looking for a set of non-dominated solutions regarding two objective functions: minimizing total completion time for in-house jobs and minimizing outsourcing cost. Since the simpler variant of our problem is NP-hard [3], we propose a genetic-based algorithm inspired from NSGA-II [4] with different dominance concept than Pareto and adaptive generation size. The proposed dominance concept is the modification of Self-Control Dominance Area of Solutions (S-CDAS) introduced by Sato et al. [5].

We show that our algorithm is able to find limited number of non-dominated solutions compared to NSGA-II but with higher quality.

2   Modified Self Control Dominance Concept

Since many years ago, Pareto-dominance concept has been integrated to different meta-heuristics to find a good estimation of set of non-dominated solutions. Recently, it has been shown that other dominance properties rather than Pareto, may help algorithms to find better estimation of non-dominated solutions, e.g., Self-Control Dominance Area of Solutions [5] and Lorenz-dominance [6].

Self-Control Dominance Area of Solutions (S-CDAS) introduced for the first time by Sato et al. [5]. They showed that by integrating S-CDAS into NSGA-II, better estimation of non-dominated solutions for multi-objective 0-1 knapsack problem could be found compared to NSGA-II with Pareto dominance. In this paper we introduce a new dominance concept inspired from S-CDAS which is called modified SCD (m-SCD). In S-CDAS the objective is to find better estimation of *whole* Pareto front so the parameters are set in a way to keep the extrema in each Pareto non-dominated front. On the contrary, in m-SCD we focus our search on the middle part of the Pareto front. The trade-off characteristic renders such area of particular interest in practical applications. So we try to find better estimation of non-dominated solutions located in this area.

In m-SCD, the concept is to make Pareto-non-dominated solutions different one from another by inducing more fine-grained ranking in order to be converged into the middle part of Pareto-front.

Definition. modified Self-Control Dominance (m-SCD)
In a bi-objective minimization problem, solution x dominates solution y based on m-SCD properties ($x \prec_{m-SCD} y$) if *one* of the following statements holds true:

- x dominates y in Pareto sense ($x \prec_P y$); or
- x and y are Pareto-equivalent and $SCD(x) \prec_P SCD(y)$ where $SCD(x) = (f_1'(x), f_2'(x))$ in which $f_1'(x)$ and $f_2'(x)$ are derived from non-orthogonal projection of point x onto xy-plane as described in more details in step 2-2.

In Fig. 2, although x_1, x_2, x_3, x_4 are all Pareto non-dominated solutions, x_1 and x_4 which are the extrema are dominated by x_2 and x_3 based on m-SCD concept. In the following, we describe step by step how we calculate the values of different parameters shown in Fig. 2 to reach the values of f_1' and f_2'.

Step 1: Consider a set of Pareto non-dominated solutions $X = \{x_1, x_2, ...x_k\}$. We define the parameters as below:

$$O = (f_1^{min} - \varepsilon, f_2^{min} - \varepsilon) \tag{1}$$

$$P_1 = (f_1^{max} - \varepsilon, f_2(O)) \tag{2}$$

$$P_2 = (f_1(O), f_2^{max} - \varepsilon) \tag{3}$$

where f_i^{min} (f_i^{max}) is the minimum (maximum) value of the i-th objective function in the set X and ε is a tiny constant.

Step 2: For each solution $x_j \in X$ ($j = 1, 2, ..., k$) we repeat the following steps:

Step 2-1: Find the slope of both lines (accordingly use SLOPE1 and SLOPE2 as the values) through two pairs of points x_j, P_1 and x_j, P_2. We have:

$$SLOPE1 = \frac{f_2(x_j) - f_2(P_1)}{f_1(x_j) - f_1(P_1)} \tag{4}$$

$$SLOPE2 = \frac{f_2(x_j) - f_2(P_2)}{f_1(x_j) - f_1(P_2)} \tag{5}$$

$$SCD(x_j) = (f_1'(x_j), f_2'(x_j)) = (f_1(P_1), f_2(P_2)) \tag{6}$$

In other words, x-value of P_1 and y-value of P_2 are the same as $f_1'(x_j)$ and $f_2'(x_j)$.

Step 2-2: For each solution $y \in X - \{x_j\}$, we calculate $f_1'(y)$ and $f_2'(y)$ by projecting point y onto the x-axis regarding SLOPE1 and onto the y-axis regarding SLOPE2.

$$f_1'(y) = f_1(y) + \frac{f_2(O) - f_2(y)}{SLOPE1} \tag{7}$$

$$f_2'(y) = f_2(y) + SLOPE2(f_1(O) - f_1(y)) \tag{8}$$

$$SCD(y) = (f_1'(y), f_2'(y)) \tag{9}$$

Step 2-3: Regarding the new values calculated for each member of set X, we use Pareto-dominance properties to find the solutions that dominate x_j.

3   Experimental Results

We conduct experiments on 7 randomly generated problems with 10, 15, 20, 40, 50, 70 and 100 jobs to test the performance of our proposed algorithm. Processing times are generated from the discrete uniform distribution within a range of [1,100] on both machines. Due dates of the jobs are generated using two parameters, T (tardiness factor) and R (due date range) as described in [7] with more details. In this paper we set T=0.3 and R=1.4. The outsourcing costs are calculated based on the function $exp(5 + \sqrt{a} \times b)$, where a is a random integer number within [1,80] and b is a random number within [0,1].

Our proposed algorithm is based on NSGA-II coupled with m-SCD dominance with adaptive generation size proposed by Tan et al. [8] in which chromosome repairing is done by eliminating the job with minimum outsourcing cost scheduled before first tardy job. This algorithm is compared to classical NSGA-II with Pareto dominance and fixed number of generation in which the first tardy job detected in schedule is eliminated for making the solution feasible. The results on 7 different instances show that on the average, 31% of solutions found by NSGA-II are dominated by m-SCD-NSGA-II while 4% of solutions found by m-SCD-NSGA-II are dominated by NSGA-II. The hypervolume ratio of NSGA-II to m-SCD-NSGA-II is 0.87 in average which implies that the area dominated by m-SCD-NSGA-II is larger than NSGA-II dominance area. In addition, the solutions found by m-SCD-NSGA-II are distributed more evenly than those found by NSGA-II. However, the spread of solutions in NSGA-II is significantly more than those solutions achieved by our proposed algorithm. This fact is completely in-line with our parameters definitions in m-SCD. Regarding the computational time, our proposed algorithm is in average 4 times slower than classical NSGA-II. The reason is first due to the additional computational effort for calculating dominance area based on m-SCD and also increasing in number of generation.

4   Conclusion

In this paper we studied a bi-objective two-machine re-etrant scheduling problem in which due to the strict due dates, outsourcing of the tardy jobs has been considered. We proposed a genetic-based algorithm with m-SCD dominance concept and adaptive generation size. We have shown that better estimation of non-dominated solutions could be achieved by comparing the results with NSGA-II regarding coverage, hypervolume and spacing metrics however, since in the proposed algorithm middle part of Pareto front was focused, less spread solutions were found. The results clearly indicate the ability of our proposed algorithm to find a good estimation of the solutions located in the middle part of the Pareto front.

Studying more general case of the problem and integrating the proposed idea into other meta-heuristics rather than genetic algorithm would be interesting for further investigations.

References

1. Danping, L., Lee, C.K.M.: A review of the research methodology for the re-entrant scheduling problem. International Journal of Production Research 49(8), 2221–2242 (2011)
2. Slotnick, S.A.: Order acceptance and scheduling: A taxonomy and review. European Journal of Operational Research 212, 1–11 (2011)
3. Gonzalez, T., Sahni, S.: Flowshop and jobshop schedules: complexity and approximations. Operations Research 26(1), 36–52 (1978)
4. Deb, K., Pratap, A., Agarwal, S., Meyarivan, T.: A fast and elitist multiobjective genetic algorithm: NSGA-II. IEEE Transaction on Evolutionary Computation 6(2), 182–197 (2002)
5. Sato, H., Aguirre, H.E., Tanaka, K.: Self-Controlling Dominance Area of Solutions in Evolutionary Many-Objective Optimization. In: Deb, K., Bhattacharya, A., Chakraborti, N., Chakroborty, P., Das, S., Dutta, J., Gupta, S.K., Jain, A., Aggarwal, V., Branke, J., Louis, S.J., Tan, K.C. (eds.) SEAL 2010. LNCS, vol. 6457, pp. 455–465. Springer, Heidelberg (2010)
6. Dugardin, F., Yalaoui, F., Amodeo, L.: New multi-objective method to solve re-entrant hybrid flow shop scheduling problem. European Journal of Operational Research 203, 22–31 (2010)
7. Choi, S.W., Kim, Y.D.: Minimizing total tardiness on a two-machine re-entrant flowshop. European Journal of Operational Research 199, 375–384 (2009)
8. Tan, K.C., Lee, T.H., Khor, E.F.: Evolutionary algorithms with dynamic population size and local exploration for multiobjective optimization. IEEE Transactions on Evolutionary Computation 5(6), 565–588 (2001)

A Tree Search Approach to Sparse Coding

Rui Rei[1,2], João P. Pedroso[1,2], Hideitsu Hino[3], and Noboru Murata[3]

[1] University of Porto, Faculty of Science
Rua do Campo Alegre, 1021/1055, 4169-007 Porto, Portugal
rui.rei@dcc.fc.up.pt,jpp@fc.up.pt
[2] INESC Porto
Rua Dr. Roberto Frias, 378, 4200-465 Porto, Portugal
[3] Waseda University, School of Science and Engineering
3-4-1 Okubo, Shinjuku-ku, Tokyo 169-8555, Japan
hideitsu.hino@toki.waseda.jp,noboru.murata@eb.waseda.ac.jp

Abstract. Sparse coding is an important optimization problem with numerous applications. In this paper, we describe the problem and the commonly used pursuit methods, and propose a best-first tree search algorithm employing multiple queues for unexplored tree nodes. We assess the effectiveness of our method in an extensive computational experiment, showing its superiority over other methods even for modest computational time.

Keywords: Sparse Coding, Tree Search.

1 Introduction

Sparse Coding is an important optimization problem in signal processing, with applications in classification, image processing, etc. Computationally, it is an NP-hard problem [2], meaning that exact methods are inapplicable in medium- and large-scale instances of the problem. Thus, we take a look at some basic, yet extremely fast heuristic methods developed specifically for this problem. Afterwards, we propose a more sophisticated tree search method which will allow considerable improvements, depending on the allowed CPU time: good results are obtained quickly (even when comparing to greedy heuristics), and improved as more time is given to the search.

The paper is structured as follows. Section 2 describes the sparse coding problem and methods commonly used in practice. Section 3 discusses tree search methods and explains our approach to sparse coding. A computational experiment and its results are discussed in Section 4, and concluding remarks and ideas for future research are presented in Section 5.

2 Sparse Coding

In the sparse coding problem, given a dictionary defined as a matrix $D \in \mathbb{R}^{n \times k}$, comprised of k prototype signals or atoms (the columns of D) of dimension

Y. Hamadi and M. Schoenauer (Eds.): LION 6, LNCS 7219, pp. 472–477, 2012.
© Springer-Verlag Berlin Heidelberg 2012

n, and a target signal $Y \in \mathbb{R}^n$, we seek the best approximation of Y using a combination of at most $m < k$ atoms. The approximation can be written as $D \cdot X$, where $X \in \mathbb{R}^k$ is a vector containing the coefficients of the atoms of D in the linear combination, i.e., X_i is the coefficient of atom i (the i-th column-vector in D, denoted as D_i). Since we can use at most m atoms, the number of non-zero entries in X, $\|X\|_0$ (using the L^0 pseudo-norm) must be less or equal to m. So, Y can be written as

$$Y = D \cdot X + R \ , \tag{1}$$

where $R \in \mathbb{R}^n$ is a residual vector, that is, the portion of Y that cannot be represented by the linear combination of atoms $D \cdot X$. The objective of the problem is then to minimize the L^2 (Euclidean) norm of R, since $\|R\|_2 = 0$ means that we have a perfect representation of Y. The problem can be formally defined as follows:

$$\begin{aligned} \text{minimize} \quad & \|Y - D \cdot X\|_2 \\ \text{subject to} \quad & \|X\|_0 \leq m \\ & X \in \mathbb{R}^k \ . \end{aligned} \tag{2}$$

The given problem is NP-hard, which means that there are no known efficient (polynomial) algorithms to solve it exactly. Thus, several approximate methods have been proposed to find good values for the coefficients X within reasonable computational time. Some of these methods are briefly described below.

First, the Matching Pursuit (MP, [3]) algorithm is a greedy heuristic method which starts with an empty solution ($X^0 = \mathbf{0}$ and $R^0 = Y$), and at each iteration $j = 0, 1, \ldots$, the atom D_i with highest correlation with R^j (the current residual), $|D_i \cdot R^j|$, is added to the sparse support (the set of atoms with non-zero coefficients). Its coefficient X_i^{j+1} is then computed such that the updated residual $R^{j+1} = R^j - X_i^j.D_i$ is orthogonal to D_i, i.e., $R^{j+1} \perp D_i$.

Another greedy heuristic, Orthogonal Matching Pursuit (OMP, [4]), improves on MP by keeping the residual orthogonal to all atoms in the sparse support at each iteration. This way, unlike MP, OMP avoids undoing work done in previous iterations. An interesting characteristic of OMP is that, for any given support, the computed coefficients are optimal. That is, given a set of atoms, OMP computes the best possible approximation using those atoms. This characteristic turns sparse coding into a purely combinatorial problem, where one must find a subset of $S \subseteq D$ to represent Y, using the coefficient calculation method employed in OMP to provide optimal coefficients for any given set of atoms.

Both MP and OMP are deterministic algorithms, since the choice of the next atom to include in the support is fixed: the atom with highest correlation with the current residual is always chosen. These can be easily turned into probabilistic algorithms, for instance, by randomly selecting an atom in the case of a tie. However, due to the unlikelihood of ties in the atom selection step, these variants would most often produce the same solutions as their original deterministic counterparts. A third method, Randomized OMP (RandOMP), is a variant of OMP where variability is introduced into the atom selection step (in every iteration, not just for tie-breaking), allowing it to generate a variety of solutions if run multiple times. The variability can be easily exploited by generating a set of solutions and selecting the best of them.

In the next section, we propose a tree search algorithm to find good sparse supports, and later compare it to OMP and RandOMP.

3   Tree Search

For combinatorial problems, one may create a tree where different branches contain alternative decisions. The leaves of such tree correspond to the problem's search space, that is, the set of complete solutions. Thus, completely exploring the tree and finding the leaves with best objective function value corresponds to solving the problem to optimality.

However, this is not always possible due to the size of the search space. For example, in the case of sparse coding, the search space corresponds to all subsets of the k atoms with size less or equal to m. In other words, the number of possible solutions is equal to the number of combinations of k atoms taken m at a time.

Since it is infeasible to completely explore the tree even for medium-sized instances, tree search is commonly stopped after some given limit (e.g. CPU time) has been reached. In this case, since the tree could not be entirely explored, the best solution found may not be optimal, and the order of exploration of the nodes in the tree becomes very important. If promising nodes are explored first, then the chance of obtaining good-quality solutions when the search is stopped increases.

There exist several schemes for the traversal order of tree nodes. Uninformed traversal methods, such as depth-first or breadth-first, are often used, but other methods like best-first, or variants of beam search, may lead to better performance under time limitation. In the next section, we describe the base elements of a tree search specific to sparse coding, and follow with the description of our multi-queue tree search traversal scheme.

3.1   Application to Sparse Coding

A complete decision tree for sparse coding may be derived as follows. Let δ be a node in the tree, S_δ a set corresponding to the sparse support in δ, and R_δ a set containing the remaining atoms in δ, which may be chosen for inclusion in the support in subsequent decisions. Note that δ is a leaf node if $|S_\delta| = m$ or $R_\delta = \emptyset$.

At the root of the tree γ, we have $S_\gamma = \emptyset$ and $R_\gamma = \{1, \ldots, k\}$. Then, given a node δ, an atom a is chosen from R_δ and two child nodes (δ' and δ'') are created. On the first child node, atom a is included in the support and removed from the remaining set, i.e.,

$$\begin{aligned} S_{\delta'} &= S_\delta \cup \{a\} \\ R_{\delta'} &= R_\delta \setminus \{a\} \ . \end{aligned} \tag{3}$$

On the second child node, δ'', atom a is simply removed from the remaining set

$$S_{\delta''} = S_\delta$$
$$R_{\delta''} = R_\delta \setminus \{a\} \ . \tag{4}$$

Using this branching method, we create two subtrees: on the first subtree all solutions will contain atom a, and on the second subtree no solution will contain a. This will be the basic branching scheme used in our tree search algorithm.

3.2 Multi-Queue Tree Search

The proposed tree search scheme is a variant of best-first search which uses multiple queues to hold the tree nodes which are waiting to be explored. The idea of using several queues comes from the fact that sparse supports of different size are not comparable, since the larger support will in general have a smaller residual. To avoid putting together nodes containing supports of different size, a sub-queue is created for each support size from 0 to $m - 1$. No queue is created for size m because all such supports are already complete solutions.

Each time the tree search iterates, a node is picked from one of the sub-queues (in round-robin fashion), its two child nodes are generated and checked, to determine if new leaves were reached. Then, non-leaf child nodes are placed in the sub-queue corresponding to the size of their support. The specific position of a node in a sub-queue is determined by its current residual norm: nodes with smaller residuals are placed in positions closer to the front of the sub-queue, meaning that these will be explored before nodes with larger residuals. Note that this results in a per-sub-queue best-first order, using the residual norm of the partial solutions as a scoring criterion.

Using this scheme allows us to quickly find complete solutions, since nodes are taken from all sub-queues in turn, and at least one leaf node is generated from each node in the $(m - 1)$-th sub-queue. The selection of the atom to include in the branching step is taken from the OMP heuristic, i.e., the atom with highest correlation with the current residual is chosen. This way, the first solution found by our Multi-Queue Tree Search (MQTS) is identical to the one produced by OMP, which guarantees a good minimum quality level even with very little time. Since nodes are taken from all sub-queues in equal number, the search will not be trapped in a part of the tree as would occur with depth-first and sometimes best-first search. One additional benefit of tree search is that no duplicate solutions are ever analyzed, as opposed to, for example, repeated RandOMP, where the same supports may be selected in different runs.

In the next section, we describe a computational experiment designed to assess the effectiveness of MQTS, comparing it to OMP and repeated RandOMP.

4   Computational Experiment

Image encoding and compression is a common application of sparse coding, traditionally using a fixed predefined dictionary. However, the use of specially designed dictionaries is known to yield better results. In our experiment, we used the K-SVD [1] dictionary learning algorithm to build specific dictionaries for color and greyscale images. K-SVD was run for 30 iterations on two images of each set, using OMP as a pursuit algorithm in its sparse coding step, producing two dictionaries with $k = 500$ atoms, for color and greyscale images, respectively.

After generating the dictionaries, the three pursuit algorithms were run on the images, broken down into manageable patches of 16×16 pixels, with a CPU limit of 1 second per patch. Note that each patch corresponds to an instance of sparse coding. For color images with three channels, $n = 16 \times 16 \times 3 = 768$, while for greyscale images each patch has $n = 256$ since there is only one channel. For both image sets the maximum support size m was set to 10 atoms.

The experiment was run on an Intel Atom 330 1.6GHz dual-core processor with 2GB of main memory. All programs were implemented in Python, version 2.6.5.

Tables 1a and 1b show the total representation error (Frobenius norm) for all color and greyscale images, respectively. The Frobenius norm of a matrix with r rows and c columns A is given by

$$\|A\|_F = \sqrt{\sum_{i=1}^{r} \sum_{j=1}^{c} A_{ij}^2} \ . \tag{5}$$

Considering the set of patches in the original image as a matrix A (each patch being a column in A), and the set of patches in the encoded image as a matrix B, the representation error of the encoded image is then obtained by $\epsilon = \|A - B\|_F$.

Table 1. Representation error on color (left) and greyscale (right) images, given by the Frobenius norm of the difference between original and encoded images

(a) Results for color images.

Image	OMP ϵ	RandOMP ϵ	Impr. %|	MQTS ϵ	Impr. %
4.1.01	5103.17	5013.60	1.76	4970.13	2.61
4.1.02	4685.48	4620.57	1.39	4560.13	2.68
4.1.03	4692.30	4609.49	1.76	4531.38	3.43
4.1.04	5650.37	5550.54	1.77	5481.10	3.00
4.1.05	6013.83	5912.02	1.69	5814.47	3.31
4.1.06	8877.23	8713.37	1.85	8623.09	2.86
4.1.07	5202.37	5119.77	1.59	5037.36	3.17
4.1.08	6400.62	6293.05	1.68	6230.79	2.65
4.2.01	9491.35	9249.17	2.55	9143.37	3.67
4.2.02	9227.53	9083.93	1.56	9001.29	2.45
4.2.03	10594.51	10531.24	0.60	10501.82	0.87
4.2.04	7642.89	7556.65	1.13	7523.30	1.56
4.2.05	11554.95	11299.07	2.21	11146.34	3.54
4.2.06	14900.17	14640.49	1.74	14475.44	2.85
4.2.07	12692.56	12458.14	1.85	12272.67	3.31
Average	8181.96	8043.41	1.67	7954.18	2.80

(b) Results for greyscale images.

Image	OMP ϵ	RandOMP ϵ	Impr. %|	MQTS ϵ	Impr. %
5.1.09	2178.22	2146.45	1.46	2104.44	3.39
5.1.10	4685.04	4577.79	2.29	4478.13	4.42
5.1.11	1966.47	1910.98	2.82	1870.58	4.88
5.1.12	3030.55	2955.66	2.47	2886.36	4.76
5.1.13	7035.98	6860.73	2.49	6635.06	5.70
5.1.14	3509.27	3428.30	2.31	3343.48	4.72
5.2.08	6319.43	6174.19	2.30	6050.19	4.26
5.2.09	8174.90	7988.63	2.28	7818.44	4.36
5.2.10	6030.22	5976.06	0.90	5912.82	1.95
5.3.01	7743.45	7627.46	1.50	7535.48	2.69
5.3.02	12019.84	11780.98	1.99	11565.14	3.78
7.1.01	3826.14	3746.98	2.07	3671.68	4.04
7.1.02	2879.91	2820.30	2.07	2756.68	4.28
7.1.03	3801.43	3743.87	1.51	3682.09	3.14
7.1.04	3360.94	3290.87	2.08	3223.30	4.10
7.1.05	5357.74	5258.38	1.85	5167.89	3.54
7.1.06	5265.03	5167.05	1.86	5075.77	3.59
7.1.07	4717.24	4636.78	1.71	4564.54	3.24
7.1.08	3202.14	3153.27	1.53	3109.18	2.90
7.1.09	4726.41	4645.52	1.71	4569.80	3.31
7.1.10	3370.25	3286.69	2.48	3229.31	4.18
7.2.01	5602.48	5552.87	0.89	5491.84	1.97
Average	4945.59	4851.35	1.93	4761.01	3.78

The individual image results indicate a consistent improvement of MQTS over both OMP and repeated RandOMP. The repeated RandOMP algorithm also proved to be better than OMP, due to its exploitation of the variability introduced in the atom selection step. As for MQTS, its superior performance even with very little CPU time indicates that the best-first search order is suitable for this problem, and the overhead of maintaining a search tree and the algorithm's additional complexity do not represent a significant burden. Additionally, this overhead should be diluted as CPU time is increased. The complete absence of symmetries in the tree (no repeated solutions) is also an advantage over RandOMP, which should manifest even more with longer run times.

A deeper analysis of the algorithm, for example by comparison with depth-first search, should allow us to conclude whether the multi-queue mechanism to avoid entrapment is effective or not.

5   Conclusion

We propose a tree search algorithm for obtaining good quality solutions to the sparse coding problem. The computational results reveal superior performance of Multi-Queue Tree Search in all test images, despite the very low CPU time budget. The algorithm quickly provides solutions of reasonable quality, improving as more time is allowed. Its performance gap over repeated Randomized Orthogonal Matching Pursuit should increase as more time is given, since solutions are analyzed exactly once.

Comparing the tree search to other metaheuristics and commercial solvers is an interesting direction for future research. Also, the performance of the algorithm could be radically improved with the use of a lower bound function, since it would allow us to discard branches of the search tree where we would be sure not to find improving solutions.

References

1. Aharon, M., Elad, M., Bruckstein, A.: K-SVD: An algorithm for designing overcomplete dictionaries for sparse representation. IEEE Transactions on Signal Processing 54(11), 4311–4322 (2006)
2. Davis, G., Mallat, S., Avellaneda, M.: Greedy adaptive approximation. Journal of Constructive Approximation 13, 57–98 (1997)
3. Mallat, S., Zhang, Z.: Matching pursuits with time-frequency dictionaries. IEEE Transactions on Signal Processing 41(12), 3397–3415 (1993)
4. Pati, Y., Rezaiifar, R., Krishnaprasad, P.: Orthogonal matching pursuit: recursive function approximation with applications to wavelet decomposition. In: 1993 Conference Record of The Twenty-Seventh Asilomar Conference on Signals, Systems and Computers, vol. 1, pp. 40–44 (November 1993)

Adaptive Control of the Number of Crossed Genes in Many-Objective Evolutionary Optimization

Hiroyuki Sato[1], Carlos A. Coello Coello[2],
Hernán E. Aguirre[3], and Kiyoshi Tanaka[3]

[1] Faculty of Informatics and Engineering,
The University of Electro-Communications 1-5-1 Chofugaoka,
Chofu, Tokyo 182-8585 Japan
[2] Departamento de Computación, CINVESTAV-IPN
Av. IPN No. 2508, México, D.F. 07360, México
[3] Faculty of Engineering, Shinshu University
4-17-1 Wakasato, Nagano, 380-8553 Japan

Abstract. To realize effective genetic operation in evolutionary many-objective optimization, crossover controlling the number of crossed genes (CCG) has been proposed. CCG controls the number of crossed genes by using an user-defined parameter α. CCG with small α significantly improves the search performance of multi-objective evolutionary algorithm in many-objective optimization by keeping small the number of crossed genes. However, to achieve high search performance by using CCG, we have to find out an appropriate parameter α by conducting many experiments. To avoid parameter tuning and automatically find out an appropriate α in a single run of the algorithm, in this work we propose an adaptive CCG which adopts the parameter α during the solutions search. Simulation results show that the values of α controlled by the proposed method converges to an appropriate value even when the adaptation is started from any initial values. Also we show the adaptive CCG achieves more than 80% with a single run of the algorithm for the maximum search performance of the static CCG using an optimal α^*.

1 Introduction

The research interest of the multi-objective evolutionary algorithm (MOEA) [1] community has rapidly shifted to develop effective algorithms for many-objective optimization problems (MaOPs) because more objective functions should be considered and optimized in recent complex applications. However, in general, MOEAs noticeably deteriorate their search performance as we increase the number of objectives [2], especially Pareto dominance-based MOEAs such as NSGA-II and SPEA2. One reason for this is that these MOEAs face difficulty to rank solutions in the population, i.e., most of the solutions become non-dominated and the same rank is assigned to them, which seriously spoils proper selection pressure required in the evolution process. To overcome this problem in

Y. Hamadi and M. Schoenauer (Eds.): LION 6, LNCS 7219, pp. 478–484, 2012.
© Springer-Verlag Berlin Heidelberg 2012

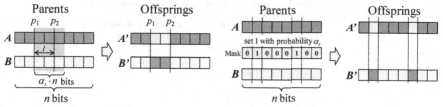

Fig. 1. Controlling crossed genes for two-point crossover (CCG$_{TX}$)

Fig. 2. Controlling crossed genes for uniform crossover (CCG$_{UX}$)

selection, several methods to improve selection pressure have been proposed [2]. Contrary to these studies focusing on selection, to realize effective genetic operation in MaOPs, the variable space of many-objective 0/1 knapsack problem has been analyzed [3]. [3] shows that variables of true Pareto optimal solutions (POS) become noticeably diverse, and true POS becomes distributed almost uniformly in variable space by increasing the number of objectives. Also, [3] shows that offspring created by conventional two-point and uniform crossovers has less chance to be selected as parents because the operators becomes too disruptive and its effectiveness decreases in MaOPs. To overcome this problem and enhance the evolution by crossover operator in MaOPs, controlling the number of crossed genes (CCG) has been proposed [3]. CCG$_{TX}$, an extension of two-point crossover, controls the maximum length of crossed genes by using an user-defined parameter α_t. Also, CCG$_{UX}$, an extension of uniform crossover, controls the number of crossed genes by using a parameter α_u. In MaOPs, CCG using a small α remarkably improves the search performance of several MOEAs [3]. However, to achieve high search performance by using CCG, we have to find out an appropriate parameter α by conducting many experiments.

To avoid time consuming parameter tuning and automatically find out an appropriate number of crossed genes in a single run of the algorithm, in this work we propose an adaptive CCG which adopts the parameter α during the solutions search. In this work, we analyze the adaptation process of α and verify the effectiveness of adaptive CCG$_{TX}$ and CCG$_{UX}$ on many-objective 0/1 knapsack problems with $m = \{4, 6, 8, 10\}$ objectives.

2   Controlling the Number of Crossed Genes

2.1   CCG for Two-Point Crossover (CCG$_{TX}$)

CCG$_{TX}$ controls the length of crossed genes by using a parameter α_t. Fig.1 shows the conceptual diagram of CCG$_{TX}$. First we select parents A and B from the parent population $\mathcal{P}$, and randomly choose the 1st crossover point p_1. Then, we randomly determine the length of the crossed genes ℓ in the range $[0, \alpha_t \cdot n]$. The 2nd crossover point is set to $p_2 = (p_1 + \ell) \mod n$. The possible range of the parameter α_t is $[0.0, 1.0]$. $\alpha_t = 1.0$ indicates the conventional two-point crossover, and the length of crossed genes becomes short by decreasing α_t.

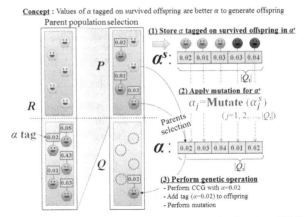

Fig. 3. Conceptual figure of the proposed adaptive CCG

2.2 CCG for Uniform Crossover (CCG$_{UX}$)

CCG$_{UX}$ controls the probability of 1 in the mask bits by using a parameter α_u. According to the generated mask bits, we perform uniform crossover as shown in Fig.2. The possible range of α_u is $[0, 0.5]$. $\alpha_u = 0.5$ indicates the typical uniform crossover, and the number of crossed genes becomes small by decreasing α_u.

3   Adaptive Control of the Number of Crossed Genes

CCG using a small α remarkably improves the search performance of several MOEAs in MaOPs [3]. However, to achieve high search performance by the conventional CCG using a static parameter α for the entire solutions search [3], we have to find out an appropriate α by conducting many experiments. To avoid time consuming parameter tuning and automatically find out an appropriate α while achieving high search performance in a single run of the algorithm, in this work we propose an adaptive CCG which adopts α so that the parameter is automatically guided to an appropriate value during the solutions search in a single run of the algorithm.

Fig.3 shows the conceptual figure of the proposed adaptive CCG. Since this method is designed based on a framework used in NSGA-II and S-CDAS [3], entire population $\mathcal{R}$ consists of parent (elite) population $\mathcal{P}$ and offspring population $\mathcal{Q}$. In the process of adaptive CCG, we use two vectors. The first one is $\boldsymbol{\alpha^s} = \{\alpha_1^s, \alpha_2^s, \cdots, \alpha_{|\mathcal{Q}|}^s\}$, in which effective values of α are kept to create superior offspring. The other one is $\boldsymbol{\alpha} = \{\alpha_1, \alpha_2, \cdots, \alpha_{|\mathcal{Q}|}\}$, which is used to generate offspring for the next generation. Also, for all solutions in the offsprings population $\mathcal{Q}$, we individually put a tag showing α used to create each solution. Before we start the solutions search, we initialize α_j^s ($j = 1, 2, \cdots, |\mathcal{Q}|$) by initial α_i. Also, we initialize the counter that measures the number of survived offspring c by 0. For each generation, we update the elements in $\boldsymbol{\alpha^s}$. After selection of new parents population $\mathcal{P}$, we pick up one survived offspring from $\mathcal{P}$. Then, we increment c and replace the element $\alpha_{1+c \bmod |\mathcal{Q}|}^s$ with the value of α tagged on

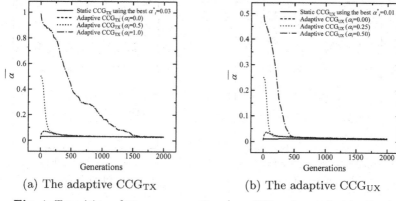

(a) The adaptive CCG$_{TX}$ (b) The adaptive CCG$_{UX}$

Fig. 4. Transition of $\overline{\alpha}$ over generation ($n = 500$ and $m = 8$ objectives)

the current offspring. We repeat this process for all survived offspring in $\mathcal{P}$. Next, we determine the value of α_j $(j = 1, 2, \cdots, |\mathcal{Q}|)$ in $\boldsymbol{\alpha}$ by applying polynomial mutation [4] to all the elements in $\boldsymbol{\alpha^s}$. Finally, we create offspring by performing CCG using the updated elements in $\boldsymbol{\alpha}$ to fill up new offsprings population $\mathcal{Q}$.

4 Experimental Results and Discussion

4.1 Problems, Parameters and Metrics

In this work we use many-objective 0/1 knapsack problems [5] with $m = \{4, 6, 8, 10\}$ objectives, $n = 500$ items, and feasibility ratio $\phi = 0.5$. We verify the effects of CCG when it is combined with S-CDAS for parents selection similar to [3]. We adopt crossovers with a crossover rate $p_c = 1.0$, and apply bit-flipping muta-tion with a mutation rate $p_m = 1/n$. In the following experiments, we show the average performance with 30 runs, each of which spent $T = 2,000$ generations. Population size is set to $N = 200$ ($|\mathcal{P}| = 100$ and $|\mathcal{Q}| = 100$). Also, we employ polynomial mutation [4] with $\eta_m = 40$ to obtain $\boldsymbol{\alpha}$ from $\boldsymbol{\alpha^s}$.

To evaluate the search performance of MOEAs, we use *Hypervolume (HV)* [6], which measures the m-dimensional volume of the region enclosed by the obtained non-dominated solutions and a dominated reference point in objective space. Here, we use $\boldsymbol{r} = (0, 0, \cdots, 0)$ as the reference point. Obtained POS showing a higher value of hypervolume can be considered as a better set of non-dominated solutions from both convergence and diversity viewpoints.

4.2 Transition of $\overline{\alpha}$ in the Proposed Adaptive CCG

First, we observe the adaptation process of $\boldsymbol{\alpha}$ by the adaptive CCG. Fig.4 shows the transition of average $\overline{\alpha} = \sum_{j=1}^{|\mathcal{Q}|} \alpha_j / |\mathcal{Q}|$ over generations. For the adaptive CCG$_{TX}$, we plot three different results by using initial $\alpha_i = \{0.0, 0.5, 1.0\}$. For the adaptive CCG$_{UX}$, we use $\alpha_i = \{0.0, 0.25, 0.5\}$. Also, we plot $\alpha_t^* = 0.03$ and $\alpha_u^* = 0.01$ maximizing HV by the static CCG$_{TX}$ and CCG$_{UX}$ as horizontal lines.

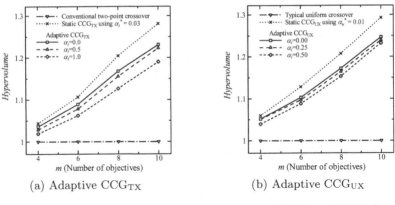

(a) Adaptive CCG$_{TX}$ (b) Adaptive CCG$_{UX}$

Fig. 5. *HV* obtained by the adaptive and static CCG (n = 500 bits)

From the result for adaptive CCG$_{TX}$ shown in Fig.4 (a), we can see that $\overline{\alpha}$ converges to a specific value even when we start adaptation from any initial α_i. Also, the converged value of $\overline{\alpha}$ is close to $\alpha_t^* = 0.03$. Convergence of $\overline{\alpha}$ by the adaptive CCG$_{TX}$ using $\alpha_i = 0.0$ is fastest among three different adaptive CCG$_{TX}$. Also, from the result for adaptive CCG$_{UX}$ shown in Fig.4 (b), we can see the similar tendency of the adaptive CCG$_{TX}$. From these results, the adaptation of α by the proposed adaptive CCG is thought to be working well.

4.3   Performance of the Proposed Adaptive CCG$_{TX}$ and CCG$_{UX}$

Next, we verify the search performance of the adaptive CCG. Fig.5 (a) shows results of *HV* obtained by the conventional two-point crossover, the adaptive CCG$_{TX}$ with $\alpha_i = \{0.0, 0.5, 1.0\}$ and the static CCG$_{TX}$ with the optimal $\alpha_t^* = 0.03$ maximizing *HV* in [3]. All results are normalized by the results of the conventional two-point crossover. Here, in the case of the static CCG$_{TX}$ with α_t^*, we only show the best result among many experiments varying α_t step by step in the possible range of $\alpha_t \in [0, 1]$. On the other hand, in the cases of the conventional two-point crossover and the proposed adaptive CCG$_{TX}$, we show results obtained by a single run of the algorithm. As De Jong mentioned in [7], note that performance comparison between EA with static parameter setting and EA with adaptive setting is unfair since it is likely that the static setting is established via preliminary parameter tuning by many experiments conducted in advance, which is not included in the comparison. Therefore, note that comparing the adaptive CCG$_{TX}$ with the static CCG$_{TX}$ using α_t^* is not fair comparison. In these graphs, we show the achievement of the search performance by the adaptive CCG$_{TX}$ between the basic performance by conventional two-point crossover and the maximum performance by the static CCG$_{TX}$ using α_t^*.

From the results of Fig.5 (a), we can see that the adaptive CCG$_{TX}$ using any α_i achieves higher *HV* than the conventional two-point crossover. Also, the adaptive CCG$_{TX}$ using smaller initial adaptation range α_i shows higher *HV*, and the adaptive CCG$_{TX}$ with $\alpha_i = 0.0$ achieves the highest *HV* among the adaptive CCG_{TX} using three different initial α_i. This is because the adaptive CCG_{TX} with $\alpha_i = 0.0$ realizes the fastest convergence of α to the optimal value as shown in Fig.4 (a). Next, by comparing results of the adaptive CCG_{TX} with the static CCG_{TX} using α^*, we can see that the adaptive CCG_{TX} using $\alpha_i = 0.0$ achieves $\{82.1, 83.7, 82.1, 82.1\}\%$ of HV for the maximum HV obtained by the static CCG_{TX} with α_t^* for $m = \{4, 6, 8, 10\}$ objectives, respectively.

Next, Fig.5 (b) shows results of HV obtained by the typical uniform crossover, the adaptive CCG_{UX} with $\alpha_i = \{0.0, 0.25, 0.5\}$ and the static CCG_{UX} with the optimal $\alpha_u^* = 0.01$ maximizing HV. All results are normalized by the results of the typical uniform crossover. From the results of Fig.5 (b), we can see the similar tendency obtained by the adaptive CCG_{TX}. The adaptive CCG_{UX} with $\alpha_i = 0.0$ achieves $\{86.6, 80.3, 83.0, 84.3\}\%$ of HV for the maximum HV obtained by the static CCG_{UX} with α_t^* for each $m = \{4, 6, 8, 10\}$ objectives problem.

5   Conclusions

To avoid parameter tuning and automatically find out an appropriate number of crossed genes in a single run of the algorithm, we have proposed the adaptive CCG which adopts the parameter α during the solutions search. Also, we have analyzed the adaptation process of α, and have verified the effectiveness of the adaptive CCG in the search performance on many-objective 0/1 knapsack problems with $m = \{4, 6, 8, 10\}$ objectives. Simulation results showed that average $\overline{\alpha}$ controlled by the adaptive CCG converges to an appropriate value even when the adaptation is started from any initial values. Also, we showed that the converged value of $\overline{\alpha}$ is close to α^* maximizing HV by the static CCG. Through performance verification, we showed the adaptive CCG_{TX} and CCG_{UX} achieve higher HV than the conventional two-point crossover and the typical uniform crossover. Also, we showed that the adaptive CCG using small initial α_i achieves more than 80% with a single run of the algorithm for the maximum HV obtained by the static CCG with α^* found through many experiments.

As future works, we want to further improve the search performance of the proposed algorithm by refining the adaptation mechanism. Also, we are planning to study on the effective crossover operators for many-objective continuous optimization problems.

References

1. Coello, C.A.C., van Veldhuizen, D.A., Lamont, G.B.: Evolutionary Algorithms for Solving Multi-Objective Problems. Kluwer Academic Publishers, Boston (2002)
2. Ishibuchi, H., Tsukamoto, N., Nojima, Y.: Evolutionary many-objective optimization: A short review. In: Proc. of 2008 IEEE Congress on Evolutionary Computation (CEC 2008), pp. 2424–2431 (2008)
3. Sato, H., Aguirre, H., Tanaka, K.: Improved S-CDAS using Crossover Controlling the Number of Crossed Genes for Many-objective Optimization. In: Proc. GECCO 2011, pp. 753–760 (2011)

4. Deb, K., Goyal, M.: A combined genetic adaptive search (GeneAS) for engineering design. Computer Science and Informatics 26(4), 30–45 (1996)
5. Zitzler, E., Thiele, L.: Multiobjective Optimization Using Evolutionary Algorithms - A Comparative Case Study. In: Eiben, A.E., Bäck, T., Schoenauer, M., Schwefel, H.-P. (eds.) PPSN 1998. LNCS, vol. 1498, pp. 292–304. Springer, Heidelberg (1998)
6. Zitzler, E.: Evolutionary Algorithms for Multiobjective Optimization: Methods and Applications. PhD thesis, Swiss Federal Institute of Technology, Zurich (1999)
7. De Jong, K.: Parameter Setting in EAs: a 30 Year Perspective. In: Parameter Setting in Evolutionary Algorithms, pp. 1–18. Springer (2007)

Counter Implication Restart for Parallel SAT Solvers

Tomohiro Sonobe and Mary Inaba

Graduate School of Information Science and Technology, University of Tokyo

Abstract. A portfolio approach has become the mainstream for parallel SAT solvers, making diversification of the search for each process more important. In the SAT Competition 2011, we proposed a novel restart method called counter implication restart (CIR), for sequential solvers and won gold and silver medals with CIR. CIR enables SAT solvers to change the search spaces drastically after a restart. In this paper, we propose an adaptation of CIR to parallel SAT solvers to provide better diversification. Experimental results indicate that CIR provides good diversification and its overall performance is very competitive with state-of-the-art parallel solvers.

1 Introduction

The Boolean satisfiability (SAT) problem asks whether an assignment of variables exists that can evaluate the given formula as true. A SAT problem is one of NP-complete problems. A formula is given in Conjunctive Normal Form (CNF), which is a conjunction of clauses. A clause is a disjunction of literals, where a literal is a positive or negative form of a variable. The solvers for this problem are called SAT solvers. The recent innovations in SAT solvers are significant and these solvers are used in many real applications, such as circuit design and software verification.

Many SAT solvers are based on the Davis-Putnam-Logemann-Loveland (DPLL) algorithm. In the last decades, conflict-driven learning and backjumping, Variable State Independent Decaying Sum (VSIDS) decision heuristic, and restart were added to DPLL, which improved the performance of DPLL solvers tremendously. These solvers are called Conflict Driven Clause Learning (CDCL) solvers. This kind of solver is now standard and it appears to be difficult to make a drastic improvement without a replacement of the fundamental algorithm.

Due to recent developments in multi-core hardware, we can easily run SAT solvers in parallel on standard PCs. However, there still appears to be a need for parallel SAT solvers. In the SAT Competition 2011, in the application category, the number of participants for the parallel category was about only ten, compared with more than 50 in the category of sequential solvers. Moreover, even though the parallel solvers were run on eight cores, the performance of the sequential solvers was very competitive with that of parallel solvers.

Many state-of-the-art parallel solvers are based on the portfolio approach [1]. In this approach, each solver runs competitively and they share learnt clauses

Y. Hamadi and M. Schoenauer (Eds.): LION 6, LNCS 7219, pp. 485–490, 2012.
© Springer-Verlag Berlin Heidelberg 2012

between them. Each solver uses a particular parameter set and conducts a differentiated but complementary search. This diversification is important for efficient searching [2]. Diversification is attained by employing, for example, differentiated restart policies [3], various strengths of saving literal polarity [4], decision heuristics, and so on.

In the SAT Competition 2011, we submitted a solver based on MiniSAT 2.2 [5] with our novel restart method, Counter Implication Restart (CIR). Our CIR enables SAT solvers to convert the search spaces by changing the decision-order after the restart, and thus enables an escape from desert search spaces [6]. This method is also valid for the diversification of the parallel SAT solver. In this paper, we propose the adaptation of CIR for use with parallel SAT solvers. Experimental results indicate that CIR also works efficiently in parallel solvers.

In Section 2, we explain the details of CIR. We show the experimental results in Section 3 and conclude the paper in Section 4.

2   Counter Implication Restart (CIR)

Existing restart policies only implement the restarting of the search from the beginning without changing anything. In many cases, this is sufficient to enable an escape from wrong branches. However, in some instances there are desert search spaces [6] where neither the solution nor the useful learnt clause exists. For such cases, it is difficult for SAT solvers to escape from these desert search spaces with a standard restart. Therefore, it is necessary to change the search activity after the restart drastically. CIR is a novel restart policy that consists of a standard restart and bumping the VSIDS scores to change the decision order after the restart. CIR traverses the implication graph [7] just before the restart, focusing on the indegrees of the variables.

A variable with a large indegree implies that this variable used to be the unit variable in a large clause. Let us consider the transformation from CSP to SAT. Suppose a variable a in the original CSP instance has a domain between 1 to n ($1 \leq a \leq n$), and its corresponding Boolean variables in the SAT instance are $a_1, a_2, ..., a_n$. There are clauses, $\prod_{1 \leq i < j \leq n} (\neg a_i \vee \neg a_j)$, that ensure at-most-one (AMO) constraint. In addition, there is one clause, $(a_1 \vee a_2 \vee ...a_n)$, that ensures at-least-one (ALO) constraint. In this setting, if any variables other than a are assigned to certain values and it causes the ALO clause for a to be unit clause by other constraints, such that only a variable $a_k (1 \leq k \leq n)$ is not assigned and the others are assigned to false, then a_k has $n - 1$ indegrees in the implication graph. Such variables like a are focused on by CIR and they are selected as decision-variables at early depth of the search tree. Before the execution of CIR, the assignments of such variables are forced by the values of other variables. However after CIR, they contribute early branching, and intuitively it enables the change of the search space.

The C-language-like pseudo code of the function of CIR is shown below. This function is called before the restart routine.

```
1. int run_count = 0;
```

```
2. CounterImplicationRestart() begin
3.   if (run_count++ % INTERVAL > 0)
4.     int indegree[nVar] = {0};
5.     int max_indegree = 0;
6.     [calculate indegree for each variable and max_indegree]
7.     for each variable var
8.       bumpScore(var, BUMP_RATIO * indegree[var] / max_indegree);
9.   restart();
10. end
```

The variable "run_count" stands for the number of times this function is executed. The main part of the function is run for every "INTERVAL" restart. In the seventh and eighth lines, all the VSIDS scores of the variables are bumped in proportion to their indegrees. To bump the VSIDS score drastically, the constant number of the "BUMP_RATIO" needs to be relatively large. So far, we have confirmed that the performance of CIR depends on the value of "INTERVAL" [8]. Fig. 1 shows the experimental result of various "INTERVAL" and fixed "BUMP_RATIO" using 200 instances from SAT Race 2008. From this result, We have found that small "INTERVAL" such as 3 is relatively better and it affects the total performance.

In the SAT Competition 2011, in the application category, we submitted MiniSAT 2.2 [5] with CIR, and won a gold medal in the minisat-hack track and a silver medal in the satisfiable problem track. Our solver could solve 202 instances in total - eight more than the original MiniSAT 2.2. The source code of this solver is available at http://www.cril.univ-artois.fr/SAT11/solvers/SAT2011-sources.tar.gz.

3 Experimental Results

We conducted experiments to confirm the performance of CIR for parallel solvers. In these experiments, the number of threads was set to four. As the first step, we implemented parallel settings of MiniSAT 2.2, called "para_minisat2.2", by using OpenMP. We modified the base number of Luby restart and the initial VSIDS scores of the variables for "para_minisat2.2", and added the function of learnt clause sharing. Then, we added the CIR top to "para_minisat2.2", called "para_cir_minisat". Three of the four threads used the CIR (the other ran as the default MiniSAT 2.2). In consideration of the previous results [8], the value of "INTERVAL" was set to 1, 2 and 3 respectively, and the "BUMP_RATIO" was fixed to 10000 for all of them.

The experiments were conducted on a Linux machine with an Intel Xeon quad-core CPU, running at 2.67 GHz and 24 GB of RAM. The benchmarks were 200 instances from SAT Race 2010. Timeout was set to 5000 seconds. We used six solvers: "para_minisat2.2", "para_cir_minisat", the latest version of Cryptominisat (denoted as "cryptominisat2.9.1"), the latest version of Plingeling (denoted as "plingeling276"), MiniSAT 2.2 in single thread (denoted as "minisat2.2_single"), and MiniSAT 2.2 with CIR whose "INTERVAL" is 3 in single thread (denoted as "cir_minisat_single").

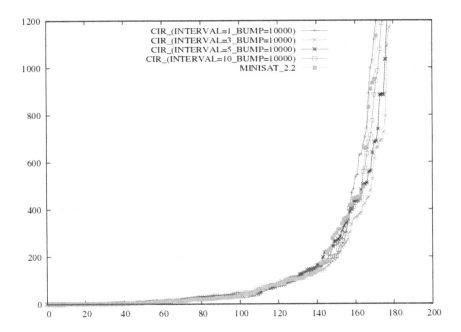

Fig. 1. The experimental result of various "INTERVAL" using 200 instances from SAT Race 2008

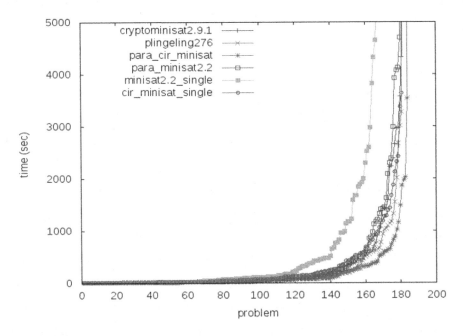

Fig. 2. The cactus plot of the experimental result using 200 instances from SAT Race 2010

Table 1. The number of solved instances for each solver

	SAT	UNSAT	total
cryptominisat2.9.1	65	116	181
plingeling276	70	111	181
para_cir_minisat	69	115	184
para_minisat2.2	69	111	180
minisat2.2_single	66	106	172
cir_minisat_single	70	111	181

The results are shown in Fig. 2 as a cactus plot and Table 1. Despite the naive and minimum configuration for parallel searching, "para_minisat2.2" provided good performance. In addition, "cir_minisat_single" is competitive with parallel solvers. The proposed solver, "para_cir_minisat" displayed relatively better performance than both "cryptominisat2.9.1" and "plingeling276", which won the first and second places in the SAT Competition 2011. This result indicates that CIR can also work in a parallel context and that CIR encourages the diversification of search activity in a portfolio approach.

4   Conclusion

CIR can convert the search space drastically after a restart. We propose the adaptation of CIR for parallel solvers in order to achieve good diversification. Experimental results show that CIR performed well for parallel settings, even though only simple functions were implemented. The vigorous conversion of the decision-order by CIR accelerates the diversification of search spaces. As future work, we will consider learnt clause sharing, such as clause length control [9] and arrange the scoring system of VSIDS so that it combines better with CIR.

Acknowledgment. We appreciate the insightful comments from the reviewers in LION 6.

References

1. Hamadi, Y., Sais, L.: Manysat: a parallel sat solver. Journal on Satisfiability, Boolean Modeling and Computation, JSAT (2009)
2. Guo, L., Hamadi, Y., Jabbour, S., Sais, L.: Diversification and Intensification in Parallel SAT Solving. In: Cohen, D. (ed.) CP 2010. LNCS, vol. 6308, pp. 252–265. Springer, Heidelberg (2010)
3. Huang, J.: The effect of restarts on the efficiency of clause learning. In: Proceedings of the International Joint Conference on Artificial Intelligence, pp. 2318–2323 (2007)
4. Pipatsrisawat, K., Darwiche, A.: A Lightweight Component Caching Scheme for Satisfiability Solvers. In: Marques-Silva, J., Sakallah, K.A. (eds.) SAT 2007. LNCS, vol. 4501, pp. 294–299. Springer, Heidelberg (2007)

5. Sorensson, N.: Minisat 2.2 and minisat++ 1.1. A short description in SAT Race 2010 (2010)
6. Crawford, J.M., Baker, A.B.: Experimental results on the application of satisfiability algorithms to scheduling problems. In: Proceedings of the Twelfth National Conference on Artificial Intelligence, pp. 1092–1097 (1994)
7. Zhang, L., Madigan, C.F., Moskewicz, M.H.: Efficient conflict driven learning in a boolean satisfiability solver. In: ICCAD, pp. 279–285 (2001)
8. Tomohiro Sonobe, M.I., Nagai, A.: Counter implication restart. In: Pragmatics of SAT 2011 (2011)
9. Hamadi, Y., Jabbour, S., Sais, L.: Control-based clause sharing in parallel sat solving. In: Proceedings of the 21st International Joint Conference on Artificial Intelligence, pp. 499–504. Morgan Kaufmann Publishers Inc, San Francisco (2009)

Learning the Neighborhood with the Linkage Tree Genetic Algorithm

Dirk Thierens[1,2] and Peter A.N. Bosman[2]

[1] Institute of Information and Computing Sciences
Universiteit Utrecht, The Netherlands
[2] Centrum Wiskunde & Informatica (CWI)
Amsterdam, The Netherlands

Abstract. We discuss the use of online learning of the local search neighborhood. Specifically, we consider the *Linkage Tree Genetic Algorithm (LTGA)*, a population-based, stochastic local search algorithm that learns the neighborhood by identifying the problem variables that have a high mutual information in a population of good solutions. The LTGA builds each generation a *linkage tree* using a hierarchical clustering algorithm. This bottom-up hierarchical clustering is computationally very efficient and runs in $O(n^2)$. Each node in the tree represents a specific cluster of problem variables. When generating new solutions, these linked variables specify the neighborhood where the LTGA searches for better solutions by sampling values for these problem variables from the current population. To demonstrate the use of learning the neighborhood we experimentally compare iterated local search (ILS) with the LTGA on a hard discrete problem, the nearest-neighbor NK-landscape problem with maximal overlap. Results show that the LTGA is significantly superior to the ILS, proving that learning the neighborhood during the search can lead to a considerable gain in search performance.

1 Introduction

Stochastic local search (SLS) is a powerful class of algorithms to solve discrete optimization problems. Although many variations of SLS can be found in the literature, all of them assume that the neighborhood to be explored is given before the search starts. However, the choice of the neighborhood is a critical factor which can make all the difference between success and failure of the search. The limitations of a fixed neighborhood are sometimes recognized, and this has led to the development of algorithms like variable neighborhood search (VNS) [3], where a nested set of neighborhoods is explored when the search stagnates in its current neighborhood. However, VNS has still a static, predefined neighborhood structure. In this paper, we show that it can be beneficial to search in a dynamically changing neighborhood structure. Our aim is to learn what neighborhood to use during the search. This neighborhood learning is guided by structural similarities present in a set of good solutions found so far during the search process. In the next section we briefly review the linkage tree genetic algorithm

Y. Hamadi and M. Schoenauer (Eds.): LION 6, LNCS 7219, pp. 491–496, 2012.
© Springer-Verlag Berlin Heidelberg 2012

that precisely does this. Section 3 specifies our benchmark function, the so called nearest-neighbor NK-landscape problem with maximal overlap. In Section 4 we compare the LTGA with iterated local search to see how much can be gained from dynamically learning the neighborhood to explore.

2   Linkage Tree Genetic Algorithm

The linkage tree genetic algorithm [7,8,1] is a population-based, stochastic local search algorithm that aims to identify which variables should be treated as a dependent set during the exploration phase. The LTGA represents this dependence information in a hierarchical cluster tree of the problem variables. The linkage tree of a population of solutions is the hierarchical cluster tree of the problem variables using an agglomerative hierarchical clustering algorithm. The problem variables - or cluster of variables - that are most similar are merged first. Similarity is measured by the mutual information between individual variables, or by the average linkage clustering between clusters of variables $X_{Fi} and X_{Fj}$. This average linkage clustering is equal to the unweighted pair group method with a arithmetic mean (UPGMA) and is defined by:

$$I^{UPGMA}(X_{Fi}, X_{Fj}) = \frac{1}{|X_{Fi}||X_{Fj}|} \sum_{X \in X_{Fi}} \sum_{Y \in X_{Fj}} I(X, Y).$$

This agglomerative hierarchical clustering algorithm is computationally very efficient. Only the mutual information between pairs of variables needs to be computed which is a $O(\ell^2)$ operation. The bottom-up hierarchical clustering can also be done in $O(\ell^2)$ computation by using the *reciprocal nearest neighbor chain* algorithm [2]. For a problem of length ℓ the linkage tree has ℓ leaf nodes (the clusters having a single problem variable) and $\ell - 1$ internal nodes. Each node divides the set of problem variables into two mutually exclusive subsets. One subset is the cluster of variables at that node, while the other subset is the complementary set of problem variables. The LTGA uses this division of the problem variables as a set of variables whose values are sampled simultaneously. Sample values are obtained by taking a random solution of the current population and copying the corresponding variable values. Each generation the LTGA builds a linkage tree of the current population. New solutions are generated by greedily exploring the neighborhood of each solution in the current population. This neighborhood is defined by the linkage tree. The LTGA traverses the tree in the opposite order of the merging of the clusters by the hierarchical clustering algorithm. Therefore, LTGA first samples the variables which are the least dependent on each other. New solutions are only accepted when they have a better fitness value than the original solution. When the tree is completely traversed, the solution obtained is copied to the population of the next generation. This tree traversal process is done for each solution in the current generation.

3   Nearest-Neighbor NK-Landscape with Maximal Overlap

To show that learning the neighborhood during the search can be very useful, we compare ILS and LTGA on the nearest-neighbor NK-landscape problem with maximum overlap [5,6]. The nearest-neighbor NK-problem is a subclass of the general NK-landscape problem where the interacting bit variables are restricted to groups of variables of length k. The fitness contributions of each group of k-bits are specified in a table of 2^k random numbers between 0 and 1. Formally, the nearest-neighbor NK-problem is specified by its length ℓ, the size of the subproblems k, the amount of overlap between the subproblems o, and the number of subproblems m. The first subproblem is defined at the first k string positions. The second subproblem is defined at the last o positions of the first subproblem and the next $(k-o)$ positions. All remaining subproblems are defined in a similar way. The relationship between the problem parameters is $\ell = k + (m-1)(k-o)$. In this paper, we look at NK-problems with maximal overlap between the groups of interacting variables, thus $o = k - 1$. For instance, for $\ell = 100$, $k = 5$, $0 = 4$, and $m = 96$ we get the subproblems: (0 1 2 3 4) (1 2 3 4 5) (2 3 4 5 6) ... (95 96 97 98 99) (96 97 98 99 100). An interesting property of the nearest-neighbor NK-problem is that we can compute the global optimal solution using dynamic programming when we use the structural knowledge of the position of the subfunctions. However, when used as benchmark function for ILS and LTGA the position of the bit variables are randomly shuffled and unknown to these algorithms. ILS has no mechanism to figure out what bits are possibly related. The LTGA however is be able to learn some important interactions and take advantage of this during the search process.

4   Experiments

The neighborhood explored by ILS is defined by single bit flips of the current solution. ILS alternates local search with stochastic perturbation of the current local optimum. When after perturbation and subsequent local search a better local optimum is found the search will proceed from that new solution. If not, the newly found solution is rejected and ILS continues from the old local optimum. The perturbation should be large enough such that the local search does not return to the same local optimum in the next iteration. However the perturbation should not be too large, otherwise the search characteristics will resemble those of a multi-start local search algorithm. To investigate the impact of the perturbation size p_m of ILS on a large set of NK-problem instances we first run a series of experiments on 100 randomly generated problems. The top subfigures from Figure 1 show results for perturbation sizes varying from 2 to 10 (left), and from 5 to 20 (right). The Y-axis shows the number of runs (out of 100) that successfully found the global optimal solution to a randomly generated NK-problem instance. The X-axis shows the number of function evaluations needed. The plots are cumulative thus a value (10000, 20) means that 20 of the 100 runs have found the global optimum in less (or equal) than 10000 function evaluation [4].

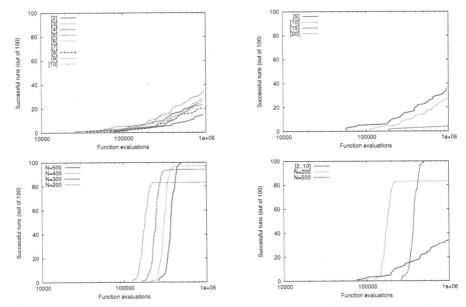

Fig. 1. RLDs of single run on 100 NK-problem instances for ILS with perturbation sizes varying from 2 to 20 and LTGA with population sizes varying from 200 to 500

Whereas in Figure 1 single independent runs are performed on 100 different random NK-problem instance, Figure 2 shows results for 100 independent runs on 10 specific NK-problem instances. The top left subfigure shows ILS with perturbation size randomly chosen between 2 and 10. The top right subfigure shows results for LTGA with population size 500. The two bottom subfigures show results for ILS (perturbation size between 2 and 10) and LTGA (population size 300 and 500) for the NK_3 and NK_2 problem instances, which are respectively the easiest and second easiest problem to solve.

4.1   Discussion

A number of observations can be made from the experiments:

- The top subfigures of Figure 1 show that the perturbation size for ILS does not seem to be a very sensitive parameter for the NK-problems considered. For $p_m = 2$ the success rate is a bit lower than for values between 3 and 10. Closer inspection of the actual runs shows that this is due to the much higher percentage of local searches that return to the starting solution after the perturbation. Performance of ILS only drops when p_m becomes larger than 10. For $p_m = 20$ none of the 100 runs did find the global solution, showing that multi-start local search is hopelessly inefficient for these nearest-neighbor NK landscape problems with maximal overlap.

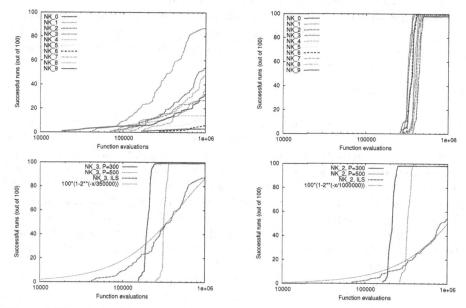

Fig. 2. RLDs of 100 runs on 10 NK-problem instances for ILS with random perturbation sizes between 2 and 10, and LTGA with population size 500

- In the bottom right subfigure of Figure 1 we have also plotted the results for ILS with a perturbation size randomly chosen from the interval [2..10]. Each time ILS perturbs a local optimum, p_m is randomly selected. This way a robust parameter setting is obtained that returns good overall performance.
- LTGA finds the global optima very reliably when the population size $N = 500$. Population sizes 300 and 400 result in a success rate well above 90%, while for $N = 200$ the success rate becomes 83%. Of course a smaller population size means that we need less function evaluations to reach a certain level of performance. For $N = 200$ we could ideally cut the search after 200000 evaluations and then restart the search. This would result in a combined success rate of $1 - (1 - 0.83)^2 = 97\%$ after 400000 function evaluations. Interestingly, this is about the same performance and efficiency result as with a population size $N = 500$, so here there is no gain in restarting the LTGA using a smaller population.
- The top subfigures of 2 show the success rate (out of 100 runs) of ILS ($p_m = \text{random}[2..10]$) and LTGA ($N = 500$) for 10 different NK-landscape problems. Clearly, the LTGA is not only more performant and more efficient, it is also more consistent. There is much less variance in the required function evaluations to reach a certain performance level.
- ILS's result varies significantly with the specific instances of the NK-landscape problem. The bottom subfigures of 2 compare the results on the two easiest instances for ILS with the LTGA. We have also drawn a tangent

cumulative exponential distribution function to ILS's success rate. In both cases the success rate of ILS is not dropping lower than the cumulative exponential - rather on the contrary - showing that here also nothing can be gained from restarting ILS after a certain number of function evaluations.

5 Conclusions

We have shown that learning a neighborhood structure during the search process can be very beneficial in terms of performance, efficiency, and consistency. To do so, we have compared iterated local search (ILS) using optimally tuned perturbation sizes with the linkage tree genetic algorithm (LTGA) on a set of nearest-neighbor NK-landscape problems with maximal overlap. The LTGA builds each generation a linkage tree using a hierarchical clustering algorithm. Each node in the tree represents a specific groups of problem variables that are linked together. When generating new solutions, these linked variables specify the neighborhood where the LTGA searches for better solutions by sampling values for these problem variables from the current population. In future work, we plan to investigate the use of neighborhood learning for other problem types.

References

1. Bosman, P.A.N., Thierens, D.: The roles of local search, model building and optimal mixing in evolutionary algorithms from a BBO perspective. In: Krasnogor, N., Lanzi, P.L. (eds.) GECCO (Companion), pp. 663–670. ACM Press (2011)
2. Gronau, I., Moran, S.: Optimal implementations of UPGMA and other common clustering algorithms. Inf. Process. Lett. 104(6), 205–210 (2007)
3. Hansen, P., Mladenovic, N., Moreno-Pérez, J.A.: Variable neighbourhood search: methods and applications. Annals OR 175(1), 367–407 (2010)
4. Hoos, H.H., Stützle, T.: Evaluating Las Vegas algorithms: pitfalls and remedies. In: Proc. of the 14th UAI, pp. 238–245. Morgan Kaufmann (1998)
5. Pelikan, M., Sastry, K., Butz, M.V., Goldberg, D.E.: Performance of Evolutionary Algorithms on Random Decomposable Problems. In: Runarsson, T.P., Beyer, H.-G., Burke, E.K., Merelo-Guervós, J.J., Whitley, L.D., Yao, X. (eds.) PPSN IX. LNCS, vol. 4193, pp. 788–797. Springer, Heidelberg (2006)
6. Pelikan, M., Sastry, K., Goldberg, D.E., Butz, M.V., Hauschild, M.: Performance of evolutionary algorithms on NK landscapes with nearest neighbor interactions and tunable overlap. In: Raidl, G. (ed.) GECCO, pp. 851–858. ACM (2009)
7. Thierens, D.: The Linkage Tree Genetic Algorithm. In: Schaefer, R., Cotta, C., Kołodziej, J., Rudolph, G., et al. (eds.) PPSN XI. LNCS, vol. 6238, pp. 264–273. Springer, Heidelberg (2010)
8. Thierens, D., Bosman, P.A.N.: Optimal mixing evolutionary algorithms. In: Krasnogor, N., Lanzi, P.L. (eds.) GECCO, pp. 617–624. ACM (2011)

A Comparison of Operator Utility Measures for On-Line Operator Selection in Local Search

Nadarajen Veerapen[1], Jorge Maturana[2], and Frédéric Saubion[1]

[1] LUNAM Université, Université d'Angers, LERIA, Angers, France
firstname.lastname@univ-angers.fr
[2] Instituto de Informática, Universidad Austral de Chile, Valdivia, Chile
jorge.maturana@inf.uach.cl

Abstract. This paper investigates the adaptive selection of operators in the context of Local Search. The utility of each operator is computed from the solution quality and distance of the candidate solution from the search trajectory. A number of utility measures based on the Pareto dominance relationship and the relative distances between the operators are proposed and evaluated on QAP instances using an implied or static target balance between exploitation and exploration. A refined algorithm with an adaptive target balance is then examined.

1 Introduction

An increasing number of solving techniques have been proposed to address larger and more complex optimization problems but they are often difficult to adapt and to tune for a given problem. In fact, efficient solving tools have become out of reach for practitioners. Among the possible solving techniques, metaheuristics are now widely to efficiently solve optimization problems. Nevertheless, attempting to design increasingly efficient metaheuristics often results in highly complex systems which require a non-negligible amount of expert knowledge to use, for instance to wisely choose the method's required parameters.

A relatively recent avenue of research is the design of generic high level control strategies in an attempt to make optimization techniques more user-friendly [3]. A classification of these different approaches can be found in [5]. In general only one criterion, solution quality, is considered. Concerning the control of parameters, the most advanced techniques were first developed in the context of evolutionary computation [6]. A number of operator selection strategies for genetic algorithms (adaptive operator selection) are presented in [4]. In [7], operator selection techniques were proposed to handle simultaneously two criteria in the evaluation of the operators: quality and diversity of the population.

We focus on local search algorithms to solve combinatorial optimization problems. In a previous work [10], we have proposed a general framework to control dynamically the search process of a local search algorithm targeted at problems that can be modeled as permutations. In this paper, we improve this mechanism by introducing new performance evaluation techniques and more sophisticated and dynamic control features.

Y. Hamadi and M. Schoenauer (Eds.): LION 6, LNCS 7219, pp. 497–502, 2012.
© Springer-Verlag Berlin Heidelberg 2012

The paper has five more sections. Section 2 looks at operator control in local search. Section 3 describes the different utility values used in operator selection. Section 4 deals with the experiments. Section 5 looks at an attempt at adaptively changing the utility value parameter. Finally the last section is the conclusion.

2   Operator Control for Local Search

The defining feature of a good local search algorithm is the efficient exploration of the search space in order to find the optimal solution. This requires striking a balance between two generally conflicting objectives: exploitation (converging towards a local optimum) with exploration (suitably sampling different areas of the search space). One way of achieving this balance is by controlling the basic operations (moves or operators) which drive the solution around the search space.

Our aim is to select an appropriate operator out of a set of operators to be applied at each iteration of the local search algorithm. In order to determine the likeliness of an operator to be useful, in terms of both exploitation and exploration, its previous behavior needs to be recorded and analyzed. Figure 1 shows how the operator control interacts with the local search algorithm.

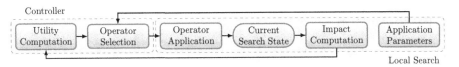

Fig. 1. Overview : Operator control (left) and local search algorithm (right)

Impact Computation. Quality is measured directly as the change in the objective function.

Measuring diversity is relatively straightforward in evolutionary algorithms but less clear in single point algorithms. Here we consider a sliding window containing the last solutions in the search path and measure the difference between them and the current candidate solution $c = op(s)$. This difference is computed at the variable-value couple level: the less frequent the occurrences of the candidate solution's variable-value couples in the path, the greater the distance between them. The following equation formalizes this notion. Let $P_{i,j}$ be the path from iteration i through j, $i \leq j$. Then $d^P(c, P_{i,j}) = 1/n \times \sum_{k_c=1}^{n} (1 - occ(P_{i,j}, (k_c, \pi_{k_c}))/|P_{i,j}|)$ where $occ(P_{i,j}, (a, b))$ returns the number of times the variable-value couple (a, b) is found in $P_{i,j}$.

3   Operator Selection and Utility

An operator is selected with a probability proportional to some utility value which is meant to be a reflection of its previous performance. Each operator has its own fixed-length sliding window which keeps track of the quality difference and distance value for its m last applications. The sliding windows are initialized by one application of each operator at the beginning of the search. The utility of an operator is based on the average of the quality and distance values.

Given two vectors u and v of equal cardinality p and considering a maximization problem, u *dominates* v if $u_k \geq v_k, \forall k \in \{1, \ldots, p\}$ with at least one strict inequality. This is often referred to as Pareto dominance and denoted by $u \succ v$.

The Pareto-dominance-based utility U_P of operator o among the set O of n operators would then be defined as $U_P(o) = |\{o'|o' \in O, o \succ o'\}| + \epsilon$ where ϵ ensures a non-zero utility value. This utility assignment scheme (used in [10]) does not allow for a commanded balance between exploitation and exploration.

We now propose a number of ways of introducing a weight, α, in order to influence the balance. Given two operators o_1 and o_2 defined by the quality-distance couples (q_1, d_1) and (q_2, d_2), we define the weighted rectilinear displacement from o_1 to o_2 as $d_\alpha(o_1, o_2) = \alpha(d_1 - d_2) + (1 - \alpha)(q_1 - q_2)$.

Based on this metric, we define three utilities: the sum of displacements to all other operators $U_\alpha^\Sigma(o) = max(0, \sum_{i=1}^n d_\alpha(o, o_i))$, the sum of positive displacements to all other operators $U_\alpha^{\Sigma+}(o) = \sum_{i=1}^n max(0, d_\alpha(o, o_i))$, the sum of displacements to all other dominated operators $U_\alpha^{\Sigma\succ}(o) = \sum_{i=1, o \succ o_i}^n d_\alpha(o, o_i)$. We also use the very simple weighted sum of operator quality and distance $U_\alpha(o) = \alpha d + (1 - \alpha)q$.

Displacements, and their sum, are useful in the sense that they are a means of describing quantitatively (the magnitude) and, to a lesser extent, qualitatively (the sign or direction) the relationship between each operator and the rest. In addition, the weight naturally introduces a quantifiable bias towards either exploration or exploitation.

4 Experiments for Weighted Operator Utility

The different utility values described in the previous section are tested on a very classic permutation problem: the Quadratic Assignment Problem (QAP) which models the problem of finding a minimum cost allocation of N facilities into N locations, taking the costs as the sum of all possible distance-flow products.

Experimental Settings. Each operator is a combination of a neighborhood and a selection function. We use a single basic neighborhood which swaps the values of two variables. The selection functions used are random selection, first improving neighbor, best neighbor, random selection among the 5 best neighbors, and best among k neighbors.

A population of twelve operators is defined. Ten of these do not change the solution configuration much (at most 6 variables are affected): half are intensification oriented, half are exploration oriented. The last two are extremely perturbative operators which randomly swap 25% and 50% of the variables.

For each instance, each algorithm is run thirty times starting with the same thirty different random solutions and a maximum of 40 000 iterations are allowed per run. All sliding windows have an arbitrary fixed length of 100.

Analysis of Results. Table 1 reports the results for all the experiments on the QAP (instances from QAPLIB [2]) in this paper. In this section all columns except the next-to-last one is of interest to us. They report the instances, their best known value and the results for the different utility values with fixed weights: U_α with $\epsilon = 0.1$ (the values 0.001, 0.05, 0.2 and 1 were also tested but 0.1 proved the best across almost all instances) and U_α^Σ, $U_\alpha^{\Sigma+}$, $U_\alpha^{\Sigma\succ}$ and U_α with α taking values 0.2, 0.5 or 0.8. The last column gives, as a comparison, the results obtained with Robust Tabu (RoTS) [9], a dedicated local search algorithm for the QAP. The values for each algorithm express the average percentage difference above the best known values. Bold font indicates the best values and italics indicate results which are within 0.05 % of the best (RoTS results are not considered because they are always better or equal).

From Table 1 it is clear that $U_\alpha^{\Sigma+}$ is the best among the displacement-based utility values tested and it provides consistently good results over various α. Pareto-dominance-based selection, U_P, also provides good results with respect to all other selection methods. Our previous work had shown that it worked slightly better than uniform selection if the operators were relatively "good" and outperformed it when a single "very good" operator was added. Here we can see that it remains useful despite the presence of highly disruptive operators.

When considering U_P and $U_\alpha^{\Sigma+}$ selections side-by-side, the latter is generally equivalent to the former for lower α values and only really better on tai*xx*a instances. Following this observation, and also because different classes of instances seem to require different values of α to obtain the best results, a method for adapting α and obtaining better results is described in the next section.

If we consider U_α^Σ, the problem is that accepting negative displacements often negates the other positive displacements thus producing results that are, usually, worse than uniform selection. The poor performance of $U_\alpha^{\Sigma\succ}$ may be explained by the fact that since it is based on Pareto dominance and there is no ϵ to ensure a minimum selection probability, the operators at both ends of the exploration-exploitation spectrum have no real chance of being selected because they usually do not dominate any other operator. Finally, it appears that U_α might be too simple and shows that the relationship between operators can be useful when used appropriately (in U_P and $U_\alpha^{\Sigma+}$).

The next section looks at how the weight α can be varied during the search.

5   Adaptive Parameter Values for Operator Control

We consider the "correct" diversity (CD) strategy [8]. The CD strategy uses the value of the quality of the solutions in the population as a means to assess the diversity of population. If the number of solutions having the same quality is above a certain threshold T_{max} then it is assumed that the population is too homogeneous and the commanded diversity is incremented by some step s_{inc}. Symmetrically, if the number of solutions having the same quality is below another threshold T_{min} the commanded diversity is decremented by some step s_{dec} (solutions are haphazardly distributed; exploitation is not strong enough).

Experiments. To tune the CD parameters we use F-Race [1] an off-line tuning algorithm. We first tested 320 parameter combinations. The winning parameters ($T_{max} = 0.3$, $T_{min} = 0.25$, $s_{inc} = 0.0001$, $s_{dec} = 0.1$) were at either end of the available domain for each parameter. One could thus assume that a better combination of parameters might be obtained by extending their domain. We therefore ran a second race with new parameter domains (144 combinations) and obtained a new winner ($T_{max} = 0.35$, $T_{min} = 0.15$, $s_{inc} = 0.0001$, $s_{dec} = 0.01$) which was relatively different from the previous one and did not benefit from the new values at the extremities of each domain. For both winners the distributions of results were statistically equivalent. This leads us to believe that the strategy parameters need only be within some tight domain (and not one specific value) to obtain the best results.

Results and Discussion. The results are presented in Table 1 under the $U_\alpha^{\Sigma+}CD$ column. It seems clear that CD is better than the other selection methods, or within 0.05 % of the best, on most instances in terms of raw results. This superiority is further confirmed by a Wilcoxon signed-rank test with 95 % confidence level. If we compare $U_\alpha^{\Sigma+}CD$ with the best values across the different α for $U_\alpha^{\Sigma+}$ both distributions are statistically equivalent. This leads us to conclude that the CD strategy is good enough to produce results equivalent to the best results of $U_\alpha^{\Sigma+}$ with fixed α values.

Table 1. Results for QAP instances

Instance	BKV	Uniform	U_P	$U_\alpha^{\Sigma+}$	U_α^{Σ}	$U_\alpha^{\Sigma\succ}$	U_α	$U_{\alpha CD}^{\Sigma+}$	RoTS
chr25a	3796	20.18	**10.67**	13.78	33.53	34.05	31.34		
				14.94	30.75	28.68	30.18	12.45	7.09
				12.11	28.70	29.75	16.44		
kra30a	88900	2.49	0.79	1.57	5.14	4.67	5.24		
				1.63	4.39	4.55	4.97	**0.61**	0.06
				0.89	3.75	4.42	2.20		
kra30b	91420	1.11	0.21	*0.16*	3.06	3.32	3.25		
				0.45	2.78	3.03	2.63	**0.13**	0.02
				0.32	2.50	2.37	0.68		
nug20	2570	0.12	*0.01*	**0.00**	1.02	1.09	1.74		
				0.03	0.56	0.89	0.65	*0.01*	0.00
				0.00	1.15	1.45	0.07		
nug30	6124	1.24	0.20	0.31	1.67	1.27	1.86		
				0.19	1.43	1.54	1.75	**0.11**	0.01
				0.39	1.60	1.71	0.68		
sko42	15812	2.28	0.29	*0.19*	1.91	1.65	1.93		
				0.28	1.38	1.63	1.59	**0.16**	0.03
				0.67	2.01	2.14	1.48		
sko49	23386	2.48	0.36	**0.21**	1.37	1.46	1.57		
				0.27	1.34	1.60	1.42	*0.24*	0.13
				0.81	2.31	1.74	1.91		
tai30a	1818146	2.59	1.26	1.17	2.05	2.16	3.31		
				1.27	1.86	1.78	1.87	**0.91**	0.51
				1.68	3.18	3.41	2.33		
tai50a	4941410	4.20	2.16	**1.58**	2.13	2.27	3.40		
				1.59	2.61	2.80	2.34	1.66	1.39
				2.83	4.11	4.11	3.82		
tai30b	637117113	0.43	**0.13**	0.44	6.65	5.21	5.27		
				0.35	3.90	3.53	4.83	*0.15*	0.03
				0.16	3.49	1.74	0.34		
tai50b	458821517	2.36	0.25	0.30	4.14	4.33	5.42		
				0.39	3.13	3.92	4.33	**0.18**	0.15
				0.37	2.56	2.66	1.60		

6 Conclusion

In this paper we have presented different alternatives for the selection of operators in Local Search. The main contribution of the paper was the investigation of weighted utilities which allow a target balance to be set between exploration and exploitation. Using static weights the best of them was competitive when compared to the previously proposed Pareto-dominance-based utility. An adaptive strategy for setting the weight was investigated and proved to provide improved results.

In future works we wish to look at more advanced on-line parameter setting strategies. Another avenue of research is testing the existing proposed methods with academic problems such as the One-MAX and long-path problems, whose properties are well understood, in order to have a better theoretical understanding of the methods.

Acknowledgments. This work was supported by Microsoft Research through its PhD Scholarship Programme and by CONICYT–ECOS-Sud (Project C10E07).

References

1. Birattari, M.: The Problem of Tuning Metaheuristics as seen from a machine learning perspective. Ph.D. thesis, Université Libre de Bruxelles, Brussels, Belgium (December 2004)
2. Burkard, R.E., Karisch, S.E., Rendl, F.: QAPLIB – a quadratic assignment problem library. Journal of Global Optimization 10, 391–403 (1997)
3. Burke, E., Kendall, G., Newall, J., Hart, E., Ross, P., Schulenburg, S.: Hyperheuristics: An emerging direction in modern search technology. In: Glover, F., Kochenberger, G. (eds.) Handbook of Metaheuristics. International Series in Operations Research & Management Science, vol. 57, pp. 457–474. Springer, New York (2003)
4. Fialho, Á.: Adaptive Operator Selection for Optimization. Ph.D. thesis, Université Paris-Sud 11, Orsay, France (December 2010)
5. Hamadi, Y., Monfroy, E., Saubion, F.: What is autonomous search? In: van Hentenryck, P., Milano, M. (eds.) Hybrid Optimization. Springer Optimization and Its Applications, vol. 45, pp. 357–391. Springer, New York (2011)
6. Lobo, F.G., Lima, C.F., Michalewicz, Z. (eds.): Parameter Setting in Evolutionary Algorithms. SCI, vol. 54. Springer, Heidelberg (2007)
7. Maturana, J., Lardeux, F., Saubion, F.: Autonomous operator management for evolutionary algorithms. J. Heuristics 16(6), 881–909 (2010)
8. Maturana, J., Saubion, F.: On the Design of Adaptive Control Strategies for Evolutionary Algorithms. In: Monmarché, N., Talbi, E.-G., Collet, P., Schoenauer, M., Lutton, E. (eds.) EA 2007. LNCS, vol. 4926, pp. 303–315. Springer, Heidelberg (2008)
9. Taillard, É.D.: Robust taboo search for the quadratic assignment problem. Parallel Computing 17(4-5), 443–455 (1991)
10. Veerapen, N., Saubion, F.: Pareto Autonomous Local Search. In: Coello, C.A.C. (ed.) LION 5. LNCS, vol. 6683, pp. 392–406. Springer, Heidelberg (2011)

Monte Carlo Methods for Preference Learning

Paolo Viappiani

Department of Computer Science, Aalborg University, Denmark
paolo@cs.aau.dk

Abstract. Utility elicitation is an important component of many applications, such as decision support systems and recommender systems. Such systems query the users about their preferences and give recommendations based on the system's belief about the utility function.

Critical to these applications is the acquisition of prior distribution about the utility parameters and the possibility of real time Bayesian inference. In this paper we consider Monte Carlo methods for these problems.

1 Bayesian Utility Elicitation

Utility elicitation is a key component in many decision support applications and recommender systems, since appropriate decisions or recommendations depend critically on the preferences of the user on whose behalf decisions are being made. Since full elicitation of user utility is prohibitively expensive in most cases (w.r.t. time, cognitive effort, etc.), we must often rely on partial utility information. This is the case of *interactive preference elicitation.*

As a user's utility function will not be known with certainty, following recent models of Bayesian elicitation [3,2,6,8], the system's knowledge about the user preferences is represented as probabilistic *beliefs*. Interactive elicitation must selectively decide which queries are most informative relative to the goal of making good or optimal recommendations and then, following user responses, update the distribution. An important requirement for utility elicitation is that inference can be made *real-time*, the system needs to output a recommendation or ask a query in no more than a few seconds.

While there are a variety of query types that can be used, comparison queries are especially natural, asking a user if she prefers one option to another. As the number of items in a dataset can be extremely large, iterating over all possible comparison is unfeasible. Recently [8] we showed that, under very general assumptions, the optimal choice query w.r.t. the expected value of information (EVOI) coincides with optimal recommendation set, that is, a set maximizing expected utility of the user selection (a simpler and submodular problem). Based on this, we can provide algorithms that select near-optimal comparison queries with worst-case guarantees; we also considered a local search technique that can select the query to ask in a fraction of a second, even for datasets with several hundreds of items.

Y. Hamadi and M. Schoenauer (Eds.): LION 6, LNCS 7219, pp. 503–508, 2012.
© Springer-Verlag Berlin Heidelberg 2012

These strategies for query optimization (and similar approximated strategies [6]) rely on the assumptions that prior utility information is available and that inference is fast enough so that user answers can be used as additional knowledge for further elicitation. As users are not usually willing to wait more than a couple of seconds, effective inference methods are therefore crucial.

Probabilistic inference is challenging because common prior distributions are not closed under Bayesian update for most types of preference queries (in particular for comparison queries). In this paper we consider how Monte Carlo methods can be used in utility-based recommendation systems with the following two purposes:

1. to make inference about the user's possible utility function given the user's query responses, and
2. to acquire a prior distribution about utility parameters given preference statements from previous users

In addition to decision support systems, recent applications of preference learning and elicitation include preference-based reinforcement learning and preference-based robotic systems [4,1].

The Underlying Decision Problem. The system is charged with the task of recommending an option to a user in some multi-attribute space, for instance, the space of possible product configurations from some domain (e.g., computers, cars, apartment rental, etc.). Products are characterized by a finite set of attributes $\mathcal{X} = \{X_1, ...X_n\}$, each with finite domain $Dom(X_i)$. For instance, attributes may correspond to the features of various cars, such as color, engine size, fuel economy, etc., with $\mathbf{X}$ defined either by constraints on attribute combinations. The user has a *utility function* $u : Dom(\mathcal{X}) \to \mathbf{R}$. The precise form of u is not critical, but we assume that $u(\mathbf{x}; \mathbf{w})$ is parametric in $\mathbf{w}$ (a vector of utility "weights"). We often refer to $\mathbf{w}$ as the user's "utility function" for simplicity, assuming a fixed form for u. For sake of presentation, we assume a linear model $u(\mathbf{x}; \mathbf{w}) = \mathbf{w} \cdot \mathbf{x}$, so that the parameter vector $\mathbf{w}$ effectively represents the importance of the different features (but our framework easily extend to richer utility models such as generalized additive utilities [5]). Given a choice set S with $x \in S$, let $S \triangleright \mathbf{x}$ denote that x has the greatest utility among the items in S (for a given utility function $\mathbf{w}$).

The system's uncertainty about the user preferences is reflected in a distribution, or *beliefs*, $P(\mathbf{w}; \theta)$ over the space W of possible utility functions. Here θ denotes the parameterization of our model, and we often refer to θ as our *belief state*. Given $P(\cdot; \theta)$, we define the *expected utility* of an option $\mathbf{x}$ to be

$$EU(\mathbf{x}; \theta) = \int_W u(\mathbf{x}; \mathbf{w})P(\mathbf{w}; \theta)dw = \int_W (\mathbf{w} \cdot \mathbf{x})\, P(\mathbf{w}; \theta)\, dw \qquad (1)$$

If required to make a recommendation given belief θ, the optimal option $\mathbf{x}^*(\theta)$ is that with greatest expected utility

$$EU^*(\theta) = \max_{\mathbf{x} \in X} EU(\mathbf{x}; \theta) \qquad (2)$$

with $x^*(\theta) = \arg\max_{\mathbf{x} \in X} EU(\mathbf{x}; \theta)$. When the user selects an option x in a choice set S, the belief is updated to $P(\mathbf{w}; \theta | S \rightsquigarrow x)$. For query selection strategies, we refer to [8].

Probabilistic Response Model. In utility elicitation, the user's response to a choice set tells us something about her preferences; but this depends on the *user response model.* For any choice set S with $\mathbf{x}_i \in S$, let $S \rightsquigarrow \mathbf{x}_i$ denote the event of the user selecting $\mathbf{x}_i$. A response model R dictates, for any choice set S, the probability $P_R(S \rightsquigarrow \mathbf{x}_i; \mathbf{w})$ of any selection given utility function w. We consider three possible response models for choice queries.

In the *noiseless response model,* the user is always able to identify the preferred item in a choice query set; thus $P_R(S \rightsquigarrow \mathbf{x}; \mathbf{w}) = 1$ if $\mathbf{w}$ is such that x has higher utility than any other in the choice set, 0 otherwise. The set of feasible utility functions is refined by imposing $k - 1$ linear constraints of the form $\mathbf{w} \cdot \mathbf{x}_i \geq \mathbf{w} \cdot \mathbf{x}_j$, $j \neq i$, and the new belief state is obtained by restricting θ to have non-zero mass only on $W \cap S \rhd \mathbf{x}_i$ and renormalizing. The *constant noise model* instead assumes each option $\mathbf{x}$, apart from the most preferred option $\mathbf{x}_\mathbf{w}^*$ relative to $\mathbf{w}$, is selected with (small) constant probability

$$P_C(S \rightsquigarrow \mathbf{x}; w) = \beta \quad ; \quad \mathbf{x} \neq \mathbf{x}_\mathbf{w}^* \tag{3}$$

with β independent of $\mathbf{w}$. Finally the *logistic* response model R_L is commonly used in choice modeling, and is variously known as the *Luce-Sheppard, Bradley-Terry,* or *mixed multinomial logit* model. Selection probabilities are given by

$$P_L(S \rightsquigarrow \mathbf{x}; \mathbf{w}) = \frac{e^{\gamma\,(\mathbf{w} \cdot \mathbf{x})}}{\sum_{y \in S} e^{\gamma\,(\mathbf{w} \cdot \mathbf{y})}} \tag{4}$$

where γ is a temperature parameter. For comparison queries (i.e., $|S| = 2$), P_L is the logistic function of the difference in utility between the two options. It models the fact that is easier to make a correct choice between two items that greatly differ in utility, rather than between two items whose utility is very close.

2   Monte Carlo Methods

Expected Utility. The belief θ is represented by a set L of l particles. Each particle is a complete instantiation of the utility weights. Expected utility is approximated by considering the particles and summing up the utility associated with each particle: $EU(\mathbf{x}; L) = \frac{1}{l} \sum_{\mathbf{q} \in L} u(x; \mathbf{q})$. Obviously the accuracy of the estimation depends on the number of particles.

Inference. In an online interaction with the user, the system needs to update the belief taking into account user responses (for instance, the user select x as the preferred outcome in the set S), resulting in a new distribution $P(\mathbf{w}; \theta | S \rightsquigarrow \mathbf{x})$. Similarly, when learning a preference model from data (preference learning), the distribution can be updated incrementally in a batch process. We now consider two different techniques to update our discrete approximation of the belief distribution: importance sampling and Gibbs sampling; both methods generate a new set of particles implementing a Bayesian update. These methods can also be used (as shown below) to learn a prior from data.

Importance Sampling. Our methods use particles to represent assignments to the utility parameters $\mathbf{w}$, initially generated according to the given prior. In online settings, every time the user answers a query we can propagate the particles with importance sampling. Particle weights are determined by applying the response model to observed responses: $P_R(S \rightsquigarrow \mathbf{x}; \mathbf{w})$(the selection probability) where S is the choice set and $\mathbf{x}$ the outcome selected. In other words, the *response model* is used directly as a likelihood function for importance sampling.

To overcome the problem of particle degeneration (most particles eventually have low or no weight), we use slice-sampling [7] to regenerate particles w.r.t. to the response-updated belief state θ whenever the *effective number of samples* drops significantly. The choice of the best mixing between importance sampling and slice sampling is an open question, as the number of necessary particles. With 50000 particles in standard elicitation problems, importance sampling requires less than 1 second, but particle regeneration requires around 30 seconds.

Gibbs Sampling. In the case of noiseless responses it is possible to use Gibbs-sampling in a quite efficient way. Since responses are noiseless, a statement such that *"x is preferred to y"* imposes a linear constraint $\mathbf{w} \cdot \mathbf{x} \geq \mathbf{w} \cdot \mathbf{y}$ and the region of feasible utilities can be represented by a convex region. We call *Feasible(W)* the region of feasible $\mathbf{w}$.

Gibbs sampling generates a set of utility vectors, consistent with the user's feedback, in the following way. Given an initial feasible weight vector $\mathbf{w} = (w_1, .., w_m)$, we pick a dimension i (between 1 and m). We identify the lower bound $w_j^{\perp}$ (fixing all other values according to $\mathbf{w}$) solving a linear program

$$\min_{\bar{w}_j} \quad \bar{w}_j \tag{5}$$

$$s.t \ (w_1, .., w_{j-1}, \bar{w}_j, w_{j+1}, .., w_m) \in \text{Feasible(W)} \tag{6}$$

and we similarly find the upper bound $w_j^{\top}$ (by considering a maximization as objective). We now sample a value $\bar{w}_j \sim U(w_j^{\perp}, w_j^{\top})$ uniformly in the interval between $w_j^{\perp}$ and $w_j^{\top}$ and update $\mathbf{w} := (w_1, .., w_{j-1}, \bar{w}_j, w_{j+1}, .., w_m)$. We repeat the process alternating the dimension j and storing the retrieved samples $\mathbf{w}$.

Learning Utility Priors from Data. We want to acquire a prior distribution for factored utilities in multi attribute domain, to be used for utility elicitation. We are given as input a number of *preference statements* for several users, of the type $S_i^j \rightsquigarrow x_i^j$ (answers to preference queries from previous users) where S_i^j is the j-th query choice set shown to user i, and x_i^j his selection.

As before, we assume a given response model R for the users, that dictates for any choice set S, the probability of selection $P_R(S \rightsquigarrow x_i; \mathbf{w})$. Our algorithm

Algorithm:

1. Sample n particles uniformly from $U[0,1]^m$ and call D_0 this initial particle set
2. For each user i:
   - *Importance Sampling* step: re-sample n particles from D_0, with weights according to: $\prod_j P_R(S_i^j \leadsto x_i^j; w)$
   - Call D^i the resulting user-specific distribution of particles
3. Merge the particles obtained with all the users: $D^\top = \bigcup D^i$

Fig. 1. The algorithm for learning utility priors with Importance Sampling

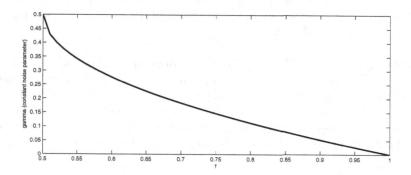

Fig. 2. The estimated $\hat{\gamma}$ as a function of f (constant noise model)

(Figure 1) iterates over all the users and perform importance sampling using the joint likelihood of all preference statements. In the following, a particle is a vector of parameters uniquely identifying a utility function (n is the number of particles, m the dimensionality of the utility parameters).

3   Learning the Response Model

We are currently experimenting these approaches in a real dataset[1] of preference rankings. 5000 users have been asked to rank two (different) sets of items (sushis). Since often the second set include items that the same user has ranked before, we can learn something about the "noise" of the user's choice model. We have the problem of simultaneously learning the user's utility and the response model. The input are rankings of the form $(\mathbf{x}_1, \mathbf{x}_2, .., \mathbf{x}_k)$, ordered outcomes from most to least preferred. The first problem is to generalize the response model from choice sets to rankings. We consider the joint likelihood of all pairwise comparisons induced by the ranking (assumed to be independent); the likelihood of a particular instantiation of the utility parameters $\mathbf{w}$ (used by our sampling techniques) is $\prod_{i,j:j>i} P_R(\{\mathbf{x}_i, \mathbf{x}_j\} \leadsto \mathbf{x}_i; \mathbf{w})$.

---

[1] http://www.kamishima.net/sushi/

Constant noise model The problem is to estimate γ, the constant error ratio, from the available rankings in the dataset. We assume a shared γ (common to all users). For each user, we consider the fraction of times two items have been place in a consistent order in both the first and the second ranking (we do not observe the "ground truth", the correct order). We call f the fraction of "agreements" (averaged among users). Given f, we note that setting $\gamma = 1 - f$ would underestimate the error rater: assuming the user make selections based on the constant error model given γ, the expected number of agreements is $f = (1 - \gamma)^2 + \gamma^2 = 2\gamma^2 - 2\gamma + 1$ (either the items are correctly ordered in both rankings, or they are incorrectly ordered in both). By solving this equation with respect to f (considering the solution between 0 and 0.5) we estimate $\hat{\gamma} = 0.5 \cdot (1 - \sqrt{2f - 1})$; see Figure 2. In the sushi dataset, the average number of agreements is 0.85, thus we find $\hat{\gamma} = 0.09$.

Logistic response model. We can learn the parameter γ of the logistic response model from data. Again, a natural assumption is to consider the value of γ shared from all users. In this case, we can learn priors for different values of γ and select the value that maximizes the overall likelihood of the preference statements. An alternative is to augment utility particles with an hypothesis about the γ and re-sample based on the likelihood of responses given both w and γ.

References

1. Akrour, R., Schoenauer, M., Sebag, M.: Preference-Based Policy Learning. In: Gunopulos, D., Hofmann, T., Malerba, D., Vazirgiannis, M. (eds.) ECML PKDD 2011, Part I. LNCS, vol. 6911, pp. 12–27. Springer, Heidelberg (2011)
2. Boutilier, C.: A POMDP formulation of preference elicitation problems. In: Proceedings of the Eighteenth National Conference on Artificial Intelligence (AAAI 2002), Edmonton, pp. 239–246 (2002)
3. Chajewska, U., Koller, D., Parr, R.: Making rational decisions using adaptive utility elicitation. In: Proceedings of the Seventeenth National Conference on Artificial Intelligence (AAAI 2000), Austin, TX, pp. 363–369 (2000)
4. Cheng, W., Fürnkranz, J., Hüllermeier, E., Park, S.-H.: Preference-Based Policy Iteration: Leveraging Preference Learning for Reinforcement Learning. In: Gunopulos, D., Hofmann, T., Malerba, D., Vazirgiannis, M. (eds.) ECML PKDD 2011, Part I. LNCS, vol. 6911, pp. 312–327. Springer, Heidelberg (2011)
5. Gonzales, C., Perny, P.: GAI networks for utility elicitation. In: Proceedings of the Ninth International Conference on Principles of Knowledge Representation and Reasoning (KR 2004), Whistler, BC, pp. 224–234 (2004)
6. Guo, S., Sanner, S.: Real-time multiattribute bayesian preference elicitation with pairwise comparison queries. In: Proceedings of the 13th International Conference on Artificial Intelligence and Statistics (AISTATS 2010), Sardinia, Italy (2010)
7. Neal, R.M.: Slice sampling. The Annals of Statistics 31(3), 705–770 (2003)
8. Viappiani, P., Boutilier, C.: Optimal bayesian recommendation sets and myopically optimal choice query sets. In: Advances in Neural Information Processing Systems 23 (NIPS), Vancouver (2010)

Hybridizing Reactive Tabu Search
with Simulated Annealing

Stefan Voß[1] and Andreas Fink[2]

[1] Institute of Information Systems (IWI), University of Hamburg
Von-Melle-Park 5, 20146 Hamburg, Germany
stefan.voss@uni-hamburg.de
[2] Chair of Information Systems, Helmut-Schmidt-University / UniBw Hamburg,
Holstenhofweg 85, 22043 Hamburg, Germany
andreas.fink@hsu-hamburg.de

Abstract. Reactive tabu search (RTS) aims at the automatic adaptation of the tabu list length. The idea is to increase the tabu list length when the tabu memory indicates that the search is revisiting formerly traversed solutions. Once too many repetitions are encountered, an escape mechanism constituting a random walk is an essential part of the method. We propose to replace this random walk by a controlled simulated annealing (SA). Excellent results are presented for various combinatorial optimization problems.

1 Introduction

The basic paradigm of tabu search is to use information about the search history to guide local search approaches to overcome local optimality. In general, this is done by a dynamic transformation of the local neighborhood. RTS aims at the automatic adaptation of the tabu list length [1,2]. A possible specification can be described as follows: Starting with a tabu list length s of 1, it is increased to $\min\{\max\{s + 2, s \times 1.2\}, b_u\}$ every time a solution is repeated, taking into account an appropriate upper bound b_u (to guarantee at least one admissible move). If there has been no repetition for some iterations, we decrease it to $\max\{\min\{s - 2, s/1.2\}, 1\}$. To accomplish detecting repetitions of solutions, we apply a trajectory based memory using hash codes.

For RTS it is appropriate to include means for diversifying moves whenever the tabu memory indicates that we are trapped in a certain region of the search space. As a trigger mechanism one may use, e.g., the combination of at least three solutions each having been traversed three times. The standard escape strategy is to perform randomly a number of moves (depending on the average of the number of iterations between solution repetitions) [1,2]. As termination criterion one may consider a given time limit. In this paper we propose to replace this random walk by a controlled SA.

The next section provides details of the specific hybridization that we propose. In Section 3 we sketch the set of problems that we have currently looked at. The paper closes with some conclusions.

Y. Hamadi and M. Schoenauer (Eds.): LION 6, LNCS 7219, pp. 509–512, 2012.
© Springer-Verlag Berlin Heidelberg 2012

2 The Hybrid Method

SA extends basic local search by allowing moves to worse solutions. Starting from an initial solution, successively a candidate move is randomly selected; this move is accepted if it leads to a solution with a better objective function value than the current solution, otherwise the move is accepted with a probability that depends on the deterioration Δ of the objective function value. The acceptance probability is computed according to the Boltzmann function as $e^{-\Delta/T}$, using a temperature T as control parameter. Various authors describe robust realizations of this general SA concept. Following [4], the value of T is initially high, which allows many worse moves to be accepted, and is gradually reduced through multiplication by a parameter α according to a geometric cooling schedule.

Instead of using random walk as the escape mechanism within RTS, we propose to apply SA, which performs, depending upon the parameter setting, diversification as well as intensification to some degree. In the computational experiments described in this paper, we examine the effect of adapting the SA parameter values in accordance with its primary role as diversification mechanism [4]. We stick to using $\alpha = 0.95$, whereas *frozenLimit* is set to 1 in order to terminate earlier. Instead of *initialAcceptanceFraction* = 0.4 we also use the value 0.1 which means less diversification; instead of *sizeFactor* = 16 we also use the value 1 which speeds up the cooling process; instead of *frozenAcceptanceFraction* = 0.02 we also use the value 0.1 which eventually means less intensification. (Whenever a SA run is performed while an overall time limit is reached we finish that run before terminating the approach.)

3 Computational Results

We have considered various optimization problems to emphasize the impact of the RTS/SA-hybridization proposed above. In the sequel we provide results for the *Ring Load Balancing Problem* (RLB) and mention other problems where implementations and results are available. All implementations have been performed by using our HOTFRAME software [3] on an average PC.

3.1 Ring Load Balancing Problem

The RLB is an NP-hard telecommunications problem where we are given a ring of nodes with a set of communication demands between node pairs [5]. Assuming that the communication demands occur simultaneously, the task is to decide for each demand whether to route it clockwise or counterclockwise, minimizing the maximum bandwidth requirement on any of the links between adjacent nodes. That is, given a set of n nodes and a set of demands between pairs of nodes, find a direction for each of the demands so that the maximum of the loads on the links in the network is as small as possible.

The solution space consists of all possible routing directions for the demands. We employ a straightforward neighborhood that is defined by switching the

Table 1. Computational results for the RLB

		opt	Esc.=RW (1s)	Esc.=SA (1s)	Esc.=SA (max(1,D/90)s)
ss	(5;6)	131.5	131.5	131.5	131.5
	(10;12)	231.6	231.6	231.6	231.6
	(15;25)	507.5	507.5	507.5	507.5
	(20;40)	734.5	734.5	734.5	734.5
	(25;60)	1013.4	1013.3	1013.3	1013.3
	(30;90)	1435.8	1434.9	1434.2	1434.2
mm	(5;8)	173.6	173.6	173.6	173.6
	(10;23)	422.5	422.5	422.5	422.5
	(15;50)	883.8	882.2	882.2	882.2
	(20;95)	1457.5	1457.4	1455.8	1455.8
	(25;150)	2253.3	2241.0	2234.3	2234.2
	(30;200)	3013.2	3019.1	3006.8	3006.8
ll	(5;10)	186.0	186.0	186.0	186.0
	(10;45)	728.6	728.3	728.3	728.3
	(15;105)	1605.1	1602.2	1599.9	1599.9
	(20;190)	2742.3	2736.2	2721.0	2720.6
	(25;300)	4243.5	4238.7	4225.2	4221.3
	(30;435)	5982.0	5987.7	5968.2	5956.9
ce	(5)	155.0	155.0	155.0	155.0
	(10)	349.4	349.4	349.4	349.4
	(15)	530.7	530.7	530.7	530.7
	(20)	721.8	721.8	721.8	721.8
	(25)	991.8	991.8	991.8	991.8
	(30)	1101.7	1101.7	1101.7	1101.7
	Average:		0.12%	0.03%	0.01%

routing direction for one demand (node pair). The quality of such a local search move is assessed by the implied change of the objective function value. The RLB has been used as a testbed as optimal solutions are available and we are yet able to show the impact of our approach. We report results for problem instances of the RLB proposed in [5].

Table 1 provides a detailed view on the characteristics of the data. In the first column we describe the different scenarios together with $(n; D)$, the number of nodes n and number of demands D. The first three blocks are non-centralized demands while the last block gives centralized demands, i.e., $D = n - 1$. Each row refers to an average of ten runs. Correspondingly, column 'opt' provides (in each row) the average of the optimal solution values for these ten runs. We provide results for the case where diversification is performed by using the original random walk as an escape mechanism (Esc.=RW) with time limit 1 second. On the right side of the table we consider the case where the escape mechanism is performed by applying a SA run with the standard parameter setting as described above. The time limit is the same as before as well as one with a possible instance-dependent extension based on the given data. While all approaches provide small deviations from optimality the hybrid approach is able to considerably improve on its pure counterpart.

3.2 Other Problems

Additional problems where we have applied our ideas include, among others, the *Minimum Weight Vertex Cover Problem* and the *Minimum Labelling Spanning Tree Problem*. For all considered problems we show that the hybridization improves the numerical results of the pure RTS and the SA.

4 Conclusions

In this paper, we have presented a very simple and yet very effective modification of the well-known reactive tabu search. As a conclusion we may deduce that randomness helps in metaheuristics, though a controlled way of incorporating randomness might be more successful than pure randomness. The number of successful implementations of RTS in literature provides an option to revisit those implementations to crosscheck whether our idea also holds in those applications that have not been looked at in this paper.

References

1. Battiti, R.: Reactive search: Toward self-tuning heuristics. In: Rayward-Smith, V.J., Osman, I.H., Reeves, C.R., Smith, G.D. (eds.) Modern Heuristic Search Methods, pp. 61–83. Wiley, Chichester (1996)
2. Battiti, R., Tecchiolli, G.: The reactive tabu search. ORSA Journal on Computing 6, 126–140 (1994)
3. Fink, A., Voß, S.: HOTFRAME: A heuristic optimization framework. In: Voß, S., Woodruff, D.L. (eds.) Optimization Software Class Libraries, pp. 81–154. Springer (2002)
4. Johnson, D.S., Aragon, C.R., McGeoch, L.A., Schevon, C.: Optimization by simulated annealing: An experimental evaluation; part I, graph partitioning. Operations Research 37, 865–892 (1989)
5. Myung, Y.-S., Kim, H.-G., Tcha, D.-W.: Optimal load balancing on SONET bidirectional rings. Operations Research 45, 148–152 (1997)

Author Index